D0847504

High energy astrophysics is one of the most exciting areas of contemporary astronomy, covering the most energetic phenomena in the universe. The highly acclaimed first edition of Professor Longair's book immediately established itself as an essential text book on high energy astrophysics. In this complete revision, the subject matter has expanded to the point where two volumes are desirable. In the first a thorough treatment is given of the physical processes that govern the behaviour of particles in astrophysical environments such as interstellar gas, neutron stars, and black holes. Special emphasis is placed on how observations are made in high energy astrophysics and the limitations imposed on them. The tools of the astronomer and high energy astrophysicist are introduced in the context of specific astronomical problems. The material in Volume 1 leads to a study of all kinds of high energy phenomena in the Galaxy and the universe, given in the second volume.

This book assumes that readers have some knowledge of physics and mathematics at the undergraduate level, but no prior knowledge of astronomy is required. The pair of books covers all aspects of modern high energy astrophysics to the point where current research can be understood.

High energy astrophysics

Volume 1

Particles, photons and their detection

High energy astrophysics

Volume 1
Particles, photons and their detection

Second edition

M. S. Longair

Jacksonian Professor of Natural Philosophy, Cavendish Laboratory, University of Cambridge, Cambridge

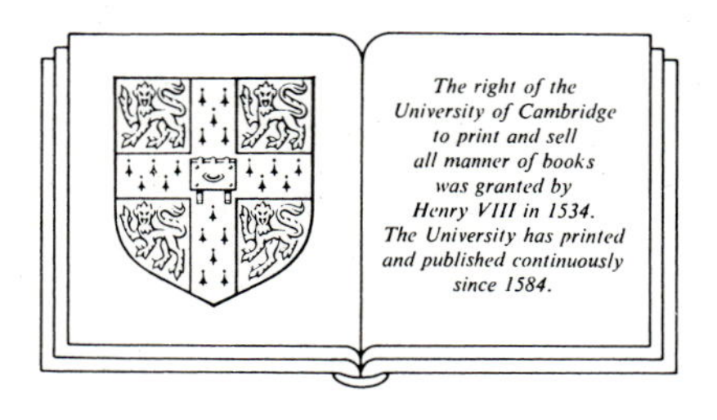

CAMBRIDGE UNIVERSITY PRESS

Cambridge

New York Port Chester Melbourne Sydney

Published by the Press Syndicate of the University of Cambridge
The Pitt Building, Trumpington Street, Cambridge CB2 1RP
40 West 20th Street, New York, NY 10011-4211, USA
10 Stamford Road, Oakleigh, Victoria 3166, Australia

First published 1981

Second edition 1992

Printed in Great Britain at the University Press, Cambridge

British Library cataloguing in publication data
Longair, M. S. (Malcolm S.)
High energy astrophysics.
Vol. 1 Particles, photons and their detection – 2nd ed
1. High energy astrophysics
I. Title
523.01976

Library of Congress cataloguing in publication data

ISBN 0 521 38374 9 hardback
ISBN 0 521 38773 6 paperback

UP

For Deborah

Contents

Preface

It is a pleasure to produce a second edition of *High energy astrophysics*. Writing the first edition was great fun and it corresponded rather closely to the lecturing style in which I had presented high energy astrophysics in the period 1973–7. Although I updated the material of the first edition to 1980 when the manuscript was sent to the press, it still remained in essence a lecture course. The reception of the book was encouraging and the time is now ripe for revising the contents considerably in the light of more recent developments and of changing perspectives about what should be included in an introduction to high energy astrophysics.

In preparing the revised edition, I have aimed to include a much broader range of astrophysical topics into the text. Whilst the first edition contained many useful tools for high energy astrophysics, many more examples can be given of their practical application in astrophysical problems. There is now a need to give more thorough treatments of phenomena such as accretion discs and the astrophysics of extragalactic radio sources, of active galactic nuclei and of compact objects. For much of the new material I have taken as a basis my recent review article 'The new astrophysics' which was published in the volume *The new physics* (Cambridge University Press, 1989) but again brought up to date.

I have, however, aimed to maintain as many of the positive features of the first edition as possible. In particular, I maintain the informal style and have no hesitation about using the first person singular or expressing my personal opinion about the material under discussion. I will emphasise strongly physical principles and the discussion of general results rather than particular models which may have only ephemeral appeal.

My general approach to astrophysics was set out in the first edition. Physics and astrophysics have a symbiotic relation. On the one hand, the astrophysical sciences are concerned with the application of the laws of physics to phenomena on a large scale in a Universe. On the other hand, new laws of physics are discovered through astronomical observations and their astrophysical interpretation. In these ways, the new astrophysics, of which high energy astrophysics is perhaps the most important ingredient, is just as much a part of modern physics as laboratory physics.

My other aim is similar to that already expressed in the preface to the first edition and in my book *Theoretical concepts in physics* (Cambridge University Press 1984, 1986, 1987) and that is to give undergraduates a feeling for what it is like to undertake research at the limits of present understanding. Astrophysics is fortunate in that many of the fundamental problems can be understood without a great deal of new physics or new physical concepts. Thus, the text may be considered valuable as an introduction to the way in which research is carried out in the astrophysical context.

Unfortunately, in preparing the revised edition, it became obvious that the material could not be contained within one reasonably-sized volume and I have therefore, with considerable reluctance, split the text into two volumes. *Volume 1* is concerned with establishing most of the basic tools needed for the high energy astrophysics which can be tried and tested within the Solar System. We will continually make forward reference to the use of these tools in purely astrophysical situations. There is considerable emphasis upon techniques of observations of high energy particles and photons, both from the surface of the Earth and from above the Earth's atmosphere. Some attention is given to the properties of cosmic ray particles since these are the only particles which we can detect on Earth which originate in astrophysical sources outside our own Solar System. We have a unique opportunity to understand the behaviour of these particles in magnetic fields through their dynamics in the interplanetary magnetic field and in collisionless shocks when they encounter the shock wave which surrounds the Earth's magnetosphere. In my opinion, this is an area which has not received its due attention from those astrophysicists who wish to understand the behaviour of charged particles in the somewhat more exotic circumstances of active galactic nuclei and black holes. There is a strong emphasis upon observation in this volume, my belief being that it is important to understand the observational limitations which exist in each astronomical waveband in order to assess the feasibility of some of the more demanding observations which one would certainly wish to carry out. I do not shrink from tackling topics which do not normally appear in text books on astrophysics, such as the economics and politics of high energy astrophysics.

In *Volume 2*, we will adopt a much more straightforward approach to the 'traditional' problems of *high energy astrophysics* – I use this term in the way in which it is normally used nowadays, although I would hate to have to try to define it.

Malcolm Longair
Edinburgh, Scotland and Cambridge, Massachusetts, USA.

May 1990.

Acknowledgements

There are many people whom it is a pleasure to thank for help and advice during the preparation of this volume. Just as the first edition was begun during a visit to the Osservatorio Astronomico di Arcetri in Florence, so the second edition could not have been completed without the Regents' Fellowship of the Smithsonian Institution which I held at the Harvard-Smithsonian Astrophysical Observatory during the period April–June 1990. I am particularly grateful to Professors Irwin Shapiro and Giovanni Fazio for sponsoring this visit to Harvard during which time the final drafts of Chapters 1–10 were completed. During that period, I had particularly helpful discussions with Drs Eugene Avrett, George Rybicki, Giovanni Fazio, Margaret Geller and many others. I am particularly grateful to them for their advice.

Much of the preliminary rewriting was completed while I was at the Royal Observatory, Edinburgh. Among the many colleagues with whom I discussed the contents of this volume, I must single out Dr John Peacock who provided deep insights into many topics. In completing the final chapter on the high energy astrophysics of the Solar System, I greatly benefitted from the advice of Professors John Brown, Carole Jordan and Eric Priest. Not only did they point me in the correct directions but they also reviewed my first drafts of that chapter. I am especially grateful to them for this laborious task. Many colleagues made helpful suggestions about corrections and additions to the first addition among whom Dr Roger Chevalier provided an especially useful list.

To all of these friends and colleagues I make the usual disclaimer that any misrepresentations of the material presented in this book is entirely my responsibility and not theirs. Finally, I acknowledge the unfailing support of my family, Deborah, Mark and Sarah who have contributed much more than they will ever know to the completion of this book.

1

High energy astrophysics – two approaches

1.1 Introduction – the author's dilemma

There has been a revolution in astronomy and astrophysics over the last 40 years. The prime reason for this has been the opening up of the whole of the electromagnetic spectrum for astronomical observations. This revolution would not have been possible without the development of new techniques for making astronomical observation both from the ground and from space. Hand in hand with these developments have been major advances in laboratory physics and the development of high speed computers. It is the combination of all these factors which has lead to the enormous advances in the astrophysical sciences.

Among the most important of the new disciplines which have developed through this exciting period has been that of *high energy astrophysics*. I would hate to have to provide a definition of the subject because it is an all-embracing title covering those areas of research undertaken by astrophysicists, physicists and astronomers interested in high energy phenomena of all types. Roughly speaking, it is the astrophysics of high energy processes and their application in astrophysical contexts. The object of this book is to describe these processes and their application in astrophysics. There is no question but that this is one of the most exciting areas of modern astrophysical research and involves some of the most difficult problems of contemporary physics. A few examples include the study of massive black holes in active galactic nuclei, the acceleration of high energy particles in astronomical environments, the origins of enormous fluxes of high energy particles from active galaxies, the physical processes in the interiors and environments of neutron stars. In these examples we find processes taking place in physical conditions which cannot be reproduced in the laboratory. Indeed, in many cases, the astrophysical environment is the only one in which the problems can be addressed. The aim of this book is to set out the logical sequence of steps by which the astronomers can address these problems meaningfully.

The aim of the astrophysical science is two-fold – the understanding of the application of the laws of physics in the extreme physical conditions encountered in astronomical systems and the discovery of new laws of physics from observation. This second aspect has a long and distinguished pedigree. For example, Kepler's discovery of the laws of planetary motion resulted from Tycho Brahe's magnificent set of astronomical observations which set new standards of precision and reliability. Kepler's laws lead directly to Newton's law of gravity and his laws of motion. Balmer's discovery of the formula for the wavelengths of the spectral lines of what we now call the Balmer series of hydrogen involved spectroscopic observations of the violet and ultraviolet lines of hydrogen in white dwarf stars. These were not observable in the laboratory at that time. The Balmer formula was the key to Bohr's theory of the atom and consequently to the unravelling of atomic structure. Good modern examples of the same process are provided by some of the best tests of the General Theory of Relativity which involve astronomical observations. One of the most spectacular of these has been the demonstration of gravitational radiation loss in the case of a binary pulsar system. There are countless other examples of astronomical observations leading directly or indirectly to advances in fundamental and applied physics.

The dilemma of the title of this section is how to reconcile all of these different demands into a single text. My own view is that, in contemporary astrophysics, it is as important to understand the means by which the observations are made as it is to understand the astrophysical tools and the astronomical context within which the tools are used. Equally, a firm understanding of the underlying physical principles of all aspects of this story seems to me to be essential if creative astrophysical research is to be undertaken. I have a long-standing goal of making the disciplines of astronomy and astrophysics as exact a science as laboratory physics. Equally, it is essential to demonstrate the firm physical underpinning of all our astrophysical research endeavours. I would like the material to be regarded as the natural extension of laboratory physics to the larger scale problems of astrophysics. I therefore put a great emphasis upon dealing with very concrete physical problems and then looking at their application in the astrophysical context. Throughout the text, I will emphasise those aspects of high energy astrophysics in which the astrophysical applications are reasonably secure. I will try to avoid ephemera and exotica and if they have to appear in the text they will be clearly signalled.

It was these considerations which led me originally to approach the subject of high energy astrophysics through the study of cosmic rays. My reasons were very simple and mundane. The cosmic ray particles which we detect directly on Earth, at the top of the atmosphere and in the interplanetary medium are the only high energy particles of genuinely cosmic origin which we can detect. As such they are presumably examples or close relatives of the relativistic particles and plasmas which are found in extreme astrophysical environments. Of course, cosmic ray physics is a major discipline in its own right but, in the present text, it should be regarded in addition as a vehicle for introducing the techniques and processes involved in high energy astrophysics. The cosmic ray particles which reach the

Earth from interstellar space exhibit many of the features of the fluxes of high energy particles assumed to be present in extraterrestrial sources such as supernovae, radio galaxies, quasars and active galactic nuclei. I therefore make no apology for approaching the subject of high energy astrophysics from a somewhat non-traditional point-of-view.

At the same time, we have to describe the astrophysical environment in which this discussion is to take place and hence we have to provide a modern view of 'conventional' astrophysics. In fact, there is almost no aspect of modern astronomy which has not in some way been strongly influenced by discoveries in high energy astrophysics. I will try to draw out these connections as the narrative develops. This also means that we have to introduce a wide range of diagnostic tools for analysing the properties of plasmas and lower temperature gases in astrophysical environments.

Finally, we should not neglect the social, intellectual and political environment in which these studies are carried out. In many cases, we will describe very large and expensive projects which are directed towards answering basic purely scientific questions. Any big science can easily price itself out of what could conceivably be funded by any national or international agency. We should not neglect the role of science, including high energy astrophysics, as part of the intellectual, cultural and economic life of nations. Scientists are more and more being asked to account for the expenditure of scarce resources on fundamental sciences and it is a challenge which should be answered convincingly on the basis of the achievements of these disciplines and their wide ramifications for technology and society. I will not shrink from including these aspects in this story.

1.2 Historical perspective I – a history of cosmic ray physics

I make no apology for beginning with a few historical notes about the history of cosmic ray physics and of high energy astrophysics. This is much more than simply the recounting of the key events in the history of these disciplines. Many of the key ideas and experimental procedures have a long and distinguished history which reflect the insight and ingenuity of the great scientists of the past. These are our legacy and the foundation of modern scientific practice.

1.2.1 *The discovery of sub-atomic particles – 1890–1910*

The story begins in the late nineteenth century. The unification of electricity and magnetism in Maxwell's theory of the electromagnetic field was one of the great triumphs of nineteenth century science. The theory appeared in its full modern guise in 1864 and much effort was devoted in the succeeding years to testing it and comparing its predictions with those of other thoeries. The prediction that light is a form of electromagnetic radiation was fully confirmed by the ingenious experiments of Hertz more than 20 years later. I have told this remarkable story elsewhere (Longair 1987). From the point of view of the present

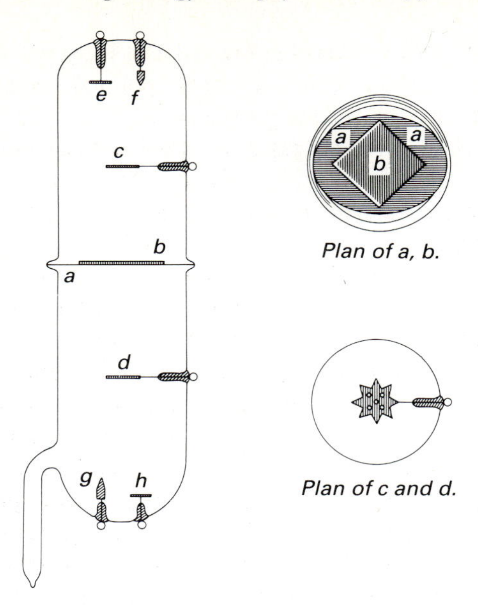

(*a*)

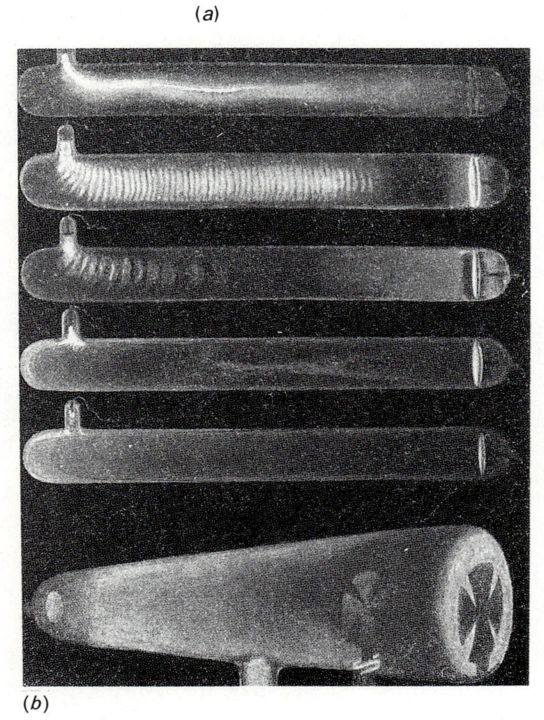

(*b*)

Figure 1.1. (*a*) Crookes' original experiment in which he demonstrated the projection of 'molecular shadows' when a vacuum tube (or 'Crookes tube') is operated with a high vacuum. The electrodes *e* and *h* can be used as cathodes and *f* and *g* as anodes. The mica screen *a* had attached to it a flat plate of uranium glass onto which the shadows of the stars *c* and *d* were projected. (W. Crookes (1879). *Phil. Trans. Roy. Soc.*, **175**, 135.)

study, the important point is that the late nineteenth century was a period when the full ramifications of Maxwell's theory were being explored and experiments in electricity and magnetism were a major growth area in physics.

Among the most interesting of these experiments were those concerning the conduction of electricity through gases. It was already known in the early nineteenth century that an electric current can flow through gases at low pressure. These studies were put on a firm experimental basis by William Crookes who began his famous series of investigations in 1879. The key to the success of his programme was the production of good vacuum tubes and high voltages between the positive and negative electrodes. These technological advances can be attributed to two great nineteenth century inventors. The first was Geissler who was an expert glass blower and who made his own vacuum tubes which were renowned throughout Europe for their high quality. His great invention was the Geissler pump for evacuating glass vessels to produce excellent vacua. The second was Rümhkorff who invented the Rümhkorff coil which enabled large voltages to be generated across the electrodes of the vacuum tubes. Typically voltages of a few thousands volts could be generated by his devices. These tools were used in many of the great experiments of the late nineteenth century including the electromagnetic experiments of Hertz, Röntgen's X-ray experiments, Zeeman's experiments in which he discovered the splitting of spectral lines and Thomson's experiments on the mass-to-charge ratio of the electron.

Crookes tubes glow when a current is passed through the gas and display beautiful colours and patterns which made them popular scientific toys. They are the forerunners of the sodium and mercury arc lights which are used in street lighting. As the pressure in the tubes was lowered, the glow from the gas throughout the tube decreased but there remained a glow from the end of the tube beyond the anode (Fig. 1.1). By placing obstacles in the tube, Crookes demonstrated that their shadows were projected onto the end of the tube beyond the anode. This was convincing evidence that the glow in the tube was caused by rays emitted by the cathode and these were naturally referred to as *cathode rays*. By 1895, it had been shown that the cathode rays could be deflected by a magnetic field and then, in one of the great experiments of modern science, J. J. Thomson succeeded in measuring the charge-to-mass ratio, e/m, of the cathode rays in 1897. In his famous experiment, the cathode rays were accelerated down the Crookes tube by the electric field between the anode and cathode and the beam could then be deflected by crossed electric and magnetic fields. An elementary calculation shows that the deflection produced by the electric field alone is a measure of the quantity (e/mv^2) where v is the velocity of the cathode ray between the crossed

---

(*b*) Examples of the effects observed in a Crookes tube. The top series of five sketches shows the phenomena observed as the tube is evacuated. In the fifth sketch, the inner surface of the tube shines with a greenish glow. The last sketch shows the image cast by a Maltese cross placed in the tube when it is operated under a high vacuum. (From F. Close, M. Marten and C. Sutton (1987). *The particle explosion*, page 23, Oxford: Oxford University Press.)

fields which can be found by balancing exactly the electric and magnetic forces acting on the rays, $v = E/B$.

This part of the story is well known. What is less well known is the fact that this experiment was only possible because of Thomson's great experimental skill in being able to produce a better vacuum than anyone else at that time. If there had been a significant amount of ionised gas in the tube, ion–electron pairs would have neutralised the electric field. The value of e/m was found to be about 2000 times that of the hydrogen atom, the lightest of all the chemical elements. Thomson measured independently the charge of the electron e and he inferred that the cathode rays were particles with mass about 1/1000 that of the hydrogen atom. This was the discovery of the first sub-atomic particle. Thomson immediately concluded that there is structure within the atom and that the process of ionisation and the flow of electric currents were associated with the tiny electrically charged particles which can be split off atoms. The identification of the cathode rays with electrons, the particles which were assumed to carry the elementary electric charge, was complete. Soon after, Thomson went on to show that the particles emitted in the photoelectric effect discovered some years earlier by Hertz, had the same value of e/m as the cathode rays, confirming the identity of the photoelectric particles with electrons.

Many physicists experimented with Crookes tubes about this period and among them was Röntgen. In 1895, he discovered by accident that wrapped unexposed photographic plates left close to Crookes tubes were darkened. In addition, fluorescent materials left close to Crookes tubes glowed in the dark. Röntgen came to the correct conclusion that both phenomena were associated with some new form of radiation emitted by the Crookes tube and he named these rays *X-rays*. This discovery caused an immediate sensation when the first X-ray photographs showing the bones of the body were published. Overnight, X-rays became a matter of the greatest public interest and were very rapidly incorporated into the armoury of the doctor's surgery. The nature of the X-rays was not clear. They were found to be more penetrating than the cathode rays since they could blacken photographic plates at a considerable distance from the hot spot on the Crookes tube which was known to be their source. Their identification with 'ultra-ultraviolet' radiation was only convincingly demonstrated when, in 1906, Barkla found that the X-radiation was polarised and, even more convincingly, when von Laue had the inspiration of looking for their diffraction by crystals in 1912.

The association of X-rays with fluorescent materials led to the search for other sources of X-radiation, the idea being that other substances known to be fluorescent might also be sources of X-rays. In 1896 Becquerel tested several known fluorescent substances before he investigated some samples of potassium uranyl disulphate. The standard procedure was to wrap the photographic plate in several sheets of black paper, expose the phosphorescent material to sunlight and then develop the plate to find if it had been darkened by X-rays. Becquerel's remarkable discovery was that the plates became darkened even when the phosphorescent material was not exposed to light. This was the discovery of *natural radioactivity*. Further experiments in that year by Becquerel showed that the amount of radioactivity was proportional to the amount of uranium in the substance and that the radioactive flux of radiation was constant in time. Another key discovery was the fact that the radiation from the uranium compounds discharged electroscopes.

Other radioactive substances were soon identified. Thorium was discovered in 1898 and then followed the Curies' isolation of polonium and radium, both of them much stronger sources of radioactivity than uranium. Rutherford's first publication on radioactivity appeared in 1898 and in it he established that there are at least two separate components in the radiation emitted by radioactive substances from a study of the absorption properties of the radiation. The component which is most easily absorbed he called α-radiation (or α-rays) and the much more penetrating component was called β-radiation (or β-rays). It took another ten years before it was conclusively demonstrated that the α-radiation consisted of what we now know as the nuclei of helium atoms. In contrast, the β-radiation was quickly shown to have the same mass-to-charge ratio as the recently discovered electron. γ-radiation was discovered by Villard in 1900 as an extremely penetrating form of radiation emitted in radioactive decays. The γ-rays displayed no deviation in a magnetic field and they became thought of as extremely high energy X-rays. Fourteen years later, they were conclusively identified as electromagnetic waves when Rutherford and Andrade observed the reflection of γ-rays from crystal surfaces.

These were the only particles known which could cause ionisation of air. The characteristic property which distinguished them so far as their ionisation properties were concerned was their penetrating power. In quantitative terms, they were distinguished as follows:

> The α-*particles* ejected in radioactive decays produce a dense stream of ions and are stopped in air within about 0.05 m. This is called the *range* of the particles. We now understand that, in this type of radioactive decay in which a single species of particle is ejected from the nucleus, the range of the α-particle is directly related to the difference in binding energies of the parent and product nuclei.
>
> The β-*particles* have greater ranges but there is not a well-defined value for any particular radioactive decay. We now understand that the spread in range is due to the fact that the electrons are emitted as part of a three-body process involving the emission of a neutrino as well as an electron.
>
> The γ-*rays* were found to have by far the longest ranges, a few centimetres of lead being necessary to reduce their intensity by a factor of 10.

1.2.2 *The discovery of cosmic rays*

The cosmic ray story begins about 1900 when it was found that electroscopes discharged even if they were kept in the dark well away from sources of natural radioactivity. The electroscope was a key instrument in many of the early experiments in radioactivity because the rate at which the leaves of the

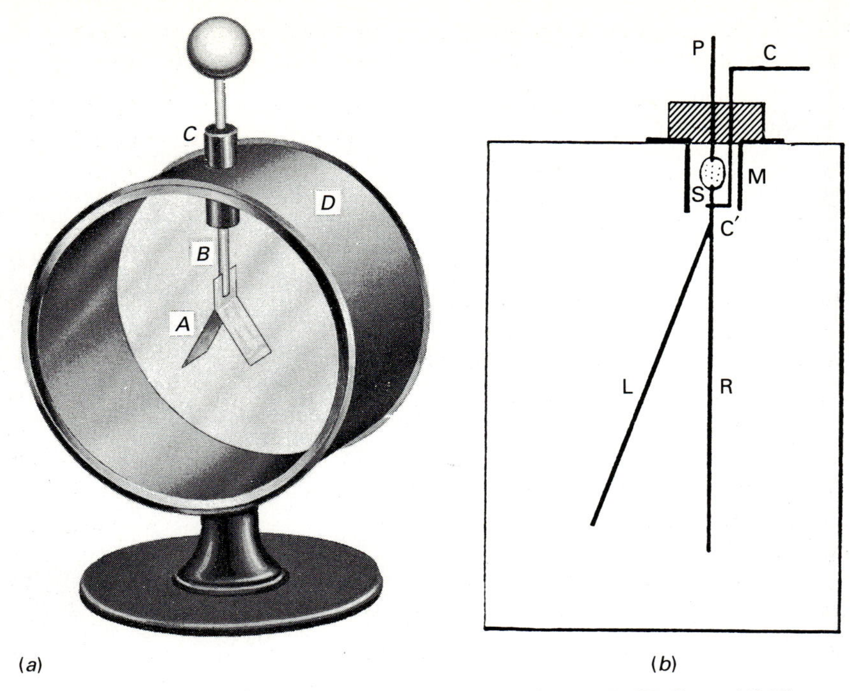

Figure 1.2. (*a*) An example of a traditional electroscope. (From F. W. Sears (1958). *Electricity and magnetism*, page 5, London: Addison-Wesley.) (*b*) The gold-leaf electroscope developed by C. T. R. Wilson and used by Rutherford in his early experiments. In Rutherford's words, 'A brass cylindrical vessel is taken of about 1 litre capacity. The gold-leaf system, consisting of a narrow strip of gold-leaf L attached to a flat rod R, is insulated inside the vessel by the small sulphur bead or piece of amber S, supported from the rod P. In a dry atmosphere a clean sulphur bead or piece of amber is almost a perfect insulator. The system is charged by a light bent rod CC′ passing through an ebonite cork. The rod C is connected to one terminal of a battery of small accumulators of 200–300 volts. If these are absent, the system can be charged by means of a rod of sealing-wax. The charging rod CC′ is then removed from contact with the gold-leaf system. The rods P and C and the cylinder are then connected with earth.' The amount of ionisation was measured by observing the rate at which the charged gold-leaf strip L moved towards the rod R as the gold-leaf discharged due to leakage currents associated with the ionisation. (From E. Rutherford (1905). *Radio-activity*, pages 86–7, Cambridge: Cambridge University Press.)

electroscope came together provided a measure of the amount of ionisation (Fig. 1.2). The origin of this behaviour was a major puzzle and, as might be expected, various ingenious experiments were carried out to discover the origin of the ionising radiation. A good example is this quotation from C. T. R. Wilson (Wilson 1901)

> The experiments with this apparatus were carried out at Peebles. The mean rate of leak when the apparatus was in an ordinary room amounted to 6.6 divisions of the micrometer scale per hour. An experiment made in the Caledonian Railway tunnel near Peebles (at night after the traffic had ceased) gave a leakage of 7.0 divisions per hour.... There is thus no evidence of any falling off of the rate of

production of ions in the vessel, although there were many feet of solid rock overhead.

It was later shown by Rutherford, however, that most of the ionisation was due to natural radioactivity, either in rocks or from radioactive contamination of the equipment. Other experiments included a tantalising experiment in 1910 by Wulf who was responsible for the construction of the best electrometers used by most workers. He found that the ionisation fell from 6×10^6 ions m^{-3} to 3.5×10^6 ions m^{-3} as he ascended the Eiffel Tower, a height of 330 m. γ-rays were the most penetrating of the ionising radiations known at that time and their absorption coefficient in air was known. If the ionisation had been due to γ-rays originating at the surface of the Earth, the intensity of ions should have halved in only 80 m and would have been negligible at the top of the Eiffel Tower.

The big breakthrough came in 1912 and 1913 when first Hess and then Kolhörster made manned balloon ascents in which they measured the ionisation of the atmosphere with increasing altitude. By late 1912 Hess had flown to 5 km and then Kolhörster by 1914 had made ascents to 9 km, all of these experiments being carried out in open balloons. These were experiments of the greatest danger and the great success of the experiments matched the courage of the experimenters. It was Hess who discovered the first definite evidence that the source of the ionising radiation was extraterrestrial. Fig. 1.3(*b*) is a photograph of him after one of his successful flights in 1912.

Hess and Kolhörster found the startling result that the average ionisation increased with respect to the ionisation at sea-level above about 1.5 km (see Table 1.1). This is clear evidence that the source of the ionising radiation must be located above the Earth's atmosphere. From the observed decrease in the number of ions $n(l)$ at a distance l through the atmosphere, the attenuation constant α, defined by $n(l) = n_0 e^{-\alpha l}$, can be found. From the data in Table 1.1 it was found that the values of α correspond to 10^{-3} m^{-1} or less. This can be compared with the absorption coefficient for the γ-rays from radium C, which in air has a value of 4.5×10^{-3} m^{-1}, i.e. the γ-rays from radium C are at least five times less penetrating than the ionising radiation which must originate from above the atmosphere. Hess (1912) made the immediate inference:

> The results of the present observations seem to be most readily explained by the assumption that a radiation of very high penetrating power enters our atmosphere from above, and still produces in the lower layers a part of the ionisation oberved in closed vessels.

Even at sea-level there is a residual ionisation due to the extraterrestrial ionising radiation, amounting to about 1.4×10^6 ion pairs m^{-3}.

It was not too much of an extrapolation to assume that the *cosmic radiation* or *cosmic rays*, as they were named by Millikan in 1925, were γ-rays with greater penetrating power than those observed in natural radioactivity. In 1929, Skobeltsyn, working in his father's laboratory in Leningrad, constructed a cloud chamber to study the properties of the β-rays emitted in radioactive decays. The experiment involved placing the chamber within the jaws of a strong magnet so

(*a*)

(*b*)

Figure 1.3. The balloon flights of Victor F. Hess. (*a*) Preparation for one of his flights in the period 1911–12. (*b*) Hess after one of the successful balloon flights in which the increase in ionisation with altitude through the atmosphere was discovered. (From Y. Sekido and H. Elliot (eds) (1985). *Early history of cosmic ray studies*, Dordrecht: D. Reidel Publishing Company.)

that the curvature of their tracks could be measured. Among the tracks, he noted some which were hardly deflected at all and which looked like electrons with energies greater than 15 MeV (Fig. 1.4). He identified them with secondary electrons produced by the 'Hess ultra γ-radiation'. Although the interpretation was not correct, these were the first pictures of the tracks of cosmic rays.

The year 1929 also saw the invention of the *Geiger–Müller detector* which enabled individual cosmic rays to be detected. The distinctive feature of this type of counter is that it has a very fast response time so that not only is it possible to identify individual events but their arrival times can be determined very precisely. We will have much more to say about this type of detector later. The problem in using such counters is that they are sensitive to contaminating radioactivity. In the same year, Bothe and Kolhörster performed one of the key experiments in cosmic ray physics and in the process introduced the important concept of coincidence counting to eliminate background events. In their experiment, Bothe and Kolhörster recorded the events on film and were able to measure coincidences to about 0.01 s. The object of the experiment was to determine whether the cosmic

Table 1.1. *The variation of ionisation with altitude from the observations of Kolhörster (data from Hillas* (1972))

Altitude (km)	Difference between observed ionisation and that at sea-level ($\times 10^6$ ions m^{-3})
0	0
1	−1.5
2	+1.2
3	+4.2
4	+8.8
5	+16.9
6	+28.7
7	+44.2
8	+61.3
9	+80.4

radiation consisted of high energy γ-rays or charged particles. By using two counters, one placed above the other, they found that simultaneous discharges of the two detectors occurred very frequently, even when a strong absorber was placed between the detectors, indicating that charged particles of sufficient penetrating power to pass through both of them were very common (Fig. 1.5). This coincidence technique is now standard practice in many different types of cosmic ray and X and γ-ray experiments and we will meet it over and over again.

In the crucial experiment they placed slabs of lead and then gold up to 4 cm thick between the counters and measured the decrease in the number of coincidences when the absorber was introduced. The gold block was, in fact, on very temporary loan from a local bank and its security was a subject of considerable concern. The mass absorption coefficient agreed very closely with that of the atmospheric attenuation of the cosmic radiation. The experiment strongly suggested that the cosmic radiation consists of charged particles. If the cosmic radiation were γ-rays, the particles detected in the Geiger counters would have had to be secondary electrons and two separate events would have to take place in the two detectors initiated by a single γ-ray. It is most unlikely that such separate secondary electron events would occur in the two detectors. As they put in their classic paper (1929):

> One can perhaps summarise the whole discussion in a single argument: the mean free path of a γ-ray between two electron ejecting processes would be $1/\mu = 10$ m in water $1/\mu = 0.9$ m in lead and $1/\mu = 0.52$ m in gold for the high latitude radiation. Hence one can see that a quite exceptional accident must be supposed to happen if two electrons produced by the same γ-ray should display the necessary penetrating power and the correct direction to strike both counters directly.

They also showed that the flux of these particles could account for the observed intensity of cosmic rays at sea-level. Finally, they noted that the particles would

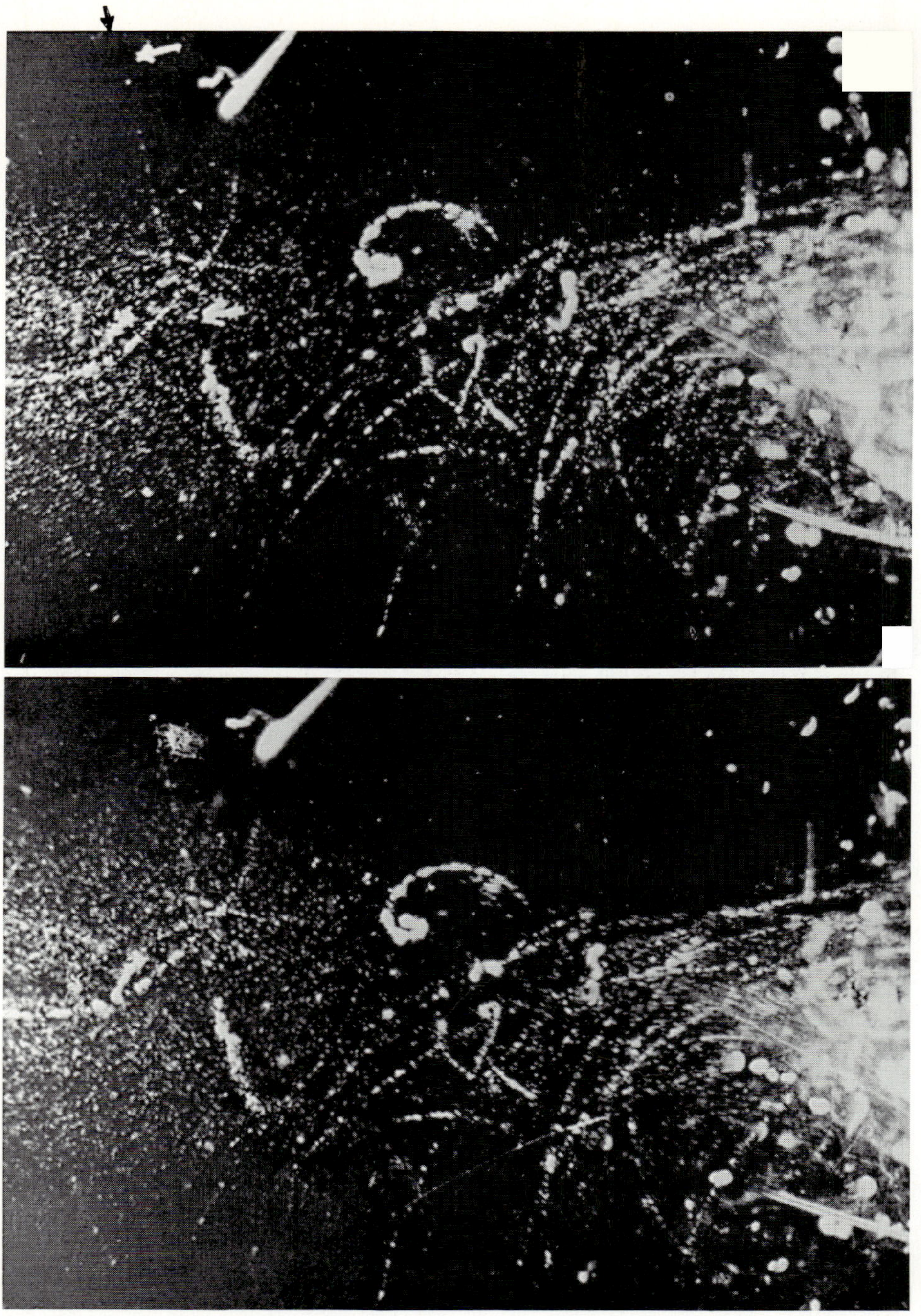

Figure 1.4. Stereographic images of the first photographic record of the arrival of a cosmic ray particle by Skobeltsyn in 1929. The track of the particle in the cloud chamber is indicated by the two white and one black arrows in the upper picture. (From D. V. Skobeltsyn (1985). *Early history of cosmic ray studies*, eds Y. Sekido and H. Elliot, page 47, Dordrecht: D. Reidel Publishing Company, 1985.)

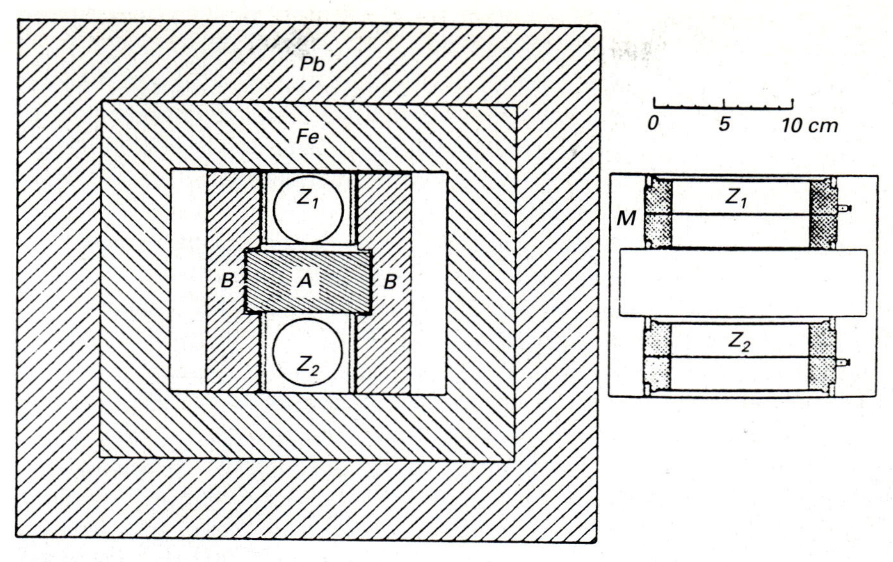

Figure 1.5. The experimental arrangement used by Bothe and Kolhörster to demonstrate that cosmic rays are charged particles and not high energy γ-rays. Z_1 and Z_2 are Geiger-Müller detectors and A is the absorbing slab – lead and gold were used in the key experiments. (W. Bothe and W. Kolhörster (1929). *Zeitschrift fur Physik*, **56**, 751; translation – A. M. Hillas (1972). *Cosmic Rays*, page 161, Oxford: Pergamon Press.)

have to be very energetic because of their long ranges in matter. They estimated the energies of the particles to be about 10^9–10^{10} eV.

1.2.3 *Cosmic rays and the discovery of elementary particles*

In Section 1.2.1 we took the story of the development of understanding of the internal structure of the atom and its nucleus up to about 1910. The unravelling of the nature of the atomic nucleus continued throughout the period 1910–30, a particularly significant development being the invention of the cloud chamber by C. T. R. Wilson. It was soon established that typical nuclei have a mass about two or more times that which can be attributed to the protons. The commonly held explanation for this difference was that the nucleus itself was composed of electrons and protons, the 'inner' electrons neutralising the extra protons. The fact that certain nuclei ejected electrons in radioactive β decays supported this point of view. Rutherford had speculated in the early 1920s that the neutral mass in the nucleus might be in the form of some new type of particle, similar to the proton but with no electric charge. During the 1920s Rutherford and his colleagues, particularly Chadwick, made a number of unsuccessful attempts to find evidence for these particles which became known as *neutrons*.

In 1930, Bothe and Becker in Germany and, in 1932, Irène Joliot-Curie and her husband Frédéric Joliot in France discovered that neutral penetrating radiation was emitted when light elements were bombarded by α-particles. Both groups believed that the radiation was some form of γ-radiation. Chadwick guessed that the penetrating radiation was a flux of the elusive neutrons. He rapidly performed a classic series of experiments in which the neutral radiation collided with different substances, including hydrogen and nitrogen, and then, from the recoil effects of the collisions between the unseen particles and the ambient gas, he could estimate the mass of the particles. This measurement of the mass of the neutral radiation showed conclusively that it could not be γ-radiation but rather neutral particles ejected from the nucleus with mass roughly the same as that of the proton.

From the 1930s to the early 1950s, the cosmic radiation provided a natural source of very high energy particles which were energetic enough to penetrate into the nucleus. This procedure turned out to be the principal technique by which new particles were discovered until the early 1950s. The first discoveries came from extensions of Skobeltsyn's experiments with cloud chambers. In 1930, Millikan and Anderson used an electromagnet ten times stronger than that used by Skobeltsyn to study the tracks of particles passing through the cloud chamber. Anderson observed curved tracks identical to those of electrons but corresponding to particles with positive electric charge. This discovery was confirmed by Blackett and Occhialini in 1933 using an improved technique in which the cloud chamber was only triggered after it was certain that a cosmic ray had passed through. They achieved this by placing Geiger counters above and below the chamber and only triggering the chamber and the cameras if both counters were triggered simultaneously. The chamber could be triggered within about 0.01 s of the passage of the cosmic ray. In this way, they achieved a very high success rate in photographing the tracks of cosmic rays passing through the chamber. Blackett and Occhialini obtained many excellent photographs of the positive electrons. On many occasions showers containing equal numbers of positive and negative electrons were observed. In both sets of experiments, showers containing both positive and negative electrons were created by cosmic ray interactions with the body of the apparatus.

The discovery of the positive electron or *positron* coincided almost exactly with Dirac's theory of the electron. In one of the great theoretical extensions of quantum mechanics, Dirac succeeded in deriving the relativistic wave equation for the electron which not only predicted its spin and magnetic moment but also the existence of what we would now call the *antiparticle* to the electron, the positron.

There were more surprises in store, however. Anderson noted that there often seemed to be much more penetrating positive and negative particle tracks in the cloud chamber pictures. These particles were bent much less than the electrons and positrons in the magnetic field and displayed little evidence of interaction with the gas in the chamber. By 1936, Anderson and Neddermeyer were sufficiently confident of their results to announce the discovery of particles with mass intermediate between that of the electron and the proton. The 'mesotrons' as they were called has mass in the range about 50–400 times the mass of the electron, the best estimate being about 200 m_e. This discovery coincided rather nicely with another theoretical prediction, this time based upon Yukawa's theory of the strong force which binds neutrons and protons together in the nucleus. According to

Yukawa's theory, published in 1936, the strong short-range force which binds the constituents of the nucleus together could be understood in terms of the exchange of particles about 250 times as massive as the electron. The mesotrons seemed to be remarkably like the predicted particles. In fact, the particles discovered by Anderson and Neddermeyer, nowadays known as *muons*, are not the particles which bind nuclei together. The identification was somewhat unsatisfactory because the mesotrons showed so little interaction with the nuclei in the chamber whereas the exchange particle is expected to show a strong interaction with nuclei. Nonetheless, a new type of particle, the muon, had been discovered.

The same procedures were used immediately after the Second World War by Rochester and Butler who constructed a new cloud chamber to use with a large electromagnet obtained by Blackett before the War. In 1947 they reported the discovery of two cases of particle tracks in the form of 'V's with apparently no incoming particle. They suggested, correctly as it turned out, that the Vs resulted from the spontaneous decay of an unknown particle whose mass could be estimated from the decay products. One of the particles was neutral and the other charged and both had mass about half that of the proton. To obtain higher fluxes of cosmic radiation, the experiments were repeated at much higher altitudes. Two years later, the experiments were carried out by Blackett's group working at the Pic du Midi Observatory in the Pyranees and by Anderson and Cowan on White Mountain in California. Many more examples of Vs were found and this class of particle became known as *strange particles*. Both neutral and charged strange particles were discovered. Most of them had mass about half that of the proton and are what are now referred to as charged and neutral *kaons* (K^+, K^-, K^0). There were a few examples, however, of neutral particles with mass greater than the mass of the protons – these are now known as *lambda* particles (Λ). What puzzled physicists was their long lifetimes – 10^{-8} and 10^{-10} s which is many orders of magnitude greater than the timescale associated with the strong interactions.

Meanwhile another powerful tool for the study of particle collisions and interactions had been developed by Powell at Bristol University. Photographic plates had played a key role in the discovery of X-rays and radioactivity in the 1890s. As we will discuss in more detail later, photographic emulsions are activated by the passage of charged particles which leave tracks showing their paths through the emulsion. Powell, in collaboration with the Ilford company, developed special 'nuclear' emulsions which were sufficiently sensitive to register the tracks of protons, electrons and all the other classes of charged particle which had been discovered. Powell and his colleagues mastered the techniques of producing thick layers of emulsion by stacking layer upon layer of emulsion which could then be separated and developed. The result was a three-dimensional picture of the interactions taking place in the emulsion. Among the first discoveries using this high precision technique was that of the *pion* (π) in 1947 which was the particle predicted by Yukawa in 1936. The photographic emulsions showed clearly the production of charged pions (π^+, π^-) in cosmic ray interactions and then their decay within a few tenths of a millimetre into muons which subsequently decayed into electrons (or positrons) and invisible neutrinos (Fig. 1.6). The nuclear

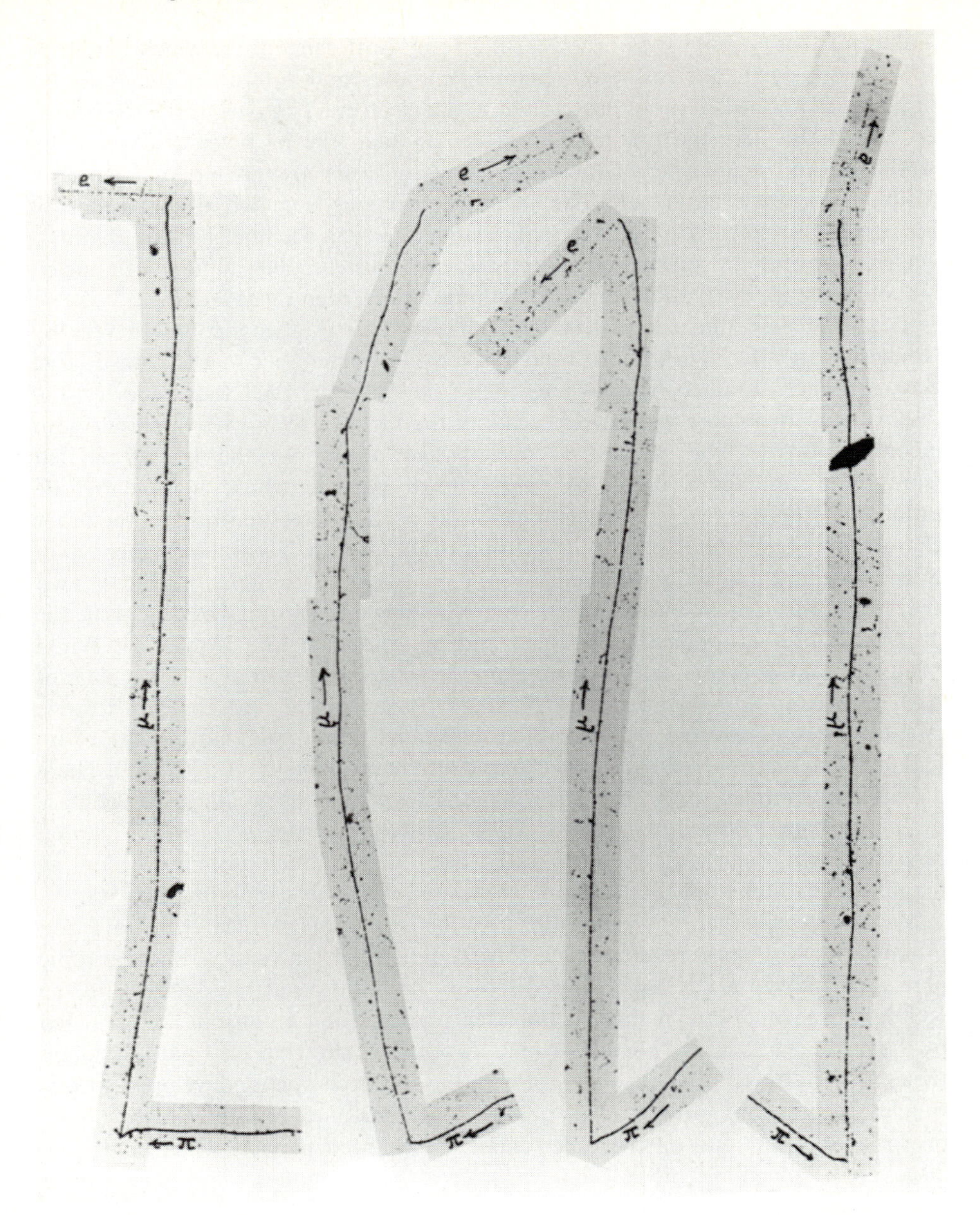

Figure 1.6. Four examples of the decay of a pion into a muon, followed by the subsequent decay of the muon into an electron. These processes were discovered by Powell and his collaborators using nuclear emulsions. (From C. F. Powell, P. H. Fowler and D. H. Perkins (1959). *The study of elementary particles by the photographic method*, page 245, Plate 8-5, Oxford: Pergamon Press.)

emulsions enabled the whole sequence of interactions and decays to be studied in a single photographic image.

Two more types of particle were discovered through cosmic ray studies. The Ξ^- particle left a clear signature in a bubble chamber photograph recorded by the Manchester group at the Pic du Midi Observatory in 1952. The decay took place within the cloud chamber and showed a decay chain which ultimately ended up with the production of a proton. The Σ particle was discovered by a group of Italian physicists in 1953.

By 1953, accelerator technology had developed to the point where energies comparable to those available in the cosmic rays could be produced in the laboratory. These 'artificial' cosmic rays had the great advantage that the beams could be produced with known energies and directed precisely onto the target. After about 1953, the future of high energy physics lay in the accelerator laboratory rather than in the use of cosmic rays. The interest in cosmic rays shifted to the problems of their origin and their propagation in astrophysical environments from their sources to the Earth.

1.2.4 *Cosmic ray astrophysics from space and from the ground*

The experiments carried out using cloud chambers showed that showers of cosmic ray particles are often observed. Most of the cosmic ray particles observed at the surface of the Earth are, in fact, secondary, tertiary or higher products of very high energy cosmic rays entering the top of the atmosphere. There had been much work from balloons and from the tops of high mountains but the whole subject was revolutionised with the development of rockets and, more important, satellites which could stay outside the atmosphere for long periods.

The astrophysical study of the origin and propagation of the cosmic ray particles had to await the 1960s when cosmic ray particle detectors were flown in satellites. These observations established many crucial facts about the particles detected in the cosmic radiation. This is the story which we will tell in more detail in the following chapters but let us summarise some of the key results which will have to be built into our astrophysical picture. First of all, the energy spectra of the particles are almost exactly the same as the typical spectrum of high energy particles inferred to be present in both Galactic and extragalactic non-thermal radio sources. In the region of the energy spectrum which is unaffected (or, to put it technically, unmodulated) by the propagation of the particles to the Earth through the Solar Wind ($E \geqslant 10^9$ eV), the energy spectra of the cosmic ray particles can be described by

$$N(E)\,\mathrm{d}E = KE^{-x}\,\mathrm{d}E$$

with $x \approx 2.5$–2.7. This relation is found to be applicable for protons, electrons and nuclei with energies in the range 10^9–10^{14} eV. The relation of this flux of particles to the relativistic gas inferred to be present in the interstellar gas is through two types of observation. First, the synchrotron radiation of cosmic ray electrons is detected in the radio waveband and, second, the Galactic γ-ray emission at energies $E \gtrsim 100$ MeV is attributed to the decay of neutral pions π^0 created in collisions between cosmic ray protons and nuclei and the nuclei of atoms, ions and molecules in the interstellar gas. The fact that these very different types of astronomy can be brought successfully to bear on these problems indicates that the cosmic ray particles observed at the top of the atmosphere are only part of a population of high energy particles pervading the whole Galaxy.

The chemical composition of the cosmic rays is similar to the abundances of the elements in the Sun, with some important exceptions, particularly for the light elements such as lithium, beryllium and boron which appear in relatively high abundances in the cosmic rays. These observations provide evidence about the chemical composition of the cosmic rays as they left their sources and also about the modifications which take place during propagation from their sources to the Earth. These observations are significant for high energy astrophysics because these are the only *particles* which we can detect which have traversed a considerable distance through the interstellar medium and which were accelerated in events such as supernovae in the relatively recent past, probably within the last 10^7 years.

At the very highest energies, cosmic rays are detected by large air shower arrays on the surface of the Earth. The arrival rate of the most energetic particles is very low indeed but particles with energies up to about 10^{20} eV have been detected. One important puzzle is the origin of these very high energy particles. Their arrival directions seem to be reasonably isotropic and, at these very high energies, these should not be significantly influenced by the magnetic field in our own Galaxy. It may be that these very high energy cosmic rays are of extragalactic origin. The acceleration mechanism for these particles is uncertain and poses a real problem for high energy astrophysicists.

1.3 Historical perspective II – the origin of high energy astrophysics

Up till 1945, astronomers could only study the Universe at large in the optical waveband. Since that time there has been an enormous expansion of the wavebands available for astronomical study. The new disciplines of radio, millimetre, infrared, ultraviolet, X and γ-ray astronomies combined with optical astronomy have led to the growth of many new areas of astrophysics. None of these developments has been more dramatic than the rise of high energy astrophysics which barely existed in 1945. The rise of high energy astrophysics has been intimately connected with the development of astronomy as a whole through this period which has unquestionably been one of the golden ages of astronomy and astrophysics.

What is the reason for this great upsurge in modern astrophysics? To put it at its very simplest, it is a question of the range of *temperatures* which are accessible for astronomical study. Fig. 1.7(*a*) shows a plot of the temperature of a black-body against the frequency (or wavelength) at which most of the radiation is emitted. Fig. 1.7(*b*) is a representation of the transparency of the atmosphere to radiation

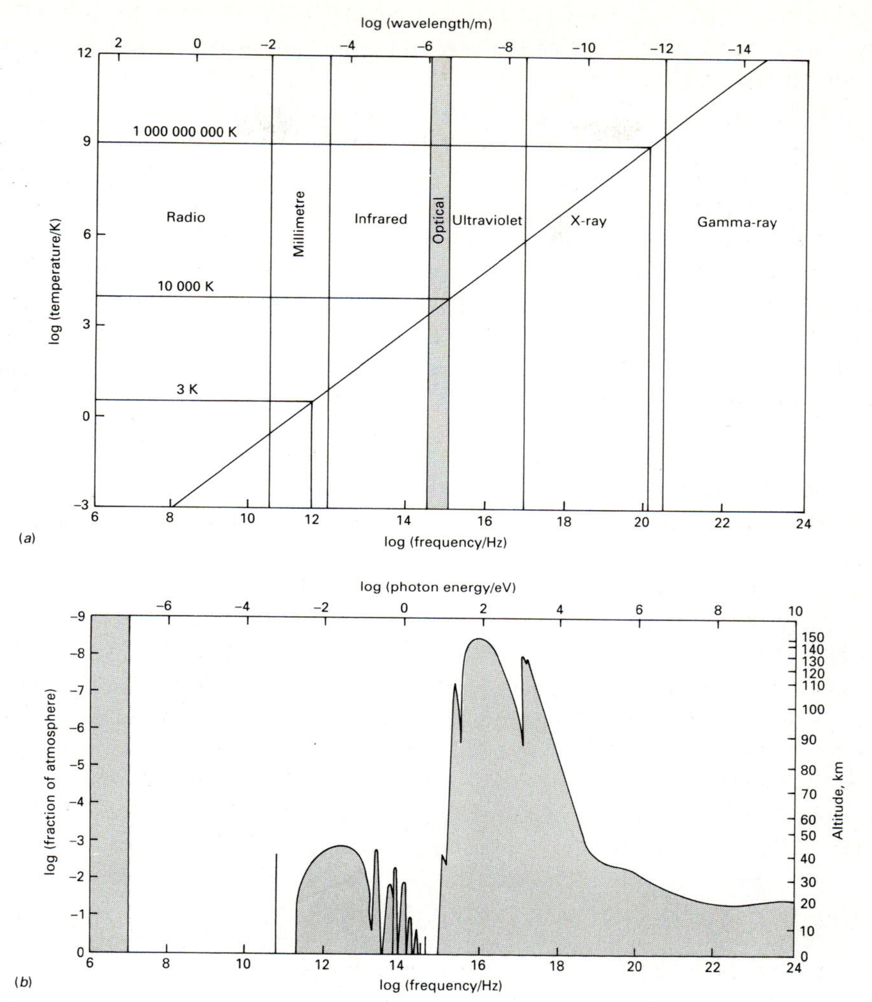

Figure 1.7. (*a*) The relation between the temperature of a black-body and the frequency (or wavelength) at which most of the energy is emitted. The frequency (or wavelength) plotted is that corresponding to the maximum of a black-body at temperature T. Convenient expressions for this relation are:

$$\nu_{\max} = 10^{11}(T/\mathrm{K})\ \mathrm{Hz}; \quad \lambda_{\max} T = 3 \times 10^{6}\ \mathrm{nm\ K}$$

The ranges of wavelength corresponding to the different wavebands – radio, millimetre, infrared, optical, ultraviolet, X and γ-rays – are shown. (*b*) The transparency of the atmosphere for radiation of different wavelengths. The solid line shows the height above sea-level at which the atmosphere becomes transparent for radiation of different wavelengths. (After R. Giacconi, H. Gursky and L. P. van Speybroeck (1968). *Ann. Rev. Astr. Astrophys.*, **6**, 373. Both diagrams are from M. S. Longair (1988). *The new physics*, ed. P. C. W. Davies, page 94, Cambridge: Cambridge University Press.)

as a function of wavelength and shows how high a telescope must be placed above the surface of the Earth for the atmosphere to become transparent to radiation of different wavelengths.

Until 1945, astronomy meant optical astronomy and Fig. 1.7(*a*) shows that this corresponds to studying the Universe in the rather small wavelength interval 300–800 nm, and hence to black-body temperatures in the range 3000–10000 K. Of course, a somewhat wider range of temperatures can be studied since bodies at temperatures outside this range emit some radiation in the optical waveband but this range of temperatures is a fair representation of the temperatures of most of the objects observed at optical wavelengths. It is from these observations that most people derive their intuitive picture of the Universe, a picture dominated by stars, galaxies and hot gas, most of these components emitting at temperatures between 3000 and 10000 K (Fig. 1.8(*d*)).

1.3.1 *Radio astronomy*

The first of the new astronomies was *radio astronomy*. Radio waves of extraterrestrial origin were discovered by Karl Jansky in the early 1930s but this caused no great stir in the astronomical world. After the Second World War, radio astronomy developed very rapidly as major advances were made in electronics, radio techniques and digital computers. Radio emission was discovered from a wide range of different astronomical objects. Some of the radio emission processes could be associated directly with phenomena observed at optical wavelengths – for example, the thermal radiation (or, more precisely, the free–free emission or bremsstrahlung) of hot electrons from regions of ionised hydrogen – but others were totally new. The radio emission did not possess the spectrum of the thermal radiation of hot gas. It was soon established that, in most of the sources, the radio emission was synchrotron radiation, the emission of ultrarelativistic electrons spiralling in magnetic fields (Fig. 1.8(*a*)). Contrary to what might have been expected from Fig. 1.7(*a*), these radio observations provide information about the very hottest, indeed relativistic, plasmas in the Universe.

Observations made in the 1950s established that most of the discrete radio sources are extragalactic objects and two features were of particular significance. First, a number of the most massive galaxies known were found to be extremely powerful sources of radio waves. They are so powerful that it is easy to detect them as radio sources at cosmological distances i.e. at distances such that the Universe was much younger than it is now when the radio waves were emitted. Simple calculations of the amount of energy necessary to power these radio sources showed that they must contain an energy in relativistic matter equivalent to the rest mass energy of about 100 million solar masses (i.e. about $10^8 M_\odot c^2 \approx 2 \times 10^{55}$ J). These galaxies must be able to convert mass of this order into relativistic particle energy. The second key fact is that the radio emission does not generally originate from the galaxy itself but comes from giant radio lobes which extend far beyond the confines of the parent galaxy. In the 1960s and 1970s it was established that the sources of these vast energies are the active nuclei of these galaxies and the

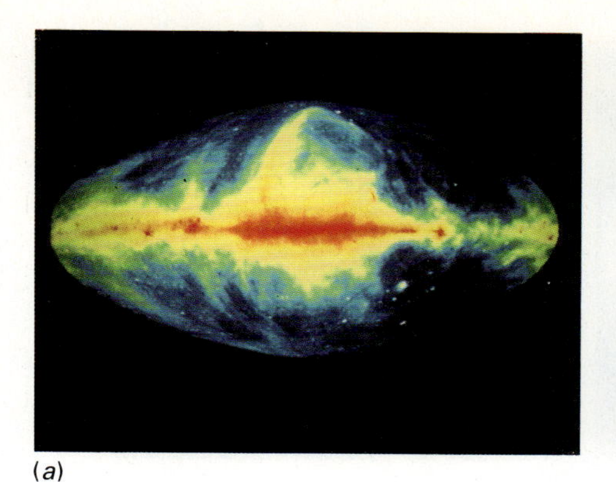

(*a*)

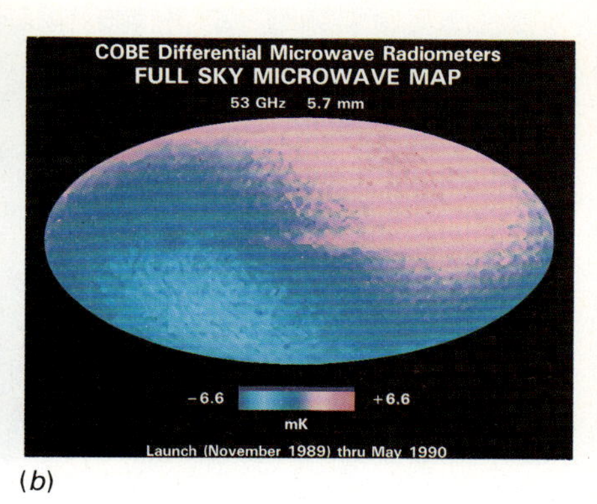

(*b*)

Figure 1.8. The whole sky as observed in different astronomical wavebands. In each image, the plane of our Galaxy, the Milky Way, lies along the centre of the image and the direction of the centre of our Galaxy lies in the centre of each diagram. This form of projection is known as an Aithof projection and in it equal areas on the surface of the celestial sphere are preserved but the orthogonality of the coordinate system is distorted away from the Galactic equator. The coordinate system shown is known as *Galactic coordinates*. The north and south galactic poles ($b = \pm 90°$) are the top and bottom of each diagram. The scale of Galactic longitude runs from 0° at the centre, through +180° at the left of each image and then from +180° at the right of each image to 360° (or 0°) at the centre.

Before 1945, only the optical waveband (*d*) was available for astronomical study. All the others have been opened up since that time as outlined in the text.

(*a*) *Long radio wavelengths* (408 MHz; 73 cm). This image is dominated by the radio emission of relativistic electrons gyrating in the interstellar magnetic field, the process known as *synchrotron radiation*. The radiation is most intense in the plane of the Galaxy but it can be seen that there are extensive 'loops' and filaments of radio emission extending far out of the plane. At high Galactic latitudes, there is a radio background component, most of it associated with the Galactic disc and the halo of the Galaxy but some of it associated with an isotropic component of diffuse radiation. In addition, at high Galactic latitudes, there are many discrete radio sources, some of which are visible on this image. The vast majority of these sources are of small angular diameter. If a survey of the discrete radio sources is made, their distribution is found to be isotropic and the integrated intensity of these sources can account for the isotropic background radiation at long radio wavelengths. (Courtesy of Dr Glyn Haslam, Max-Planck-Institut für Radioastronomie, Bonn.)

(*b*) *Millimetre wavelengths* (53 GHz; 5.7 mm). This image of the sky was made by the COBE satellite which made a complete map of the sky at this wavelength. The image is a differential map in the sense that all the intensities are measured relative to the mean temperature of the Microwave Background Radiation at a fixed point in the sky which has brightness temperature 2.736 K. The colour coding of the temperature scale corresponds to a difference between the maximum and minimum temperatures of about 6 mK. The Galactic plane can be seen, the radiation being associated with the free–free emission (or bremsstrahlung) of extensive regions of ionised hydrogen. The image is, however, dominated by the 'dipole' component which is hottest in the direction $l = 270°$, $b = 30°$ and coolest in the direction $l = 90°$, $b = -60°$. The amplitude of this dipole component amounts to a temperature fluctuation of $\Delta T/T \approx 10^{-3}$. This can be wholly attributed to the motion of the Earth through the frame of reference in which the Microwave Background Radiation is 100% isotropic at a velocity of about 350 km s^{-1}. (Courtesy of Dr M. Hauser and NASA.)

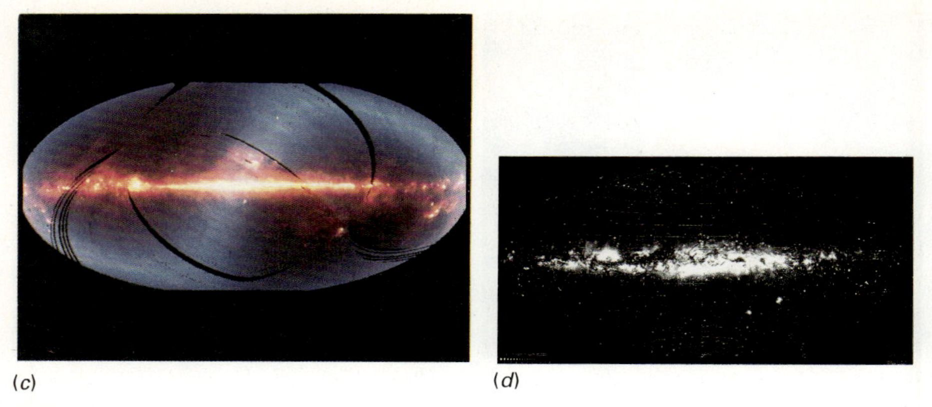

Figure 1.8. (*c* and *d*).

(*c*) *Far-infrared wavelengths* (60–100 μm). This image of the sky was generated from the IRAS all-sky survey and shows the distribution of radiation at 60 μm. The picture is dominated by emission from the Galactic plane, the radiation being the reradiated emission of heated dust grains. This map therefore delineates regions in which active star formation is proceeding. In addition, a broad band of radiation can be observed stretching across the map from top right to bottom left. This is the thermal radiation of zodiacal dust which is dust lying in the ecliptic plane of our own Solar System and which is heated by the Sun. (Courtesy of NASA, the Jet Propulsion Laboratory and the Rutherford-Appleton Laboratory.)

(*d*) *Optical wavelengths* (500–600 nm). A painting of the whole sky as observed at optical wavelengths by M. and T. Keskula of the Lund Observatory. The painting reproduces accurately the optical images of nebulae and the observed brightnesses of stars down to tenth magnitude. This image is convincing evidence that we live in a disc-shaped galaxy. The nearby dwarf companion galaxies to our own Galaxy, the Large and Small Magellanic Clouds, are seen in the Southern Galactic Hemisphere at about Galactic longitudes 290° and 310° respectively. (Courtesy of the Lund Observatory.)

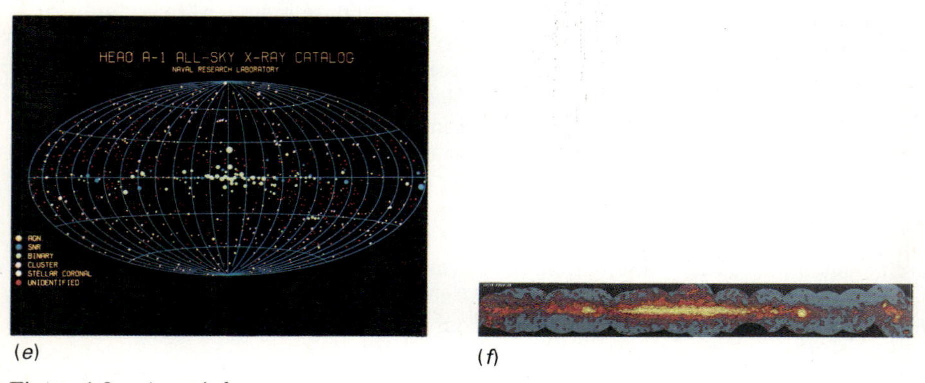

Figure 1.8. (*e* and *f*).

(*e*) *X-ray wavelengths* (1–5 keV). This distribution of bright X-ray sources was derived from the HEAO-1 survey of the X-ray sky. The image shows the distribution of different types of X-ray source seen among the brightest objects. There is a concentration of the brightest sources towards the plane of the Galaxy and towards the Galactic Centre but at high galactic latitudes the distribution of sources is isotropic extended sources result from the expulsion of this energy from the nuclei in the form of jets of relativistic plasma.

These discoveries revealed a major new component of the Universe which had been previously unrecognised – *relativistic plasma*. It is present in all galaxies and in intergalactic space. Of particular interest from the perspective of our approach to high energy astrophysics is the fact that the cosmic rays which we measure at the top of the atmosphere are a sample of the relativistic plasma present in the interstellar medium of our own Galaxy. These astronomical discoveries were the touchstone for the explosive growth of high energy and relativistic astrophysics over the last 25 years. The basic result of these new discoveries was that the energy demands and the timescales over which the energy had to be released were so extreme that conventional astrophysical sources of energy, in particular nuclear energy, were inadequate and relativistic phenomena associated with compact energy sources had to be investigated in detail.

The study of these radio sources led to further discoveries. Amongst the earliest of these was the fact that supernovae, or exploding stars, are very powerful sources of relativistic plasma. The study of the radio galaxies led to the discovery in the late 1950s of a class of galaxies known as *N-galaxies* which have very bright star-like nuclei in which there is a great deal of high energy activity as demonstrated by the observation of strong, broad emission lines in their spectra and their strongly varying optical continuum radiation. The culmination of these studies was the discovery of the *quasi-stellar radio sources*, or *quasars*, in the early 1960s, in which the starlight of the galaxy is completely overwhelmed by the intense non-thermal optical radiation from the nucleus. In some cases, the optical emission from the nuclear regions can be more than 1000 times greater than that of the parent galaxy (Fig. 1.9). These objects and their close relatives, the *BL-Lacertae* or *BL-Lac objects*, which were discovered in 1968, are the most powerful energy sources known in the Universe. Because of this, these objects can be observed at very great distances and provide important diagnostic tools for cosmology. The most distant quasars now known emitted their light and radio waves when the Universe was less than one-fifth of its present age. Because they can be observed at such great distances, the quasars can provide important information about the way in which high energy astrophysical activity in galaxies has changed as the Universe grows older.

---

within the limits of the available statistics. There is, in addition, an intense background component of diffuse X-ray emission. Its origin is as yet uncertain but discrete sources such as active galaxies must make up a significant fraction of the background. (Courtesy of NASA.)

(*f*) *γ-ray wavelengths* (70 MeV–5 GeV). This map of the γ-ray emission from the Galactic plane was made by the COS-B satellite. The emission consists of diffuse γ-ray emission from the interstellar gas, most of it probably associated with γ-rays produced in the decay of π^0 particles generated in collisions between cosmic ray protons and nuclei and the interstellar gas. In addition, 25 discrete sources of γ-rays have been detected including the pulsars in the Crab and Vela supernova remnants and the quasar 3C 273. (Courtesy of Dr K. Bennett and the European Space Agency.)

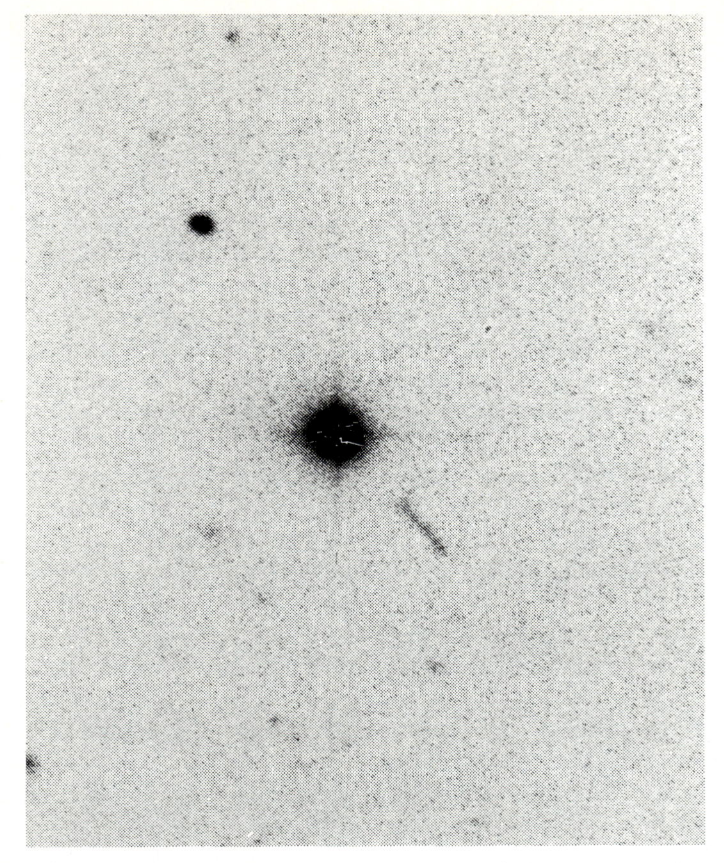

Figure 1.9. The quasar 3C 273. This deep image taken by Dr Halton Arp shows the quasar and its famous jet pointing towards the bottom right-hand corner of the image. The faint smudges to the south of the quasar are galaxies at the same distance as the quasar. (H. A. Arp (1981). *Optical jets in galaxies*, eds B. Battrick and J. Mort, page 55, ESA Publications SP-162.)

As studies of active galactic nuclei were mushrooming in the 1960s, two other great discoveries were made in radio astronomy. The first was the discovery of the *Microwave Background Radiation* by Penzias and Wilson in 1965. Subsequent studies showed that this radiation is remarkably isotropic in the sense that it has the same intensity in all directions on the sky to a very high degree of precision (Fig. 1.8(*b*)) and has an almost perfect black-body spectrum (Fig. 1.10). This radiation is the cool remnant of the equilibrium radiation spectrum formed early in the hot, dense phases of the expanding Universe.

The second was the discovery of *pulsars*. In 1967 Bell and Hewish constructed a radio telescope to study very short timescale fluctuations imposed upon the intensities of compact radio sources by density fluctuations in the interplanetary plasma streaming out from the Sun, what is known as the Solar Wind. Early in these studies, sources consisting entirely of pulsed radio emission with very stable

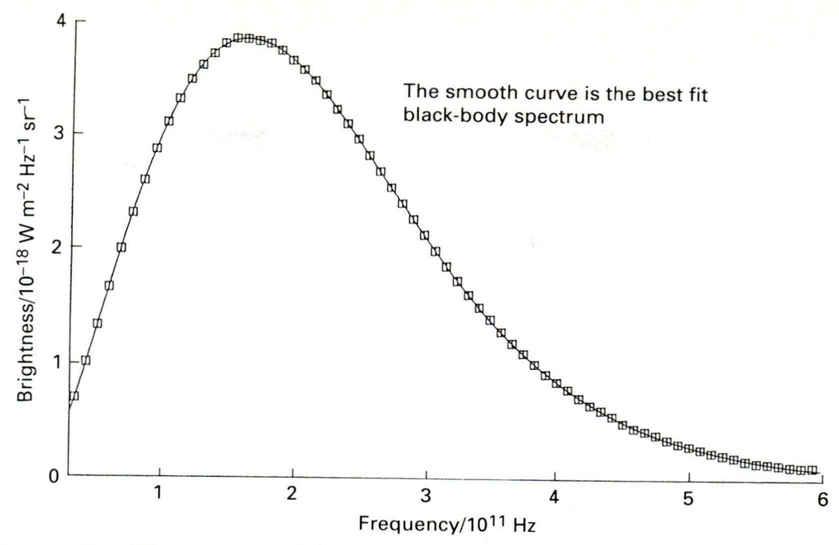

Figure 1.10. The spectrum of the Microwave Background Radiation as measured by the Cosmic Background Explorer (COBE) in the direction of the North Galactic Pole. The error boxes on each point are 1% errors. The best fitting black-body curve has radiation temperature 2.735 ± 0.06 K. (Mather *et al.* (1990). *Astrophys. J.*, **354**, 37–41.)

periods of about 1 s were discovered. They were soon identified conclusively as rotating, magnetised *neutron stars* and thus provided the first definite proof of the existence of these highly compact stars in which the central densities are as high as 10^{18} kg m^{-3}. A key point from the perspective of relativistic astrophysics was the fact that solar mass objects had been discovered with radii within a factor of 3 of the Schwarzschild radius of solar mass black holes. Thus, in these compact objects, general relativity is no longer simply a small correction term to the equations of motion. The effects of general relativity are strong and these objects provide laboratories for the study of matter in strong gravitational fields.

1.3.2 *X-ray and γ-ray astronomy*

X-ray astronomy can only be carried out at very high altitudes because of photoelectric absorption of X-rays by the atoms and molecules of the Earth's atmosphere (see Fig. 1.7(*b*)). Thus, it was only after rockets capable of lifting scientific payloads above the atmosphere became available that the exploration of the X-ray sky was possible. These rocket flights provided only about five minutes of observation above the atmosphere but this was enough, even in the first rocket flights of 1962 and 1963, to show that the X-ray sky was rich for astrophysical study. As in the case of the radio waveband, the sources which were first observed had not been predicted by astrophysicists. Amongst the earliest detections in the 1–10 keV waveband were the supernova remnant the Crab Nebula, the nearby radio galaxy M87, a number of stellar X-ray sources, which seemed to be highly variable, and the diffuse X-ray background radiation.

The full scope of X-ray astronomy became clear in the early 1970s with the launch of the first dedicated X-ray satellite, the UHURU satellite Observatory, which mapped the X-ray sky and provided systematic monitoring of variable X-ray sources (Fig. 1.8(*e*)). Some remarkable discoveries resulted from this mission. The variability of some of the Galactic X-ray sources was found to be due to the fact that the compact X-ray emitter is a member of an eclipsing binary star system. In a number of these cases, the X-ray binaries were found to contain 'pulsating' X-ray sources and these were soon identified with magnetised rotating neutron stars but, in the cases of the X-ray sources, the source of energy is the infall of matter transferred from the primary star, the process known as *accretion*. The X-ray emission is basically the thermal emission of very hot gas heated up as it falls onto the magnetic poles of the neutron star. In the case of the pulsating X-ray sources, the inferred masses are consistent with their being neutron stars but, intriguingly, there are a few X-ray sources in which the mass of the invisible secondary is inferred to be greater than the upper limit for stable neutron stars. These are all candidates for *black holes* in binary systems and as such are objects of the greatest astrophysical interest.

In extragalactic astronomy, the nuclei of active galaxies were found to be intense and often variable X-ray sources. It is likely that this emission is associated with the emission of ultrarelativistic gas generated close to the nucleus itself but the precise nature of the emission is not clear. One class of extragalactic source in which the emission processes are understood is the X-ray emissions of diffuse hot gas in clusters of galaxies. The mass of the cluster forms a deep gravitational potential well in which the gas must be very hot if it is to form a stable extended atmosphere. This is what is observed in a number of great clusters. The X-ray emission is extended, filling the core of the cluster and the gas responsible for the emission has temperature in the range 10^7–10^8 K. The emission process is thermal free–free emission (or bremsstrahlung) as is confirmed by the observation of very highly ionised iron lines from the intracluster gas.

In 1978, the Einstein X-ray Observatory was launched. It provided the first high resolution images of many X-ray sources and made very deep surveys of small areas of sky. Many different classes of astronomical object were detected as X-ray sources including regions of star formation and normal galaxies. Perhaps most significant of all was the fact that X-ray emission was detected from all types of star and not just from the binary X-ray sources where there are special reasons why there should be strong X-ray emission.

γ-ray emission from the plane of our Galaxy was first detected by the OSO III satellite in 1967. This was followed by the SAS-2 satellite which discovered the diffuse γ-ray background and by the COS-B satellite which provided a detailed map of the Galactic γ-ray emission and discovered about 25 discrete γ-ray sources. These included the pulsars in the Crab and Vela supernova remnants and the quasar 3C 273. The γ-ray image of the sky is dominated by the emission from the Galactic plane (Fig. 1.8(*f*)). At high photon energies, $\varepsilon \geqslant 100$ MeV, the principal emission mechanism is the decay of neutral pions generated in collisions between the nuclei of atoms and molecules of the interstellar gas and cosmic ray protons and nuclei. At lower energies, non-thermal processes, in particular inverse Compton scattering and bremsstrahlung, can make important contributions to the background γ-ray emission.

The first evidence of γ-ray line emission came from balloon observations in the early 1970s by the Rice University Group. In 1977 definitive observations of the electron–positron annihilation line at 511 keV in the direction of the Galactic Centre were made by Leventhal and MacCallum in balloon observations. Since then observations have also been made of the 1.809 MeV line of radioactive ^{26}Al by the HEAO-C satellite, this line also being detected from the direction of the Galactic Centre. Another unexpected discovery was that of γ-ray bursts which were detected by the US Vela and also by Soviet satellites. Over 100 of these events have now been observed over the last ten years but their nature is not yet established.

Very high energy γ-rays ($\varepsilon \approx 10^{11-12}$ eV) have recently been detected by a remarkable ground-based technique. γ-rays of these energies initiate small electron–photon cascades in the upper atmosphere. The electrons are of such high energy that their velocities exceed the speed of light in air and consequently they emit optical Cherenkov radiation. The optical light emitted by these showers is detected at sea-level by simple telescope arrays. Several well-known sources, including Cygnus X-3, Hercules X-1 and the Crab pulsar have been detected as γ-ray emitters at about 10^{12} eV. An important result claimed by the Durham group is the discovery of a 12.6 ms pulsar at γ-ray energies in the binary X-ray source Cygnus X-3. Finally, at even higher γ-ray energies, $\varepsilon \approx 10^{16}$ eV, the Kiel and Leeds University groups have claimed to detect γ-rays from Cygnus X-3 using ground-based cosmic ray air-shower detector arrays. At these very high energies, the fluxes of γ-rays are very low indeed, typically only about one γ-ray per month being detected by the Haverah Park team.

1.3.3 *Ultraviolet and infrared astronomy*

As soon as observations from above the atmosphere became possible, one of the obvious developments was the extension of the classical techniques of optical astronomy to the ultraviolet spectral regions, 120–320 nm. Ultraviolet spectrographs were flown on rockets in the mid-1960s and were followed by the series of orbiting astrophysical observatories culminating in the launch of the International Ultraviolet Explorer (IUE) in 1978.

As expected, a wide range of hot objects could be studied but perhaps of most importance was the fact that a wide range of the common elements could be studied because their strong resonance transitions fall in the ultraviolet spectral region. Active galaxies and quasars are particularly strong emitters in the ultraviolet waveband because the non-thermal radiation in the nucleus observed in the optical waveband extends to far-ultraviolet wavelengths. This continuum radiation excites a wide range of ions and atoms which emit strong resonance lines in the ultraviolet waveband. These lines have proved to be particularly valuable diagnostic tools for the astrophysics of active galactic nuclei.

The waveband 120–300 nm has now been extensively studied spectroscopically by the IUE but the shorter ultraviolet wavelengths, $\lambda < 120$ nm, remain relatively unexplored. There are two reasons why this is a difficult waveband for astronomical observations. First, there is the problem of constructing an efficient telescope because most materials are strongly absorbant for normal incidence optics at wavelengths shorter than about 120 nm. Second, the interstellar gas is expected to be opaque to radiation of wavelength less than 91.2 nm corresponding to the wavelength at which the Lyman continuum begins, and this restricts the targets available for study. Fortunately, it appears that the interstellar medium is sufficiently clumpy for there to be 'holes' through which the more distant Universe can be observed.

The development of infrared astronomy has been largely determined by the availability of detector materials for the wavelength range 1 μm–1 mm. There are certain wavelength 'windows' which can be successfully exploited from the surface of the Earth in the ranges 1–30 μm and 350 μm–1 mm (see Fig. 1.7(*b*)). In the range 30–350 μm, however, the Earth's atmosphere is opaque and observations can only be made from high flying aircraft, balloons and satellites.

The distinctive problem to be overcome in infrared astronomy is the fact that the telescope and the Earth's atmosphere are strong thermal emitters of infrared radiation. For example, the radiation of a black-body at room temperature, say 300 K, peaks at a wavelength of about 10 μm. Therefore, normally, the strength of the signal from an astronomical source is very much weaker than the background due to the telescope and the atmosphere.

Besides the ability to study cool objects in the temperature range roughly 3000–1 K, there are two other important features of the infrared waveband. First, there is a great deal of *dust* in the interstellar gas in galaxies but the dust grains become transparent in the infrared waveband and hence it is possible to look deep inside regions which are obscured at optical wavelengths. Second, at wavelengths longer than about 3 μm, dust grains become strong emitters rather than absorbers of radiation. They emit more or less like little black-bodies at the temperature to which they are heated by the radiation they absorb. They do not radiate at shorter wavelengths because, if the grains were heated to temperatures greater than about 1000 K, they would evaporate.

The first complete survey of the far infrared sky was carried out by the Infrared Astronomical Satellite (IRAS) in 1983–4. It revealed intense emission from regions of star formation in our own Galaxy and nearby galaxies as well as a host of new detections of stars, galaxies, active galaxies and quasars (Fig. 1.8(*c*)).

1.3.4 *Neutral hydrogen and molecular line astronomy*

One of the great predictions of modern astronomy was made during the Second World War by van de Hulst who worked out which emission and absorption lines of atoms, ions and molecules might be detectable from astronomical sources in the radio waveband. The most significant prediction was that neutral hydrogen should emit line radiation at a wavelength of about 21 cm because of the minute change in energy when the relative spins of the proton and electron in a hydrogen atom change. Although this is a highly forbidden transition with a spontaneous transition probability for a given hydrogen atom of only once every 12 million years, there is so much neutral hydrogen present in the Galaxy that it should be detectable. In 1951, the 21 cm line of neutral hydrogen was discovered by Ewan and Purcell and it has proved to be one of the most powerful tools for diagnosing not only the properties of the interstellar gas but also the dynamics of galaxies since the line is so narrow that it provides an excellent measure of the velocity fields inside galaxies.

Molecules had been known to exist in the interstellar medium from the absorption bands seen in the optical spectra of stars. The real significance of molecular line astronomy only became apparent, however, with the development of high precision radio telescopes and line-receivers working in the centimetre and millimetre wavebands. In 1967, the hydroxyl radical, OH, was first detected by radio techniques. In many ways, this was an unexpected detection because the signals were very strong indeed and variable in intensity. In fact, the inferred temperature of the source regions was greater than 10^9 K, indicating that some form of *maser* action must be pumping the energy levels of the molecules. As in cases of masers and lasers, the populations of the energy levels of the molecules must be far from equilibrium in such a way that enormous intensities in the lines are observed, far exceeding those expected from the thermodynamic temperature of the source region.

Soon, many more molecules were discovered through their molecular line emission in the centimetre, millimetre and sub-millimetre wavebands, including species such as ammonia, water vapour and even ethanol. Most of the molecular lines which have been discovered are associated with the rotational transitions of molecules, linear molecules with up to 11 carbon atoms having now been detected. The importance of these studies is that they have led to the development of the new discipline of *interstellar chemistry*. For the molecules to survive, it is essential that they should be shielded from the intense interstellar ultraviolet radiation. It is therefore not surprising that they are found in large abundances in dusty star-formation regions into which ultraviolet radiation cannot penetrate. These observations are of special importance because star formation is one of the key areas of contemporary astronomy which is most poorly understood.

1.3.5 *Optical astronomy*

It is perhaps ironic that a survey of developments in modern observational astronomy should end rather than begin with the oldest astronomical discipline – optical astronomy. We first note the great technical developments in instrumental technique which have taken place over the last 20 years. Until 1945, essentially all astronomy was undertaken by photographic techniques. The period 1960–90 has seen the gradual replacement of photographic plates by electrooptical detectors. Photographic plates typically have a quantum efficiency of only a few per cent so that most of the light incident upon the plate is not detected. The most recent electronic detectors such as charge-coupled devices (CCDs) have quantum efficiencies of about 60–70 % at the red end of the optical spectrum (500–1000 nm). Furthermore, detectors such as CCDs have a linear response to the intensity of the incident radiation whereas photographic plates are non-linear in the sense that they saturate at high brightness levels.

The great advance represented by these new devices is that nowadays astronomers deal directly with digital data and can perform much more readily procedures such as sky subtraction and intensity and wavelength calibration of their data. This development would not have been possible without the continued development of electronic computers. Nowadays, it is routine to deal with arrays of, say, 1000×1000 picture elements and each element can be separately processed by computer. This has resulted in a great increase in the quantity and quality of the data which the astronomer can study astrophysically. The only present limitation in the use of these electrooptical devices is that they have only a small field of view. Thus, the photographic plate still has advantages when very wide field images or very long-slit spectra have to be obtained.

What is the present status of optical astronomy in the era of the new astrophysics? There is no question but that it continues to play a central role in all astrophysics. The fundamental reason for this is that a large fraction of the matter in the Universe is locked up in stars with masses within about a factor of 10 of that of the Sun and these emit a large fraction of their energy in the optical waveband. Since they have long lifetimes, they are the most readily observable objects in the Universe. The stars are assembled into galaxies and these are the basic building blocks of the Universe. Furthermore, the evolution of stars is one of the most exact of the astrophysical sciences and so stars provide among the best probes of the evolutionary history of any stellar system.

Thus, whenever observations are made in the new wavebands, it is important to relate the objects detected to what is observed optically in the same region of space. One particularly important aspect of the procedure of associating optical objects with objects detected in other wavebands is that very often distances can only be estimated from optical observations, for example from the properties of the associated star or the redshift of a galaxy. The key point to be appreciated is the richness of the optical spectrum in emission and absorption features.

1.3.6 *Theoretical astronomy*

Unquestionably the years since 1945 have been one of the most exciting periods in theoretical astronomy. There are three points which are worth noting. First, the theoretical developments have been largely stimulated by the great observational discoveries of the period. What I find particularly striking is the completely new range of astrophysical tools which have had to be developed by theoretical astronomers before the interpretation of the observations can be undertaken. All the discoveries of the new astronomies have resulted in entirely new pieces of theoretical astrophysics and these, in turn, have deepened our understanding of the behaviour of matter in circumstances which are not found within terrestrial laboratories. Some obvious examples are the interiors and environments of neutron stars, the nuclei of galaxies, X-ray emission from accretion processes in X-ray binaries and so on. It is questionable whether or not advances in understanding general relativity would have been so rapid without the stimulus of active galaxies and the discoveries of modern astrophysical cosmology.

A second important feature has been the use of high speed computers in essentially all aspects of theoretical astrophysics. This has been of special importance in some of the most exact of the astrophysical sciences. An excellent example is the study of the internal structure and evolution of stars. The development of powerful computer codes for stellar models has resulted in precise predictions of the surface properties of most classes of star. Detailed comparison between theory and observation is now possible and is leading to more and more precise understanding of the processes of stellar evolution. This is, however, but one example and it is fair to say that, in essentially all branches of modern astrophysics where well-defined astrophysical problems have been formulated, theorists have provided the observers with good predictions which enable sensible astrophysical questions to be asked.

The third aspect is the impact of advances in theoretical physics upon theoretical astronomy. Essentially all the major advances in theoretical physics have had an important impact upon some aspect of astronomy. The importance of nuclear physics in understanding the processes going on in the centres of stars is obvious. But there are many perhaps less obvious examples – superconductivity and superfluidity in the centres of neutron stars, the role of the electroweak theory of elementary particles in understanding how the energy of collapse may result in a supernova explosion, the role of plasma physics in understanding the dynamics of clouds and jets of relativistic particles. It is remarkable how many of these theories are playing a role in modern astrophysics and by extension how these new applications provide further tests of the theory. We will find numerous examples of this as the story unfolds.

In a spectacular example of the inverse process, Hawking's discovery of the evaporation of black holes by radiation is a fundamental piece of theoretical physics which developed from astronomical and cosmological studies. The most ambitious of these endeavours is the application of Grand Unified Theories of elementary particles and quantum gravity to the very earliest phases of the hot big bang. The relevant energies are so high that these theories can probably only be tested by using the very early Universe itself as a laboratory for ultrahigh energy physics. The fact that these ideas are being taken seriously by the best workers in the field is some measure of the advance in understanding and ambitions of astrophysicists in the 1990s.

1.4 Units, basic definitions and wavebands

1.4.1 *SI units*

We will use SI units throughout this text, although this is not at all common practice in the astronomical literature. Historically, astronomers and high energy particle physicists have used convenient non-SI units and, indeed, many professionals will not recognise some of the units we will employ! For example, the literature is full of ergs, gauss, janskies, etc. as well as specifically astronomical terms such as astronomical units, parsecs, magnitudes, colours, etc. The SI units are not always the most convenient units to use since they may result in enormously large or small numbers. Unfortunately, there is little likelihood of standardisation in the near future and one just has to learn to live with it. My policy is as follows. All formulae and derivations will be given using strict SI units and, wherever possible, I will quote numerical values in SI units. I will relax my procedures slightly when observational results are quoted but relate the units to SI units where appropriate. I will introduce non-standard units at the appropriate points in the text. It is, however, useful to pull all of them together in one place and these are displayed in Table 1.2.

Energies will play an important part in the story which follows. Normally, particle energies are quoted in *electron-volts* (eV) with the usual SI modifications for greater orders of magnitude:

$$1 \text{ kiloelectron-volt} = 1 \text{ keV} = 10^3 \text{ eV},$$
$$1 \text{ megaelectron-volt} = 1 \text{ MeV} = 10^6 \text{ eV},$$
$$1 \text{ gigaelectron-volt} = 1 \text{ GeV} = 10^9 \text{ eV},$$
$$1 \text{ teraelectron-volt} = 1 \text{ TeV} = 10^{12} \text{ eV},$$
$$1 \text{ petaelectron-volt} = 1 \text{ PeV} = 10^{15} \text{ eV},$$
$$1 \text{ exaelectron-volt} = 1 \text{ EeV} = 10^{18} \text{ eV, etc.}$$

The following conversion factors will also be found to be very useful:

$$(\text{proton rest mass}) \times c^2 = m_p c^2 = 938.3 \text{ MeV} \approx 10^9 \text{ eV} = 1 \text{ GeV}$$
$$(\text{electron rest mass}) \times c^2 = m_e c^2 = 0.511 \text{ MeV} \approx 5 \times 10^5 \text{ eV} = 0.5 \text{ MeV}$$
$$1 \text{ eV} = 1.602 \times 10^{-19} \text{ J} = 1.602 \times 10^{-12} \text{ erg}$$

Other useful energy conversion factors for photons are as follows:

$$E = h\nu = hc/\lambda = 1.2399/\lambda = 4.136 \times 10^{-15}\nu \text{ eV}$$

where the frequency ν is measured in hertz (or cycles s^{-1}) and the wavelengths λ in microns (i.e. 10^{-6} m or 10^3 nm). Although it is usual to use angstroms (Å) as wavelength units in astronomy, I will prefer to use microns or micrometres (μm) and nanometres (nm), recalling that 1 Å = 0.1 nm = 10^{-4} μm.

Very often, we will be interested in the radiation from hot bodies and then a useful conversion factor is:

$$E = kT = 1.380 \times 10^{-23} T \text{ J} = 8.617 \times 10^{-5} T \text{ eV}$$

where the temperature T is measured in degrees Kelvin.

Table 1.2. *Physical constants*

Velocity of light	$c = 2.997925 \times 10^{8}$ m s^{-1}
Gravitational constant	$G = 6.6726 \times 10^{-11}$ N m^{2} kg^{-2}
Planck's constant	$2\pi\hbar = h = 6.6261 \times 10^{-34}$ J s
Electron charge	$e = 1.6022 \times 10^{-19}$ C
Mass of electron	$m_e = 9.109 \times 10^{-31}$ kg
Mass of proton	$m_p = 1.6726 \times 10^{-27}$ kg
Boltzmann's constant	$k = 1.3807 \times 10^{-23}$ J K^{-1}
Gas constant	$R = 8.315$ J K^{-1} mol^{-1}
Avogadro's number	$N_0 = 6.0221 \times 10^{23}$ mol^{-1}
Permittivity of free space	$\varepsilon_0 = 8.8542 \times 10^{-12}$ C^{2} m^{-2} N^{-1}
Permeability of free space	$\mu_0 = 4\pi \times 10^{-7}$ H m^{-1}
Fine structure constant ($\alpha = e^2/4\pi\varepsilon_0 \hbar c$)	$\alpha = 7.297 \times 10^{-3}$
	$\alpha^{-1} = 137.04$
Classical electron radius ($r_e = e^2/4\pi\varepsilon_0 m_e c^2$)	$r_e = 2.818 \times 10^{-15}$ m
Magnetic moment of 1 Bohr magneton ($\mu_B = eh/4\pi m_e$)	$\mu_B = 9.274 \times 10^{-24}$ J T^{-1}
Radiation density constant ($a = 8\pi^5 k^4/15c^3h^3$)	$a = 7.5660 \times 10^{-16}$ J m^{-3} K^{-4}
Stefan–Boltzmann constant ($\sigma = ac/4$)	$\sigma = 5.6705 \times 10^{-8}$ W m^{-2} K^{-4}

Astronomical units

1 astronomical unit (AU)	$= 1.496 \times 10^{11}$ m
1 parallax-second (parsec, pc)	$= 3.0856 \times 10^{16}$ m
1 light year (ly)	$= 9.4605 \times 10^{15}$ m
1 solar mass ($M_\odot$)	$= 1.989 \times 10^{30}$ kg
1 solar radius ($R_\odot$)	$= 6.9598 \times 10^{8}$ m
Luminosity of Sun ($L_\odot$)	$= 3.90 \times 10^{26}$ W

1.4.2 *Special relativity, four-vectors and basic energy relations*

Everyone has their own favourite conventions for calculations in special relativity. Throughout this book, I use the notation and conventions for special relativity as presented by Rindler in his excellent book *Essential relativity* (1977). According to his conventions, the Lorentz transformations are written:

$$\left.\begin{aligned} x' &= \gamma(x - Vt) \\ y' &= y \\ z' &= z \\ t' &= \gamma(t - Vx/c^2) \end{aligned}\right\} \tag{1.1}$$

where the Lorentz factor γ is defined to be

$$\gamma = (1 - V^2/c^2)^{-\frac{1}{2}} \tag{1.2}$$

V is the relative velocity of the inertial frames of reference S and S'. The velocity of light is always written explicitly in all calculations. The components of four-vectors are defined to correspond to the components of the primitive displacement four-vector **R** which is taken to be the quantity

$$\mathbf{R} \equiv [x, y, z, t] \tag{1.3}$$

The norm of the four-vector is then the quantity

$$\text{norm}(\mathbf{R}) = c^2t^2 - x^2 - y^2 - z^2 \quad (1.4)$$

and is a Lorentz invariant in any inertial frame of reference. The components of other four-vectors are defined to be the quantities which transform like x, y, z, t. In this way, it is possible to define:

$$\left.\begin{array}{l} \text{Four-velocity } \mathbf{U} \equiv [\gamma u_x, \gamma u_y, \gamma u_z, \gamma] \\ \text{Four-momentum } \mathbf{P} \equiv m_0 \mathbf{U} \equiv [\gamma m_0 u_x, \gamma m_0 u_y, \gamma m_0 u_z, \gamma m_0] \\ \text{Four-frequency } \mathbf{K} \equiv [\hbar \mathbf{k}, \hbar\omega/c^2] \end{array}\right\} \quad (1.5)$$

and so on. For more details of the derivations of the forms of these and other four-vectors and their uses, the reader may consult Rindler's text or my own version in *Theoretical concepts in physics* (1984).

It is useful to recall the various types of energy which we will meet in dealing with relativistic particles. We will deal with the *total energies, kinetic energies and momenta* of particles. We recall the standard results. If m_0 is the rest mass of the particle, v its velocity and c the velocity of light,

Total energy = (rest energy + kinetic energy) = $\gamma m_0 c^2$
Kinetic energy = total energy − rest mass energy = $(\gamma - 1)m_0 c^2$
Relativistic three-momentum $p = \gamma m v$

We will try to use these formulae in their proper relativistic form as often as possible. However, the limiting non-relativistic and ultrarelativistic forms are frequently needed.

In the *non-relativistic case*, $v/c \ll 1$, $\gamma \to 1$

Kinetic energy = $\frac{1}{2} m_0 v^2$
Non-relativistic three-momentum = $m_0 v$

In the *ultrarelativistic case*, $\gamma \gg 1$, $v \approx c$, we find

Total energy ≈ Kinetic energy
Kinetic energy ≈ $\gamma m_0 c^2$
Relativistic three-momentum ≈ $\gamma m_0 c$
Total energy ≈ kinetic energy ≈ (relativistic three-momentum) c

1.4.3 *Astronomical wavebands*

All accessible astronomical wavebands will contribute essential information to the story we have to tell and so it is useful to define what is conventionally meant by these wavebands. The nomenclature is a mixture of history, tradition, instrumental technique and the influence of the atmosphere and ionosphere upon the detectibility of extraterrestrial sources of radiation. As a result, the boundaries between wavebands are often somewhat fuzzy.

Radio waveband $3 \times 10^6 \leqslant \nu \leqslant 3 \times 10^{10}$ Hz; 3 MHz $\leqslant \nu \leqslant$ 30 GHz; 100 m $\geqslant \lambda \geqslant$ 1 cm

The first radio astronomical observations were made at metre wavelengths but, as radio technology developed through the 1960s and 1970s, observations became possible up to the shortest centimetre wavelengths. In these wavebands,

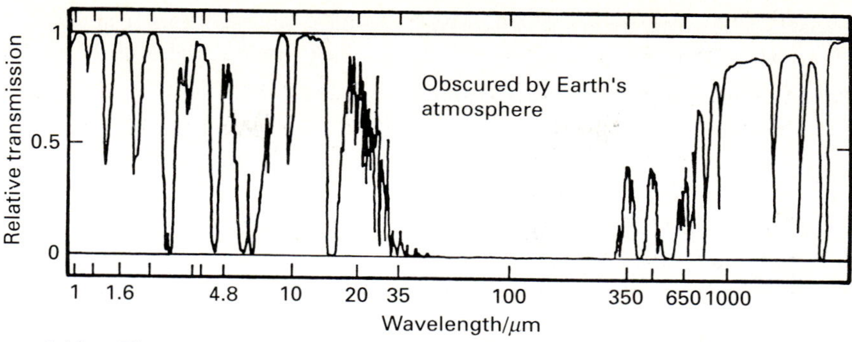

Figure 1.11. The transmission of the atmosphere as a function of wavelength in the infrared and sub-millimetre wavebands. The precipitable water vapour content of the atmosphere is assumed to be 1 mm. (Courtesy of the Royal Observatory, Edinburgh.)

observations are made with a single radio antenna or with a number of them combined into arrays which use the principle of aperture synthesis to provide high angular resolution images. At the low frequency end of this range, 1–10 MHz, observations of extraterrestrial sources become very difficult because of the reflection of radio waves by the plasma of the ionosphere. Observations are difficult at these wavelengths although there are certain favourable sites close to the auroral zones at which the sky can be observed. Even if the telescope is located above the ionosphere, observations at frequencies lower than about 1 MHz are difficult because of the same plasma reflection effects occurring in the interplanetary and interstellar plasma. The upper frequency end of the radio waveband merges into the millimetre waveband.

Millimetre and sub-millimetre waveband $3 \times 10^{10} \leqslant \nu \leqslant 3 \times 10^{12}$ Hz; 30 GHz $\leqslant \nu \leqslant$ 3000 GHz; $10 \geqslant \lambda \geqslant 0.1$ mm

Millimetre wave astronomy is similar in many respects to radio astronomy but there are important astronomical and technical differences. A distinct astronomical feature of these wavebands is the presence of a wealth of molecular lines in cool sources. The technology of the detectors is also different in that solid state mixers and bolometric detectors are used for heterodyne and continuum receivers. The transparency of the atmosphere varies dramatically with wavelength in this waveband (Fig. 1.11). At wavelengths less than about 1 mm, there are very strong absorption bands due to water vapour, carbon dioxide and other molecules in the atmosphere. The transparency of the atmosphere is particularly sensitive to the amount of water vapour in the atmosphere. To have a reasonable chance of making observations at wavelengths less than 1 mm, it is essential to observe from a high, dry site. Thus, a site such as Mauna Kea in Hawaii at 4200 m is ideal for observations in the sub-millimetre waveband ($1 \leqslant \lambda \leqslant 0.1$ mm) where it is found that there is less than about 1 mm of precipitable water vapour about 35% of the time. Notice, however, that there are only a few windows available in the sub-millimetre waveband at 0.8, 0.65, 0.45 and 0.35 mm. To make observations in the other parts of the waveband, it is necessary to make observations from above the Earth's atmosphere, either from high flying aircraft, such as the Kuiper Airborne Observatory or from satellites.

Infrared waveband $3 \times 10^{12} \leqslant \nu \leqslant 3 \times 10^{14}$ Hz; $100 \geqslant \lambda \geqslant 1$ μm

There is a distinction between those parts of the infrared waveband which can be observed from high ground-based sites and those which can only be successfully observed from above the Earth's atmosphere. Fig. 1.11 shows the transparency of the atmosphere in the waveband $1 \leqslant \lambda \leqslant 100$ μm in which it can be seen that there are 'windows' at a number of wavelengths. The centres of these windows are at wavelengths of 1.2, 1.65, 2.2, 3.5, 5, 10, 20 and 30 μm The last two windows are only accessible from high, dry sites and even observations at 10 μm are often difficult except under the best observing conditions. Observations outside these windows have to be undertaken from balloons, high flying aircraft or satellites. There is thus a complementarity between the types of observation attempted from the ground and from above the atmosphere. The waveband is often separated into *near* and *thermal* infrared wavelengths. The distinction is related to those parts of the waveband at which the observations are detector noise limited (the near-infrared) and those at which the thermal background radiation from the sky and the telescope are the dominant source of noise (the thermal infrared). The distinction thus depends upon the type of observation being undertaken. For broad-band observations using the present generation of detectors, observations at wavelengths longer than 3 μm are thermal background limited whereas those at shorter wavelengths are normally detector noise limited. In making observations in the thermal infrared waveband, the observer is almost always searching for very faint signals against an enormous thermal background.

Optical waveband $3 \times 10^{14} \leqslant \nu \leqslant 10^{15}$ Hz; 1 μm $\geqslant \lambda \geqslant 300$ nm

This waveband corresponds to the classical optical region of the spectrum in which virtually all astronomy was carried out until 1945. The wavelength range to which our eyes are sensitive is roughly 400–700 nm, corresponding to the blue and red ends of the optical spectrum respectively. Consequently the waveband from 700 to 1000 nm is often referred to as 'infrared' wavelengths. To distinguish these wavelengths from the waveband $\lambda > 1$ μm, I often refer to the waveband $700 < \lambda < 1000$ μm as the 'optical infrared' region. At the short wavelength end of the waveband, the atmosphere becomes opaque because of absorption by ozone in the upper atmosphere. This has the beneficial effect of protecting us from the Sun's hard ultraviolet radiation. The absorption sets in rather suddenly with decreasing wavelength so that observations from ground-based observatories at wavelengths less than about 320 nm are generally impossible. A characteristic feature of the optical waveband is the ability to carry out astronomical photography. Photographic emulsions have been developed which are sensitive throughout the optical waveband, although at the longest wavelengths very special emulsions are needed and require long exposure times. Photographic emulsions have been developed which extend well into the ultraviolet waveband. Nowadays, however, photographic plates have been largely replaced by opto-electronic detectors such as charge-coupled devices (CCDs) which use silicon as the detector material and which have quantum efficiencies about 50 times those of photographic plates. The band-gap in silicon corresponds to a limiting maximum wavelength of 1 μm. As a result, for many purposes, CCDs based upon silicon technology have become a dominant research tool for optical astronomy.

Ultraviolet waveband $10^{15} \leqslant \nu \leqslant 3 \times 10^{16}$ Hz; $300 \geqslant \lambda \geqslant 10$ nm

The atmosphere is opaque to radiation in this waveband because of ozone and molecular absorption and therefore astronomy in these wavebands has to be carried out from above the atmosphere, preferably from satellites. The band divides rather naturally into two regions. The region $300 \geqslant \lambda \geqslant 120$ nm can be studied using techniques similar to those used in the optical waveband. At shorter wavelengths, however, it is difficult to find materials which reflect radiation at normal incidence. Rather, the incident radiation is simply absorbed by the mirror material with little or no reflection. One solution is to use grazing incidence rather than normal incidence optics and then the ultraviolet radiation can be focussed in a similar manner to optical radiation. However, the telescopes look rather different from optical telescopes. Another problem is that at wavelengths shorter than 91.2 nm, the Lyman limit for hydrogen, it is expected that the interstellar gas becomes opaque because of photoelectric absorption by neutral hydrogen in the Lyman continuum. There is evidence that this is indeed important but, fortunately for astronomers, it appears that the neutral hydrogen is sufficiently clumpy for there to be holes through the interstellar gas which enable the more distant Universe to be observed.

X-ray waveband $3 \times 10^{16} \leqslant \nu \leqslant 3 \times 10^{19}$ Hz; $10 \geqslant \lambda \geqslant 0.01$ nm; $0.1 \leqslant E \leqslant 100$ keV

As in the case of the far-ultraviolet waveband, the atmosphere is opaque to X-rays because of photoelectric absorption by the atoms which make up the molecular gases of the atmosphere. The detectors of these energetic photons begin to resemble the detectors used in particle physics experiments. Proportional counters and scintillation detectors are used as well as other devices such as CCDs in which the total energy deposited by the X-ray on entering the detector can be measured. X-ray astronomy is wholly carried out from above the atmosphere. The photons are of such high energy that they behave like particles and the telescopes for high energy X-rays are essentially collimators in which the resolution of the telescope is determined by the geometric design of the collimator. At low X-ray energies, $0.1 < E < 1$ keV, grazing incidence optics can still be used to image the X-rays to a focal plane, but at higher energies the grazing incidence angles are so small that enormously long telescopes would be needed to focus the X-ray image.

γ-ray waveband $\nu \geqslant 3 \times 10^{19}$ Hz; $\lambda \leqslant 0.01$ nm; E $\geqslant$ 100 keV

All photons with energies greater than about 100 keV are referred to as γ-rays. Except at the very highest energies, these studies have to be carried out from above the atmosphere. Between 100 keV and 1 Mev, photoelectric absorption is the dominant absorption mechanisms but at higher energies Compton scattering and then electron–positron pair production become the dominant absorption processes. The detectors used in satellite experiments are very similar to the detectors used in particle physics experiments but of course they have to be made as light as possible since they have to be flown in orbit. At the very highest energies, $E \geqslant 10^{11}$ eV, γ-rays from extraterrestrial sources are so energetic that they initiate electromagnetic cascades in the upper atmosphere and the Cerenkov radiation of the ultrahigh energy electrons and positrons in the showers can be detected at ground level. Thus indirectly, the very highest energy γ-rays can be detected at ground level.

2

Interaction of high energy particles with matter I – Ionisation losses

2.1 Introduction

We consider first of all the interaction of high energy particles with matter. When they pass through a solid, liquid or gas, they can cause considerable wreckage to the constituent atoms, molecules and nuclei. There are three basic types of process which can occur:

(i) ionisation and excitation of the atoms and molecules of the material. In the process of ionisation, electrons are torn off atoms by the electrostatic forces between the charged high energy particle and the electrons. This is not only a source of ionisation but also a source of heating of the material because of the transfer of kinetic energy to the electrons;
(ii) the destruction of crystal structures and molecular chains;
(iii) nuclear interactions between the high energy particles and the nuclei of the atoms of the material.

All three types of interaction find applications in the construction of high energy particle detectors. In this chapter we will be solely concerned with the first of the processes, in particular with that known as *ionisation losses*. This process is important for many reasons. First of all, the ionisation caused by high energy particles in detectors can be used to measure the flux of particles and some of their other properties. Second, the process influences the propagation of high energy particles under cosmic conditions. Third, these losses provide an effective mechanism for heating the interstellar gas, in particular the cool giant molecular clouds in galaxies. One feature of my approach in this book will be that elementary physical processes will find a wide variety of applications, both in the methods of detecting particles and radiation and in astrophysics. I will also point out how these results can be adapted to apparently quite different physical problems. This

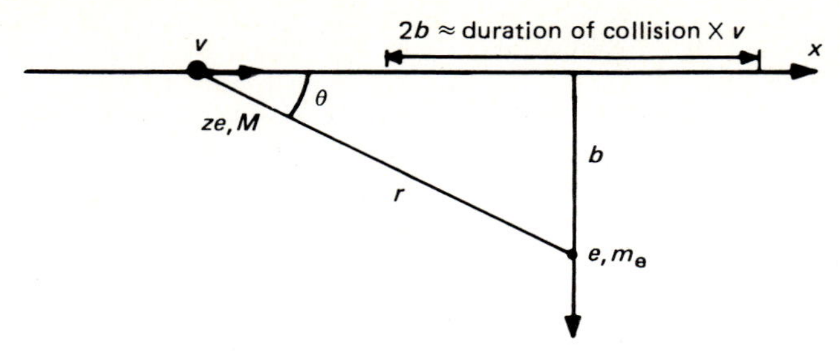

Figure 2.1. The geometry of the collision of a high energy particle with a stationary electron illustrating the definition of the collision parameter b.

is part of the fun of my approach – we will build up many powerful astrophysical tools as we go along.

We will undertake the analysis of ionisation losses reasonably carefully so that the problems of performing a more complete calculation can be appreciated.

2.2 Ionisation losses – non-relativistic treatment

Before going into the detailed analysis, it is useful to look at some general features of the physical process. We consider first the collision of high energy protons and nuclei with stationary electrons. We can show that only a very small fraction of the kinetic energy of the high energy particle is transferred to the electron. Consider a head-on collision between a high energy proton or nucleus of mass M and velocity v with an electron of mass m_e. If we take each of them to be solid spheres, the maximum velocity which the electron can acquire must be less than $2v$ in a perfectly elastic collision. It is a simple calculation to show that the maximum velocity acquired by the electron in a non-relativistic collision is $[2M/(M+m_e)]v$. Recalling that $m_e \ll M$, this is approximately $2v$.

Therefore, the loss of kinetic energy of the high energy particle is less than $\frac{1}{2}m_e(2v)^2 = 2m_e v^2$ and its fractional kinetic energy loss is less than $\frac{1}{2}m_e(2v)^2/\frac{1}{2}Mv^2 = 4m_e/M$. Thus, except for electron–electron collisions, the fractional loss of energy per collision is very small. This means that in real collisions in which the interaction is mediated through the electrostatic fields of the particles, the *incident high energy particle is effectively undeviated*. All that happens is that the electrons of the material receive a small kick through the electrostatic attraction or repulsion of the particle. We can now give the first treatment which is non-relativistic and in which the high energy particle is assumed to move sufficiently fast for the electron in its orbit to remain effectively stationary during the interaction. Methodologically, it is interesting that, as our story unfolds, there will be several occasions on which we will follow the same line of approach in dealing with the interaction of individual particles with the atoms, ions and electrons of a diffuse medium.

The interaction is illustrated in Fig. 2.1. The charge of the high energy particle is ze and its mass M, and it is assumed that it is undeviated in the interaction; b, the distance of closest approach of the particle to the electron, is called the *collision parameter* of the interaction. Then, the total *momentum impulse* given to the electron in this encounter is $\int F\,dt$. By symmetry, the forces parallel to the line of flight of the high energy particle cancel out and therefore we need only work out the component of force perpendicular to the line of flight. Then,

$$F_{\perp} = \frac{ze^2}{4\pi\varepsilon_0 r^2}\sin\theta\,; \mathrm{d}t = \frac{\mathrm{d}x}{v}$$

Changing variables to θ: $b/x = \tan\theta$, $r = b/\sin\theta$ and therefore $\mathrm{d}x = (-b/\sin^2\theta)\mathrm{d}\theta$; v is effectively constant and therefore

$$\int_{-\infty}^{\infty} F_{\perp}\,\mathrm{d}t = -\int_0^{\pi} \frac{ze^2}{4\pi\varepsilon_0 b^2}\sin^2\theta\,\frac{b\sin\theta}{v\sin^2\theta}\,\mathrm{d}\theta = -\frac{ze^2}{4\pi\varepsilon_0 bv}\int_0^{\pi}\sin\theta\,\mathrm{d}\theta$$

Therefore

$$\text{momentum impulse } p = \frac{ze^2}{2\pi\varepsilon_0 bv} \tag{2.1}$$

Therefore, the kinetic energy transferred to the electron is

$$\frac{p^2}{2m_e} = \frac{z^2e^4}{8\pi^2\varepsilon_0^2 b^2v^2m_e} = \text{energy loss by high energy particle}$$

We now want the *average energy loss per unit length* and so we have to work out the number of collisions with collision parameters in the range b to $b+\mathrm{d}b$ and integrate over collision parameters. From the geometry of Fig. 2.2, it can be seen that the total energy loss of the high energy particle, $-\mathrm{d}E$, in length $\mathrm{d}x$ is:

(number of electrons in volume $2\pi b\,\mathrm{d}b\,\mathrm{d}x$) $\times$ (energy loss per interaction)

$$= \int_{b_{\min}}^{b_{\max}} N_e\,2\pi b\,\mathrm{d}b\,\frac{z^2e^4}{8\pi^2\varepsilon_0^2 b^2v^2m_e}\,\mathrm{d}x \tag{2.2}$$

where N_e is the number density, or concentration, of electrons. Notice that I have included limits $b_{\max}$ and $b_{\min}$ to the range of collision parameters in this integral. Let us complete the integral:

$$-\frac{\mathrm{d}E}{\mathrm{d}x} = \frac{z^2e^4N_e}{4\pi\varepsilon_0^2 v^2m_e}\ln\left(\frac{b_{\max}}{b_{\min}}\right) \tag{2.3}$$

Notice how the logarithmic dependence on $b_{\max}/b_{\min}$ comes about. The closer the encounter, the greater the momentum impulse, $p \propto b^{-2}$. However, there are more electrons at large distances ($\propto b\mathrm{d}b$) and hence, when we integrate, we obtain only a logarithmic dependence of energy loss upon the range of collision parameters. We will encounter the same phenomenon in the case of bremsstrahlung (Section 3.4) and of working out the conductivity of a plasma (Section 10.4). You may well ask, 'Why introduce the limits $b_{\max}$ and $b_{\min}$, rather than work out the answer

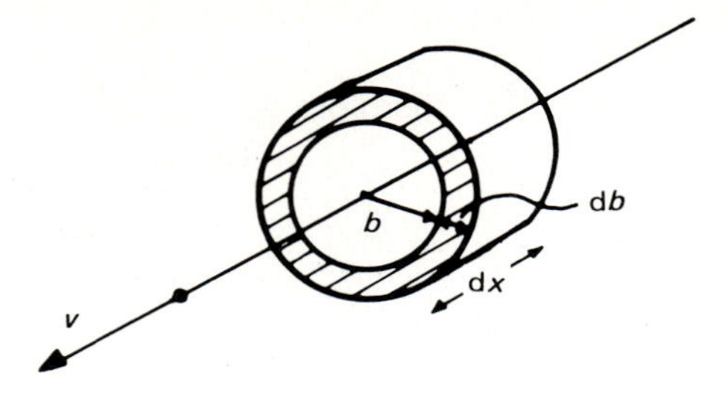

Figure 2.2.

properly?' The simple reason is that the proper sum is very much more complicated and would take account of the acceleration of the electron by the high energy particle and include a proper quantum mechanical treatment of the interaction. Our approximate methods give rather good answers, however, because the limits b_{max} and b_{min} only appear inside the logarithm and hence need not be known very precisely.

Upper limit b_{max}

An upper limit to the range of integration over collision parameters occurs when the duration of the collision is of the same order as the period of the electron in its orbit. When this occurs, the interaction is no longer impulsive. In the limit in which the duration of the collision is much greater than the period of the orbit, the electron feels a slowly varying weak field and, in terms of the dynamics of particles which we will discuss later, it 'conserves its motion adiabatically' during the perturbation and no ionisation takes place. What do we mean by the *duration* of the collision? The energy transfer to the electron can be derived in the following way: if we take the time during which the particle experiences a strong interaction with the electron to be $\tau = 2b/v$ (see Fig. 2.1) and multiply by the electrostatic force at the distance of closest approach to the electron b, then

$$F = ze^2/4\pi\varepsilon_0 b^2; \text{ impulse } p = F\tau = 2ze^2/4\pi\varepsilon_0 bv$$

Therefore

$$p = ze^2/2\pi\varepsilon_0 bv$$

This is the same answer as before (expression (2.1)). In other words, we can think of the encounter as lasting a time $\tau = 2b/v$. If the collision time is the same as the orbital period of the electron, we obtain an order of magnitude estimate for b_{max}. Hence, a good upper limit for b_{max} is

$$2b_{max}/v \approx 1/\nu_0$$

where ν_0 is the orbital frequency of the electron. Writing $\omega_0 = 2\pi\nu_0$,

$$b_{max} \approx \frac{v}{2\nu_0} = \frac{\pi v}{\omega_0} \tag{2.4}$$

Lower limit b_{min}

There are two possibilities:

(i) According to classical physics, the closest distance of approach corresponds to that collision parameter at which the electrostatic potential energy of interaction of the high energy particle and the electron is equal to the maximum possible energy transfer which, according to our first calculation, is $2m_e v^2$. Thus,

$$ze^2/4\pi\varepsilon_0 b_{\text{min}} \approx 2m_e v^2$$

$$b_{\text{min}} = ze^2/8\pi\varepsilon_0 m_e v^2 \tag{2.5}$$

We can easily show that, if this amount of energy is transferred during the interaction, then the electron moves roughly a distance b_{min} during the encounter and hence the assumptions on which the calculation is based break down. To demonstrate this, we note that the average velocity of the electron perpendicular to its initial line of flight during the encounter is $\approx p/m_e$. Therefore the distance moved in the collision time $\tau = 2b/v$ is $(p/m_e) \times (2b/v) = ze^2/\pi\varepsilon_0 m_e v^2$, which is of the same order of magnitude as b_{min}.

(ii) A second possible value of b_{min} comes from the fact that we should really use quantum physics to describe close encounters between the atomic system and the high energy particle. The maximum velocity acquired by the electron in the encounter is $\Delta v \approx 2v$ and hence its change in momentum is $\Delta p = 2m_e v$. There is therefore a corresponding uncertainty in the position Δx, according to the Heisenberg Uncertainty Principle, $\Delta x \approx \hbar/2m_e v$. Therefore,

$$b_{\text{min}} = \hbar/2m_e v \tag{2.6}$$

If this turns out to be the appropriate value of b_{min}, then we ought to have undertaken a proper quantum mechanical calculation. Granted this defect in our calculation, the value of b_{min} still tells us the smallest meaningful value of b for the purposes of our integration.

We must choose whichever of these values of b_{min} is the larger according to the physical conditions of the problem. We must consider the ratio of possible values of b_{min}:

$$\frac{b_{\text{min}}(\text{quantum})}{b_{\text{min}}(\text{classical})} = \frac{\hbar}{2m_e v}\,\frac{8\pi\varepsilon_0 m_e v^2}{ze^2} = \frac{4\pi\varepsilon_0 v\hbar}{ze^2}$$

$$= \frac{1}{z\alpha}\left(\frac{v}{c}\right) = \frac{137}{z}\left(\frac{v}{c}\right) \tag{2.7}$$

where $\alpha = e^2/4\pi\varepsilon_0 c\hbar = 1/137$ is the fine structure constant. Thus, if the particles have $v/c \gtrsim 0.01z$ we should use the quantum restriction. For high energy particles, we will use this value in the subsequent chapters. Note, however, that the same calculation applies for ionisation losses involving 'thermal' matter, for example, the gas in a cloud of hot hydrogen. In this case, the typical velocities of the particles are less than $0.01c$ and so the classical limit is the relevant one.

Substituting the appropriate limits into equation (2.3), we obtain the following loss rate per unit length for high energy particles in the non-relativistic limit:

$$-\frac{\mathrm{d}E}{\mathrm{d}x} = \frac{z^2 e^4 N_\mathrm{e}}{4\pi\varepsilon_0^2 v^2 m_\mathrm{e}} \ln\left(\frac{2\pi m_\mathrm{e} v^2}{\hbar\omega_0}\right) \tag{2.8}$$

The next step is to express the angular frequency ω_0 of the electron in its orbit in terms of its atomic binding energy. For simplicity, we adopt the simple primitive Bohr model for the atom in which ω_0 is genuinely the orbital angular frequency of the electron in its ground state and then the binding energy ε of the electron can be shown to be

$$|\varepsilon| = \tfrac{1}{2}\hbar\omega_0$$

This energy is also the ionisation potential I of the atom and so we can write

$$-\frac{\mathrm{d}E}{\mathrm{d}x} = \frac{z^2 e^4 N_\mathrm{e}}{4\pi\varepsilon_0^2 v^2 m_\mathrm{e}} \ln\left(\pi\frac{m_\mathrm{e} v^2}{I}\right) \tag{2.9}$$

In practice, I should be some properly weighted mean over all states of the electrons in the atom i.e. we should write $\bar{I}$ not I. This value of $\bar{I}$ takes account of the fact that there are electrons in many different energy levels in the atoms of the substance which can be ejected by the high energy particle. The value of $\bar{I}$ cannot be calculated exactly except for the simplest atoms and has to be found by experiment. Conventionally, the loss rate is written,

$$-\frac{\mathrm{d}E}{\mathrm{d}x} = \frac{z^2 e^4 N_\mathrm{e}}{4\pi\varepsilon_0^2 v^2 m_\mathrm{e}} \ln\left(\frac{2m_\mathrm{e} v^2}{\bar{I}}\right) \tag{2.10}$$

where we recognise $2m_\mathrm{e} v^2$ as an old friend, the maximum kinetic energy E_max which can be transferred to the electron.

Another way of obtaining the same result is to work out the energy spectrum of the ejected electrons. It is easy to show that the energy spectrum is of power-law form:

$$N(E)\mathrm{d}E = \frac{z^2 e^4 N_\mathrm{e}}{8\pi\varepsilon_0^2 v^2 m_\mathrm{e}} \frac{\mathrm{d}E}{E^2}\mathrm{d}x \tag{2.11}$$

This calculation is left as an exercise to the reader. Integration over all energies from $\bar{I}$ to E_max gives the same logarithmic term, $\ln(E_\mathrm{max}/\bar{I})$, derived above.

Notice what we have learned already. Inspection of formula (2.11) shows that *the ionisation losses are independent of the mass of the high energy particle.* If we measure the loss rate per unit length, $-\mathrm{d}E/\mathrm{d}x$, we obtain information about $(z/v)^2$. Another interesting point is that the ionisation losses are proportional to m_e^{-1} and therefore we infer that 'ionisation' losses due to electrostatic interactions of the high energy particles with protons and nuclei can be safely neglected.

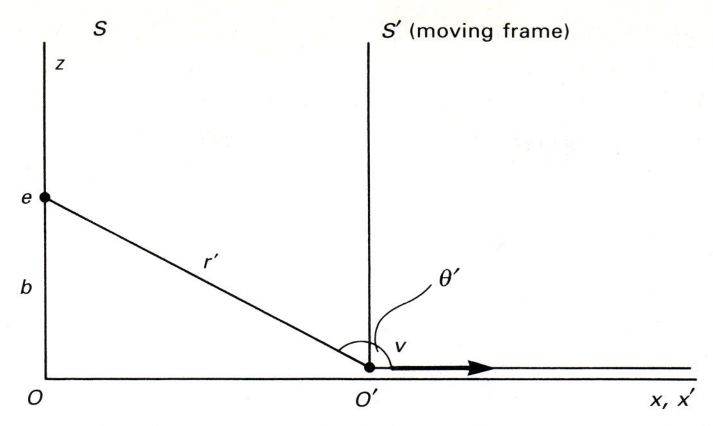

Figure 2.3. The reference frames S and S' for evaluating the strength of the electric field of a relativistic charged particle as observed at the origin of the frames S and S'.

2.3 The relativistic case

The extension of the above analysis of the relativistic domain is straightforward. Although the high energy particle is relativistic, the electron still feels a non-relativistic force acting upon it. What we need is the electric field of the relativistic particle as experienced by the stationary observer.

2.3.1 *The relativistic transformation of an inverse square law Coulomb field*

We orient our reference frames S and S' in standard configuration with the high energy particle moving along the positive x-axis and the electron located at a distance b along the z-axis (Fig. 2.3). We set up the coordinate systems so that $t = t' = 0$ and $x = x' = 0$ when the high energy particle is at its distance of closest approach. At time t, the particle is at the point x in S. In S', the coordinates of the electron (or its displacement four-vector) are $[-vt', 0, b, t']$. Furthermore, in S' the electric field E of the particle is spherically symmetric about the origin $0'$ and hence, at the electron,

$$E_{x'} = \frac{ze}{4\pi\varepsilon_0 r'^2}\cos\theta' = -\frac{ze}{4\pi\varepsilon_0}\frac{x'}{r'^3}$$

$$E_{z'} = \frac{ze}{4\pi\varepsilon_0 r'^2}\sin\theta' = \frac{ze}{4\pi\varepsilon_0}\frac{b}{r'^3}$$

where $r'^2 = (vt')^2 + b^2$ and θ' is the angle between the positive x-axis and the direction of the electron. We now relate time measured by the stationary observer on the electron in S to that measured by the observer moving with high energy particle

$$t' = \gamma(t - vx/c^2)$$

But, by our choice of coordinates, $x = 0$ for the electron in S and hence $t' = \gamma t$. Therefore

$$E_{x'} = -\frac{ze(\gamma vt)}{4\pi\varepsilon_0[b^2+(\gamma vt)^2]^{\frac{3}{2}}}$$

$$E_{z'} = \frac{zeb}{4\pi\varepsilon_0[b^2+(\gamma vt)^2]^{\frac{3}{2}}}$$

Now using the inverse transforms for E and B from S' to S,

$$\left.\begin{aligned} E_x &= E_{x'} & B_x &= B_{x'} \\ E_y &= \gamma(E_{y'}+vB_{z'}) & B_y &= \gamma\left(B_{y'}-\frac{v}{c^2}E_{z'}\right) \\ E_z &= \gamma(E_{z'}-vB_{y'}) & B_z &= \gamma\left(B_{z'}+\frac{v}{c^2}E_{y'}\right) \end{aligned}\right\}$$

Since $B_{x'} = B_{y'} = B_{z'} = 0$, we obtain

$$\left.\begin{aligned} E_x &= -\frac{\gamma zevt}{4\pi\varepsilon_0[b^2+(\gamma vt)^2]^{\frac{3}{2}}} & B_x &= 0 \\ E_y &= 0 & B_y &= -\frac{\gamma vzeb}{4\pi\varepsilon_0 c^2[b^2+(\gamma vt)^2]^{\frac{3}{2}}} \\ E_z &= \frac{\gamma zeb}{4\pi\varepsilon_0[b^2+(\gamma vt)^2]^{\frac{3}{2}}} & B_z &= 0 \end{aligned}\right\} \quad (2.12)$$

We notice that

$$B_y = -\frac{v}{c^2}E_z$$

2.3.2 *Relativistic ionisation losses*

The expressions (2.12) for the electric and magnetic fields associated with a relativistically moving charge are rather useful. It will be noted that, in the non-relativistic limit, $v/c \ll 1$, the expressions for the electric field revert to the standard form of Coulomb's law as would be expected. When the particle is relativistic, however, the electric field at the electron is much enhanced but it is experienced by the electron for a much shorter period of time (see Fig. 2.4). At its distance of closest approach, $x = 0$, $t = 0$, E_z is greater in the relativistic case by a factor γ, whereas the width of the pulse E_z, or the collision time, is shorter by a factor of $1/\gamma$. The magnitude of the E_x component is smaller by a factor of $1/\gamma$, as compared with E_z. In the ultrarelativistic limit, $v \to c$, the pulse looks very like an electromagnetic wave, with $E_z = cB_y$ propagating in the positive x-direction.

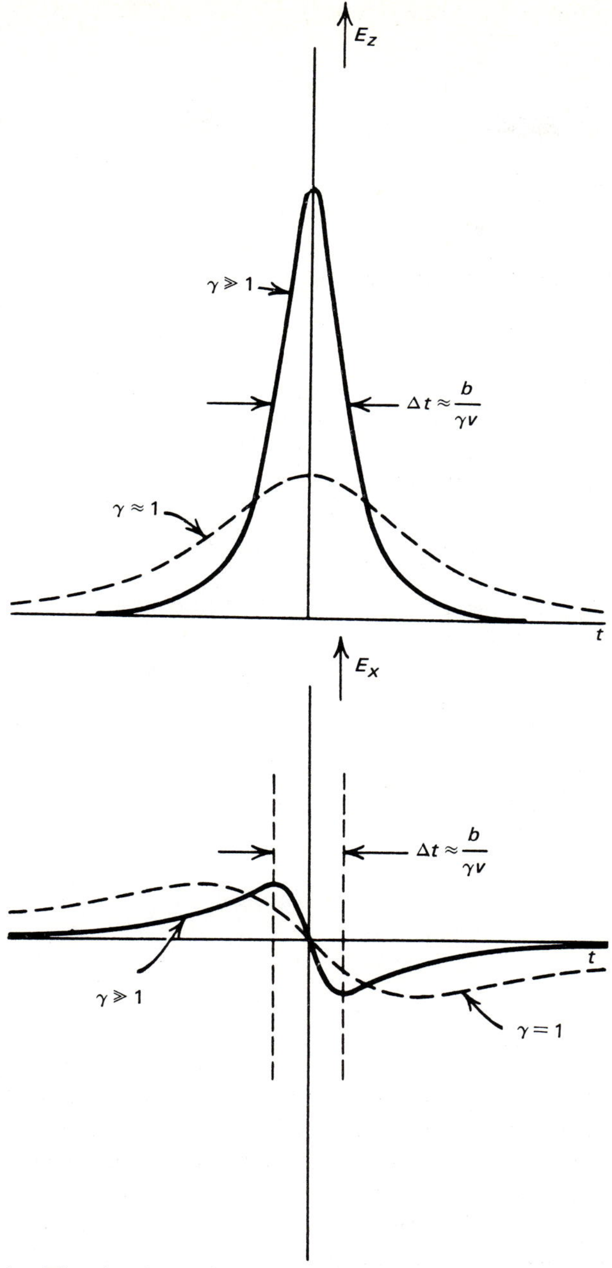

Figure 2.4. The electric fields E_x and E_z of a relativistically moving charged particle as observed from the laboratory frame of reference S. The cases of a non-relativistic particle, $\gamma = 1$ (dashed line) and a relativistic particle, $\gamma \gg 1$ (solid line), are compared. (From J. D. Jackson (1975). *Classical electrodynamics*, page 554, New York: Wiley and Sons, Inc.)

Because of the symmetry of the E_x field about $t = 0$, it again gives no net impulse to the electron. From the E_z field, we get

$$\int F_{\perp}\,\mathrm{d}t = \int eE_z\,\mathrm{d}t = \frac{ze^2\gamma b}{4\pi\varepsilon_0}\int_{-\infty}^{\infty}\frac{\mathrm{d}t}{[b^2+(\gamma vt)^2]^{\frac{3}{2}}}$$

Changing variables to $x = \gamma vt/b$,

$$\int F_{\perp}\,\mathrm{d}t = \frac{ze^2\gamma b}{4\pi\varepsilon_0}2\frac{1}{\gamma vb^2}\int_0^{\infty}\frac{\mathrm{d}x}{(1+x^2)^{\frac{3}{2}}}$$

$$= \frac{ze^2}{2\pi\varepsilon_0\,vb} \qquad (2.13)$$

exactly the same as expression (2.1). This should not be unexpected because the simple argument given in Section 2.2 indicates that it is the product of E_z and the collision time which determines the magnitude of the momentum impulse and the first term increases by a factor γ while the second decreases by the same amount.

The integration over collision parameters proceeds as in the non-relativistic case and thus all we need worry about are the values of b_{max} and b_{min} to put inside the logarithm. The correct form may be found either by asking how the values of b_{max} and b_{min} change in the relativistic case, or by making a relativistic generalisation of the logarithmic form $\ln(E_{\mathrm{max}}/\bar{I})$ when the high energy particle is relativistic.

In the first approach, b_{max} is greater by a factor γ because the duration of the impulse is shorter by this factor. In the case of b_{min}, the transverse momentum of the electron is greater by a factor γ and hence, because of the Heisenberg Uncertainty Principle,

$$\Delta x \approx b_{\mathrm{min}} = \hbar/\Delta p \propto \gamma^{-1}$$

Thus, we expect the logarithmic term to have the form $\ln(2\gamma^2 m_e v^2/\bar{I})$. The second approach is a useful exercise in relativity.

2.3.3 *Relativistic collision between a high energy particle and a stationary electron*

We use Rindler's notation to write the momentum four-vectors of the high energy particle and the electron in the laboratory frame of reference (see Section 1.4.2);

high energy particle	$[\gamma M\mathbf{v}, \gamma M]$
electron	$[0, m_e]$

We transform both four-vectors into a frame of reference moving at velocity V_F, for which the Lorentz factor $\gamma_F = (1 - V_F^2/c^2)^{-\frac{1}{2}}$ and $V_F \parallel \mathbf{v}$. Therefore

$$(\gamma Mv)' = \gamma_F(\gamma Mv - V_F\gamma M)$$

$$p_e' = \gamma_F(0 - V_F m_e)$$

In the centre of momentum frame $(\gamma Mv)' + p'_e = 0$ and hence,

$$V_F(m_e + \gamma M) = \gamma Mv$$

$$V_F = \frac{\gamma Mv}{m_e + \gamma M} \tag{2.14}$$

In this frame of reference, the relativistic three-momentum of the electron is $-\gamma_F V_F m_e$ i.e. the particle is travelling in the negative x'-direction. We obtain the maximum energy exchange if the electron is sent back along the positive x'-direction. Since the collision is elastic, the three-momentum is $+\gamma_F V_F m_e$ and the fourth component of the four-vector, the total energy, is the same. Now transform the four-momentum $[\gamma_F V_F m_e, 0, 0, \gamma_F m_e]$ back to the laboratory frame of reference. To find the energy of the particle, we need only consider the fourth component of the momentum four-vector. Using the inverse Lorentz transformation.

$$(\gamma m_e)_{\text{in} S} = \gamma_F(\gamma_F m_e + \frac{V_F}{c^2}\gamma_F V_F m_e)$$

Therefore the total energy in S is $\gamma_F^2 m_e c^2(1 + V_F^2/c^2)$. Correspondingly, the maximum kinetic energy of the electron is

$$\gamma_F^2 m_e c^2(1 + V_F^2/c^2) - m_e c^2 = 2(V_F^2/c^2)\gamma_F^2 m_e c^2$$

Now, $m_e \ll \gamma M$ and hence

$$V_F \approx v; \gamma_F \approx \gamma$$

In the ultrarelativistic limit, the maximum energy transfer to the electron is

$$E_{\max} = 2\gamma^2 m_e v^2 \tag{2.15}$$

If we use this expression for $E_{\max}$, we recover the same logarithmic factor as before.

$$\ln(2\gamma^2 m_e v^2/\bar{I})$$

2.4 Relativistic ionisation losses and practical forms of the formulae

2.4.1 *Relativistic ionisation losses*

The exact result derived from relativistic quantum theory is known as the Bethe–Bloch formula

$$-\frac{dE}{dx} = \frac{z^2 e^4 N_e}{4\pi\varepsilon_0^2 m_e v^2}\left[\ln\left(\frac{2\gamma^2 m_e v^2}{\bar{I}}\right) - v^2/c^2\right] \tag{2.16}$$

We have succeeded in deriving this formula except for the final correction factor $-v^2/c^2$ which is always small. As discussed earlier, $\bar{I}$ is treated as a parameter to be fitted to laboratory experimental data.

The Bethe–Bloch formula indicates that the energy loss rate depends only upon the velocity of the particle and its charge. The velocity (or energy) dependence of this loss rate is shown in Fig. 2.5. Up to velocities of $v \approx c$ or kinetic energies

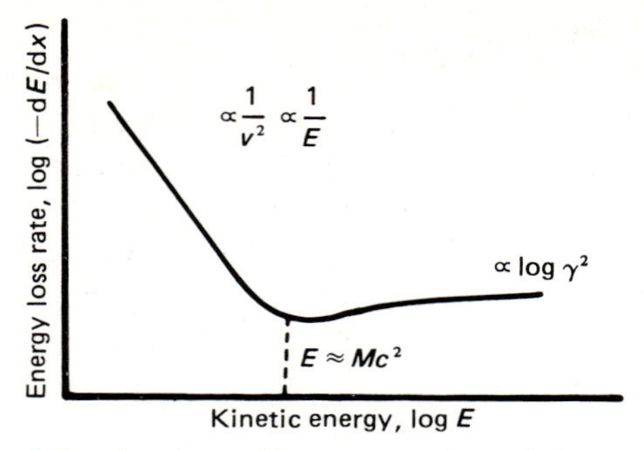

Figure 2.5. A schematic representation of the energy loss rate due to ionisation losses.

$E \approx Mc^2$, the ionisation loss rate decreases as v^{-2} or E^{-1}. At higher energies, the loss rate increases only logarithmically with increasing energy, as $\ln\gamma^2$ according to our analysis. For energies $E \approx Mc^2$, there is a minimum loss rate.

These results are found to be satisfactory for not-too-relativistic high energy particles in not-too-dense materials. There is another modification which has to be made for very high energies and dense media since the Bethe–Bloch formula overestimates the losses of the highest energy particles. In our treatment, we have assumed that we can add together the energy transfers to the electrons incoherently, i.e. we assume that there is no net reaction of the electrons back on the field of the high energy particle, which is equivalent to saying that we have neglected the polarisation of the medium. So far the interactions have been assumed to take place in free space and this holds good for interactions which do not extend to many atomic diameters. For highly relativistic particles, however, the upper limit to the range of collision parameters is $\gamma v/4v_0$ and we cannot neglect these collective effects for the most energetic particles. The simplest thing to do is to split up the range of collision parameters at a value b_0 into near and distant encounters and then treat the distant ones as if they took place in a medium having a refractive index, i.e.

$$-\frac{dE}{dx} = \frac{z^2e^4N_e}{4\pi\varepsilon_0^2 m_e v^2}\left[\ln\left(\frac{\gamma m_e v}{\hbar}b_0\right) + \ln\left(\frac{b(\gamma,\varepsilon)}{b_0}\right) - \frac{v^2}{c^2}\right] \tag{2.17}$$

Since b_0 appears in both logarithms, it is not too important to use an exact value for it. This phenomenon is known as the *density effect* and was first discussed by Fermi.

Jackson (1975) gives a nice treatment of this topic and derives suitable forms for modifications at large collision parameters. He shows that in the extreme relativistic limit, the second term inside the square brackets becomes

$$\ln(1.123c/b_0\,\omega_p)$$

where ω_p is the plasma frequency, $\omega_p = (N_e e^2/\varepsilon_0 m_e)^{\frac{1}{2}}$; this may be compared with the value from the previous treatment $\ln(1.123\gamma c/b_0\,\omega)$. The net effect is that the losses are somewhat smaller at the highest energies as compared with the Bethe–Bloch formula.

2.4.2 *Practical forms of the ionisation loss formulae*

The energy loss formulae do not involve explicitly the mass of the high energy particle but only its velocity v, or equivalently its Lorentz factor $\gamma = (1 - v^2/c^2)^{-\frac{1}{2}}$, and its charge z. The mass of the high energy particle can be written $M \approx N_N m_N$, where N_N is the number of nucleons in the nucleus and m_N is the average nucleon mass, which is roughly that of the proton or neutron, i.e. $m_N = (m_p + m_n)/2 \approx m_p \approx m_n$. Therefore, since the kinetic energy of the particle is $(\gamma - 1)Mc^2$, the kinetic energy per nucleon is

$$(\gamma - 1)\, Mc^2/N_N = (\gamma - 1)\, m_N\, c^2 \tag{2.18}$$

Thus, if we have some way of measuring the charge of the particle z, the ionisation losses measure its *kinetic energy per nucleon* and this is generally the most convenient way of describing the energies of the high energy particles.

Suppose that the atomic number of the medium through which the high energy particle passes is Z and that the number density of atoms is N. Then, $N_e = NZ$. We may therefore write

$$\begin{aligned} -\frac{\mathrm{d}E}{\mathrm{d}x} &= \frac{z^2 e^4 NZ}{4\pi\varepsilon_0^2 m_e v^2}\left[\ln\left(\frac{2\gamma^2 m_e v^2}{\bar{I}}\right) - v^2/c^2\right] \\ &= z^2 NZ f(v) \end{aligned} \tag{2.19}$$

$\mathrm{d}E/\mathrm{d}x$ is often referred to as the *stopping power* of the material. It is convenient to express the stopping power not in terms of length but in terms of the total mass per unit cross-section traversed by the particle. Thus, if a particle travels a distance x through material of density ρ, we say that it has traversed ρx kg m^{-2} of material. Then, writing $\rho x = \xi$,

$$-\frac{\mathrm{d}E}{\mathrm{d}\xi} = z^2 f(v)\frac{NZ}{\rho} = z^2 f(v)\frac{Z}{m} \tag{2.20}$$

where m is the mass of a nucleus of the material. The benefit of expressing the losses in this way is that Z/m is rather insensitive to Z for all the stable elements. For light elements Z/m is $1/2m_N$ while for uranium, it decreases slightly to about $1/2.4m_N$. Thus, the only cause of variation of the energy loss rate from element to element is the variation in $\bar{I}$.

The energy loss rate, expressed as $-(\mathrm{d}E/\mathrm{d}\xi)/z^2$, for high energy particles passing through different materials is shown in Fig. 2.6 which is taken from Enge (1966). Despite the wide range of values of $\bar{I}$ for those materials, it can be seen that the curves lie remarkably close together and this is because the mean ionisation potential only appears inside the logarithm in the expression (2.19). The corresponding diagram presented by Hillas (1972) extends to much higher energies and also shows the energy loss rate as a function of momentum (Fig. 2.7). We begin to see how these curves can be used to measure the charges of high energy particles. If we can measure simultaneously the energy loss $\mathrm{d}E/\mathrm{d}\xi$ and the kinetic energy of the particle E, we define a single point on these loss rate diagrams and

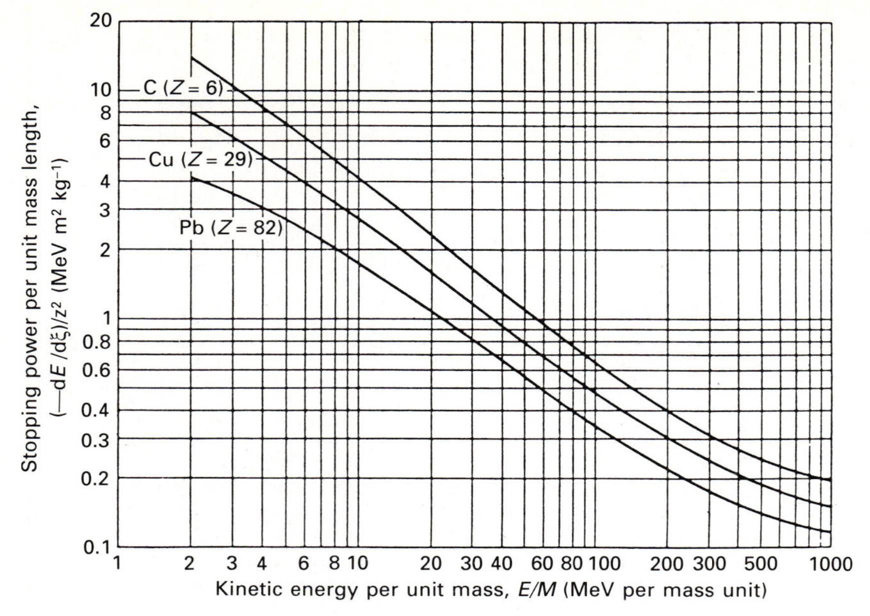

Figure 2.6. The energy loss rate for protons in carbon, copper and lead expressed in terms of their stopping powers. The scales are given in units such that the graph can be used for particles other than protons. The particle's charge is ze and its mass M. (From H. A. Enge (1966). *Introduction to nuclear physics*, page 184, London: Addison-Wesley Publishing Co.)

the only remaining variable is the charge z. Since the loss rate increases as z^2, we see that the loss rate at a given kinetic energy is a sensitive measure of z.

Another useful feature of these curves is the fact that a minimum ionisation loss rate occurs at Lorentz factors $\gamma \approx 2$, corresponding to kinetic energies $E \approx Mc^2$. A good approximation is that the minimum ionisation loss rate for any species in any medium is roughly

$$0.2z^2 \text{ MeV (kg m}^{-2})^{-1}$$

Indeed, if we measure this ionisation loss rate, we can be sure that the particle is relativistic.

One way of estimating the total initial energy of the particle is to measure how far it travels in the medium until it is brought to rest. This distance is called the *range* of the cosmic ray and it is found by integrating the energy loss rate from the particle's initial energy until it is brought to rest.

$$R = \int_E^0 \mathrm{d}x = \int_0^E \frac{\mathrm{d}E}{(\mathrm{d}E/\mathrm{d}x)} \tag{2.21}$$

We know that the calculation breaks down at the very smallest kinetic energies but the particle travels only a very short distance once its kinetic energy falls below that at which our calculation is valid. To be strictly accurate, we can integrate down to some energy E_1, at which energy we know the loss rate can be accurately represented by the expression (2.19) and then add on the little bit of residual range.

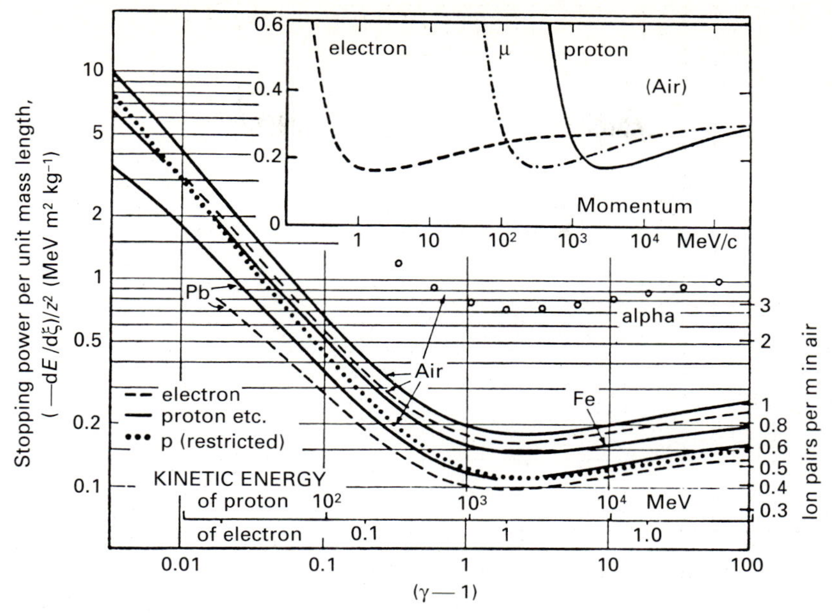

Figure 2.7. The energy loss rate due to ionisation losses in various materials. In contrast to Fig. 2.6, these curves extend into the relativistic regime, $\gamma \gg 1$. The diagram shows both the values of the Lorentz factor γ and the kinetic energies of the particles. The inset shows the loss rates in air as a function of the momentum of the particles. (From A. M. Hillas (1972). *Cosmic rays*, page 30, Oxford: Pergamon Press.)

Let us demonstrate how the range provides information about the initial kinetic energy per nucleon of the particle. We write

$$-\mathrm{d}E/\mathrm{d}\xi = z^2 f(v)\, Z/m$$

where, as discussed above, Z/m is roughly constant. The particle enters the material with energy E_0 and hence

$$R = \int_0^{E_0} \frac{\mathrm{d}E}{z^2 f(v)\, Z/m}$$

Now

$$E = (\gamma - 1)\, Mc^2; \quad \mathrm{d}E = \mathrm{d}(\gamma Mc^2) = Mv\gamma^3 \mathrm{d}v$$

Therefore,

$$\frac{Rz^2}{M} = \frac{m}{Z} \int_0^{E_0} \frac{v\gamma^3 \mathrm{d}v}{f(v)} \tag{2.22}$$

which is a function of only v_0 or γ_0 which, in turn, is a measure of the initial kinetic energy per nucleon of the particle. Thus, if we fire arbitrary types of high energy particle into a material, the range gives information about the initial kinetic energy per nucleon, the charge z and the mass M of the particle. Enge (1966) has evaluated this integral and shown how insensitive the range R, expressed as Rz^2/M, is to the material into which the particle is injected (Fig. 2.8).

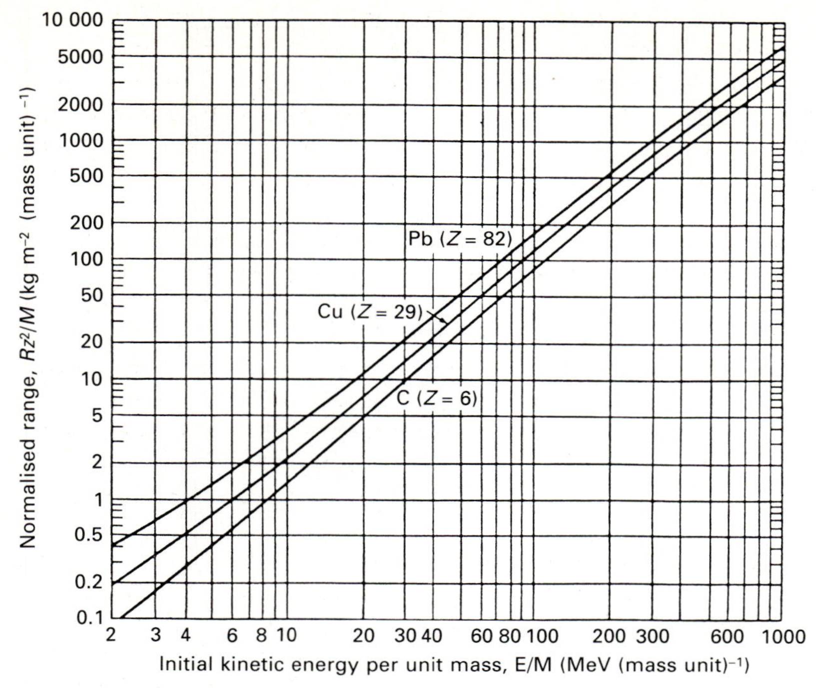

Figure 2.8. The ranges for protons in carbon, copper and lead measured in terms of their path lengths in kg m^{-2}. The scales are such that the graph may be scaled for particles other than protons. (From H. A. Enge (1966). *Introduction to nuclear physics*, page 186, London: Addison-Wesley.)

In the ideal experiment, the energy loss rate $-(dE/d\xi)$ would be measured all along the path of the particle. Then, curves of the form shown in Figs. 2.6 and 2.7 would be reproduced with distance plotted along the x-axis rather than kinetic energy. This can be achieved in certain types of experiment, in particular, in the use of nuclear emulsions as particle detectors (Section 6.2).

It is important to remember that the process of ionisation energy loss is statistical in nature and that the high energy particle makes random encounters with the electrons of the atoms of the material. There will therefore be a spread in the ranges of identical high energy particles which enter the material with the same kinetic energies because some particles make more encounters than others. This phenomenon is known as *straggling* and imposes a fundamental limitation upon the accuracy with which the initial kinetic energy can be measured. For particles of a given kinetic energy, an approximately Gaussian distribution of path lengths is expected.

Another statistical phenomenon which influences estimates of the instantaneous loss rate of the particle occurs if a very close encounter takes place between the high energy particle and an electron. A high energy electron can be ejected in such an encounter which may be energetic enough to produce secondary ionisation of its own. These secondary tracks are called *delta-rays* and examples of them are found in nuclear emulsion photographs (Section 6.2). They start off at right angles to the track and then cause extensive ionisation. The worry is that a few of these energetic delta-rays in the section of track studied can cause large fluctuations in the measured value of dE/dx. One may be able to spot them in a nuclear emulsion but one would never know whether or not one or two such encounters had taken place in a solid-state detector in which only the total energy loss dE in dx is measured.

Finally, what happens to the energy that is deposited in the material? A trail of ions is left behind and those electrons that were sufficiently energetic ionise further atoms of the substance. Thus, for a given energy loss rate, a mean number of ion–electron pairs is produced which is almost independent of the material. To do the sum properly is very difficult because one must estimate how much energy goes into excitation without ionisation, elastic collisions and so on. The observed values are that one ion–electron pair in air is produced for every 34 eV, in hydrogen for every 36 eV and in argon for every 26 eV. Thus, measuring the number of ion pairs produced in the material in length dx enables the energy dE deposited in the material to be found.

This process has found an important astrophysical application in the heating and ionisation of cold, dense molecular clouds in the interstellar medium. It is known that, inside giant molecular clouds, a great deal of interstellar chemistry takes place despite the low temperature of the gas, $T \approx 10$–50 K. At these low temperatures, the gas should be completely neutral. However, the clouds are permeated by the interstellar flux of high energy particles and therefore their ionisation losses can ionise and heat the material of the clouds. This is believed to be the process responsible for the production of the low levels of ionisation inferred to be present in molecular clouds. Estimating the ionisation rate due to the interstellar flux of high energy particles is not straightforward because it depends upon the spectrum of the particles down to low energies and upon their ability to penetrate into the cold clouds. There are considerable uncertainties associated with both of these quantities as we will discuss later.

2.5 A diversion – ionisation losses and dynamical friction

One of the great delights of modern astrophysics is how apparently different physical processes turn out to be related. It is the similarities and differences which add to one's physical understanding and I will indulge in these excursions whenever they will prove useful in the development of the story at some point in the future. The following arguments, developed by my colleague Rashid Sunyaev and myself some years ago, are in no sense original but they show how powerful these methods can be.

Dynamical friction is the gravitational analogue of ionisation losses. In a strict analogy with the development of Section 2.2., the physical process describes the interaction of a massive, fast-moving star with a cluster of stars or a rapidly-moving massive galaxy interacting with other members of a cluster of galaxies. In both cases, the idea is the same as that of ionisation losses – the star or galaxy transfers kinetic energy to the other stars or galaxies in the cluster and so loses energy itself. The big difference between the electrostatic and gravitational cases is the fact that gravity is so much weaker than the electromagnetic force. For simplicity, we will discuss first the case of a massive star interacting gravitationally with the stars in a galaxy and then specialise to the case of a star cluster. We expect to find exactly the same type of formula for the loss of kinetic energy of the massive star as those derived above (relations (2.3)). To convert from the electrostatic to the gravitational case, we compare the forms of the inverse square laws of electrostatics and gravitation:

$$F = \frac{(ze)\,e}{4\pi\varepsilon_0 r^2}; \quad F = \frac{GMm}{r^2}$$

We have to replace $(ze)e/4\pi\varepsilon_0$ by GMm where M is the mass of the fast-moving star and m is the mass of each of the swarm of less massive stars. Clearly we have to make the identifications $ze/(4\pi\varepsilon_0)^{\frac{1}{2}} \equiv G^{\frac{1}{2}}M$ and $e/(4\pi\varepsilon_0)^{\frac{1}{2}} \equiv G^{\frac{1}{2}}m$. The number density of particles is N. We therefore find that the energy loss rate due to gravitational interactions is:

$$-\frac{\mathrm{d}E}{\mathrm{d}x} = \frac{4\pi G^2 M^2 m N}{v^2} \ln\left(\frac{b_{\max}}{b_{\min}}\right) \tag{2.23}$$

If we wish, we can write this relation in terms of the mass density $\rho = Nm$ through which the particle moves and then

$$-\frac{\mathrm{d}E}{\mathrm{d}x} = \frac{4\pi G^2 M^2 \rho}{v^2} \ln\left(\frac{b_{\max}}{b_{\min}}\right) \tag{2.24}$$

This energy loss rate describes the losses due to the force of dynamical friction acting upon the massive particle.

We may therefore define a loss-time τ during which the massive particle loses all its initial kinetic energy $E = \frac{1}{2}Mv^2$ in accelerating light particles

$$\tau = \frac{E}{(-\mathrm{d}E/\mathrm{d}t)} = \frac{\frac{1}{2}Mv^2}{v(-\mathrm{d}E/\mathrm{d}x)} = \frac{v^3}{8\pi G^2 MmN \ln(b_{\max}/b_{\min})} \tag{2.25}$$

This loss-time τ is closely related to the *gravitational relaxation time* of a star in the cluster. By this we mean the time it takes to change the energy of a typical star in the cluster by roughly a factor of 2 due to distant random encounters. This is also roughly the time to establish equipartition of kinetic energy with the other stars in the cluster. Obviously, a much more complete analysis is needed to describe the interaction of particles of the same mass which are all in motion. It is apparent that a number of the assumptions made in Section 2.2 break down in this case but the more detailed analyses find the same type of functional dependence which we have found already. The proper expression for the gravitational relaxation time τ_r is

$$\tau_r = \frac{3\sqrt{2}}{32\pi} \frac{v^3}{G^2 m^2 N \ln(b_{\max}/b_{\min})} \tag{2.26}$$

The similarity of this relation with the one we derived above may be seen by setting $M = m$ in equation (2.25).

To go just a little further, let us apply this result to a cluster of stars or galaxies which has already come into *gravitational equilibrium* through their mutual gravitational interactions but has not yet come into *equipartition*. We will suppose that there are N_s stars in the cluster which has radius R. Then, a natural upper bound to the range of collision parameters, b_{max}, is the radius of the cluster since there will not be gravitational interactions at greater distances. The lower limit is set by the same restriction which we described in the low temperature case of ionisation losses – namely that the particles cannot exchange more than their kinetic energies (see the expression (2.5)):

$$\tfrac{1}{2}mv^2 \approx Gm^2/b_{min}; \quad b_{min} \approx 2Gm/v^2$$

Therefore, we find

$$b_{max}/b_{min} \approx Rv^2/2Gm$$

It is now convenient to rewrite b_{max}/b_{min} in terms of the numbers of particles in the cluster, N_c. We do this in the following way. First of all, we write the number of particles in the system as $N_c = \frac{4}{3}\pi R^3 N$. Then, we use the virial theorem which states that for a cluster in dynamical equilibrium, the total kinetic energy of the particles in the cluster is one-half of the gravitational potential energy of the cluster (see Section 22.3). To order of magnitude we can write

$$U = 2T; \quad \frac{1}{2}\frac{GM_c^2}{R} \approx N_c mv^2 \tag{2.27}$$

where the mass of the cluster M_c is just Nm. We therefore find

$$\tfrac{1}{2}\frac{GN_c^2 m^2}{R} \approx N_c mv^2; \quad v^2 \approx \frac{GN_c m}{2R} \tag{2.28}$$

and hence, from (2.27),

$$b_{max}/b_{min} \approx N_c/4 \tag{2.29}$$

Thus, the thermalisation time may be written

$$\tau_r = \frac{3\sqrt{2}v^3}{32\pi G^2 m^2 N \ln(N_c/4)} \tag{2.30}$$

Let us make two simple applications of these results. Specialists may well be horrified by the simplicity of my arguments but I find them useful. In the first case, let us enter typical parameters for a globular star cluster in our Galaxy into the expression (2.30) – for demonstration purposes, I adopt the following values: $R = 10$ pc, $M = 0.3M_\odot$, $v = 8$ km s^{-1}, $N_c = 10^6$. These figures are self-consistent according to the virial theorem (2.28). According to the relation (2.30), the gravitational relaxation time is 10^{11} years. This may look like bad news because the age of our Galaxy is probably about $(1.5\text{–}2) \times 10^{10}$ years old. However, we have greatly oversimplified the problem. The most important simplifications are to assume that the stars are uniformly distributed within a sphere of radius 10 pc. This is certainly not the case since the space density of stars increases rapidly towards the centre where values as large as 10^4 pc^{-3}, compared with a value $N = 200$ pc^{-3} in the above sum, are found. Second, there is certainly a distribution of stellar masses in the cluster and the more massive stars have shorter relaxation times than the less massive stars. These modifications mean that dynamical friction can be important in globular clusters and certainly warrants a much improved calculation over the crude argument given here. One obvious point is that one must look to dense environments like globular clusters before the effects of dynamical friction become important in exchanging kinetic energy between stars. In the general interstellar medium, the effects are not expected to be important during the lifetime of the Galaxy which has important consequences for the orbits of stars in our Galaxy.

Let us perform the same calculation for the members of a cluster of galaxies. In this case, the approximations should be even worse because the galaxies are not point masses but are significantly extended. We adopt $R = 2.5$ Mpc, $N = 1000$, $M = 10^{11} M_{\odot}$, $v = 10^3$ km s^{-1}, parameters again consistent with the virial theorem (2.28). In this case, the gravitational relaxation time τ_r is 2.25×10^{12} years. This means that, in general, the galaxies in a cluster will not have come into equipartition although they must have attained gravitational equilibrium according to the virial theorem. The same remarks as before apply, however. We have to remember that the clusters are centrally concentrated and that the most massive galaxies, $M \approx 10^{13} M_{\odot}$, will have relaxation times with the lighter members and with each other which are much shorter than the above estimate. Indeed, simple extensions of the above arguments show that the most massive galaxies can relax in about 10^{10} years and this may be the explanation of the observation that the most massive galaxies are normally found in the centres of clusters, having given up their kinetic energy to the lighter members.

2.6 Comments

The most important result which we have obtained in this chapter is the Bethe–Bloch formula for ionisation losses. As we have indicated above, this formula will find at least three important uses in what follows. First, it provides one of the most important techniques by which particle diagnostics can be carried out in experimental studies of high energy particles. This will be explored in much more detail in Chapter 6. Second, the losses will be important in considering the energy loss of particles as they make their way from their sources through the interstellar medium. Third, the heating and ionisation caused by high energy particles will be important in the heating and ionisation of molecular clouds.

In addition to these practical applications, I hope I have indicated the approach I am to take to the understanding of physical processes. I will endeavour to set out the fundamental physics as clearly as possible, emphasising the physical bases of the processes, if necessary, at the expense of strict mathematical rigour. I will give extensive references to the more detailed texts where much fuller treatments are given. In fact, I have treated ionisation losses in considerably more detail than I will often need to treat other processes. I have been concerned in this presentation to indicate how it is possible to reduce most of the processes to relatively straightforward pieces of basic physics. Note also the extensive approach to the physical processes. This example of the close relation between dynamical friction and ionisation losses will only be the first of a number of such analogies.

3

Interactions of high energy particles with matter II – Electrons: ionisation losses and bremsstrahlung

3.1 Introduction

We will deal briefly with the ionisation loss formula for electrons and then proceed to the subject of bremsstrahlung which is very important in many different astronomical contexts as well as in studies of high energy particles. Wherever there is hot ionised gas in the Universe, it emits *free–free radiation* or *bremsstrahlung*. Some of the astrophysical applications include the radio emission from compact regions of ionised hydrogen at temperatures $T \approx 10^4$ K, the X-ray emission of binary X-ray sources for which $T \approx 10^7$ K and the diffuse X-ray emission from the hot intergalactic gas in clusters of galaxies, which may be as hot as $T \approx 10^8$ K. There is also the application of bremsstrahlung as a loss mechanism for relativistic electrons which plays a key role in cosmic rays physics. We will develop useful formulae for astrophysical and high energy particle studies as well as a number of important general radiation formula in this chapter.

3.2 Ionisation losses of electrons

There are two important differences between the ionisation losses of electrons and those of protons and nuclei given in Chapter 2. First, the interacting particles, the high energy electron and the 'thermal' electrons are identical and second, the electrons suffer much larger deviations than the high energy protons and nuclei, which are effectively undeviated in the electrostatic encounters with the cold electrons. The net result is, however, not very different from what was found before. The formula for the ionisation losses of electrons is as follows:

$$-\frac{\mathrm{d}E}{\mathrm{d}x} = \frac{e^4 NZ}{8\pi\varepsilon_0^2 m_e v^2}\left[\ln\frac{\gamma^2 m_e v^2 E_{\max}}{2I^2} - \left(\frac{2}{\gamma} - \frac{1}{\gamma^2}\right)\ln 2 + \frac{1}{\gamma^2} + \frac{1}{8}\left(1 - \frac{1}{\gamma}\right)^2\right] \tag{3.1}$$

where $E_{\max}$ is the maximum kinetic energy which can be transferred to the stationary electron. It is a useful exercise to perform a more exact version of the calculation performed in Section 2.3.3. Making no approximations at all, the maximum kinetic energy transfer is

$$E_{\text{max}} = \frac{2\gamma^2 M^2 m_e v^2}{m_e^2 + M^2 + 2\gamma m_e M} \tag{3.2}$$

Notice that the previous result (2.15) is obtained in the limit $M \gg m_e$ and $M \gg 2\gamma m_e$. In the case of electron–electron collisions, $M = m_e$ and E_{max} takes the value

$$E_{\text{max}} = \frac{\gamma^2 m_e v^2}{1+\gamma} \tag{3.3}$$

The resulting ionisation loss formula is of very similar form to that given in Section 2.4 as may be observed by setting $z = 1$ in the loss rate (2.19) and from inspection of Fig. 2.7. The only important differences are found when the loss rates are compared for protons and electrons *of the same kinetic energy*. It can be seen that the loss rate of the protons is then *greater* than that of the electrons, until the particles become relativistic. The physical reason for this is that a proton of the same kinetic energy as an electron moves more slowly past the electrons in the atom and hence there is a larger momentum impulse acting on the electrons. However, when both the proton and the electron are relativistic, they move past the stationary electrons at the speed of light resulting in the same impulse.

3.3 The radiation of accelerated charged particles

In the 1930s, Anderson noticed that the ionisation loss rate given by equation (3.1) underestimates the energy loss rate of electrons when they become relativistic. It was soon realised that an additional energy loss mechanism was present associated with the radiation of electromagnetic waves because of the acceleration of the electron in the electrostatic field of the nucleus. This radiation was called 'braking radiation' or, in German, *bremsstrahlung*. The physical reason for the radiation is straightforward – whenever a charged particle is accelerated or decelerated, it emits electromagnetic radiation and bremsstrahlung is the radiation emitted in the encounter between the electron and the nuclei of the substance through which it passes. This process is identical to that known as *free–free emission* in the language of atomic physics in the sense that the radiation corresponds to transitions between unbound states in the field of the nucleus.

We will study the theory of this process in some detail for a number of reasons. First of all, bremsstrahlung is one of the most important physical processes in modern high energy astrophysics and it finds applications in many different astronomical environments as mentioned in Section 3.1. Second, we need to understand the loss processes which are important in the detection of high energy particles and, in particular, the interaction of high energy electrons with the atoms and molecules of our atmosphere. Third, the study enables us to introduce a number of useful theoretical tools for studying the radiation of accelerated charged particles. We begin with a discussion of some of the more powerful results concerning the radiation of accelerated electrons which we will use on several occasions as the story develops.

3.3.1 *A useful relativistic invariant*

Let us establish first of all a useful relativistic invariant. Very often we have to transform the energy loss rate by radiation, dE/dt, from one inertial frame of reference to another.

We can show that dE/dt is a Lorentz invariant between inertial frames. To the expert in relativity, this is obvious. The total energy emitted in the form of radiation is the fourth component of the momentum four-vector $[\mathbf{p}, E/c^2]$ and dt is the fourth component of the displacement four-vector $[\mathbf{r}, t]$. Therefore, both the energy dE and the time interval dt transform in the same way between inertial frames of reference and their ratio dE/dt is also an invariant.

Let us obtain this same result in a slightly gentler fashion. In the instantaneous rest frame of the accelerated charged particle, the radiation is emitted with zero net momentum, as may be seen in the case of dipole radiation (see equation (3.8) below), and therefore its four-momentum can be written $[0, dE'/c^2]$. This radiation is emitted in the interval of proper time dt which has four vector $[0, dt']$. We may now use the inverse Lorentz transformation to relate dE' and dt' to dE and dt.

$$\left.\begin{aligned} dE &= \gamma dE' \\ dt &= \gamma dt' \end{aligned}\right\}$$

and hence

$$dE/dt = dE'/dt' \tag{3.4}$$

3.3.2 *The radiation of an accelerated charged particle – J. J. Thomson's treatment*

We will derive the relevant formulae in two quite different ways. The normal derivation proceeds from Maxwell's equations and involves writing down the retarded potentials for the electric and magnetic fields at some distant point r from the accelerated charge. We will follow this procedure in our second derivation. It is, however, instructive to begin with a remarkable argument due to J. J. Thomson which indicates very clearly the origin of the radiation from an accelerated charged particle. I have given this argument in another context but it is such a pleasant argument that it is worth repeating.

We consider a charge q stationary at the origin O of some inertial frame of reference S at time $t = 0$. The charge then suffers a small acceleration to velocity Δv in the short interval of time Δt. Thomson visualised the resulting field distribution in terms of the electric field lines attached to the accelerated charge.

After a time t, we can distinguish between the field configuration inside and outside a sphere of radius $r = ct$ centred on the origin of S. Outside this sphere, the field lines do not yet know that the charge has moved away from the origin because information cannot travel faster than the speed of light and therefore the field lines are radial and centred on O. Inside this sphere, the field lines are radial about the origin of the frame of reference which is centred on the moving charge. Between these two regions, there is a thin shell of thickness $c\Delta t$ in which we have to join up corresponding electric field lines. This configuration is indicated schematically in Fig. 3.1(*a*). Geometrically, it is clear that there must be a component of the electric field in the circumferential direction, i.e. in the $\mathbf{i}_\theta$-direction. This 'pulse' of electromagnetic field is propagated away from the charge at the speed of light and consequently represents an energy loss from the accelerated charged particle.

Let us work out the strength of the electric field in the pulse. We assume that the increment in velocity Δv is very small, i.e. $\Delta v \ll c$, and therefore it is safe to assume that the field lines are radial at $t = 0$ and also at time t in the frame of reference S. There will, in fact, be small aberration effects associated with the velocity Δv but these are second-order compared with the gross effects we are discussing here. We may therefore consider a small cone of electric field lines at angle θ with respect to the acceleration vector of the charge at $t = 0$ and at some later time t when the charge is moving at a constant velocity Δv (Fig. 3.1(b)). We now have to join up the field lines between the two cones through the thin shell of thickness cdt as shown in the diagram. The strength of the E_θ component of the field is just given by number of field lines per unit area in the $\mathbf{i}_\theta$-direction. From the geometry of Fig. 3.1(b), which exaggerates the discontinuities in the field lines, the E_θ field component is given by the relative sizes of the sides of the rectangle $ABCD$ i.e.

$$E_\theta / E_r = \Delta v t \sin\theta / c\Delta t \tag{3.5}$$

But, E_r is just given by Coulomb's law,

$$E_r = q/4\pi\varepsilon_0 r^2; \quad r = ct$$

and therefore

$$E_\theta = q(\Delta v/\Delta t)\sin\theta/4\pi\varepsilon_0 c^2 r$$

$\Delta v/\Delta t$ is just the acceleration $\ddot{r}$ of the charge and hence we can write the result

$$E_\theta = q\ddot{r}\sin\theta/4\pi\varepsilon_0 c^2 r \tag{3.6}$$

Notice that the radial component of the field decreases as r^{-2} according to Coulomb's law but the field in the pulse decreases only as r^{-1} because the field lines become more and more stretched in the E_θ-direction (see (3.5)). Alternatively we can write $p = qr$ where p is the dipole moment of the charge with respect to some origin and hence

$$E_\theta = \ddot{p}\sin\theta/4\pi\varepsilon_0 c^2 r \tag{3.7}$$

This is a pulse of electromagnetic radiation and hence the energy flow per unit area per second at distance r is given by the Poynting vector $\mathbf{E}\times\mathbf{H} = E^2/Z_0$ where

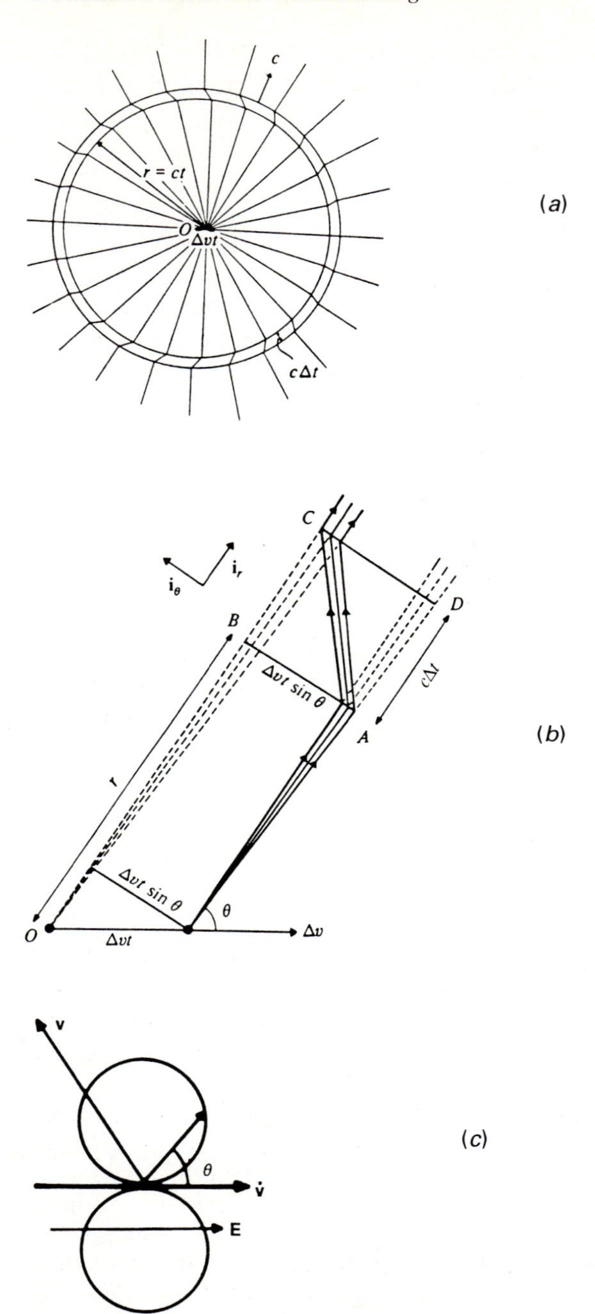

Figure 3.1. (*a*) Illustrating J. J. Thomson's method of demonstrating the radiation of an accelerated charged particle. The diagram shows schematically the configuration of electric field lines at time t due to a charge accelerated to a velocity Δv in time Δt at $t = 0$. (From M. S. Longair (1984). *Theoretical concepts in physics*, page 192, Cambridge: Cambridge University Press.)

$Z_0 = (\mu_0/\varepsilon_0)^{\frac{1}{2}}$ is the impedance of free space. The rate of loss of energy through the solid angle $\mathrm{d}\Omega$ at distance r from the charge is therefore

$$-\left(\frac{\mathrm{d}E}{\mathrm{d}t}\right)_{\mathrm{rad}} \mathrm{d}\Omega = \frac{|\ddot{\mathbf{p}}|^2 \sin^2\theta}{16\pi^2 Z_0 \varepsilon_0^2 c^4 r^2} r^2 \,\mathrm{d}\Omega$$

$$= \frac{|\ddot{\mathbf{p}}|^2 \sin^2\theta}{16\pi^2 \varepsilon_0 c^3} \mathrm{d}\Omega \qquad (3.8)$$

To find the total radiation rate, we integrate over all solid angles, i.e. we integrate over θ with respect to the direction of the acceleration. We recall that integrating over solid angle means integrating over $\mathrm{d}\Omega = 2\pi \sin\theta \mathrm{d}\theta$.

$$-\left(\frac{\mathrm{d}E}{\mathrm{d}t}\right)_{\mathrm{rad}} = \int_0^\pi \frac{|\ddot{\mathbf{p}}|^2 \sin^2\theta}{16\pi^2 \varepsilon_0 c^3} 2\pi \sin\theta \,\mathrm{d}\theta$$

Performing this integral, we find

$$-\left(\frac{\mathrm{d}E}{\mathrm{d}t}\right)_{\mathrm{rad}} = \frac{|\ddot{\mathbf{p}}|^2}{6\pi\varepsilon_0 c^3} = \frac{q^2|\ddot{\mathbf{r}}|^2}{6\pi\varepsilon_0 c^3} \qquad (3.9)$$

Exactly the same result, often called *Larmor's formula*, comes out of the full theory. Notice that it is valid for any form of acceleration $\ddot{\mathbf{r}}$. These formulae embody the three essential properties of the radiation of an accelerated charged particle.

(i) The total radiation rate is given by Larmor's formula (3.9). Notice that, in this formula, the acceleration is the *proper acceleration* of the charged particle and that the radiation loss rate is that measured in the instantaneous rest frame of the particle.
(ii) The *polar diagram* of the radiation is of *dipolar* form, i.e. the electric field strength varies as $\sin\theta$ and the power radiated per unit solid angle as $\sin^2\theta$ with respect to the acceleration vector of the particle (Fig. 3.1(*c*)). Notice that there is no radiation along the acceleration vector and the field strength is greatest at right angles to the acceleration vector.
(iii) The radiation is *polarised* with the electric field vector lying in the direction of the acceleration vector of the particle as projected onto the sphere at distance r from the charged particle (see Fig. 3.1(*b*)).

(*b*) An expanded version of (*a*) used to evaluate the strength of the azimuthal component of the electric field due to the acceleration of the electron. (From M. S. Longair (1984). *Theoretical concepts in physics*, page 192, Cambridge: Cambridge University Press.)
(*c*) The polar diagram of the radiation emitted by an accelerated electron. The polar diagram shows the magnitude of the electric field strength as a function of polar angle θ with respect to the instantaneous acceleration vector $\mathbf{a}$. The polar diagram $E \propto \sin\theta$ corresponds to circular lobes with respect to the acceleration vector.

These are very useful rules which enable us to understand the radiation properties of particles in different astrophysical situations. It is important to remember that these rules are applicable in the *instantaneous rest frame* of the particle and therefore we will have to look carefully at what an external observer sees when the particle is moving at a relativistic velocity.

3.3.3 *The radiation of an accelerated charged particle – improved treatment*

I really enjoy arguments such as that given in Section 3.3.2 above but, in my experience, students feel that they have been short-changed, if not cheated – why have they been learning all these complex things like Maxwell's equations, vector potentials, retarded potentials and so on if you can obtain the same answers by messing around with Coulomb's law? The answer is that one can make the clever arguments work if you call upon your accumulated experience and hindsight which students, by definition, do not yet possess. Another point is that one is happy to use these arguments because one already knows the answer from more rigorous treatments. On the other hand, these simple arguments are very often the clue to understanding what is likely to happen in more complex situations before the detailed analysis is carried out. In any case, let us try to do a little better – if you were perfectly satisfied with Section 3.3.2, you may advance to Section 3.3.4.

We begin with Maxwell's equations in free space in their standard form.

$$\nabla \times \mathbf{E} = -\partial \mathbf{B}/\partial t \tag{3.10a}$$

$$\nabla \times \mathbf{B} = \mu_0 \mathbf{J} + \frac{1}{c^2}\frac{\partial \mathbf{E}}{\partial t} \tag{3.10b}$$

$$\nabla \cdot \mathbf{B} = 0 \tag{3.10c}$$

$$\nabla \cdot \mathbf{E} = \rho_e/\varepsilon_0 \tag{3.10d}$$

We introduce the *scalar* and *vector potentials*, ϕ and $\mathbf{A}$ respectively, in order to simplify the evaluation of the vector fields $\mathbf{E}$ and $\mathbf{B}$ at distance $\mathbf{r}$ from the accelerated charge.

$$\mathbf{B} = \nabla \times \mathbf{A} \tag{3.11a}$$

$$\mathbf{E} = -\partial \mathbf{A}/\partial t - \nabla \phi \tag{3.11b}$$

We do this because the electric and magnetic fields $\mathbf{E}$ and $\mathbf{B}$ turn out to be the components of a four-tensor. It is much easier to work out the components of the four-vector potential $[\mathbf{A}, \phi/c^2]$ and then take its derivatives as expressed by the relations (3.11). We therefore substitute for $\mathbf{E}$ and $\mathbf{B}$ in Maxwell's equation using the scalar and vector potentials of (3.10). Taking equation (3.10b) first,

$$\nabla \times (\nabla \times \mathbf{A}) = \mu_0 \mathbf{J} - \frac{1}{c^2}\frac{\partial}{\partial t}\left(\frac{\partial \mathbf{A}}{\partial t} + \nabla \phi\right)$$

We recall that

$$\nabla\times(\nabla\times\mathbf{A}) = \nabla(\nabla\cdot\mathbf{A})-\nabla^2\mathbf{A}$$

and therefore, substituting and interchanging the order of the time and spatial derivatives,

$$\nabla(\nabla\cdot\mathbf{A})-\nabla^2\mathbf{A} = \mu_0\mathbf{J}-\frac{1}{c^2}\frac{\partial^2\mathbf{A}}{\partial t^2}-\frac{1}{c^2}\frac{\partial}{\partial t}(\nabla\phi)$$

$$\nabla^2\mathbf{A}-\frac{1}{c^2}\frac{\partial^2\mathbf{A}}{\partial t^2} = -\mu_0\mathbf{J}+\nabla\left[\nabla\cdot\mathbf{A}+\frac{1}{c^2}\frac{\partial\phi}{\partial t}\right] \tag{3.12}$$

We now make the same substitutions into equation (3.10d).

$$\nabla\cdot(-\partial\mathbf{A}/\partial t-\nabla\phi) = \rho_e/\varepsilon_0$$

$$\frac{\partial}{\partial t}(\nabla\cdot\mathbf{A})+\nabla^2\phi = -\rho_e/\varepsilon_0$$

Now we add $-(1/c^2)(\partial^2\phi/\partial t^2)$ to both sides of the equation and then

$$\nabla^2\phi-\frac{1}{c^2}\frac{\partial^2\phi}{\partial t^2} = -\frac{\rho_e}{\varepsilon_0}-\frac{\partial}{\partial t}\left[\nabla\cdot\mathbf{A}+\frac{1}{c^2}\frac{\partial\phi}{\partial t}\right] \tag{3.13}$$

It will be observed that equations (3.12) and (3.13) have remarkably similar forms and that, if we were able to set the quantities in the square brackets of each equation equal to zero, we would obtain two remarkably simple inhomogeneous wave-equations for $\mathbf{A}$ and ϕ separately. Fortunately, we are able to do this because there is considerable freedom in the definition of the vector potential $\mathbf{A}$. In classical electrodynamics, $\mathbf{A}$ only appears as the object which, when curled, results in the magnetic field $\mathbf{B}$ which is what we measure in the laboratory. We can always add to $\mathbf{A}$ the gradient of any scalar quantity and it will be guaranteed to disappear upon curling. Therefore, let us see what we have to do to $\mathbf{A}$ and ϕ in order to ensure that the same values of $\mathbf{E}$ and $\mathbf{B}$ will be found. If we write $\mathbf{A}' = \mathbf{A}+\operatorname{grad}\chi$, then we know from expression (3.11a) that the resulting value of $\mathbf{B}$ will be as before. What about $\mathbf{E}$? Substituting for $\mathbf{A}$ in equation (3.11b),

$$\mathbf{E} = -\partial\mathbf{A}/\partial t-\nabla\phi$$
$$\mathbf{E} = -\partial\mathbf{A}'/\partial t-\nabla(\phi-\dot{\chi})$$

Thus, we need to replace ϕ by $\phi' = \phi-\dot{\chi}$. Therefore, we can express the condition that $\nabla\cdot\mathbf{A}+(1/c^2)(\partial\phi/\partial t)$ should vanish in the following way

$$\nabla\cdot\mathbf{A}+\frac{1}{c^2}\frac{\partial\phi}{\partial t} = 0$$

$$\nabla\cdot(\mathbf{A}'-\nabla\chi)+\frac{1}{c^2}\frac{\partial}{\partial t}(\phi'+\dot{\chi}) = 0$$

$$\nabla\cdot\mathbf{A}'+\frac{1}{c^2}\frac{\partial\phi'}{\partial t} = \nabla^2\chi-\frac{1}{c^2}\frac{\partial^2\chi}{\partial t^2} \tag{3.14}$$

Thus, provided we can find a suitable function χ to satisfy (3.14), we obtain the following pair of equations separately for **A** and ϕ:

$$\nabla^2\mathbf{A} - \frac{1}{c^2}\frac{\partial^2\mathbf{A}}{\partial t^2} = -\mu_0\mathbf{J} \tag{3.15a}$$

$$\nabla^2\phi - \frac{1}{c^2}\frac{\partial^2\phi}{\partial t^2} = -\frac{\rho_e}{\varepsilon_0} \tag{3.15b}$$

In fact, it turns out that it is possible to obtain these equations with the more restrictive requirement

$$\nabla^2\chi - \frac{1}{c^2}\frac{\partial^2\chi}{\partial t^2} = 0$$

This process is known as selecting the gauge and this particular choice is known as the *Lorentz gauge* (see Jackson (1975) for more details).

This pair of equations have standard forms of solution. I have given a simple derivation of these in *Theoretical Concepts in Physics* (Longair 1984).

$$\mathbf{A}(\mathbf{r}) = \frac{\mu_0}{4\pi}\int\frac{\mathbf{J}(\mathbf{r}', t-|\mathbf{r}-\mathbf{r}'|/c)}{|\mathbf{r}-\mathbf{r}'|}\,\mathrm{d}^3\mathbf{r}' \tag{3.16a}$$

$$\phi(\mathbf{r}) = \frac{1}{4\pi\varepsilon_0}\int\frac{\rho_e(\mathbf{r}', t-|\mathbf{r}-\mathbf{r}'|/c)}{|\mathbf{r}-\mathbf{r}'|}\,\mathrm{d}^3\mathbf{r}' \tag{3.16b}$$

The point at which the fields are measured is **r** and the integration is over the electric current and charge distributions throughout space. The terms in $|\mathbf{r}-\mathbf{r}'|/c$ take account of the fact that the current and charge distributions should be evaluated at retarded times. We now make a number of simplifications in order to obtain the results we are seeking. First of all, we note that, in the case of an accelerated charged particle, the integral of the product of the current density **J** and the volume element $\mathrm{d}^3\mathbf{r}'$ is no more than the product of its charge times its velocity

$$\mathbf{J}\left(\mathbf{r}', t-\frac{|\mathbf{r}-\mathbf{r}'|}{c}\right)\mathrm{d}^3\mathbf{r}' = q\mathbf{v}\delta(\mathbf{r})$$

where $\delta(\mathbf{r})$ is the Dirac delta-function. The expression for the vector potential is therefore

$$\mathbf{A} = \frac{\mu_0}{4\pi}\frac{q\mathbf{v}}{r} \tag{3.17}$$

We now take the time derivative of **A** in order to find **E**

$$\mathbf{E} = -\frac{\partial\mathbf{A}}{\partial t} = -\frac{\mu_0}{4\pi}\frac{q\ddot{\mathbf{r}}}{r} = -\frac{q\ddot{\mathbf{r}}}{4\pi\varepsilon_0 c^2 r}$$

This is exactly the same expression for **E** as before (3.6) and so we need not repeat the rest of the argument which results in expression (3.9). We notice, however, that we have much more powerful tools available in the formulae (3.16). I leave as exercises to the reader the demonstration that the solutions represent outgoing electromagnetic waves from the accelerated charge and also that the **E** and **B** fields are orthogonal to each other and to the radial direction of propagation of the wave from the origin in the far field limit.

Another important point is to note that these results are correct provided the velocities of the charges are small. A more complete analysis results in the following expressions for the field potentials which are valid for all velocities – the *Liénard–Wiechert potentials.*

$$\mathbf{A}(\mathbf{r}, t) = \frac{\mu_0}{4\pi r}\left[\frac{q\mathbf{v}}{(1-\mathbf{v}\cdot\mathbf{n}/c)}\right]_{\text{ret}}; \quad \phi(\mathbf{r}, t) = \frac{1}{4\pi\varepsilon_0 r}\left[\frac{q}{(1-\mathbf{v}\cdot\mathbf{n}/c)}\right]_{\text{ret}} \tag{3.18}$$

$\mathbf{n}$ is the unit vector in the direction of the point of observation from the moving charge. In both cases, the potentials are evaluated at retarded times relative to the location of the observer. The reason for drawing attention to these more general potentials at this point is that the denominator in the potentials, $(1-\mathbf{v}\cdot\mathbf{n}/c)$, will reappear on a number of occasions in our treatment of charges and sources of radiation moving at high velocities. For example, in the case of a particle moving towards the point of observation at a velocity close to that of light, it represents the fact that the particle almost catches up with the radiation it emits.

3.3.4 *The radiation losses from accelerated charged particles moving at relativistic velocities*

We will often have to deal with the case of accelerated high energy particles moving at relativistic velocities. We can adapt the results we have already obtained to many of these problems. The one key assumption is that, in the particle's instantaneous rest frame, the acceleration of the particle is small and this is normally the case. This is a useful exercise in the use of four-vectors. We need the following general results. First, the norms of four-vectors are invariants in any inertial frame of reference. Second, the acceleration four-vector of the particle, $\mathbf{A}$, not to be confused with the vector potential $\mathbf{A}$ of the last section, can be written

$$\mathbf{A} = \gamma\left[\frac{\partial(\gamma\mathbf{v})}{\partial t}, \frac{\partial\gamma}{\partial t}\right] = \left[\gamma^2\mathbf{a} + \left(\frac{\mathbf{v}\cdot\mathbf{a}}{c^2}\right)\gamma^4\mathbf{v}, \gamma^4\left(\frac{\mathbf{v}\cdot\mathbf{a}}{c^2}\right)\right] \tag{3.19}$$

where the acceleration $\mathbf{a} = \ddot{\mathbf{r}}$ and the velocity of the particle $\mathbf{v} = \dot{\mathbf{r}}$ are measured in the observer's frame of reference S. In the instantaneous rest frame of the particle, S', the acceleration four-vector is $[\mathbf{a}_0, 0]$, where $\mathbf{a}_0 = (\ddot{\mathbf{r}})_0$ is the proper acceleration of the particle. We now equate the norms of the four-vectors in the reference frames S and S':

$$-\mathbf{a}_0^2 = c^2\gamma^8(\mathbf{v}\cdot\mathbf{a}/c^2)^2 - [\gamma^2\mathbf{a} + (\mathbf{v}\cdot\mathbf{a}/c^2)\gamma^4\mathbf{v}]^2$$

After a little bit of straightforward algebra, we find

$$\mathbf{a}_0^2 = \gamma^4[\mathbf{a}^2 + \gamma^2(\mathbf{v}\cdot\mathbf{a}/c)^2] \tag{3.20}$$

Now, we recall that the radiation rate $(\mathrm{d}E/\mathrm{d}t)$ is a Lorentz invariant (Section 3.3.1) and therefore

$$\left(\frac{\mathrm{d}E}{\mathrm{d}t}\right)_S = \left(\frac{\mathrm{d}E'}{\mathrm{d}t'}\right)_{S'} = \frac{q^2|\mathbf{a}_0|^2}{6\pi\varepsilon_0 c^3}$$

$$\left(\frac{\mathrm{d}E}{\mathrm{d}t}\right)_S = \frac{q^2\gamma^4}{6\pi\varepsilon_0 c^3}\left[\mathbf{a}^2 + \gamma^2\left(\frac{\mathbf{v}\cdot\mathbf{a}}{c}\right)^2\right] \tag{3.21}$$

Notice that all the quantities $\mathbf{a}$, $\mathbf{v}$ and γ are measured in S. This is a useful formula. Let us rewrite it in a slightly different form by resolving the acceleration of the particle into components parallel $a_\parallel$ and perpendicular $a_\perp$ to its velocity vector v, i.e.

$$\mathbf{a} = a_\parallel \mathbf{i}_\parallel + a_\perp \mathbf{i}_\perp \text{ and } |\mathbf{a}|^2 = |a_\parallel|^2 + |a_\perp|^2$$

Therefore,

$$\begin{aligned}\mathbf{a}^2 + \gamma^2(\mathbf{v}\cdot\mathbf{a}/c)^2 &= |a_\parallel|^2 + |a_\perp|^2 + \gamma^2(va_\parallel/c)^2 \\ &= |a_\perp|^2 + |a_\parallel|^2(1 + \gamma^2 v^2/c^2) \\ &= |a_\perp|^2 + |a_\parallel|^2\gamma^2\end{aligned} \tag{3.22}$$

Therefore, the loss rate can also be written,

$$\left(\frac{\mathrm{d}E}{\mathrm{d}t}\right)_S = \frac{q^2\gamma^4}{6\pi\varepsilon_0 c^3}(|a_\perp|^2 + \gamma^2|a_\parallel|^2) \tag{3.23}$$

These results will prove useful in the subsequent development.

3.3.5 *Parseval's theorem and the spectral distribution of the radiation of an accelerated electron*

The final tool we need before tackling bremsstrahlung is the decomposition of the radiation field of the electron into its spectral components. Parseval's theorem provides an elegant method of relating the dynamical history of the particle to its radiation spectrum.

We introduce the Fourier transform of the acceleration of the particle through the Fourier transform pair:

$$\left.\begin{aligned}\dot{\mathbf{v}}(t) &= \frac{1}{(2\pi)^{\frac{1}{2}}}\int_{-\infty}^{\infty} \dot{\mathbf{v}}(\omega)\exp(-\mathrm{i}\omega t)\, d\omega \\ \dot{\mathbf{v}}(\omega) &= \frac{1}{(2\pi)^{\frac{1}{2}}}\int_{-\infty}^{\infty} \dot{\mathbf{v}}(t)\exp(\mathrm{i}\omega t)\,\mathrm{d}t\end{aligned}\right\} \tag{3.24}$$

According to Parseval's theorem, $\dot{\mathbf{v}}(\omega)$ and $\dot{\mathbf{v}}(t)$ are related by the following integrals:

$$\int_{-\infty}^{\infty} |\dot{\mathbf{v}}(\omega)|^2\,\mathrm{d}\omega = \int_{-\infty}^{\infty} |\dot{\mathbf{v}}(t)|^2\,\mathrm{d}t \tag{3.25}$$

This is proved in all textbooks on Fourier analysis. We can therefore apply this relation to the energy radiated by a particle which has an acceleration history $\dot{\mathbf{v}}(t)$

$$\int_{-\infty}^{\infty} \frac{dE}{dt} dt = \int_{-\infty}^{\infty} \frac{e^2}{6\pi\varepsilon_0 c^3} |\dot{\mathbf{v}}(t)|^2 \, dt$$

$$= \int_{-\infty}^{\infty} \frac{e^2}{6\pi\varepsilon_0 c^3} |\dot{\mathbf{v}}(\omega)|^2 \, d\omega \tag{3.26}$$

Now, what we really want is $\int_0^\infty \ldots d\omega$ rather than $\int_{-\infty}^\infty \ldots d\omega$. Since the acceleration is a real function, there is another theorem in Fourier analysis which tells us that

$$\int_0^\infty |\dot{\mathbf{v}}(\omega)|^2 \, d\omega = \int_{-\infty}^0 |\dot{\mathbf{v}}(\omega)|^2 \, d\omega$$

and hence we find

$$\text{total emitted radiation} = \int_0^\infty I(\omega) \, d\omega$$

$$= \int_0^\infty \frac{e^2}{3\pi\varepsilon_0 c^3} |\dot{\mathbf{v}}(\omega)|^2 \, d\omega$$

Therefore

$$I(\omega) = \frac{e^2}{3\pi\varepsilon_0 c^3} |\dot{\mathbf{v}}(\omega)|^2 \tag{3.27}$$

Note that this is the total energy per unit bandwidth emitted throughout the period during which the particle is accelerated. Obviously, for a distribution of particles, this result must be integrated over all the particles contributing to the radiation at frequency ω from the system.

3.4 Bremsstrahlung

Bremsstrahlung is the radiation associated with the acceleration of electrons in the electrostatic fields of ions and the nuclei of atoms. We adopt here a classical approach to which quantum mechanical parts are added as appropriate. The quantum mechanical treatment is beyond the scope of this book but is very important in deriving the photon distributions expected in the case of high energy interactions.

Our plan of attack is clear. We already have an expression for the acceleration of an electron in the electrostatic field of a high energy proton or nucleus (see Section 2.3.1). Now the roles of the particles are interchanged – the electron is moving at a high velocity past the stationary nucleus but of course, by symmetry, the field experienced by the electron in its rest frame is exactly the same as before. The immediate result is that we know exactly the acceleration experienced by the electron. We then have to make the following calculations. First, we take the Fourier transform of the acceleration of the electron and then use the expression

(3.27) to work out the radiation spectrum of the emitted radiation. We then integrate this result over all collision parameters, just as in the case of ionisation losses, and we have to worry again about suitable limits for b_{max} and b_{min}. In the case in which the electron is relativistic, we then have to perform a transformation back into the laboratory frame of reference.

Both the relativistic and non-relativistic cases begin in the same way and so let us treat both cases simultaneously. The acceleration of the electron in its rest frame, already worked out in Section 2.3.1, can be expressed as follows. The accelerations along the trajectory of the electron, $a_{\parallel}$, and perpendicular to it, $a_{\perp}$, are given by (2.12)

$$\left.\begin{aligned} a_{\parallel} &= \dot{\mathbf{v}}_x = -\frac{eE_x}{m_e} = \frac{\gamma Ze^2 vt}{4\pi\varepsilon_0 m_e[b^2+(\gamma vt)^2]^{\frac{3}{2}}} \\ a_{\perp} &= \dot{\mathbf{v}}_z = -\frac{eE_z}{m_e} = \frac{\gamma Ze^2 b}{4\pi\varepsilon_0 m_e[b^2+(\gamma vt)^2]^{\frac{3}{2}}} \end{aligned}\right\} \tag{3.28}$$

where Ze is the charge of the nucleus. In general, these are non-relativistic perturbations. Note also that we do not need to worry about the motion of the particle in the magnetic field B since the electron is initially at rest and it is not accelerated to relativistic velocities during the interaction. This would be the condition for the magnetic field to influence the acceleration of the particle.

We now have to take the Fourier transforms of the accelerations (3.28). On this occasion, let us work through this part of the calculation properly so that it can be seen how the cruder methods can give perfectly adequate results for our purposes.

$$\dot{\mathbf{v}}_x(\omega) = \frac{1}{(2\pi)^{\frac{1}{2}}}\int_{-\infty}^{\infty} \frac{\gamma Ze^2 vt}{4\pi\varepsilon_0 m_e[b^2+(\gamma vt)^2]^{\frac{3}{2}}} \exp(\mathrm{i}\omega t)\,\mathrm{d}t \tag{3.29a}$$

$$\dot{\mathbf{v}}_z(\omega) = \frac{1}{(2\pi)^{\frac{1}{2}}}\int_{-\infty}^{\infty} \frac{\gamma Ze^2 b}{4\pi\varepsilon_0 m_e[b^2+(\gamma vt)^2]^{\frac{3}{2}}} \exp(\mathrm{i}\omega t)\,\mathrm{d}t \tag{3.29b}$$

Changing variables to $x = \gamma vt/b$,

$$\begin{aligned} \dot{\mathbf{v}}_x(\omega) &= \frac{1}{(2\pi)^{\frac{1}{2}}}\frac{Ze^2}{4\pi\varepsilon_0 m_e}\frac{1}{\gamma bv}\int_{-\infty}^{\infty}\frac{x}{(1+x^2)^{\frac{3}{2}}}\exp\left(\mathrm{i}\frac{\omega b}{\gamma v}x\right)\mathrm{d}x \\ &= \frac{1}{(2\pi)^{\frac{1}{2}}}\frac{Ze^2}{4\pi\varepsilon_0 m_e}\frac{1}{\gamma bv}I_1(y) \end{aligned} \tag{3.30a}$$

$$\begin{aligned} \dot{\mathbf{v}}_z(\omega) &= \frac{1}{(2\pi)^{\frac{1}{2}}}\frac{Ze^2}{4\pi\varepsilon_0 m_e}\frac{1}{bv}\int_{-\infty}^{\infty}\frac{1}{(1+x^2)^{\frac{3}{2}}}\exp\left(\mathrm{i}\frac{\omega b}{\gamma v}x\right)\mathrm{d}x \\ &= \frac{1}{(2\pi)^{\frac{1}{2}}}\frac{Ze^2}{4\pi\varepsilon_0 m_e}\frac{1}{bv}I_2(y) \end{aligned} \tag{3.30b}$$

where $y = \omega b/\gamma v$. The standard practice is now to look up a table of Fourier transforms (see, for example, Gradshteyn and Ryzhik (1980) 8.432.5, page 959, and

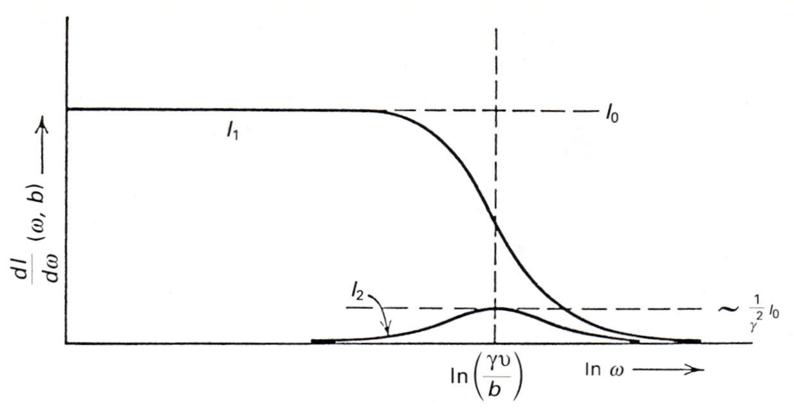

Figure 3.2. The spectrum of bremsstrahlung resulting from the acceleration of the electron parallel and perpendicular to its initial direction of motion. (From J. D. Jackson (1975). *Classical electrodynamics*, page 722, New York: Wiley and Sons, Inc.)

Abramowitz and Stegun (1965), pp. 375–8) and you will find that these integrals become

$$I_1(y) = 2iyK_0(y)$$
$$I_2(y) = 2yK_1(y)$$

where K_0 and K_1 are modified Bessel functions of order zero and one. We can therefore work out the radiation spectrum of the electron in the encounter with the charged nucleus

$$\begin{aligned} I(\omega) &= \frac{e^2}{3\pi\varepsilon_0 c^3}[|a_{\parallel}(\omega)|^2 + |a_{\perp}(\omega)|^2] \\ &= \frac{e^2}{3\pi\varepsilon_0 c^3}\frac{1}{2\pi}\left(\frac{Ze^2}{4\pi\varepsilon_0 m_e bv}\right)^2\left[\frac{1}{\gamma^2}I_1^2(y) + I_2^2(y)\right] \\ &= \frac{Z^2e^6}{24\pi^4\varepsilon_0^3 c^3 m_e^2}\frac{\omega^2}{\gamma^2 v^2}\left[\frac{1}{\gamma^2}K_0^2\left(\frac{\omega b}{\gamma v}\right) + K_1^2\left(\frac{\omega b}{\gamma v}\right)\right] \end{aligned} \tag{3.31}$$

This is the intensity spectrum which results from a single collision between an electron and a nucleus with collision parameter b. It is interesting to plot the intensity spectrum, displaying separately the terms which arise from the accelerations parallel and perpendicular to the velocity vector of the electron (Fig. 3.2). It is apparent that the impulse perpendicular to the direction of travel contributes the greater intensity, even in the non-relativistic case, $\gamma = 1$. In addition, this component results in significant radiation at low frequencies. When the particle is relativistic, the intensity due to acceleration along the trajectory of the particle decreases by a further factor of γ^{-2}. Thus, in bremsstrahlung, the dominant contribution to the radiation spectrum results from the momentum impulse perpendicular to the line of flight of the electron.

It is instructive to study the asymptotic limits of $K_0(y)$ and $K_1(y)$. These are:

$$y \ll 1 \quad K_0(y) = -\ln y; \quad K_1(y) = 1/y$$
$$y \gg 1 \quad K_0(y) = K_1(y) = (\pi/2y)^{\frac{1}{2}} \exp(-y)$$

Thus, at high frequencies, we find an exponential cut-off in the radiation spectrum

$$I(\omega) = \frac{Z^2 e^6}{48\pi^3 \varepsilon_0^3 c^3 m_e^2 \gamma v^3} \left[\frac{1}{\gamma^2} + 1\right] \exp\left(-\frac{2\omega b}{\gamma v}\right) \tag{3.32}$$

Notice the origin of the exponential cut-off. It will be recalled that the duration of the relativistic collision is roughly $\tau = 2b/\gamma v$. Thus, the dominant Fourier component in the radiation spectrum must correspond to frequencies $\nu \approx 1/\tau = \gamma v/2b$ and hence to $\omega \approx \pi v \gamma/b$ i.e. to order of magnitude $\omega b/\gamma v \approx 1$. The exponential cut-off tells us that there is little power emitted at frequencies greater than $\omega \approx \gamma v/b$.

The low frequency spectrum has the form

$$I(\omega) = \frac{Z^2 e^6}{24\pi^4 \varepsilon_0^3 c^3 m_e^2} \frac{1}{b^2 v^2} \left[1 - \frac{1}{\gamma^2}\left(\frac{\omega b}{\gamma v}\right)^2 \ln^2\left(\frac{\omega b}{\gamma v}\right)\right]$$

In this low frequency limit, $\omega b/\gamma v \ll 1$, the second term in square brackets can be neglected and hence a good approximation for the low frequency intensity spectrum is

$$I(\omega) = \frac{Z^2 e^6}{24\pi^4 \varepsilon_0^3 c^3 m_e^2 b^2 v^2} = K \tag{3.33}$$

As noted above, this means that the low frequency spectrum is almost entirely due to the momentum impulse perpendicular to the line of flight of the electron. In fact, we could have guessed that the low frequency spectrum of the emission would be flat because, so far as these frequencies are concerned, the momentum impulse is a delta-function i.e. the duration of the collision is very much less than the period of the waves. It is a standard result of Fourier analysis that the Fourier transform of the delta-function is a flat spectrum $I(\omega) = \text{constant}$. To an excellent approximation, the low frequency spectrum is flat up to frequency $\omega = \gamma v/b$ above which the spectrum falls off exponentially.

It is intriguing to note that once again the factor γ has disappeared from the intensity spectrum (3.33) even in the relativistic case. This is not so surprising when we remember that the momentum impulse is the same in the relativistic and non-relativistic cases as was demonstrated by the expression (2.13).

Finally, we have to integrate over all relevant collision parameters which contribute to the radiation at frequency ω. So far, we have performed a completely general analysis in the rest frame of the electron. If the electron is moving relativistically, the number density of nuclei it observes is enhanced by a factor γ because of relativistic length contraction. Hence, in the moving frame of the electron, $N' = \gamma N$ where N is the space density of nuclei in the laboratory frame of reference. The number of encounters per second is $N'v$ and since, properly speaking, all parameters are now measured in the rest frame of the electron, let us add superscript dashes to all the relevant parameters. The radiation spectrum in frame of the electron is therefore

$$I(\omega') = \int_{b'_{\min}}^{b'_{\max}} 2\pi b' \gamma N v K \, \mathrm{d}b'$$

$$= \frac{Z^2 e^6 \gamma N}{12\pi^3 \varepsilon_0^3 c^3 m_e^2} \frac{1}{v} \ln\left(\frac{b'_{\max}}{b'_{\min}}\right) \tag{3.34}$$

3.5 Non-relativistic and thermal bremsstrahlung

We are interested in two cases: in the first, we evaluate the total energy loss rate by bremsstrahlung of a high energy but non-relativistic electron and second, the continuum spectrum and radiation loss rate of hot ionised gas in which the velocity distribution of the electrons is Maxwellian at temperature T. In both cases, we can neglect the relativistic correction factors and hence obtain the low frequency radiation spectrum from (3.34)

$$I(\omega) = \frac{Z^2 e^6 N}{12\pi^3 \varepsilon_0^3 c^3 m_e^2} \frac{1}{v} \ln \frac{b_{\max}}{b_{\min}} \tag{3.35}$$

Again, we have to make the correct choice of limiting collision parameters $b_{\max}$ and $b_{\min}$. For $b_{\max}$, we note that we should only integrate out to those values of b for which $\omega b/v = 1$. For larger values of b, the radiation at frequency ω lies on the exponential tail of the spectrum and we obtain a negligible contribution to the intensity (see Fig. 3.2). For $b_{\min}$, we have the same options described in Section 2.2 – at low velocities, $v \leqslant (Z/137)\,c$, we use the classical limit, $b_{\min} = Ze^2/8\pi\varepsilon_0 m_e v^2$ (expression (2.5)). This would be appropriate for the bremsstrahlung of a region of ionised hydrogen at $T \approx 10^4$ K. At high velocities, $v \geqslant (Z/137)\,c$, the quantum restriction, $b_{\min} \approx \hbar/2m_e v$ (expression (2.6)), is applicable and this is the appropriate limit to describe, for example, the X-ray bremsstrahlung of hot intergalactic gas in clusters of galaxies. Thus, the choices are

$$I(\omega) = \frac{Z^2 e^6 N}{12\pi^3 \varepsilon_0^3 c^3 m_e^2} \frac{1}{v} \ln \Lambda \tag{3.36}$$

where

$$\Lambda = 8\pi\varepsilon_0 m_e v^3/Ze^2\omega \text{ for low velocities} \tag{3.37a}$$

$$\Lambda = 2m_e v^2/\hbar\omega \text{ for high velocities} \tag{3.37b}$$

Notice that we have simplified the algebra by restricting the analysis to the flat, low frequency part of the radiation spectrum. There is, as usual a cut-off at high frequencies corresponding to $b_{\min}$.

In is interesting to compare our result with the full answer which was derived by Bethe and Heitler using a full quantum mechanical treatment of the radiation process. The key aspect of their result is that the electron cannot give up more than its total kinetic energy in the radiation process and so no photons are radiated with energies greater than $\varepsilon = \hbar\omega = \frac{1}{2}m_e v^2$. In the same notation as above, the intensity of radiation from a single electron of energy $E = \frac{1}{2}m_e v^2$ in the non-relativistic limit is

$$I(\omega) = \frac{8}{3}Z^2\alpha r_e^2 \frac{m_e c^2}{E} vN \ln\left[\frac{1+(1-\varepsilon/E)^{\frac{1}{2}}}{1-(1-\varepsilon/E)^{\frac{1}{2}}}\right]$$

where $\alpha = e^2/4\pi\hbar\varepsilon_0 c \approx 1/137$ is the fine structure constant and $r_e = e^2/4\pi\varepsilon_0 m_e c^2$ is the classical electron radius. The constant in front of the logarithm in this expression is exactly the same as that in expression (3.36). In addition, in the limit of low energies $\varepsilon \ll E$, the term inside the logarithm reduces to $4E/\varepsilon$ which is exactly the same as the expression (3.37b).

3.5.1 *Non-relativistic bremsstrahlung losses*

To find the total energy loss rate of a high energy particle, we integrate the expression (3.36) over all frequencies. In practice, this means integrating from 0 to ω_{max} where ω_{max} corresponds to the cut-off, $b_{min} \approx \hbar/2m_e v$. This is approximately

$$\omega_{max} = 2\pi/\tau \sim 2\pi v/b_{min} \approx 4\pi m_e v^2/\hbar \tag{3.38}$$

i.e. to order of magnitude $\hbar\omega \approx \frac{1}{2}m_e v^2$. This is just the kinetic energy of the electron and is obviously the maximum amount of energy which can be lost in a single encounter with the nucleus. We should therefore integrate (3.36) from $\omega = 0$ to $\omega_{max} \approx m_e v^2/2\hbar$ and thus,

$$\begin{aligned} -\left(\frac{dE}{dt}\right)_{brems} &\approx \int_0^{\omega_{max}} \frac{Z^2 e^6 N}{12\pi^3\varepsilon_0^3 c^3 m_e^2} \frac{1}{v} \ln\Lambda \, d\omega \\ &\approx \frac{Z^2 e^6 N v}{24\pi^3\varepsilon_0^3 c^3 m_e \hbar} \ln\Lambda \\ &= (\text{constant})\, Z^2 N v \end{aligned} \tag{3.39}$$

We note that the total energy loss rate of the electron is proportional to v, i.e. to the square root of the kinetic energy E: $-dE/dt \propto E^{\frac{1}{2}}$. This is in contrast to the case of relativistic bremsstrahlung losses which is developed in Section 3.6 (see equation (3.71)). In practical applications of this formula, it is necessary to integrate over the energy distribution of the particles. For example, the energy spectrum may well be of Maxwellian or power-law form $N(E)\,dE \propto E^{-x}\,dE$. We will use these formulae in the context of the hard X-ray emission from solar flares in Section 12.5.1.

3.5.2 *Thermal bremsstrahlung*

In order to work out the bremsstrahlung or free–free emission of a gas at temperature T, we integrate the expression (3.36) over a Maxwellian distribution of electron velocities

$$N_e(v)\,dv = 4\pi N_e \left(\frac{m_e}{2\pi kT}\right)^{\frac{3}{2}} v^2 \exp\left(-\frac{m_e v^2}{2kT}\right) dv \tag{3.40}$$

The algebra can become somewhat cumbersome at this stage. We can find the correct order-of-magnitude answer if we write $\frac{1}{2}m_e v^2 = \frac{3}{2}kT$ in expression (3.36). Then, the emissivity of a plasma having electron density N_e becomes in the low frequency limit,

$$I(\omega) \approx \frac{Z^2 e^6 N N_e}{12\sqrt{3}\pi^3 \varepsilon_0^3 c^3 m_e^2}\left(\frac{m_e}{kT}\right)^{\frac{1}{2}} g(\omega, T) \tag{3.41}$$

where $g(\omega, T)$ is the correction factor known as the *Gaunt factor*, which is the proper form of the term in Λ integrated over velocity, which appears in formulae (3.35). At high frequencies the spectrum of thermal bremsstrahlung cuts off exponentially as $\exp(-\hbar\omega/kT)$, reflecting the population of electrons in the high energy tail of a Maxwellian distribution of energies at $\hbar\omega \gg kT$. Finally, the total energy loss rate of the plasma may be found by integrating the spectral emissivity over all frequencies. In practice, because of the exponential cut-off, we find the correct functional form by integrating (3.41) from 0 to $\omega = kT/\hbar$ as described above, i.e.

$$-(\mathrm{d}E/\mathrm{d}t) = (\text{constant})\, Z^2 T^{\frac{1}{2}} \bar{g} N N_e \tag{3.42}$$

Detailed calculations give the following answers: the spectral emissivity of the plasma is

$$\begin{aligned}\kappa_\nu &= \frac{1}{3\pi^2}\left(\frac{\pi}{6}\right)^{\frac{1}{2}} \frac{Z^2 e^6}{\varepsilon_0^3 c^3 m_e^2}\left(\frac{m_e}{kT}\right)^{\frac{1}{2}} g(\nu, T)\, N N_e \exp\left(-\frac{h\nu}{kT}\right) \\ &= 6.8\times 10^{-51} Z^2 T^{-\frac{1}{2}} N N_e g(\nu, T) \exp(-h\nu/kT)\ \mathrm{W\ m^{-3}\ Hz^{-1}}\end{aligned} \tag{3.43}$$

where the number densities of electrons N_e and of nuclei N are given in particles per cubic metre. At frequencies $\hbar\omega \ll kT$, the Gaunt factor has only a logarithmic dependence on frequency. Suitable forms at radio and X-ray wavelengths are:

$$\text{Radio} \quad g(\nu, T) = \frac{\sqrt{3}}{2\pi}\left[\ln\left(\frac{128\varepsilon_0^2 k^3 T^3}{m_e e^4 \nu^2 Z^2}\right) - \gamma^{\frac{1}{2}}\right] \tag{3.44a}$$

$$\text{X-ray} \quad g(\nu, T) = \frac{\sqrt{3}}{\pi} \ln\left(\frac{kT}{h\nu}\right) \tag{3.44b}$$

where $\gamma = 0.577\ldots$ is Euler's constant. The functional forms of both logarithmic terms (3.44a, b) can be readily derived from the corresponding expressions (3.37a, b).

For frequencies $h\nu/kT \gg 1, g(\nu, T)$ is approximately $(h\nu/kT)^{\frac{1}{2}}$. The origin of this factor may be understood from the approximations for the Bessel functions used in deriving the relation (3.32). The high frequency approximation differs from that at $h\nu \ll kT$ by a factor of roughly $y^{\frac{1}{2}} \exp(-y)$ and the form of bremsstrahlung emissivity given in (3.43) only takes account of the factor $\exp(-y)$ in the limit $y \to \infty$. Thus, the dominant term in the Gaunt factor is $y^{\frac{1}{2}} \approx (h\nu/kT)^{\frac{1}{2}}$.

The total loss rate of the plasma is

$$-\left(\frac{\mathrm{d}E}{\mathrm{d}t}\right)_{\text{brems}} = 1.435\times 10^{-40} Z^2 T^{\frac{1}{2}} \bar{g} N N_e\ \mathrm{W\ m^{-3}} \tag{3.45}$$

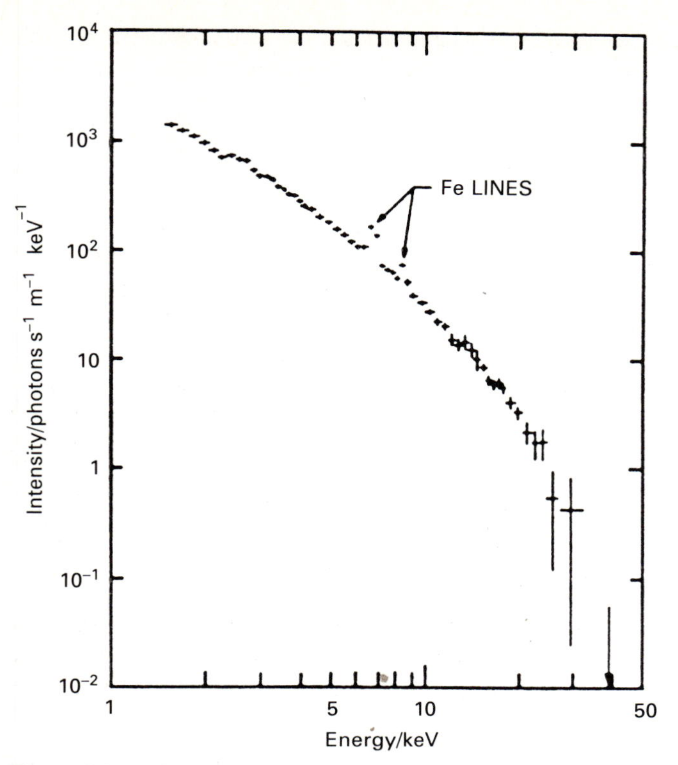

Figure 3.3. The X-ray spectrum of the Perseus cluster of galaxies observed by the HEAO-A2 instrument. The continuum emission can be accounted for by the thermal bremsstrahlung of hot intracluster gas at a temperature corresponding to $kT = 6.5$ keV, i.e. $T = 7.5 \times 10^7$ K. The thermal nature of the radiation is confirmed by the observation of the Lyα and Lyβ emission lines of highly ionised iron, Fe^{+25}, at energies of 6.7 and 7.9 keV respectively. The ionisation potential of Fe^{+24} is 8.825 keV and hence the gas must be very hot. Note also the cluster of unresolved lines of highly ionised silicon, sulphur, calcium and argon in the energy range 1.8–4 keV. (From R. Mushotzky (1980). *X-ray astronomy*, eds R. Giacconi and G. Setti, page 174, Dordrecht: D. Reidel Publishing Co.)

Detailed calculations show that the frequency averaged value of the Gaunt factor $\bar{g}$ lies in the range 1.1–1.5 and thus, to a good approximation, we can write $\bar{g} = 1.2$. The subject of suitable Gaunt factors for use in the thermal bremmsstrahlung formulae is large and complex. A compilation of a large number of useful results is given by Karzas and Latter (1961).

Fig. 3.3 shows the spectrum of the intergalactic gas in the Perseus cluster of galaxies as observed in the X-ray waveband by the HEAO-A2 experiment. The derived temperature of the emitting gas is $T = 7.5 \times 10^7$ K. Confirmation of this high temperature is provided by the observation of the Fe XXVI line at about 8 keV which can also be seen in Fig. 3.3. Since the gas is collisionally excited, the electron temperature of the hot gas must lie in the range 10^7–10^8 K. The interpretation of the diffuse X-ray emission from the cluster as the bremsstrahlung of hot gas provides an important measure of the mass of intergalactic gas in the cluster as well as providing an astrophysical tool for measuring the mass of the cluster as a whole.

3.5.3 *Thermal bremsstrahlung absorption*

To complete our study of thermal bremsstrahlung, it is useful and instructive to work out the coefficient for thermal bremsstrahlung absorption. This is important because the resulting spectrum provides a characteristic signature for compact regions of ionised hydrogen in the radio waveband. It also illustrates the general procedure for relating emission and absorption coefficients.

We write down first the transfer equation for radiation. It is most convenient to work in terms of the intensity of radiation I_ν, i.e. the radiant energy passing per second through unit area at normal incidence per steradian per unit bandwidth. In traversing dx, the decrease in intensity is $\chi_\nu I_\nu dx$ where χ_ν is the absorption coefficient, and the amount gained is $\kappa_\nu dx/4\pi$, where κ_ν is the emissivity of the plasma, i.e. the power emitted per unit volume per unit bandwidth. Therefore,

$$dI_\nu/dx = -\chi_\nu I_\nu + \kappa_\nu/4\pi \tag{3.46}$$

Now, in thermodynamic equilibrium at temperature T, dI_ν/dx is zero and every emission is balanced by an absorption by the same physical process – this is the *principle of detailed balance*. The radiation spectrum must have black-body form in thermodynamic equilibrium and therefore,

$$\chi_\nu I_\nu = \kappa_\nu/4\pi \tag{3.47}$$

where I_ν is the intensity spectrum of black-body radiation at temperature T. This equilibrium intensity spectrum for black-body radiation is given by the Planck function

$$I_\nu(T) = \frac{2h\nu^3}{c^2}\left[\exp\left(\frac{h\nu}{kT}\right) - 1\right]^{-1} \tag{3.48}$$

Substituting into (3.47), we find

$$\chi_\nu = \frac{\kappa_\nu c^2}{8\pi h\nu^3}\left[\exp\left(\frac{h\nu}{kT}\right) - 1\right]$$

The absorption coefficient for thermal bremsstrahlung is therefore

$$\chi_\nu = (\text{constant})\frac{T^{-\frac{1}{2}}NN_e}{\nu^3}g(\nu, T)\left[1 - \exp\left(-\frac{h\nu}{kT}\right)\right] \tag{3.49}$$

At high frequencies, $h\nu \gg kT$, the absorption coefficient has functional dependence

$$\chi_\nu \propto NN_e g(\nu, T)\nu^{-3}T^{-\frac{1}{2}} \tag{3.50}$$

At low frequencies, $h\nu \ll kT$, we expand the exponential for small values of $h\nu/kT$ and find

$$1 - \exp(-h\nu/kT) = h\nu/kT$$

$$\chi_\nu \propto NN_e g(\nu, T)\nu^{-2}T^{-\frac{3}{2}} \tag{3.51}$$

It is interesting to compare these expressions for the absorption coefficient with that expressed in terms of the Einstein coefficients of spontaneous emission, A_{21}, absorption B_{12} and stimulated emission B_{21}. We recall the standard relations between these coefficients for transitions between an upper energy level 2 and a lower energy level 1.

A_{21} = transition probability per unit time for spontaneous emission
$B_{12}\,I(\omega)$ = transition probability for absorption per unit time
$B_{21}\,I(\omega)$ = transition probability for induced or stimulated emission per unit time

In this formulation, we use the intensity of radiation $I(\omega)$ defined in connection with the radiative transfer equation (3.46) at frequency ω corresponding to an energy difference between the upper and lower states $\hbar\omega_{21} = E_2 - E_1$. If n_2 and n_1 are the populations of the states 2 and 1 respectively, we can write down the condition for thermodynamic equilibrium that the sum of the spontaneous and induced emissions should balance the number of absorptions

$$n_2 A_{21} + n_2 B_{21} I(\omega) = n_1 B_{12} I(\omega) \tag{3.51}$$

Therefore, solving for $I(\omega)$,

$$I(\omega) = \frac{A_{21}/B_{21}}{(n_1/n_2)(B_{12}/B_{21}) - 1}$$

Now, in thermodynamic equilibrium, n_1/n_2 is given by the Boltzmann relation $n_1/n_2 = (g_1/g_2)\exp(h\nu_{21}/kT)$ where g_1 and g_2 are the statistical weights of levels 1 and 2. Therefore

$$I(\omega) = \frac{A_{21}/B_{21}}{(g_1/g_2)(B_{12}/B_{21})\exp(h\nu_{21}/kT) - 1} \tag{3.53}$$

However, we know that this function must be the Planck function (3.48) and hence, comparing coefficients,

$$g_1 B_{12} = g_2 B_{21}; \quad A_{21} = (2h\nu^3/c^2)B_{21} \tag{3.54}$$

This remarkable argument, first given by Einstein in 1916, results in a direct relation between the elementary processes of emission and absorption. It can therefore also be used in non-equilibrium situations. Einstein's great insight was that, in writing down the transfer equation for radiation in terms of elementary processes, the induced term B_{21} must be included as well as A_{21} and B_{12}. Thus, in terms of elementary atomic processes, the emissivity of the plasma is

$$\kappa_\nu = h\nu_{21} n_2 A_{21} \tag{3.55}$$

In the transfer equation for radiation corresponding to (3.46) we have to add the terms for absorption and stimulated emission

$$\begin{aligned}\frac{dI(\omega)}{dx} &= \frac{h\nu_{21} n_2 A_{21}}{4\pi} - n_1 B_{12} h\nu_{21} I(\omega) + n_2 B_{21} h\nu I(\omega) \\ &= \frac{\kappa_\nu}{4\pi} - h\nu I(\omega)(n_1 B_{12} - n_2 B_{21})\end{aligned} \tag{3.56}$$

Thus,

$$\chi_\nu = h\nu(n_1 B_{12} - n_2 B_{21}) \tag{3.57}$$

$$= h\nu n_1 B_{12}\left(1 - \frac{n_2}{n_1}\frac{B_{21}}{B_{12}}\right)$$

$$= h\nu n_1 B_{12}\left(1 - \frac{n_2 g_1}{n_1 g_2}\right)$$

If the matter is in thermal equilibrium, but not necessarily with the radiation, we recall that

$$n_2/n_1 = (g_2/g_1)\exp(-h\nu/kT)$$

and therefore

$$\chi_\nu = h\nu n_1 B_{12}[1 - \exp(-h\nu/kT)] \tag{3.58}$$

This result is formally identical to the expression (3.49). It can be seen that the last term in square brackets is derived from the term B_{21}, i.e. the stimulated emission term. The result is that the absorption coefficient $\chi'_\nu = h\nu n_1 B_{12}$ is referred to as the absorption coefficient for bremsstrahlung *uncorrected for stimulated emission* whereas (3.58), which is what is measured in practice, is referred to as the absorption coefficient for bremsstrahlung *taking account of stimulated emission.*

The distinction is important in the sense that the different forms are encountered in different types of astronomical problem. For example, stellar astrophysicists normally use (3.50) because this is directly related to the opacity of the stellar material for photon diffusion – normally, these astronomers are interested in the opacity of the medium for photons having energies $h\nu \approx kT$. On the other hand, radio astronomers always deal with very low energy photons, $h\nu \ll kT$, and they always use the formula (3.51).

Let us apply the formula (3.51) to the spectrum of a compact region of ionised hydrogen as observed at radio wavelengths. The optical depth of the medium τ is defined to be

$$\tau = \int \chi_\nu \,\mathrm{d}x = (\text{constant}) \int N_\mathrm{e}^2 T^{-\frac{3}{2}} \nu^{-2} \mathrm{d}x \tag{3.59}$$

To evaluate the emission spectrum of the region, we integrate the transfer equation (3.46) along a column with uniform electron density

$$\int_0^{I_\nu} \frac{\mathrm{d}I_\nu}{(\kappa_\nu/4\pi - \chi_\nu I_\nu)} = \int_0^x \mathrm{d}x$$

Here we assume no background radiation is present, so that $I_\nu = 0$ at $x = 0$. The integration is straightforward

$$I_\nu = \frac{\kappa_\nu}{4\pi\chi_\nu}[1 - \exp(-\chi_\nu x)] \tag{3.60}$$

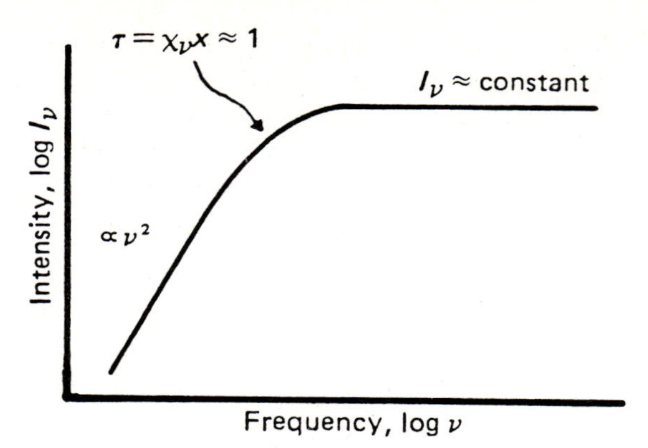

Figure 3.4. The spectrum of thermal bremsstrahlung at low radio frequencies at which self-absorption becomes important. This is the characteristic spectrum of the compact regions of ionised hydrogen found in regions of star formation.

This formula makes sense. If $\tau = \chi_\nu x \ll 1$,

$$I_\nu = \frac{\kappa_\nu}{4\pi\chi_\nu}(\chi_\nu x) = \frac{\kappa_\nu x}{4\pi} \tag{3.61}$$

If $\tau = \chi_\nu x \gg 1$,

$$I_\nu = \frac{\kappa_\nu}{4\pi\chi_\nu} = \frac{2h\nu^3}{c^2}\left[\exp\left(\frac{h\nu}{kT}\right) - 1\right]^{-1}$$

$$= \frac{2kT}{c^2}\nu^2 \quad \text{if } h\nu \ll kT \tag{3.62}$$

Thus, the spectrum of the compact region of ionised hydrogen has a characteristic shape with $I_\nu =$ constant when $\tau \ll 1$ and $I_\nu \propto \nu^2$ if $\tau \gg 1$, corresponding to the Rayleigh–Jeans tail of a black-body distribution at temperature T. This form of spectrum is found in the compact HII regions which lie close to regions of star formation (Fig. 3.4). The temperature of the region may be estimated from the intensity of radiation in the Rayleigh–Jeans region of the spectrum and a mean, temperature-weighted value of N found from the point at which the region becomes optically thick.

3.6 Relativistic bremsstrahlung

We begin with the expression (3.34) for the radiation loss rate per unit bandwidth per unit pathlength in the frame of the moving electron. Again, we must decide appropriate limits for the collision parameters b_{max} and b_{min}. We notice that these limits are linear dimensions perpendicular to the line of flight of the electron and hence take the same values in S and S'.

It would look at first sight as if the value of b_{min} should be the same as before, $\hbar/\gamma m_e v$. We are now dealing, however, with the *radiation* from the electron and we want it to radiate coherently. We may think in the following simple way about the condition for the electron to radiate coherently. We take the electron to have 'size' Δx. If the duration of the impulse is shorter than the travel time across Δx, the different bits of the 'probability distribution' of the electron will receive impulses at different times and the radiation of the electron will not be coherent. Therefore the duration of the impulse must be at least as long as the travel time across the electron

Δt	$\geqslant \Delta x/v$
duration of impulse	travel time across the electron

Therefore

$$\frac{b_{min}}{\gamma v} = \frac{\hbar}{\gamma m_e v \cdot v}; \quad b_{min} = \frac{\hbar}{m_e v} \tag{3.63}$$

Notice that, in this case, there is no Lorentz factor γ on the bottom line of the minimum collision parameter. This result makes sense in the following way. In the rest frame of the electron, this collision parameter corresponds to an angular frequency $\omega' = m_e v^2/\hbar$ and therefore to a photon energy $\hbar\omega' = m_e v^2$. Transforming to the external frame, we see that $\hbar\omega = \gamma m_e v^2$ i.e. this is the condition that the electron cannot give up more than its total kinetic energy in a collision. Notice that exactly the same physical argument applied to the derivation of the non-relativistic value of b_{min} (see Section 3.5.1).

The previous upper limit was $b_{max} = \gamma v/\omega$. The present case is slightly different in that the electron interacts with nuclei which are shielded by their electron clouds unless the collision parameter is small. We can find a suitable estimate by considering, for example, the Fermi–Thomas model of the atom (see, Leighton (1959) pp. 362–3).

The electrostatic field of the nucleus can be written approximately as

$$V(r) = \frac{Ze^2}{r}\exp\left(-\frac{r}{a}\right) \tag{3.64}$$

where

$$a = 1.4\, a_0 Z^{-\frac{1}{3}};$$
$$a_0 = 4\pi\varepsilon_0 \hbar^2/m_e e^2 = 0.53 \times 10^{-10}\ \mathrm{m}$$

Thus, for neutral atoms, a suitable value for b_{max} is

$$b_{max} = 1.4\, a_0 Z^{-\frac{1}{3}} \tag{3.65}$$

We now consider the ratio of the two collision parameters

$$\frac{b_{max}(F\text{–}T)}{b_{max}(coll)} = \frac{1.4\, a_0 \omega}{\gamma Z^{\frac{1}{3}} v} \tag{3.66}$$

We should adopt the smaller of these values of b_{max} in any particular application. In the ultrarelativistic limit $\gamma \to \infty$, we should use the Fermi–Thomas shielding limit. The ultrarelativistic limit of (3.34) is therefore

$$I(\omega') = \frac{Z^2 e^6 \gamma N}{12\pi^3 \varepsilon_0^3 c^3 m_e^2 v} \ln\left(\frac{1.4\, a_0 m_e v}{Z^{\frac{1}{3}} \hbar}\right) \tag{3.67}$$

We now have to transform this into the laboratory frame. We have already shown in Section 3.3.1 that dE/dt is a relativistic invariant. In the present case, $I(\omega')$ has dimensions (energy per unit time per unit bandwidth). Thus, we need only ask how $\Delta\omega$ transforms between frames. It is simplest to note that ω transforms in the same way as E and hence, as shown in Section 3.3.1,

$$\Delta\omega = \gamma \Delta\omega'$$

i.e. the bandwidth increases by a factor γ in S. Therefore in S, the intensity per unit bandwidth is smaller by a factor γ, i.e.

$$I(\omega) = \frac{Z^2 e^6 N}{12\pi^3 \varepsilon_0^3 c^3 m_e^2 v} \ln\left(\frac{192 v}{Z^{\frac{1}{3}} c}\right) \tag{3.68}$$

The intensity spectrum is independent of frequency up to energy $\hbar\omega = (\gamma - 1) m_e c^2$ which corresponds to the electron giving up all its kinetic energy in a single collision. Now we can find the total energy loss rate by integrating over frequency.

$$-(dE/dt) = \int_0^{E/\hbar} I(\omega)\, d\omega$$

Since $v \approx c$,

$$-\left(\frac{dE}{dt}\right) = \frac{Z^2 e^6 N E}{12\pi^3 \varepsilon_0^3 m_e^2 c^4 \hbar} \ln\left(\frac{192}{Z^{\frac{1}{3}}}\right) \tag{3.69}$$

We should compare this with the proper formula derived by Bethe and Heitler from the full relativistic quantum treatment

$$-\left(\frac{dE}{dt}\right) = \frac{Z(Z+1.3) e^6 N}{16\pi^3 \varepsilon_0^3 m_e^2 c^4 \hbar} E\left[\ln\left(\frac{183}{Z^{\frac{1}{3}}}\right) + \frac{1}{8}\right] \tag{3.70}$$

Notice that, although we have had to make a number of approximations, we have come remarkably close to the correct answer. The term $(Z+1.3)$ takes account of electron–electron interactions between the high energy electron and those bound to the atoms of the ambient material. Notice that, in contrast to the non-relativistic case (3.39), the relativistic bremsstrahlung energy loss rate is proportional to the energy of the electron.

The relativistic bremsstrahlung energy losses are of exponential form $-dE/dx \propto E$ and it is therefore possible to define a radiation length X_0 over which the electron loses a fraction $(1 - 1/e)$ of its energy, $-dE/dx = E/X_0$. As in Section 2.4.2, it is convenient to describe this length in terms of the number of kilograms

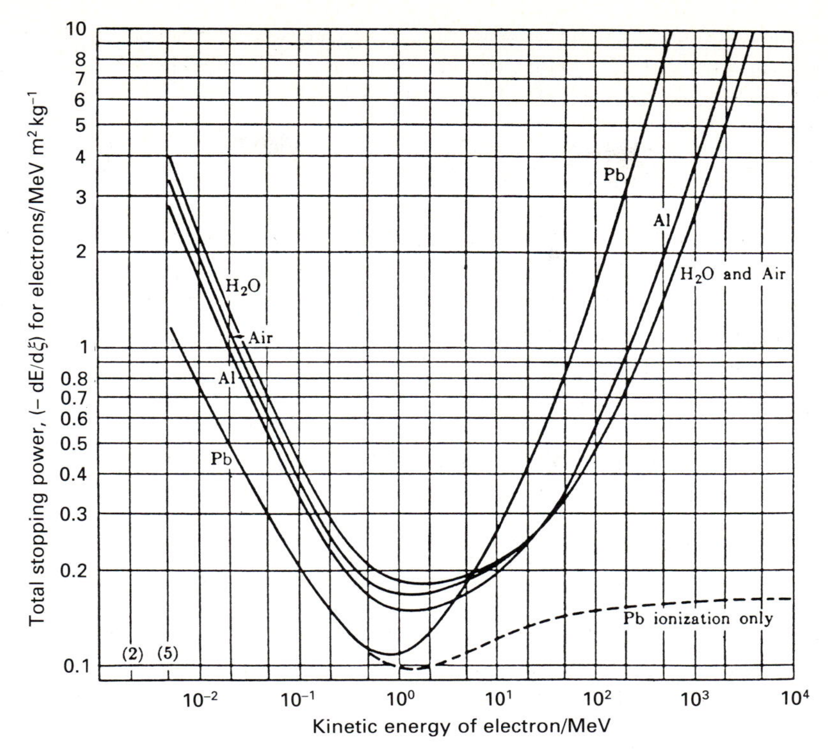

Figure 3.5. The total stopping power for electrons in air, water, aluminium and lead. At energies less than 1 MeV, the dominant loss mechanism is ionisation losses. At higher energies, the dominant loss process is bremsstrahlung. For comparison, the contribution from ionisation losses for electrons in lead is also shown. (From H. A. Enge (1966). *Introduction to nuclear physics*, page 190, London: Addison-Wesley Publishing Co.)

per square metre, $\xi_0 = \rho X_0$, through which the electron passes. In the ultrarelativistic limit

$$-\frac{dE}{d\xi} = -\frac{dE}{dt}\frac{1}{\rho c} = \frac{E}{\rho X_0} = \frac{E}{\xi_0} \tag{3.71}$$

It is also convenient to express the radiation length ξ_0 in terms of the atomic mass M_A of the material. If N_0 is Avogadro's number, $N = N_0 \rho / M_A$. We recall that M_A g of any substance contains N_0 particles. We can now write an expression for ξ_0 which is useful for numerical purposes:

$$\xi_0 = \frac{7160 M_A}{Z(Z+1.3)\,[\ln(183 Z^{-\frac{1}{3}}) + \frac{1}{8}]}\ \text{kg m}^{-2} \tag{3.72}$$

The form of the total energy loss rate $(-dE/d\xi)$ or *total stopping power* for different materials is illustrated in Fig. 3.5. Below about 1 MeV, at which the electron becomes non-relativistic, ionisation losses remain the dominant loss mechanism, but for greater energies, relativistic bremsstrahlung losses rapidly

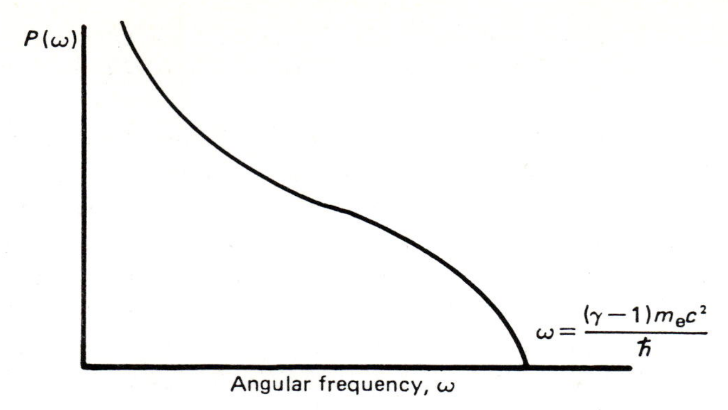

Figure 3.6. The probability per unit bandwidth of the emission of a photon by bremsstrahlung as a function of angular frequency of the emitted photon plotted on linear intensity and frequency scales.

become dominant. We can define a *critical energy* E_c as that at which bremsstrahlung losses are equal to ionisation losses. For hydrogen, air and lead the values of E_c are 340, 83 and 6.9 MeV respectively. This is an important result in the analysis of air showers initiated by cosmic rays in the atmosphere.

The radiation lengths for these materials are:

hydrogen	$\xi_0 = 580$ kg m^{-2}	$X_0 = 6.7$ km
air	$\xi_0 = 365$ kg m^{-2}	$X_0 = 280$ m
lead	$\xi_0 = 58$ kg m^{-2}	$X_0 = 5.6$ mm

The value for the atmosphere is of particular interest because its total depth is about 10000 kg m^{-2} and thus, cosmic ray electrons must suffer catastrophic bremsstrahlung losses when they enter the atmosphere.

The other crucial factor is the emitted spectrum of photon energies which we have devoted so much energy to deriving. Let us rewrite the spectrum as a probability distribution of emitting photons of energy $\hbar\omega$. Then

$$I(\omega)\,d\omega = P(\omega)\hbar\omega N_i\,d\omega \tag{3.73}$$

Therefore,

$$P(\omega) \propto 1/\omega \text{ up to } \hbar\omega = (\gamma - 1)m_e c^2$$

This means that the probability diverges at zero frequency. As indicated by Fig. 3.2, however, the intensity of radiation remains finite at zero frequency. The important point is that, although the likelihood of an energetic photon being emitted is small, when it is emitted, it takes away a significant fraction of the energy of the electron. The spectrum of bremsstrahlung plotted in terms of flux of photons per unit frequency interval is shown on a linear frequency scale in Fig. 3.6–this shows the probability distribution of energy packets being emitted. It can be seen that, on average, we expect one or two very energetic photons to be emitted in each radiation length. This is very important since it means that a very high energy cosmic ray electron deposits most of its energy into one or two high energy photons within a very short distance once it enters the atmosphere. This leads us naturally to study the interaction of high energy photons with matter in the next chapter.

In addition, relativistic bremsstrahlung is likely to be of importance astrophysically. Wherever there are relativistic electrons with energy E, these can interact with atoms and molecules to generate photons with frequencies up to $\nu = E/h$, on average their energy being about $\frac{1}{3}E$. An example of the potential application of this process to the γ-ray emission of the interstellar medium is given in Section 18.3 of Volume 2. In that case, the emission is the bremsstrahlung of a power-law distribution of electron energies $N(E) \propto E^{-x}$. In that section, it is shown that such an electron energy spectrum results in an intensity spectrum for the γ-rays of exactly the same power-law form, $N_{\gamma}(\varepsilon) \propto \varepsilon^{-x}$, provided the intensity is measured in terms of the flux density of photons $m^{-2}\,s^{-1}\,MeV^{-1}\,sr^{-1}$. This process may well be important in other astrophysical environments.

4

Interactions of high energy photons

4.1 Introduction

The three main processes involved in the interaction of high energy photons with atoms, nuclei and electrons are photoelectric absorption, Compton scattering and electron–positron pair production. Again, we need the physics of these processes, not only in diagnosing the properties of high energy particles but also in studying high energy astrophysical phenomena in a wide variety of different circumstances. The specific example we will consider in the context of the detection of particles will be the development of electron–photon cascades in which high energy electrons enter the atmosphere or a particle detector. The arrival of very high energy γ-rays at the top of the atmosphere gives rise to showers of very energetic electron–positron pairs. The latter can emit Cherenkov radiation which can be detected as optical emission at ground level. This technique has resulted in remarkable evidence for ultrahigh energy γ-rays from cosmic sources and so it is convenient to study Cherenkov radiation in this context.

These properties also find very wide applicability in high energy astrophysics. For example, photoelectric absorption is found in the spectra of most X-ray sources at energies $\varepsilon \lesssim 1$ keV. Thomson and Compton scattering appear in a myriad of guises from the processes occurring in stellar interiors, to the spectra of binary X-ray sources and the many applications of inverse Compton scattering in objects involving intense radiation fields and high energy electrons. Pair production is bound to occur wherever there are significant fluxes of high energy γ-rays – indeed, evidence for the production of positrons by this process is provided by the detection of the 511 keV electron–positron annihilation line in our own Galaxy.

We will therefore give the relations which are useful both from the point of view of particle and photon detection and of astrophysics. We will return to many of these processes in more detail later in our story.

4.2 Photoelectric absorption

At low photon energies, $\hbar\omega \ll m_e c^2$, the dominant process by which photons lose energy is photoelectric absorption. This process is familiar to stellar astronomers since photoelectric, or bound–free, absorption is one of the principal sources of opacity in stellar interiors and stellar atmospheres. We will be principally interested in the process in much more rarefied plasmas. If the energy of the incident photon is $\hbar\omega$, it can eject electrons which have binding energies $E_I \leqslant \hbar\omega$ from atoms, ions and molecules, the remaining energy $(\hbar\omega - E_I)$ being removed as the kinetic energy of the ejected electron. This is just the *photoelectric effect* first explained by Einstein in his revolutionary paper of 1905 in which he first proposed the concept of light quanta. The energy levels within the atom for which $\hbar\omega = E_I$ are called *absorption edges* because ejection of electrons from these energy levels is impossible if the photons are of lower energy. For photons with higher energies, the cross-section for photoelectric absorption from this level decreases as roughly ν^{-3}. The evaluation of this cross-section is one of the standard results of the quantum theory of radiation (see e.g. Heitler (1954)). There is an analytic solution for the absorption cross-section for photons with energies $\hbar\omega \gg E_I$ and $\hbar\omega \ll m_e c^2$ due to the ejection of electrons from the K-shells of atoms, i.e. from the 1s level

$$\sigma_K = 4\sqrt{2}\sigma_T \alpha^4 Z^5 \left(\frac{m_e c^2}{\hbar\omega}\right)^{\frac{7}{2}} = \frac{e^{12} m_e^{\frac{3}{2}} Z^5}{192\sqrt{2}\pi^5 \varepsilon_0^6 \hbar^4 c}\left(\frac{1}{\hbar\omega}\right)^{\frac{7}{2}} \tag{4.1}$$

In this cross-section, α is the fine structure constant, $\alpha = e^2/4\pi\varepsilon_0 \hbar c$ and σ_T is the Thomson cross-section, $\sigma_T = 8\pi r_e^2/3 = e^2/6\pi\varepsilon_0^2 m_e^2 c^4$. Notice that this cross-section corresponds to the removal of 2 K-shell electrons from the atom since both 1s electrons contribute to the opacity of the material. Notice also the very strong dependence of the absorption cross-section upon the atomic number Z. Thus, although heavy elements are very much less abundant than hydrogen, the combination of the ω^{-3} dependence and the fifth-power dependence upon Z means that quite rare elements make important contributions to the absorption cross-section at hard ultraviolet and X-ray energies. More detailed calculations of these cross-sections with the appropriate Gaunt factors are given by Karzas and Latter (1961).

The above results are sufficient to enable us to understand some of the characteristic features of X-ray absorption spectra seen in X-ray sources. Every atom has a characteristic X-ray term diagram showing the energies of the different stationary states for the removal of an electron from within an atom. Examples of these, the X-ray term diagrams for hydrogen, carbon, oxygen and argon, are given in Table 4.1 with the corresponding X-ray photoelectric absorption cross-sections as a function of photon energy shown in Fig. 4.1. Curves such as these are of great importance in the construction of proportional counters for the detection of X-rays since the photoelectric absorption spectra of the detector gas and the window materials determine the efficiency and sensitivity of the counter (see Section 6.4).

Table 4.1. *The X-ray energy levels for hydrogen, carbon, oxygen and argon*

	Energies in electron-volts (eV) X-ray term				
Element	K	L_I	$L_{II,III}$	M_I	$M_{II,II}$
Hydrogen	13.598				
Carbon	283.8		6.4		
Oxygen	532.0	23.7	7.1		
Argon	3202.9	320	247.3, 245.2	25.3	12.4

(From *American Handbook of Physics*, third edition, page 7.158–9, McGraw-Hill, New York, 1972.)

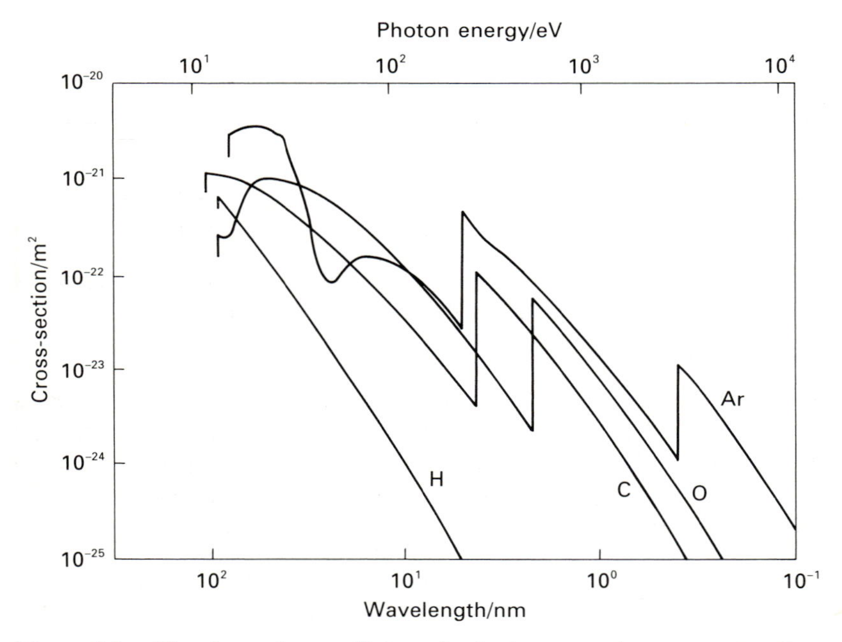

Figure 4.1. The absorption coefficients for hydrogen, carbon, oxygen and argon atoms as a function of photon energy (or wavelength). (From M. V. Zombeck (1990). *Handbook of space astronomy and astrophysics*, page 295–8, Cambridge: Cambridge University Press.)

Another important application of photoelectric absorption is in the determination of the X-ray absorption coefficient for interstellar matter. To estimate this, it is necessary to add together curves of the form shown in Fig. 4.1 for the cosmic abundance of the elements. In such a computation, the K-edges, which correspond to the ejection of electrons from the 1s shell of the atom or ion, provide the dominant source of opacity. The resulting total absorption coefficient for X-rays is shown in Fig. 4.2, the K-edges of the elements contributing to the total

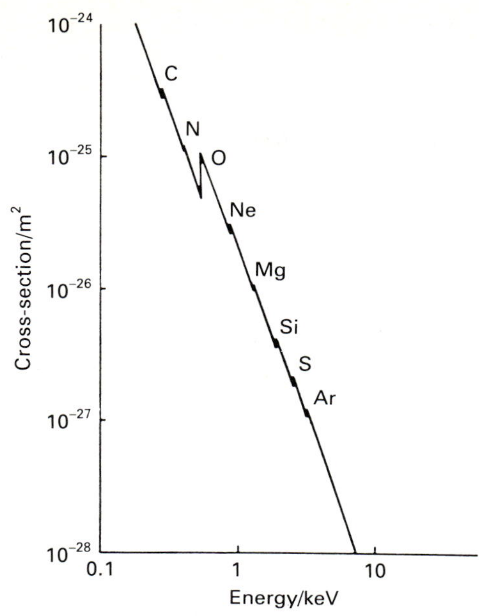

Figure 4.2. The absorption cross-section for interstellar gas with typical cosmic abundances of the chemical elements (see Section 9.2.1). The discontinuities in the absorption cross-section as a function of energy are associated with the K-shell absorption edges of the elements indicated. The optical depth of the medium is given by $\tau = \int \sigma_X(E) N_H \, dl$ where N_H is the number density of hydrogen atoms. (After R. H. Brown and R. J. Gould (1970). *Phys. Rev.*, **D1**, 2252.)

absorption coefficient being indicated. In the curve shown in Fig. 4.2, the standard cosmic abundances of the chemical elements have been assumed (see Section 9.2.1). In low resolution X-ray spectral studies, these edges cannot be resolved individually as distinct features and a useful linear interpolation formula for the X-ray absorption coefficient, σ_X, and optical depth, $\tau_X = \sigma_X N_H \, dl$ is

$$\tau_X(\hbar\omega) = 2 \times 10^{-26} \left(\frac{\hbar\omega}{1 \text{ keV}} \right)^{-\frac{8}{3}} \int N_H \, dl \tag{4.2}$$

where the column depth $\int N_H \, dl$ is expressed in particles per square metre and N_H is the number density of hydrogen atoms in particles per cubic metre. Because of the steep energy dependence of τ_X, photoelectric absorption is not generally important at energies $\hbar\omega \gtrsim 1$ keV.

4.3 Compton scattering

In 1923, Compton discovered that the wavelength of hard X-ray radiation increases when it is scattered from stationary electrons. This was the final convincing proof of Einstein's quantum picture of the nature of light according to which it may be considered to possess both wave-like and particle-like properties.

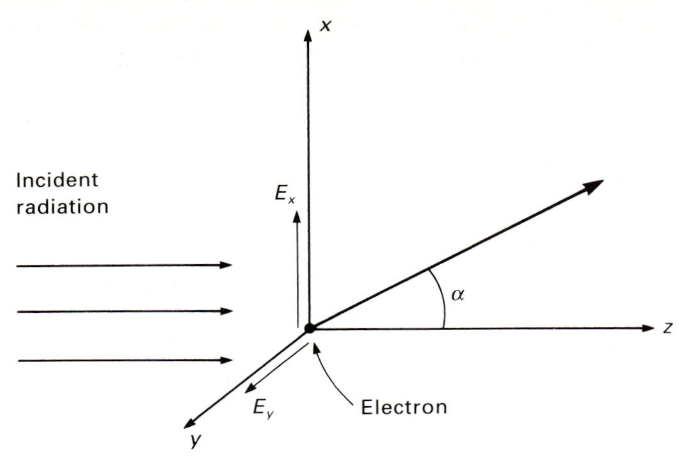

Figure 4.3. Illustrating the geometry of Thomson scattering of a beam of radiation by a free electron.

In the Compton scattering process, the incoming high energy photons collide with stationary electrons and transfer some of their energy and momentum to the electrons. Consequently, the photons come out of the collisions with less energy and momentum than they went in. Since the energy and momentum of the photons are proportional to the frequency of the radiation, $E = \hbar\omega$ and $\mathbf{p} = (\hbar\omega/c)\mathbf{i}_k$, where $\mathbf{i}_k$ is the unit vector in the direction of travel of the photon, the loss of energy of the photon corresponds to an increase in its wavelength. Let us build up a picture of some of the important aspects of Compton scattering by considering first the simpler process of Thomson scattering.

4.3.1 *Thomson scattering*

Thomson first published the formula for what is now called the Thomson cross-section in 1906. He used it in a major paper of that year to show that the number of electrons in each atom is roughly the same as the element's atomic number. He obtained this result by interpreting X-ray scattering experiments carried out by Barkla. He attributed the scattering to the reradiation of the X-rays by all the electrons in the sample which he assumed could be considered free particles. In his derivation of the cross-section for this process, he used the same Larmor formula which we derived by his method in Section 3.3.2. Let us follow some of his footsteps.

The formula we seek is that which describes the scattering of an unpolarised beam of radiation incident upon a stationary electron. The problem is to find the intensity of radiation scattered through angle α by the electron. We can give a completely classical analysis of this problem using the radiation formulae derived above. We assume that the beam of incident radiation propagates in the positive z-direction (Fig. 4.3). To simplify the analysis and without loss of generality, we arrange the geometry of the scattering so that the scattering angle α lies in the x–z

plane. We resolve the electric field strength of the unpolarised incident field into components of equal intensity with electric vectors in the $\mathbf{i}_x$ and $\mathbf{i}_y$ directions (see Fig. 4.3).

The electric fields experienced by the electron in the x and y directions, $E_x = E_{x0}\exp(i\omega t)$ and $E_y = E_{y0}\exp(i\omega t)$ respectively, cause the electron to oscillate and the accelerations in these directions are:

$$\ddot{r}_x = eE_x/m_\mathrm{e} \quad \ddot{r}_y = eE_y/m_\mathrm{e}$$

We can therefore enter these accelerations into the radiation formula (3.8) which shows the angular dependence of the emitted radiation upon the polar angle θ. Let us treat the x-acceleration first. In this case, we can use the formula (3.8) directly with the substitution $\alpha = \pi/2-\theta$. Therefore the intensity of radiation scattered through angle θ into the solid angle $\mathrm{d}\Omega$ is

$$-\left(\frac{\mathrm{d}E}{\mathrm{d}t}\right)_x \mathrm{d}\Omega = \frac{e^2|\ddot{r}_x|^2\sin^2\theta}{16\pi^2\varepsilon_0 c^3}\mathrm{d}\Omega = \frac{e^4|E_x|^2}{16\pi^2 m_\mathrm{e}^2\varepsilon_0 c^3}\cos^2\alpha\,\mathrm{d}\Omega \tag{4.3}$$

As usual, we have to take time averages of E_x^2 and we find that $\overline{E_x^2} = E_{x0}^2/2$. We sum over all waves contributing to the E_x-component of radiation and express the result in terms of the incident energy per unit area upon the electron. The latter is given by Poynting's theorem, $\mathbf{S}_x = (\mathbf{E}\times\mathbf{H}) = c\varepsilon_o E_x^2\mathbf{i}_z$. Again, we take time averages and find that the contribution to the intensity in the direction α from the x-component of the acceleration is $S_x = \sum_i c\varepsilon_0 E_{x0}^2/2$. Therefore,

$$\begin{aligned}-\left(\frac{\mathrm{d}E}{\mathrm{d}t}\right)_x \mathrm{d}\Omega &= \frac{e^4\cos^2\alpha}{16\pi^2 m_\mathrm{e}^2\varepsilon_0 c^3}\sum_i \overline{E_x^2}\,\mathrm{d}\Omega \\ &= \frac{e^4\cos^2\alpha}{16\pi^2 m_\mathrm{e}^2\varepsilon_0^2 c^4}S_x\,\mathrm{d}\Omega \end{aligned}\tag{4.4}$$

Now let us look at the scattering of the E_y-component of the incident field. From the geometry of Fig. 4.3, it can be seen that the radiation in the x–z plane from the acceleration of the electron in the y-direction corresponds to scattering at $\theta = 90°$ and therefore the scattered intensity in the α-direction is just

$$-\left(\frac{\mathrm{d}E}{\mathrm{d}t}\right)_y \mathrm{d}\Omega = \frac{e^4}{16\pi^2 m_\mathrm{e}^2\varepsilon_0^2 c^4}S_y\,\mathrm{d}\Omega \tag{4.5}$$

The total scattered radiation into $\mathrm{d}\Omega$ is the sum of these components (notice that we add the intensities of the two independent field components).

$$-\left(\frac{\mathrm{d}E}{\mathrm{d}t}\right)\mathrm{d}\Omega = \frac{e^4}{16\pi^2 m_\mathrm{e}^2\varepsilon_0^2 c^4}(1+\cos^2\alpha)\frac{S}{2}\mathrm{d}\Omega \tag{4.6}$$

where $S = S_x+S_y$ and we recall that $S_x = S_y$ for unpolarised radiation. We now express the scattered intensity in terms of a differential scattering cross-section $\mathrm{d}\sigma_\mathrm{T}$

in the following way. We define the scattered intensity in direction α by the following relation

$$\frac{d\sigma_T(\alpha)}{d\Omega} = \frac{\text{energy radiated per unit time per unit solid angle}}{\text{incident energy per unit time per unit area}} \tag{4.7}$$

Since the total incident energy is S, the differential cross-section for Thomson scattering is

$$d\sigma_T = \frac{e^4}{16\pi^2\varepsilon_0^2 m_e^2 c^4} \frac{(1+\cos^2\alpha)}{2} d\Omega \tag{4.8}$$

which we can express in terms of the classical electron radius $r_e = e^2/4\pi\varepsilon_0 m_e c^2$,

$$d\sigma_T = \frac{r_e^2}{2}(1+\cos^2\alpha)\, d\Omega \tag{4.9}$$

To find the total cross-section for scattering, we integrate over all angles α in the standard way,

$$\sigma_T = \int_0^\pi \frac{r_e^2}{2}(1+\cos^2\alpha)\, 2\pi \sin\alpha\, d\alpha$$

$$= \frac{8\pi}{3} r_e^2 = \frac{e^4}{6\pi\varepsilon_0^2 m_e^2 c^4} = 6.653 \times 10^{-29}\ \text{m}^2 \tag{4.10}$$

This is Thomson's famous result for the total cross-section for scattering by stationary free electrons and is justly referred to as the *Thomson cross-section*. This cross-section reappears in all sorts of formulae involving radiation processes as we will find as we proceed. Let us note some of the important properties of Thomson scattering.

(i) The scattering is symmetric with respect to the scattering angle α. Thus, as much radiation is scattered backwards as forwards.

(ii) Another useful calculation is to work out the scattering cross-section for 100 % polarised emission. We can work this out by integrating the scattered intensity (4.3) over all angles.

$$-\left(\frac{dE}{dt}\right)_x = \frac{e^2|\ddot{r}_x|^2}{16\pi^2\varepsilon_0 c^3} \int \sin^2\theta\, 2\pi \sin\theta\, d\theta$$

$$= \left(\frac{e^4}{6\pi\varepsilon_0^2 m_e^2 c^4}\right) S_x \tag{4.11}$$

We find the same total cross-section for scattering as before. This should not be surprising because it does not matter how the electron is forced to oscillate. The energy radiated is simply proportional to the sum of the incident intensities of the radiation field. Because of this last fact, the only important quantity so far as the electron is concerned is the total intensity incident upon it and it does not matter how anisotropic the radiation is. One convenient way of expressing this result is to write the formula for the scattered radiation in terms to the energy density of radiation at the electron

$$u_{\text{rad}} = \sum_i u_i = \sum_i S_i/c$$

and hence

$$-(\mathrm{d}E/\mathrm{d}t) = \sigma_{\mathrm{T}} c u_{\text{rad}} \tag{4.12}$$

(iii) The one distinctive feature of the process is that the scattered radiation is polarised, even if the incident beam of radiation is unpolarised. This can be seen intuitively from Fig. 4.3 because all the E-vectors of the unpolarised beam lie in the x–y plane. Therefore, in the case of observing the electron precisely in the x–y plane, the scattered radiation is 100% polarised. On the other hand, if we look along the z-direction, we observe unpolarised radiation. If we define the degree of polarisation in the standard way,

$$\Pi = \frac{I_{\text{max}} - I_{\text{min}}}{I_{\text{max}} + I_{\text{min}}} \tag{4.13}$$

we find by a simple calculation that the fractional polarisation of the radiation is

$$\Pi = \frac{1-\cos^2\alpha}{1+\cos^2\alpha} \tag{4.14}$$

This is therefore a means of producing polarised radiation from an initially unpolarised beam. A beautiful example of this phenomenon is presented in Fig. 4.4 which shows the infrared intensity and polarisation of an obscured source in Orion Molecular Cloud 1. The polarisation vectors are more or less circularly symmetric about the source to the right of the picture, showing that it is the source of excitation of the outflow which opens up towards the top left of the image and that the radiation from the flow is scattered infrared radiation.

(iv) Thomson scattering is one of the most important processes which impedes the escape of photons from any region and it is useful to write down equation (4.12) in terms of the scattering of photons out of a beam propagating in the positive x-direction. To do this we write down the expression for the energy scattered by the electron in terms of the number density N of photons of frequency ν so that

$$-(\mathrm{d}(Nh\nu)/\mathrm{d}t) = \sigma_{\mathrm{T}} c N h\nu$$

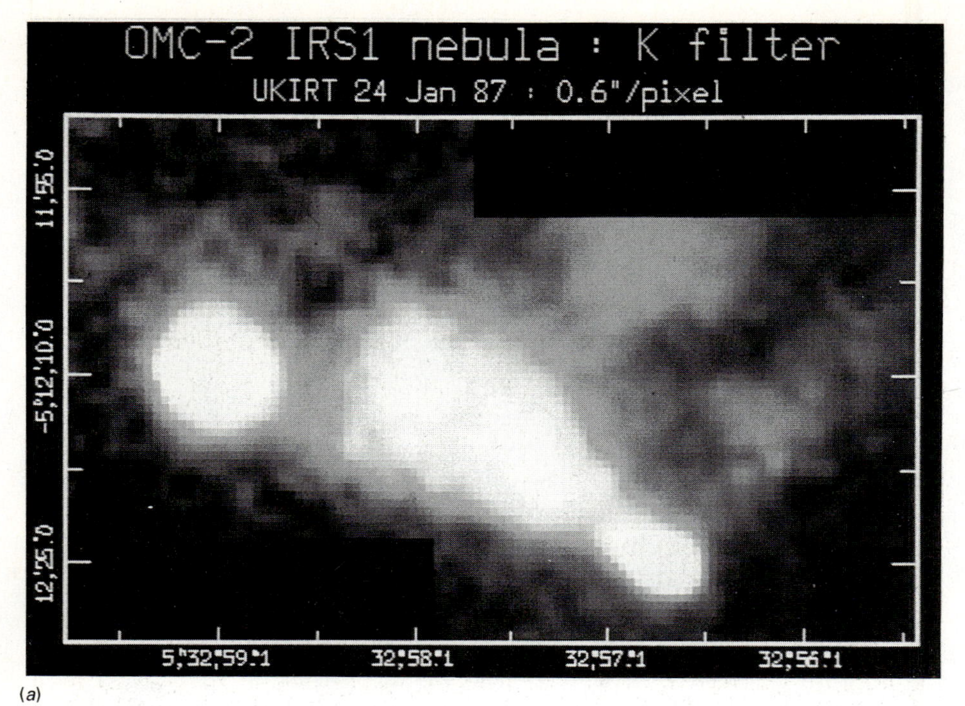

(*a*)

Figure 4.4. (*a*) For legend see facing page

There is no change of energy of the photons in the scattering process and so, if we apply the above equation to the scattering of photons from the beam and if there are now N_e electrons per unit volume, the number density of photons decreases exponentially with distance

$$-\mathrm{d}N/dt = \sigma_T c N_e N$$

$$-\mathrm{d}N/\mathrm{d}x = \sigma_T N_e N$$

$$N = N_0 \exp(-\int \sigma_T N_e \,\mathrm{d}x)$$

We can express this by stating that the *optical depth* of the medium to Thomson scattering is

$$\tau = \int \sigma_T N_e \,\mathrm{d}x$$

In this process, the photons are scattered in random directions and so they perform a random walk, each step corresponding to the *mean free path* λ_T of the photon through the electron gas where $\lambda_T = (\sigma_T N_e)^{-1}$. Thus, there is a very real sense in which the Thomson cross-section is the physical cross-section of the electron for the scattering of electromagnetic waves.

4.3.2 *Compton scattering*

In the case of Thomson scattering, there is no change in the frequency of the radiation. The electron simply acts as a radiator which scatters the incoming

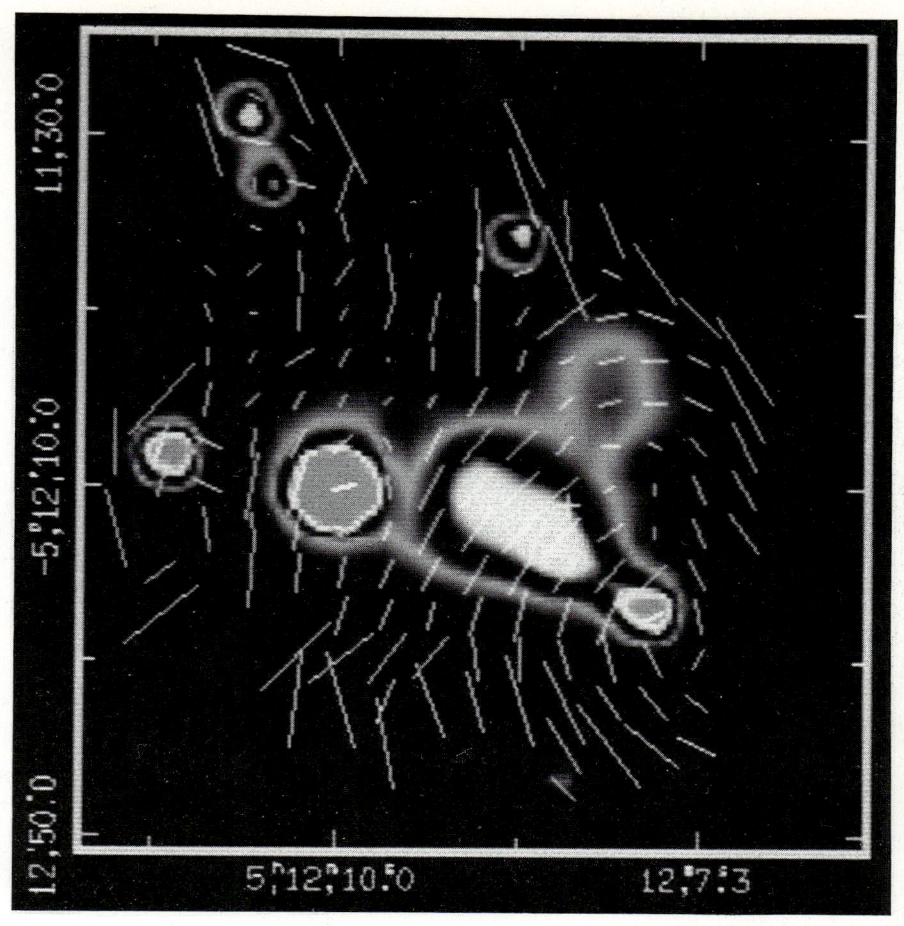

(*b*)

Figure 4.4. Infrared images at 2.2 μm of an outflow source in the Orion Molecular Cloud. (*a*) The outflow source is the object to the bottom right of the image. The cone of emission extending towards the top left of the diagram is believed to be excited by a beam of radiation from this source. (From I. S. McLean (1987). *Infrared astronomy with arrays*, eds C. G. Wynn-Williams and E. E. Becklin, page 189, Hawaii: University of Hawaii Publications. (*b*) An infrared polarisation image of the same region at 2.2 μm taken at a lower angular resolution. It can be seen that the polarisation vectors in the region of the outflow are roughly circularly symmetric about the source at the bottom right of the image, the typical signature of scattering by the free electrons in the outflow. (From J. Rayner and I. S. McLean, (1987). *Infrared astronomy with arrays*, eds C. G. Wynn-Williams and E. E. Becklin, page 277. Hawaii: University of Hawaii Publications.)

radiation. This remains a good approximation provided the energy of the photon is much less than the rest mass energy of the electron, $\hbar\omega \ll m_e c^2$. In general, as long as the energy of the photon is less than $m_e c^2$ in the centre of momentum frame of reference, the scattering may be accurately treated as Thomson scattering and we will use this in our treatment of inverse Compton scattering (Section 4.3.3). There are a number of important cases, however, in which the frequency change associated with the collision between the electron and the photon is important in high energy astrophysics. Let us establish some of the more important general results.

In classical Compton scattering the wavelengths of the photons increase. For the sake of generality, let us suppose that the electron is not at rest but moving with velocity $\mathbf{v}$ in the laboratory frame of reference S. There are several ways of tackling this problem. I confess that I normally adopt the boring approach of finding the centre of momentum frame, then treating the collision in that frame and transforming back to the laboratory frame of reference. There is, however, a very neat way of coming quickly to the complete answer using the momentum and frequency four-vectors of the electron and photon respectively. This is an excellent example of the power of four-vectors.

Let us write down the momentum four-vectors for the electron and the photon before and after the collision.

$$\text{Electron} \quad \mathbf{P} = [\gamma m_e \mathbf{v}, \gamma m_e] \quad \mathbf{P}' = [\gamma' m_e \mathbf{v}', \gamma' m_e]$$

$$\text{Photon} \quad \mathbf{K} = \left[\frac{\hbar\omega}{c}\mathbf{i}_k, \frac{\hbar\omega}{c^2}\right] \quad \mathbf{K}' = \left[\frac{\hbar\omega'}{c}\mathbf{i}_{k'}, \frac{\hbar\omega'}{c^2}\right]$$

Before After

The collision conserves four-momentum and hence

$$\mathbf{P}+\mathbf{K} = \mathbf{P}'+\mathbf{K}' \tag{4.15}$$

Now, we square both sides of this four-vector equation and recall the properties of the norms of the momentum four-vectors of the electron and the photon: $\mathbf{P}\cdot\mathbf{P} = \mathbf{P}'\cdot\mathbf{P}' = m_e^2 c^2$ and $\mathbf{K}\cdot\mathbf{K} = \mathbf{K}'\cdot\mathbf{K}' = 0$. Therefore,

$$(\mathbf{P}+\mathbf{K})^2 = (\mathbf{P}'+\mathbf{K}')^2$$

$$\mathbf{P}\cdot\mathbf{P}+2\mathbf{P}\cdot\mathbf{K}+\mathbf{K}\cdot\mathbf{K} = \mathbf{P}'\cdot\mathbf{P}'+2\mathbf{P}'\cdot\mathbf{K}'+\mathbf{K}'\cdot\mathbf{K}'$$

$$\mathbf{P}\cdot\mathbf{K} = \mathbf{P}'\cdot\mathbf{K}' \tag{4.16}$$

Now multiply (4.15) by $\mathbf{K}'$ and use the equality (4.16).

$$\mathbf{P}\cdot\mathbf{K}'+\mathbf{K}\cdot\mathbf{K}' = \mathbf{P}'\cdot\mathbf{K}'+\mathbf{K}'\cdot\mathbf{K}'$$

$$\mathbf{P}\cdot\mathbf{K}'+\mathbf{K}\cdot\mathbf{K}' = \mathbf{P}\cdot\mathbf{K} \tag{4.17}$$

This four-vector equation is the solution we seek. Let us reduce it to somewhat more familiar form by multiplying out the four-vector products. The scattering angle is given by $\mathbf{i}_k\cdot\mathbf{i}_{k'} = \cos\alpha$. We let the angle between the incoming photon and the velocity vector of the electron be θ and the angle between them after the collision be θ'. Then, $\cos\theta = \mathbf{i}_k\cdot\mathbf{v}/|\mathbf{v}|$ and $\cos\theta' = \mathbf{i}_{k'}\cdot\mathbf{v}'/|\mathbf{v}'|$. After a little algebra,

$$\frac{\omega'}{\omega} = \frac{1-(v/c)\cos\theta}{[1-(v/c)\cos\theta'+(\hbar\omega/\gamma m_e c^2)(1-\cos\alpha)]} \tag{4.18}$$

In the traditional argument, the Compton effect is described in terms of the increase in wavelength of the photon on scattering from a stationary electron i.e. $v = 0$, $\gamma = 1$.

$$\frac{\omega'}{\omega} = \frac{1}{1+(\hbar\omega/m_e c^2)(1-\cos\alpha)}$$

$$\frac{\Delta\lambda}{\lambda} = \frac{\lambda'-\lambda}{\lambda} = \frac{\hbar\omega}{m_e c^2}(1-\cos\alpha) \tag{4.19}$$

This effect of 'cooling' the radiation and transferring the energy to the electron is sometimes called the *recoil effect*. Note, however, that expression (4.18) also shows how energy can be interchanged between the electrons and the radiation field. In the limit $\hbar\omega \ll \gamma m_e c^2$, the change in frequency of the photon can be written

$$\frac{\omega'-\omega}{\omega} = \frac{\Delta\omega}{\omega} = \frac{v}{c}\frac{(\cos\theta-\cos\theta')}{[1-(v/c)\cos\theta']} \tag{4.20}$$

Thus, it can be seen that, to first order, the frequency changes are $\sim v/c$. Note, however, that, also to first order, if the angles θ and θ' are randomly distributed, a photon is just as likely to decrease as increase its energy. It can be shown that there is no net increase in energy of the photons to first order in v/c and it is only in second order, i.e. to order v^2/c^2, that there is a net energy gain. We will take this point up in a little more detail after we have dealt with inverse Compton scattering since the low energy limit of that process gives explicitly the correct expression for the net energy gain by the photons in the presence of a distribution of electrons at temperature T.

Occasions will arise when it is not adequate to use the Thomson cross-section to describe the scattering of photons by free electrons. The key consideration is whether or not the electron moves with velocity $v \sim c$ or if the photon has energy $\hbar\omega \sim m_e c^2$ in the centre of momentum frame of reference. If a photon of energy $\hbar\omega$ collides with a stationary electron, according to the analysis of Section 2.3.3, the centre of momentum frame moves at velocity

$$\frac{v}{c} = \frac{\hbar\omega}{m_e c^2 + \hbar\omega} \tag{4.21}$$

Therefore, if the photons to be scattered have energy $\hbar\omega \gtrsim m_e c^2$, we must use the proper quantum relativistic cross-section for scattering. Another case which can often arise is if the photons are of low energy $\hbar\omega \ll m_e c^2$ but the electron is moving ultrarelativistically with $\gamma \gg 1$. Then, the centre of momentum frame moves with a velocity close to that of the electron and in this frame the energy of the photon is $\gamma\hbar\omega$. Again if $\gamma\hbar\omega \sim m_e c^2$, the quantum relativistic cross-section has to be used.

The relevant total cross-section is the Klein–Nishina formula:

$$\sigma_{\text{K-N}} = \pi r_e^2 \frac{1}{\varepsilon}\left\{\left[1-\frac{2(\varepsilon+1)}{\varepsilon^2}\right]\ln(2\varepsilon+1)+\frac{1}{2}+\frac{4}{\varepsilon}-\frac{1}{2(2\varepsilon+1)^2}\right\} \tag{4.22}$$

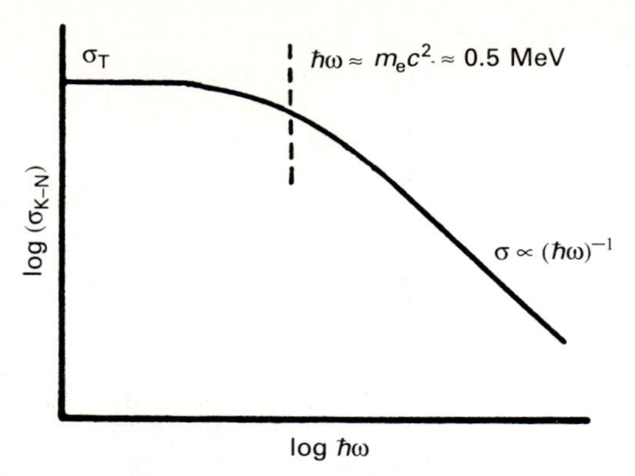

Figure 4.5. A schematic diagram showing the dependence of the Klein–Nishina cross-section upon photon energy.

where $\varepsilon = \hbar\omega/m_e c^2$ and r_e is the classical electron radius (see (4.10)). For low energy photons, $\varepsilon \ll 1$, this expression reduces to

$$\sigma_{K-N} = \frac{8\pi}{3} r_e^2(1-2\varepsilon) = \sigma_T(1-2\varepsilon) \approx \sigma_T \tag{4.23}$$

In the ultrarelativistic limit it becomes

$$\sigma_{K-N} = \pi r_e^2 \frac{1}{\varepsilon}(\ln 2\varepsilon + \tfrac{1}{2}) \tag{4.24}$$

so that the cross-section decreases roughly as ε^{-1} at the highest energies (Fig. 4.5). If the atom has Z electrons, then the total cross-section for the atom is just $Z\sigma_{K-N}$. Note that scattering by nuclei can be neglected because they cause very much less scattering than electrons, roughly by a factor of $(m_e/m_N)^2$ where m_N is the mass of the nucleus.

4.3.3 *Inverse Compton scattering*

It is convenient to continue the story of the Compton effect with a discussion of inverse Compton scattering, not only because it is one of the most important processes for high energy astrophysics but also because the results can be used in the non-relativistic limit to illustrate other aspects of the Compton scattering process. The normal way in which this subject is introduced is through consideration of the scattering of low energy photons by ultrarelativistic electrons. The high energy electrons scatter low energy photons to high energy so that in the Compton interaction the photons now gain and the electrons lose energy. The process is called *inverse Compton scattering* because the electrons lose energy rather than the photons. We will treat the case in which the energy of the photon in the centre of momentum frame of reference is much less than $m_e c^2$ and consequently the Thomson scattering cross-section can be used to describe the probability of scattering.

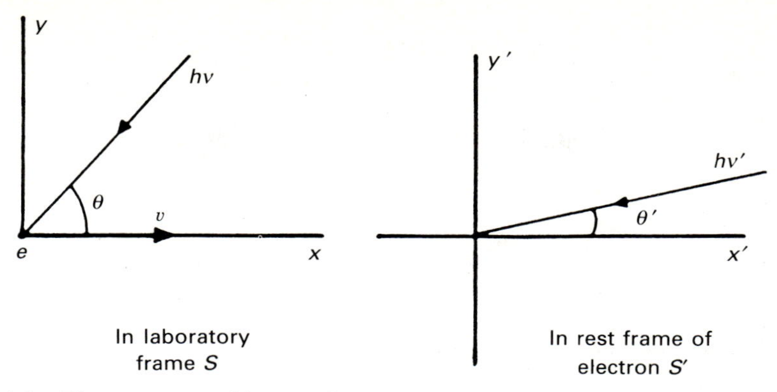

Figure 4.6. The geometry of inverse Compton scattering in the laboratory frame of reference S and that in which the electron is at rest S'.

Many of the most important results can be worked out using simple arguments (see, for example, Blumenthal and Gould (1970) and Rybicki and Lightman (1979)). We consider the geometry of inverse Compton scattering shown in Fig. 4.6 which depicts the collision between a photon and a relativistic electron as seen in the laboratory frame of reference S and in the rest frame of the electron S'. We consider the case in which $\gamma\hbar\omega \ll m_e c^2$ so that the centre of momentum frame is very closely that of the relativistic electron. If the energy of the photon is $\hbar\omega$ and the angle of incidence θ in S, its energy in the frame S' is

$$\hbar\omega = \gamma\hbar\omega[1+(v/c)\cos\theta] \tag{4.25}$$

according to the usual relativistic Doppler shift formula. Using the same arguments, the angle of incidence θ' in the frame S' is related to θ by the formulae

$$\sin\theta' = \frac{\sin\theta}{\gamma[1+(v/c)\cos\theta]}; \quad \cos\theta' = \frac{\cos\theta + v/c}{[1+(v/c)\cos\theta]} \tag{4.26}$$

Now, provided $\hbar\omega' \ll m_e c^2$, the Compton interaction in the rest frame of the electron is simply Thomson scattering and hence the energy loss rate of the electron in S' is just the rate at which energy is reradiated by the electron. According to the expression (14.12), this loss rate is

$$-(\mathrm{d}E/\mathrm{d}t)' = \sigma_T c U'_{\mathrm{rad}} \tag{4.27}$$

where U'_{rad} is the energy density of radiation in the rest frame of the electron. As shown in Section 4.2.2, it is of no importance whether or not the radiation is isotropic. The free electron oscillates in response to any incident field.

Therefore, our strategy is to work out U'_{rad} in the frame S' of the electron and then to use expression (4.27) to find out $(\mathrm{d}E/\mathrm{d}t)'$. Using the result obtained in Section 3.3.1, this is also the loss rate $(\mathrm{d}E/\mathrm{d}t)$ in the frame S.

Suppose the number density of photons in a beam of radiation incident at angle θ to the x-axis is N. Then, the energy density of these photons in S is $N\hbar\omega$. The flux density of photons incident upon an electron stationary in S is $U_{\mathrm{rad}}c = N\hbar\omega c$.

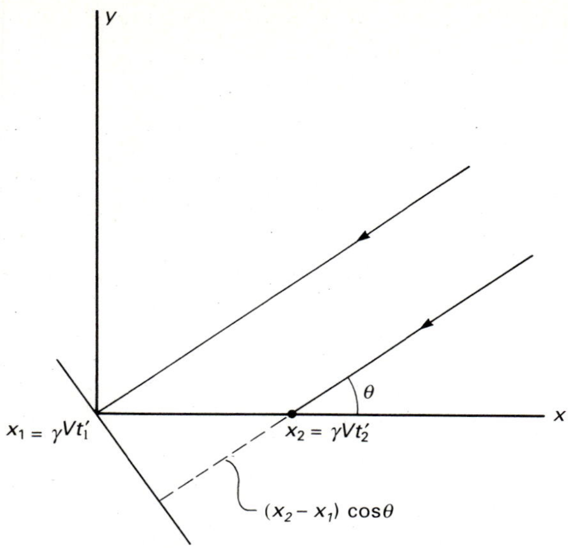

Figure 4.7. Illustrating the rate of arrival of photons at the observer in the laboratory frame of reference (see text).

Now let us work out the flux density of this beam in the frame of reference of the electron S'. We need two things: the energy of each photon in S' and the rate of arrival of photons at the electron. The first of these is easy and is given by the formula (4.25). The second factor requires a little bit of care, although the answer is obvious in the end. The beam of photons incident at angle θ in S arrives at an angle θ' according to the aberration formulae (4.26). We are interested in the rate of arrival of photons at the origin of S' and so let us consider two photons which arrive there at times t'_1 and t'_2. The coordinates of these events in S are

$$[x_1, 0, 0, t_1] = [\gamma V t'_1, 0, 0, \gamma t'_1] \quad \text{and} \quad [x_2, 0, 0, t_2] = [\gamma V t'_2, 0, 0, \gamma t'_2]$$

respectively. This calculation makes the important point that the photons in the beam are propagating along parallel but separate trajectories in S as illustrated by Fig. 4.7. From the geometry of Fig. 4.7, it is apparent that the time difference when the photons arrive at a plane perpendicular to their direction of propagation in S is

$$\begin{aligned}\Delta t &= t_2 + \frac{(x_2 - x_1)}{c} \cos\theta - t_1 \\ &= (t'_2 - t'_1)\, \gamma[1 + (v/c)\cos\theta] \qquad (4.28)\end{aligned}$$

i.e. the time interval between the arrival of photons from the direction θ' is shorter by a factor $\gamma[1 + (v/c)\cos\theta]$ in S' than it is in S. Thus, the rate of arrival of photons and correspondingly the number density of photons is greater by this factor $\gamma[1 + (v/c)\cos\theta]$ in S' as compared with S. This is exactly the same factor by which the energy of the photon has increased (expression (4.25)). On reflection, we should not be surprised about this because these are two different aspects of the same relativistic transformation between the frames S and S'.

Thus, as observed in S', the energy density of the beam is

$$U'_{\mathrm{rad}} = [\gamma(1+(v/c)\cos\theta)]^2 U_{\mathrm{rad}} \tag{4.29}$$

Now, we may think of this as the energy density associated with the photons incident at angle θ in the frame S and consequently arrives from solid angle $2\pi \sin\theta \, \mathrm{d}\theta$. We assume that the radiation field in S is isotropic and therefore we can now work out the total energy density seen by the electron in S' by integrating over solid angle in S, i.e.

$$U'_{\mathrm{rad}} = U_{\mathrm{rad}} \int_0^{\pi} \gamma^2[1+(v/c)\cos\theta]^2 \tfrac{1}{2} \sin\theta \, \mathrm{d}\theta$$

Integrating, we find

$$U'_{\mathrm{rad}} = \tfrac{4}{3} U_{\mathrm{rad}}(\gamma^2 - \tfrac{1}{4}) \tag{4.30}$$

Therefore, substituting into (4.27), we find

$$(\mathrm{d}E/\mathrm{d}t)' = \tfrac{4}{3}\sigma_{\mathrm{T}} c U_{\mathrm{rad}}(\gamma^2 - \tfrac{1}{4})$$

Because $\mathrm{d}E/\mathrm{d}t = (\mathrm{d}E/\mathrm{d}t)'$, we find

$$\mathrm{d}E/\mathrm{d}t = \tfrac{4}{3}\sigma_{\mathrm{T}} c U_{\mathrm{rad}}(\gamma^2 - \tfrac{1}{4})$$

Now, this is the energy gained by the photon field due to the scattering of the low energy photons. We have therefore to subtract the energy of these photons to find the total energy gain to the photon field in S. The rate at which energy is removed from the low energy photon field is just

$$\sigma_{\mathrm{T}} c U_{\mathrm{rad}}$$

and therefore, subtracting, we find

$$\begin{aligned} \mathrm{d}E/\mathrm{d}t &= \tfrac{4}{3}\sigma_{\mathrm{T}} c U_{\mathrm{rad}}(\gamma^2 - \tfrac{1}{4}) - \sigma_{\mathrm{T}} c U_{\mathrm{rad}} \\ &= \tfrac{4}{3}\sigma_{\mathrm{T}} c U_{\mathrm{rad}}(\gamma^2 - 1) \end{aligned}$$

We now use the identity $(\gamma^2 - 1) = (v^2/c^2)\gamma^2$ to write the loss rate in its final form

$$\mathrm{d}E/\mathrm{d}t = \tfrac{4}{3}\sigma_{\mathrm{T}} c U_{\mathrm{rad}} \left(\frac{v^2}{c^2}\right) \gamma^2 \tag{4.31}$$

This is the remarkably elegant result we have been seeking. It is exact so long as $\gamma\hbar\omega \ll m_e c^2$.

Let us complete the story of the inverse Compton scattering of high energy electrons before we return to the case of non-relativistic hot electrons. The next calculation is the determination of the spectrum of the scattered radiation. This can be found by performing two successive Lorentz transformations, first transforming the photon distribution into the frame S' and then transforming the scattered radiation back into the laboratory frame of reference S. This is not a

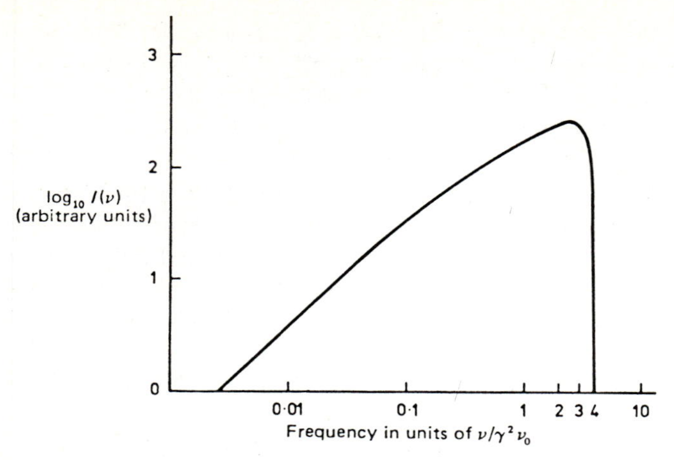

Figure 4.8. The emission spectrum of inverse Compton scattering; ν_0 is the frequency of the unscattered radiation. (From G. R. Blumenthal and R. J. Gould (1970). *Rev. Mod. Phys.*, **42**, 237.)

trivial calculation but the exact result is given by Blumenthal and Gould (1970) for an incident isotropic photon field at a single frequency ν_0 (Fig. 4.8). They show that the spectral emissivity $I(\nu)$ may be written

$$I(\nu)\,d\nu = \frac{3\sigma_T c}{16\gamma^4}\frac{N(\nu_0)}{\nu_0^2}\nu\left[2\nu\ln\left(\frac{\nu}{4\gamma^2\nu_0}\right)+\nu+4\gamma^2\nu_0-\frac{\nu^2}{2\gamma^2\nu_0}\right]d\nu \tag{4.32}$$

where the radiation field is assumed to be monochromatic with frequency ν_0; $N(\nu_0)$ is the number density of photons. At low frequencies, the term in square brackets in the expression (4.32) is a constant and hence the scattered radiation has a spectrum of the form $I(\nu) \propto \nu$.

It is an easy calculation to show that the maximum energy which the photon can acquire corresponds to a head-on collision in which the photon is sent back along its original path. In this case, it is a useful exercise to show that the maximum energy of the photon is

$$(\hbar\omega)_{\max} = \hbar\omega\gamma^2(1+v/c)^2 \approx 4\gamma^2\hbar\omega_0 \tag{4.33}$$

Another interesting result comes out of the formula for the total energy loss rate of the electron (expression (4.31)). The number of photons scattered per unit time is just $\sigma_T cU_{\text{rad}}/\hbar\omega_0$ and hence the average energy of the scattered photons is just

$$\hbar\bar{\omega} = \tfrac{4}{3}\gamma^2(v/c)^2\hbar\omega_0 \approx \tfrac{4}{3}\gamma^2\hbar\omega_0 \tag{4.34}$$

This result gives substance to the hand-waving argument that the photon gains one factor of γ in transforming into S' and then gains another on transforming back to S.

The general result that the frequency of photons scattered by ultrarelativistic electrons is $\nu \approx \gamma^2\nu_0$ is of profound importance in high energy astrophysics. We know that there are electrons with Lorentz factors $\gamma \sim 100$–1000 in various types of astronomical source and consequently they scatter any low energy photons to very much higher energies. To give some simple examples of how this might apply, consider radio, infrared and optical photons scattered by electrons with $\gamma = 1000$. The scattered radiation has frequency (or energy) 10^6 times that of the incoming photons. Thus, radio photons with $\nu_0 = 10^9$ Hz become ultraviolet photons with $\nu = 10^{15}$ Hz ($\lambda = 300$ nm); far-infrared photons with $\nu_0 = 3 \times 10^{12}$ Hz, typical of the photons seen in galaxies which are powerful far-infrared emitters, produce X-rays with frequency 3×10^{18} Hz, i.e. about 10 keV; optical photons with $\nu_0 = 4 \times 10^{14}$ Hz become γ-rays with frequency 4×10^{20} Hz i.e. about 1.6 MeV. It is apparent that the inverse Compton scattering process is a means of producing very high energy photons indeed. It also becomes an inevitable drain of energy for high energy electrons whenever they pass through a region in which there is a large energy density of photons.

When we come to use these formula in earnest in astrophysical calculations, we will have to carry out these calculations in much more detail. In addition, we will have to integrate over both the spectrum of the incident radiation and the spectrum of the relativistic electrons. For the moment, the enthusiast is urged to consult the excellent review paper by Blumenthal and Gould (1970).

4.3.4 *Comptonisation*

The calculations carried out in Sections 4.3.2 and 4.3.3 demonstrate how energy can be interchanged between photons and electrons by Compton scattering in particular limiting cases. These are only particular examples of the more general process by which electrons and photons of any energy exchange energy. If the evolution of the spectrum of a source is primarily determined by Compton scattering, this process is often referred to as *Comptonisation*. This is an enormous subject and anything other than a brief introduction to some of the key concepts is far beyond the scope of this text. For more details, the review article by Pozdnyakov, Sobol and Sunyaev (1983) and the gentler introduction by Rybicki and Lightman (1979) can be warmly recommended.

The corollary of the requirement that the evolution of the spectrum be determined by Compton scattering is that the plasma must be rarefied so that other radiation processes such as bremsstrahlung do not contribute extra photons into the system. It is also advantageous if the plasma is hot because then the exchange of energy per collision is greater if the matter is hotter than the radiation. Examples of sources in which such conditions are found include the hot gas in the vicinity of binary X-ray sources, the hot plasmas in the nuclei of active galaxies, the hot intergalactic gas in clusters of galaxies and the primordial plasma as it cools down from the hot early phases of the Hot Big Bang.

Let us build up a simple picture of the Comptonisation process. We will restrict our discussion to the non-relativistic regime in which $kT_e \ll m_e c^2$ and $\hbar\omega \ll m_e c^2$. We can use a number of the results already established to understand the process of interchange of energy between the photon field and the electrons. First of all, we have an expression for the energy transferred to stationary electrons from the photon field (expression (4.19)). If $\hbar\omega \ll m_e c^2$, we can write this in terms of the fractional change of energy of the photon

$$\frac{\Delta\varepsilon}{\varepsilon} = \frac{\hbar\omega}{m_e c^2}(1-\cos\alpha) \tag{4.35}$$

Now, in the frame of reference of the electron, the scattering is simply Thomson scattering and so the probability distribution of the scattered photons is symmetrical about their incident directions. Therefore, when averages are taken over the scattering angle α, opposite values of $\cos\alpha$ cancel out and the average energy increase of the electron is

$$\langle \Delta\varepsilon/\varepsilon \rangle = \hbar\omega/m_e c^2 \tag{4.36}$$

In the other extreme, we take the low energy limit of the energy loss rate of high energy electrons in scattering low energy photons. We recall that the derivation of equation (4.31) is correct for all values of the Lorentz factor γ and hence incorporates all the effects of aberration and Doppler shifts, even if these effects are very small. We take the low energy limit of equation (4.31) so that

$$(\mathrm{d}E/\mathrm{d}t) = \tfrac{4}{3}\sigma_T c U_{\mathrm{rad}}(v/c)^2 \tag{4.37}$$

Now the number of photons scattered per second in this loss process is simply the number of photons encountered per second by the electron which is

$$\sigma_T N_{\mathrm{phot}}\, c = \sigma_T U_{\mathrm{rad}}\, c/\hbar\omega_0$$

Therefore the average energy gain by the photons per Compton collision is

$$\left\langle \frac{\Delta\varepsilon}{\varepsilon} \right\rangle = \frac{4}{3}\left(\frac{v}{c}\right)^2 \tag{4.38}$$

It is interesting to compare this result with the expression (4.20). The net gain of energy per collision in the Compton scattering process is second order in v/c because the first-order effects cancel out. The net increase in energy is statistical because implicitly, in deriving equation (4.27), we have integrated over all angles of scattering.

Now, if the electrons have a thermal distribution of velocities at temperature T_e, we can write $\frac{1}{2}m_e\langle v^2\rangle = \frac{3}{2}kT_e$ and hence

$$\Delta\varepsilon/\varepsilon = 4kT_e/m_e c^2 \tag{4.39}$$

As a result, the equation which describes the net energy change of the photon in a Compton collision is

$$\Delta\varepsilon/\varepsilon = -\hbar\omega_0/m_e c^2 + 4kT_e/m_e c^2 \tag{4.40}$$

This equation establishes for us the condition under which energy is transferred to and from the photon field, i.e. there is no energy transfer if $\hbar\omega_0 = 4kT_e$. If $4kT_e > \hbar\omega_0$, energy is transferred to the photons whilst if $\hbar\omega_0 > 4kT_e$ energy is transferred to the electrons.

We will be primarily concerned with the case in which the electrons are hotter than the photons and so let us write down the condition that the photon distribution be significantly modified by repeated Compton scatterings. The fractional increase in energy is $4(kT_e/m_e c^2)$ per collision and hence we need to evaluate the number of collisions which the photon makes with electrons before they escape from the scattering region. If the region has electron density N_e and size l the optical depth for Thomson scattering through the region is

$$\tau_e = N_e \sigma_T l \tag{4.41}$$

If $\tau_e \gg 1$, the photons undergo a random walk in escaping from the region. By the usual stochastic arguments, the photon travels a net distance $l = N^{\frac{1}{2}}\lambda_e$ in N scatterings where $\lambda_e = (N_e \sigma_T)^{-1}$ is the mean free path of the photon. Therefore, in the limit $\tau_e \gg 1$, which is clearly necessary to alter significantly the energy of the photon, the number of scatterings is $N = (l/\lambda_e)^2 = \tau_e^2$. If $\tau_e \ll 1$, the number of scatterings is simply τ_e and hence the condition for a significant distortion of the photon spectrum by Compton scattering is given by $4y \gtrsim 1$, where

$$y = \frac{kT_e}{m_e c^2} \max(\tau_e, \tau_e^2) \tag{4.42}$$

Thus, the normal condition for Comptonisation to change significantly the spectrum of the photons is

$$y = \frac{kT_e}{m_e c^2} \tau_e^2 \gtrsim \tfrac{1}{4} \tag{4.43}$$

Let us now investigate how repeated scatterings change the energy of the photon. After one scattering, the energy of the photon relative to its initial energy is

$$\frac{\varepsilon'}{\varepsilon} = 1 + \frac{4kT_e}{m_e c^2}$$

After N scatterings, the energy is therefore

$$\frac{\varepsilon'}{\varepsilon} = \left(1 + \frac{4kT_e}{m_e c^2}\right)^N$$

Since $4kT_e \ll m_e c^2$, we can make the approximation

$$\left(1 + \frac{4kT_e}{m_e c^2}\right) \approx \exp\left(\frac{4kT}{m_e c^2}\right)$$

and hence

$$\frac{\varepsilon'}{\varepsilon} = \exp\left(\frac{4kT_e}{m_e c^2} N\right) = \exp(4y) \tag{4.44}$$

for all values of y.

Let us now investigate the case in which the effects of Comptonisation are very strong. First of all, we note the number of scatterings which must take place before the photon distribution approaches equilibrium with the electrons. From the expression (4.40), we see that net energy transfer no longer takes place from the electrons to the photon field if the photons are 'heated' to a temperature such that $\hbar\omega = 4kT_e$. The optical depth to Thomson scattering necessary for this to occur is found using the expression (4.44). Setting $\varepsilon' = 4kT_e$ we find

$$\frac{4kT_e}{\hbar\omega_0} = \exp\left[4\left(\frac{kT_e}{m_e c^2}\right)\tau_T^2\right]$$

i.e.

$$\tau_T = \left[\frac{m_e c^2}{4kT_e}\ln\left(\frac{4kT_e}{\hbar\omega_0}\right)\right]^{\frac{1}{2}} \tag{4.45}$$

If the optical depth of the medium is greater than this, the photon distribution approaches its equilibrium form determined entirely by Compton scattering. We know what this equilibrium distribution must be. Photons are bosons and, consequently, the equilibrium spectrum is given in general by the Bose–Einstein distribution, the energy density of which is given by the standard formula

$$u_\nu \, d\nu = \frac{8\pi h\nu^3}{c^3}\left[\exp\left(\frac{h\nu}{kT}+\mu\right)-1\right]^{-1} d\nu \tag{4.46}$$

μ is known as the *chemical potential* and, in the present context, is a measure of the deficit in the number of photons relative to that required for a Planck distribution at the same temperature. In the case of the Planck spectrum, $\mu = 0$ and the number and energy densities of the photons are uniquely defined by a single parameter, the thermal equilibrium temperature of the radiation and the matter T. If there is a mismatch between the number density of photons and the energy density of the matter and radiation, the equilibrium spectrum is the Bose–Einstein distribution with a finite chemical potential μ. The forms of these spectra are shown in Fig. 4.9 for different values of the chemical potential μ. In the limiting case $\mu \gg 1$, the spectrum approximates to a Wien distribution modified by the factor $\exp(-\mu)$

$$u_\nu = \exp(-\mu)\frac{8\pi h\nu^3}{c^3}\exp\left(-\frac{h\nu}{kT}\right) \tag{4.47}$$

It is useful to note that the average energy of the photons in this distribution is

$$\langle\hbar\omega\rangle = kT_e \frac{\int_0^\infty x^3 \exp(-x)\, dx}{\int_0^\infty x^2 \exp(-x)\, dx} = 3kT_e \tag{4.48}$$

It is intriguing that this is identical to the result derived by Einstein in his classic paper of 1905 in which he introduced the concept of light quanta (see Longair (1984), Chapter 11).

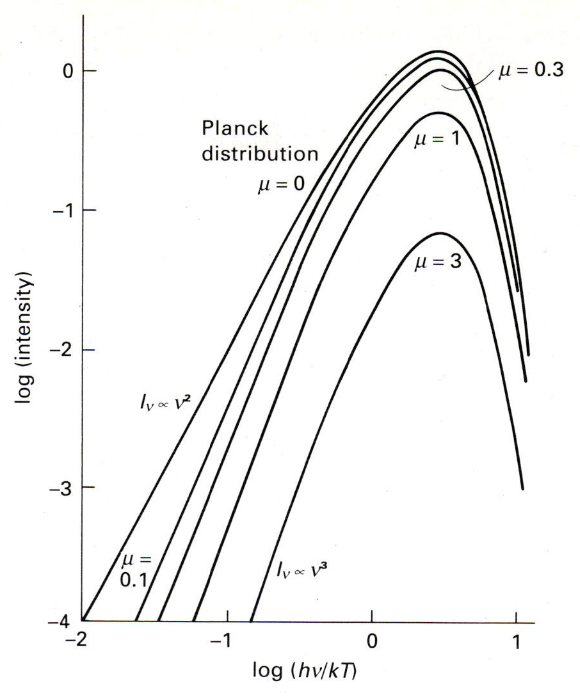

Figure 4.9. Illustrating the intensity spectra of Bose–Einstein distributions with different values of the dimensionless chemical potential μ. The distribution with $\mu = 0$ is the Planck function. At energies $h\nu \gg \mu kT$ the distributions are similar to a Planck function but with intensity reduced by a factor $\exp(-\mu)$. At energies $h\nu \ll \mu kT$, the intensity spectrum is $I_\nu \propto \nu^3$. For large values of μ, the distribution follows closely that of a Wien distribution with intensity reduced by a factor $\exp(-\mu)$.

What we now need is the equation which describes how the photon field evolves towards these equilibrium distributions. In the non-relativistic limit, this equation is known as the *Kompaneets equation* after the Soviet scientist who first derived it in 1949 (Kompaneets 1956). The derivation of this equation is distinctly non-trivial since not only does it have to take account of the interchange of energy between the photons and electrons in both directions but it also has to take account of induced effects which become important if the occupation number n of the photons in phase space becomes large. The derivation outlined by Rybicki and Lightman (1979) gives an excellent feel for what is involved.

The equation is written in terms of the evolution of the distribution of photons in phase space. It turns out to be most convenient to work in terms of the occupation number $n(\nu)$ of a volume of phase space which is defined to be the number of photons per state. We recall that the elementary volume of phase space is $(2\pi)^3$. In the case of photons there are two independent polarisations for each state and hence the number of states in the elementary volume of phase space d^3k is $2d^3k/(2\pi)^3$. If the photon distribution is isotropic, the photons which lie in the frequency range ν to $\nu+d\nu$ have wave vectors k which lie in a spherical shell of thickness dk in k-space of volume $4\pi k^2 dk$. Therefore, the number of states in this volume of phase space is

$$\frac{8\pi k^2 dk}{(2\pi)^3} = \frac{8\pi}{c^3}\nu^2 d\nu$$

The occupation number is defined to be the number of particles or photons per state. If the energy density of isotropic radiation in the frequency interval ν to $\nu + d\nu$ is $u_\nu d\nu$, then the number density of photons is $u_\nu d\nu / h\nu$ and the occupation number is

$$n(\nu) = \frac{u_\nu c^3}{8\pi h \nu^3} \quad (4.49)$$

We observe that there is a particularly simple expression for the occupation number of photons in the case of Bose–Einstein and Planck distributions

$$\text{B–E:} \quad n(\nu) = [\exp(x+\mu) - 1]^{-1};$$
$$\text{Planck:} \, n(\nu) = [\exp x - 1]^{-1}$$

where $x = h\nu/kT$. It is the occupation number $n(\nu)$ which determines when it is necessary to include stimulated emission terms in the expressions for interactions of photons. If $n > 1$, then the effects of stimulated emission cannot be neglected.

With this introduction, we write down the Kompaneets equation

$$\frac{\partial n}{\partial y} = \frac{1}{x^2}\frac{\partial}{\partial x}\left[x^4\left(n + n^2 + \frac{\partial n}{\partial x}\right)\right] \quad (4.50)$$

where y has its usual meaning, $y = \int (kT_e/m_e c^2)\sigma_T N_e \, dl$, and $x = h\nu/kT_e$. The terms in square brackets have the following meanings. The term in $\partial n/\partial x$ represents the diffusion of photons along the frequency axis in both directions with respect to their initial energies. As noted above, there is a statistical increase in their energy associated with this diffusion. The first term in n represents the cooling of the photons by the recoil effect and the second term in n^2 describes the effects of induced Compton scattering which also contributes to the cooling of the photons if the occupation number is large. It is a simple exercise to show that the term in square brackets is zero for a Bose–Einstein distribution for which the occupation number n takes the value $n = [\exp(x+\mu) - 1]^{-1}$.

Generally speaking, the solutions have to be found numerically but some useful limiting cases are described by Rybicki and Lightman. Rather than quote these cases, I will simply give some examples of the formation of spectra which are likely to be of interest for high energy astrophysics. In the case $y \gg 1$, in which saturation by the Compton effect takes place, the spectrum is expected to follow the Wien distribution. Fig. 4.10 shows an example in which photons are created by bremsstrahlung and are then strongly Comptonised.

In the intermediate cases in which $y > 1$ but the process does not reach saturation, the computations of Pozdnyakov *et al.* (1983) show how the spectrum changes as y increases. If energy losses by induced scattering are neglected, analytic

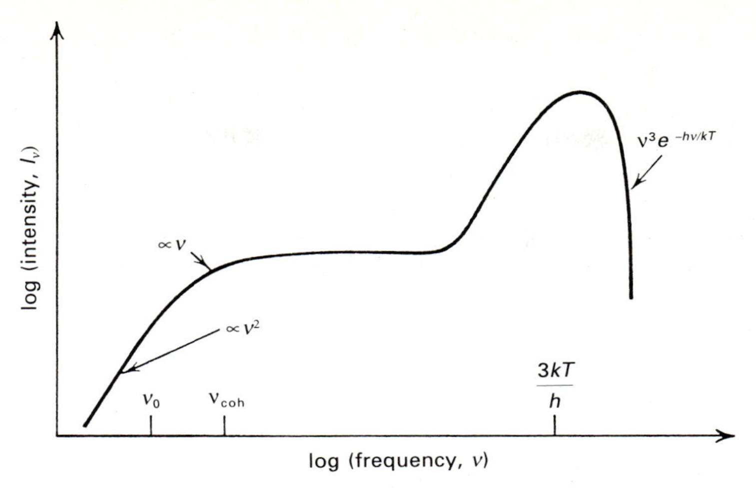

Figure 4.10. The intensity spectrum of a thermal, non-relativistic medium in which bremsstrahlung is the dominant process at low frequencies and in which Compton scattering is saturated so that the characteristic Wien distribution is formed at high frequencies. At low frequencies, the characteristic spectrum of self-absorbed thermal bremsstrahlung is shown (from G. B. Rybicki and A. P. Lightman (1979). *Radiative processes in astrophysics*, page 220, New York: John Wiley and Sons.)

solutions can be found and compared with the results of Monte Carlo calculations of the evolution of the photon spectrum. These are shown in Fig. 4.11 in which it is assumed that the input photons are of very low energy and the electrons have temperature $kT_e = 25$ keV. The figure illustrates how the spectrum changes as the Thomson scattering optical depth increases. The spectrum evolves so that for large optical depths the beginnings of the formation of the Wien peak can be observed. At smaller optical depths, the spectrum mimics very closely a power-law spectrum up to energies $h\nu \approx kT_e$. There is an analytic solution indicated by the solid lines which has the following form

$$n(\nu) \propto x^m \tag{4.51}$$

where $m = -\frac{3}{2} \pm \left(\frac{9}{4} + \frac{1}{y}\right)^{\frac{1}{2}}$ and hence $u(\nu) \propto \nu^{3+m}$. This is an intriguing example in which a power-law spectrum is created through 'thermal' processes rather than having to be ascribed to some 'non-thermal' radiation mechanism involving ultrarelativistic electrons. Processes of this type may be important in forming the spectra of X-ray sources if large amounts of very hot gas are known to be present. Pozdnyakov *et al.* give a number of possible applications of these results, for example, in explaining the X-ray spectrum of the X-ray source Cygnus X-I (Fig. 4.12).

One very important application of the Kompaneets equation is in describing distortions of the spectrum of the Microwave Background Radiation if the background photons are scattered by hot electrons. Two cases are of special interest. In the first, if for some reason the intergalactic gas were heated to a very

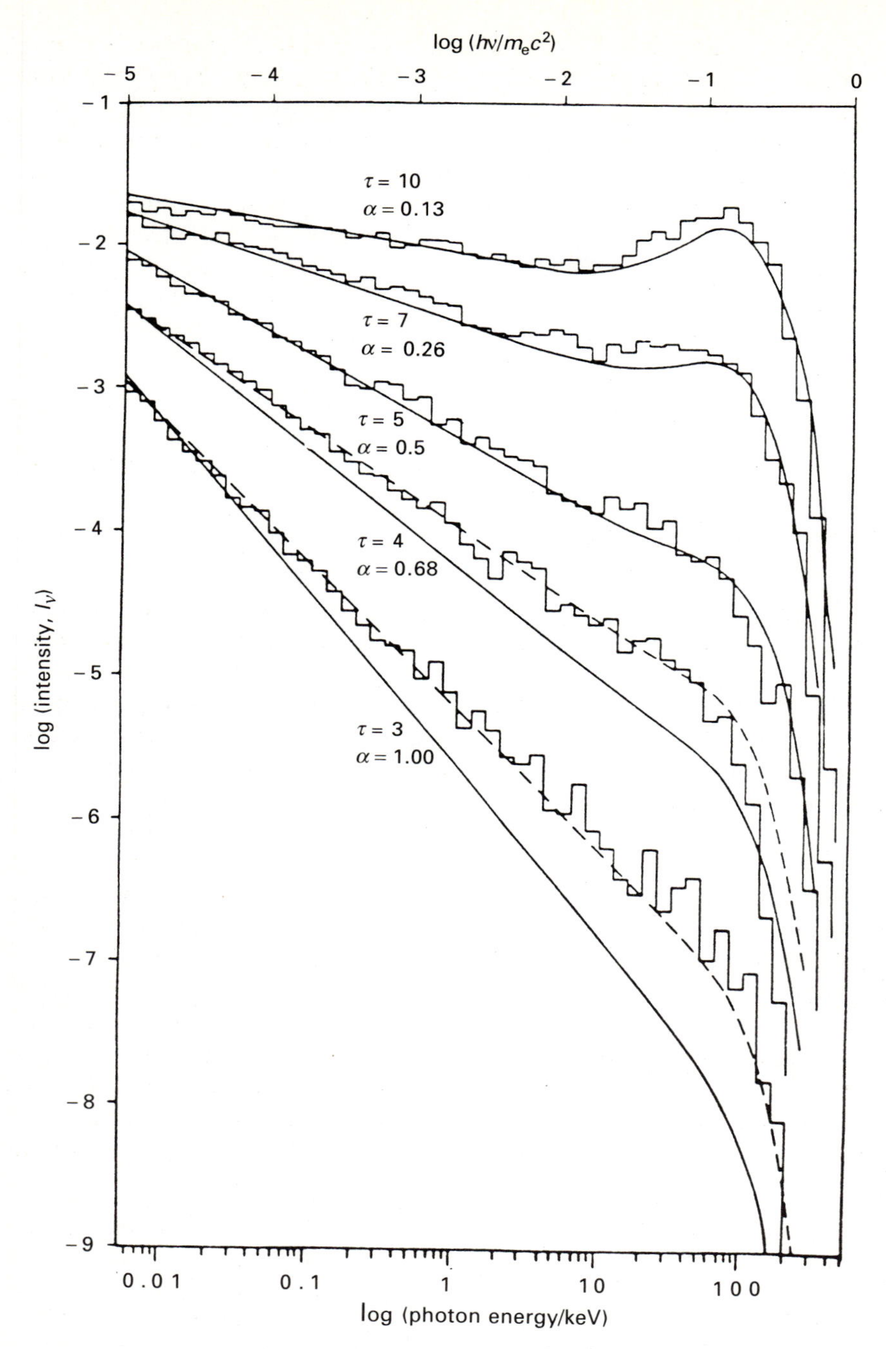

Figure 4.11. The Comptonisation of low frequency photons in a spherical plasma cloud having $kT_e = 25$ keV. The solid curves are analytic solutions of the Kompaneets equation (see Pozdnyakov *et al.* (1983)); the results of Monte Carlo simulations of the Compton scattering process are shown by the histograms and there is good agreement

high temperature such that $T_e \gg T_r$. Compton scattering would increase the energy of the photons of the Microwave Background Radiation. The net result would be a mismatch between the energy density of the photons of the Microwave Background Radiation and their number density. If this process took place in the early Universe before the epoch of recombination, there would be many scatterings and there would be time to set up an equilibrium Bose–Einstein distribution with finite chemical potential. If the heating took place after the epoch of recombination, there would not be time to set up the equilibrium distribution and the predicted spectrum would be found by solving the Kompaneets equation without the terms describing the cooling of the photons, i.e.

$$\frac{\partial n}{\partial y} = \frac{1}{x^2}\frac{\partial}{\partial x}\left(x^4\frac{\partial n}{\partial x}\right) \qquad (4.52)$$

Zeldovich and Sunyaev have given a solution to this equation in the form

$$\frac{\Delta u_\nu}{u_\nu} = y\frac{x \exp x}{\exp x - 1}\left(x\frac{\exp x + 1}{\exp x - 1} - 4\right) \qquad (4.53)$$

The predicted spectrum is shown in Fig. 4.13. As can be seen from the figure, the intensity of the background radiation in the Rayleigh–Jeans region of the spectrum *decreases* while the intensity in the Wien region *increases*. Thus, a limit to the amount of heating of the intergalactic gas which could have taken place in the Universe is provided by the very high degree of precision with which the black-body nature of the spectrum of the Microwave Background Radiation is now known. The observations from the COBE (Fig. 1.10) provide limits

$$y \leqslant 10^{-3}; \quad \mu \leqslant 10^{-2}$$

The second application is the Compton scattering of photons of the Microwave Background Radiation as they propagate to the Earth through regions of very hot ionised gas. The most famous example of this is the decrement in the Microwave Background Radiation expected in the direction of those rich clusters of galaxies which possess large amounts of hot gas, for example the Perseus cluster of galaxies (Fig. 3.3). This is known as the *Sunyaev–Zeldovich effect* (Sunyaev and Zeldovich 1970). It is a useful exercise to solve the Kompaneets equation in the form of equation (4.52) in the Rayleigh–Jeans approximation and for small values of the parameter y. Using a trial solution of the form $I_\nu \propto \nu^2$, it is straightforward to obtain the solution

$$\Delta I_\nu / I_\nu = -2y \qquad (4.54)$$

---

with the analytic solutions. A slightly better fit to the Monte Carlo calculations is found for the cases $\tau = 3$ and $\tau = 4$ if the analytic formula is fitted to the spectral index found from the Monte Carlo simulations (dashed curve). These computations illustrate the development of the Wien peak at energies $h\nu \approx kT_e$. (From L. A. Posdnyakov, I. M. Sobol and R. A. Sunyaev (1983). *Astrophysics and Space Physics Reviews*, **2**, 263. Soviet Scientific Reviews, Harwood Academic Publishers).

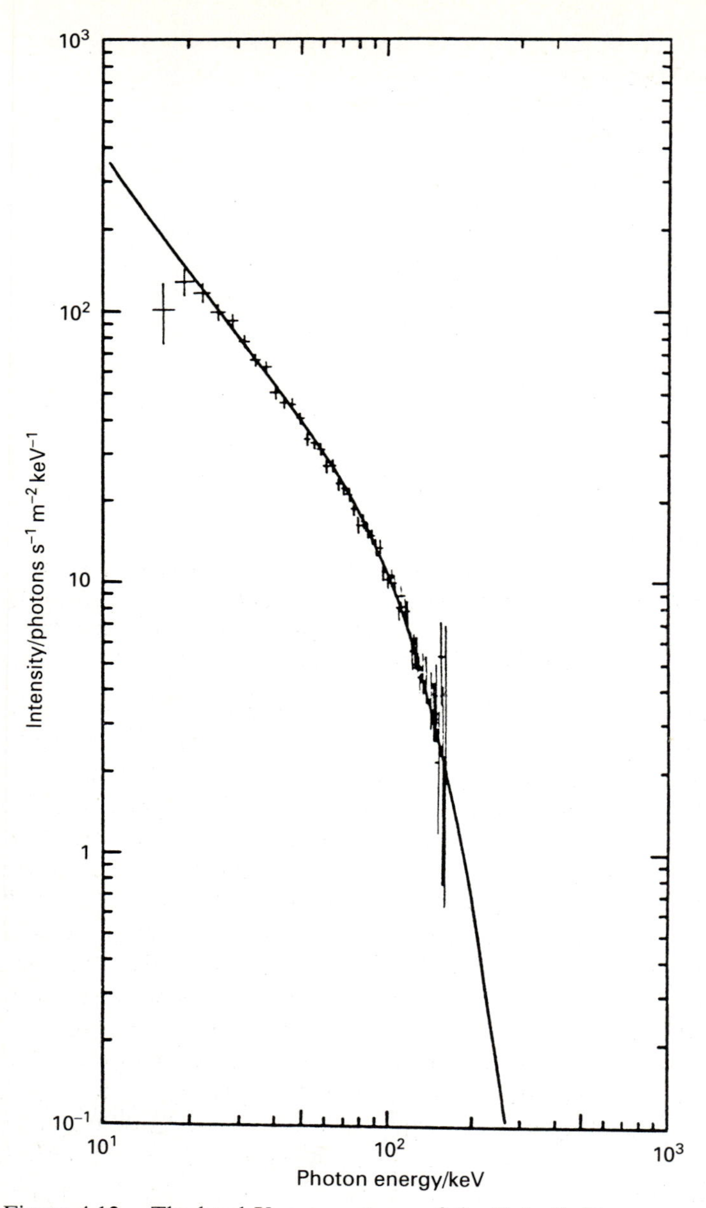

Figure 4.12. The hard X-ray spectrum of the Galactic X-ray source Cygnus X-1 observed in a balloon flight of the Max Planck Institute for Extraterrestrial Physics, on 20 September 1977 compared with the analytic solution of the Kompaneets equation with parameters $\tau_0 = 5$, $kT_e = 27$ keV. (R. A. Sunyaev and L. G. Titarchuk (1980). *Astron. Astrophys.*, **86**, 121.)

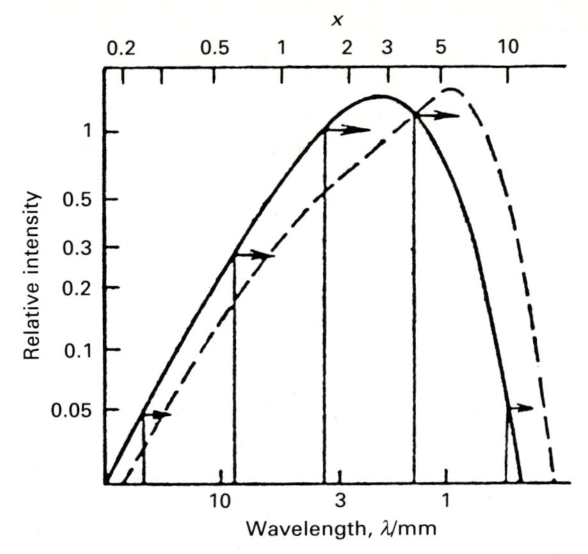

Figure 4.13. Illustrating the Compton scattering of a Planck distribution by hot electrons in the case in which the Compton optical depth $y = \int (kT_e/m_e c^2)\sigma_T N_e \, dl = 0.15$. The intensity decreases in the Rayleigh–Jeans region of the spectrum and increases in the Wien region. (From R. A. Sunyaev (1980). *Soviet Astronomy Letters*, **6**, 213.)

The reader may be intrigued to try to derive this result using the expression for the average increase in energy of the photons in a single Compton scattering given by the expression (4.39). Naive application of the result (4.39) will result in the wrong answer in that a temperature (or intensity) decrement of $-8y$ is predicted. The reason for this discrepancy is intimately connected with the statistical nature of the Compton scattering process. Fig. 4.14 taken from Sunyaev (1980) shows the probability distribution of scattered photons for a single Compton scattering. It can be observed that the average increase in energy is very small compared with the broad wings of the scattering function. In fact, this is no more than the point we emphasised before, namely that the first-order Compton collisions result in as many energy gains as losses and the breadth of the scattering function is of order v/c. Therefore, in addition to the increase in energy due to the second-order effect in $(v/c)^2$, we have also to take account of the scattering of photons by first order Compton scatterings. In the Rayleigh–Jeans limit in which the spectrum is $I_\nu \propto \nu^2$, there are more photons scattered down in energy to frequency ν than are scattered up from lower frequencies. It is left to the reader to show that simple Doppler scatterings increase the intensity at frequency ν by an increment $+6y$ so that the net decrement is $-2y$ as given by the Kompaneets equation. The moral of this fairly major digression is to note the power of the Kompaneets equation in automatically taking account of all aspects of the diffusion of photons in phase space in the one equation!

As an example of the use of this formula, Birkinshaw (1990) adopts a core radius for the hot gas in the Coma cluster of galaxies of 500 kpc, an electron temperature corresponding to $kT_e = 7.9$ keV ($T_e = 10^8$ K) and an electron density of

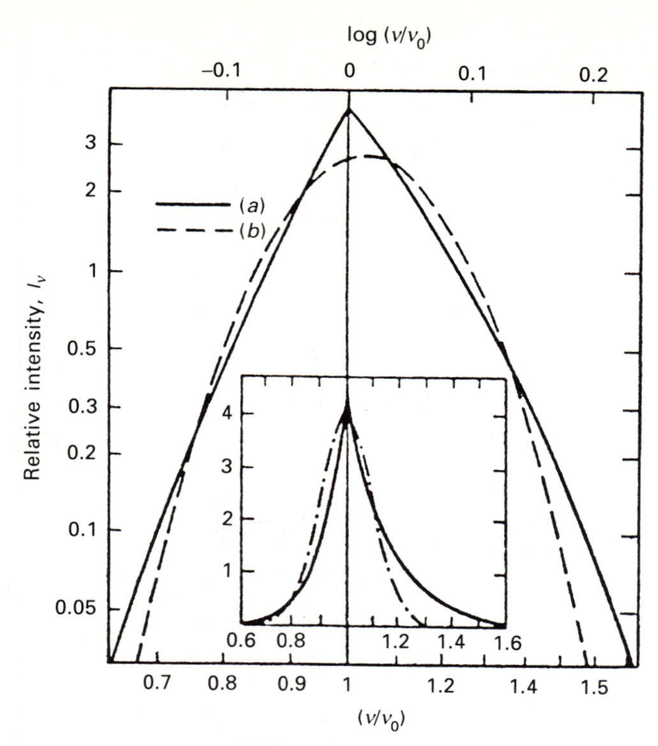

Figure 4.14. The probability distribution of photons scattered in a single Compton scattering (*a*) using the exact expression for Compton scattering (solid line) and (*b*) using the diffusion term in the Kompaneets equation (dashed line) for the case in which the hot gas has temperature $kT_e = 5.1$ keV, i.e. $kT_e/m_e c^2 = 0.01$. The insert shows these distributions on a linear scale. It can be seen that the distributions are broad with half-widths $\sigma \sim (kT_e/m_e c^2)^{\frac{1}{2}}$, i.e. $\Delta\nu/\nu \sim 0.1$. The average increase in the energy of the photon is $\Delta\nu/\nu = 4(kT_e/m_e c^2) = 0.04$. (From R. A. Sunyaev (1980). *Soviet Astronomy Letters*, **6**, 214.)

$N_e = 3 \times 10^3$ m^{-3}. Therefore, to order of magnitude the parameter y has the value 10^{-4} through the centre of the cluster. Birkinshaw's more detailed calculations for the observed distribution of hot gas in the cluster imply that the decrement in the Microwave Background Radiation in the direction of the Coma cluster should amount to $\Delta I_\nu/I_\nu = -4.8 \times 10^{-4}$ corresponding to a change in brightness temperature of the radiation in this direction of $\Delta T = -1.4$ mK in the Rayleigh–Jeans region of the spectrum. This is a very difficult experiment but the decrement has now been observed in a number of clusters which are known to have large quantities of hot diffuse gas (Fig. 4.15).

There are some interesting features of these observations. First of all, it will be noted that the measurement of a temperature decrement in the Microwave Background Radiation immediately provides an estimate of the quantity $N_e T_e$ in the cluster, i.e. the pressure of the hot gas. Second, the sense of the Sunyaev–Zeldovich effect is different on either side of the maximum of the Planck

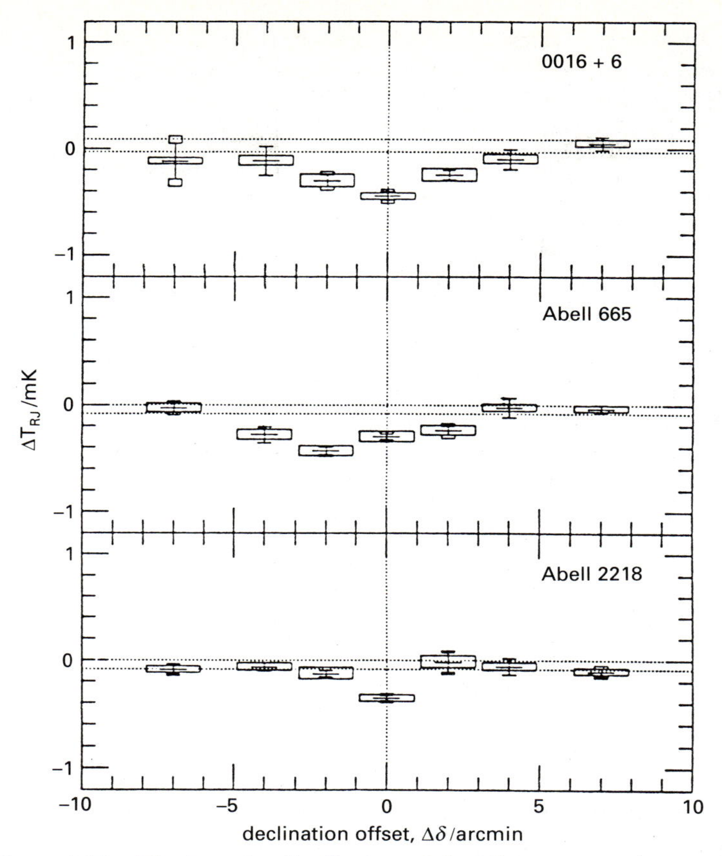

Figure 4.15. The intensity distribution of the Microwave Background Radiation in the vicinity of three rich clusters of galaxies which are known to contain large quantities of hot gas from X-rays observations of the bremsstrahlung of the intracluster gas. The observations are scans of the intensity of the Microwave Background Radiation across the clusters at a radio frequency of 20.3 GHz. The error boxes and the dotted lines represent the sum of the random and systematic uncertainties in the observations. It can be seen that there is definite evidence for a decrement in the intensity of the Microwave Background Radiation in the directions of these clusters. (From M. Birkinshaw (1990). *The cosmic microwave background: 25 years later*, eds N. Mandolesi and N. Vittorio, Dordrecht: Kluwer.)

spectrum as illustrated in Fig. 4.13 and so a corresponding increase in intensity is expected in the Wien region of the spectrum. Third, in principle, it is possible to measure the distance of the cluster by combining the information provided by the bremsstrahlung emission of the hot gas with the decrement in the Microwave Background Radiation. The temperature is determined from the shape of the bremsstrahlung spectrum and hence one can solve for both the electron density and the physical size of the emitting gas cloud. By measuring its angular extent, the distance to the cluster can be measured. If the redshift of the cluster is known, Hubble's constant can be measured (see e.g. Birkinshaw (1990)).

It has to be emphasised again that this section is no more than a mild introduction to what is an enormous subject. We have restricted attention to a number of simple cases which have practical applications. We have not considered the intriguing cases in which Comptonisation as well as other processes of emission and absorption are taking place simultaneously. Again, there is no better introduction than the exposition of Rybicki and Lightman and, for those of a robust constitution, the papers by Sunyaev and his colleagues.

4.4 Electron–positron pair production

If the photon has energy greater than $2m_e c^2$, pair production is possible in the field of the nucleus. Pair production cannot take place in free space because momentum and energy cannot be conserved simultaneously. To demonstrate this, consider a photon of enery $\hbar\omega$ decaying into an electron–positron pair, each of which has kinetic energy $(\gamma-1)m_e c^2$. The best one can do to try to conserve both energy and momentum is if the created electron–positron pair moves parallel to the original direction of the photon, then we find

Conservation of energy: energy of photon $= \hbar\omega = 2\gamma m_e c^2$

momentum of pair $= 2\gamma m_e v = (\hbar\omega/c)(v/c)$

But,

initial momentum of photon $= \hbar\omega/c$

Since v cannot be equal to c, we cannot conserve both energy and momentum in free space and this is why we need a third body, such as an ambient nucleus, which can absorb some of the energy or momentum.

Let us simply quote some of the useful results for electron-positron pair production (see e.g. Chupp (1976), Ramana Murthy and Wolfendale (1986)).

Intermediate photon energies In the case of no screening, the cross-section for photons with energies in the range $1 \ll \hbar\omega/m_e c^2 \ll 1/\alpha Z^{\frac{1}{3}}$ can be written

$$\sigma_{\text{pair}} = \alpha r_e^2 Z^2\left[\frac{28}{9}\ln\left(\frac{2\hbar\omega}{m_e c^2}\right)-\frac{218}{27}\right] \text{m}^2\ \text{atom}^{-1} \tag{4.55}$$

As usual, r_e is the classical electron radius and α is the fine structure constant.

Ultrarelativistic limit In the case of complete screening and for photon energies $\hbar\omega/m_e c^2 \gg 1/\alpha Z^{\frac{1}{3}}$, the cross-section becomes

$$\sigma_{\text{pair}} = \alpha r_e^2 Z^2\left[\frac{28}{9}\ln\left(\frac{183}{Z^{\frac{1}{3}}}\right)-\frac{2}{27}\right] \text{m}^2\ \text{atom}^{-1} \tag{4.56}$$

Notice that, in both cases, the cross-section for pair production is $\sim \alpha\sigma_T$. Notice also that the cross-section for the creation of pairs through interactions with electrons is very much smaller than the above values and can be neglected.

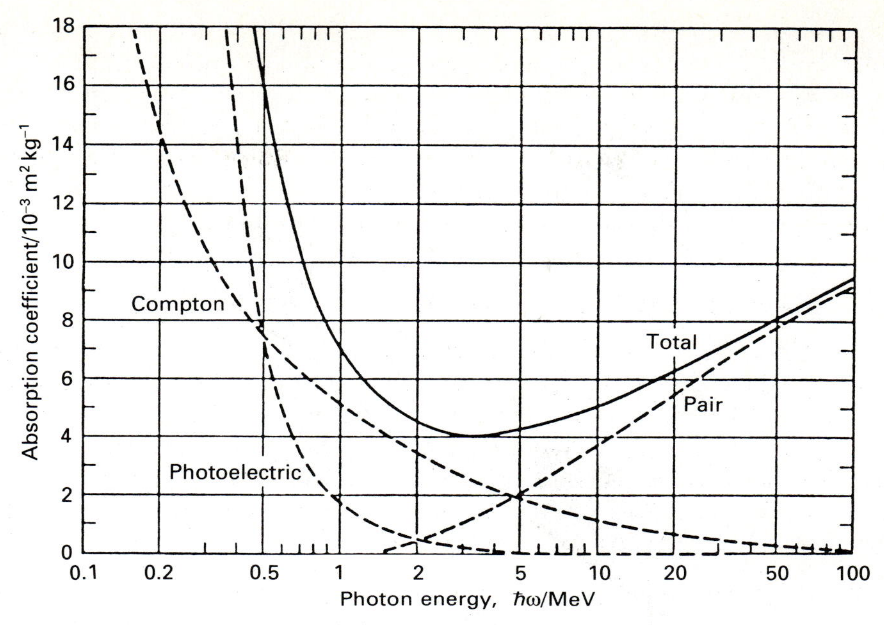

Figure 4.16. The total mass absorption coefficient for high energy photons in lead, indicating the contributions associated with the photoelectric absorption, Compton scattering and electron–positron pair production. (From H. A. Enge (1966). *Introduction to nuclear physics*, page 193, London: Addison-Wesley Publishing Co.)

Exactly as in Section 3.6, we can define a radiation length ξ_{pair} for pair production

$$\xi_{\text{pair}} = \rho/N_{\text{i}}\sigma_{\text{pair}} = M_{\text{A}}/N_0\sigma_{\text{pair}} \tag{4.57}$$

where M_{A} is the atomic mass, N_{i} is the number density of nuclei and N_0 is Avogadro's number. If we compare the radiation lengths for pair production and bremsstrahlung by ultrarelativistic electrons, we find that $\xi_{\text{pair}} \approx \xi_{\text{brems}}$. This reflects the similarity of the bremsstrahlung and pair production mechanisms according to quantum electrodynamics. More details of the physical similarities of these processes using Feynman diagrams may be found in Leighton's discussion ((1959) pp. 667–76).

We can now put the three main loss processes for high energy photons together – ionisation losses, Compton scattering and electron–positron pair production – to obtain the total mass absorption coefficient for X-rays and γ-rays. Fig. 4.16 shows how each of these processes contributes to the total absorption coefficient in lead. Notice that the energy range 500 keV $\lesssim \hbar\omega \lesssim$ 5 MeV is a difficult energy range for experimental study of photons from cosmic sources because all three processes make a significant contribution to the absorption coefficient for γ-rays. This means that this is a particularly difficult energy range for the design and construction of γ-ray telescopes. To make matters worse, the fluxes of photons from astrophysical sources are known to be low in this energy range.

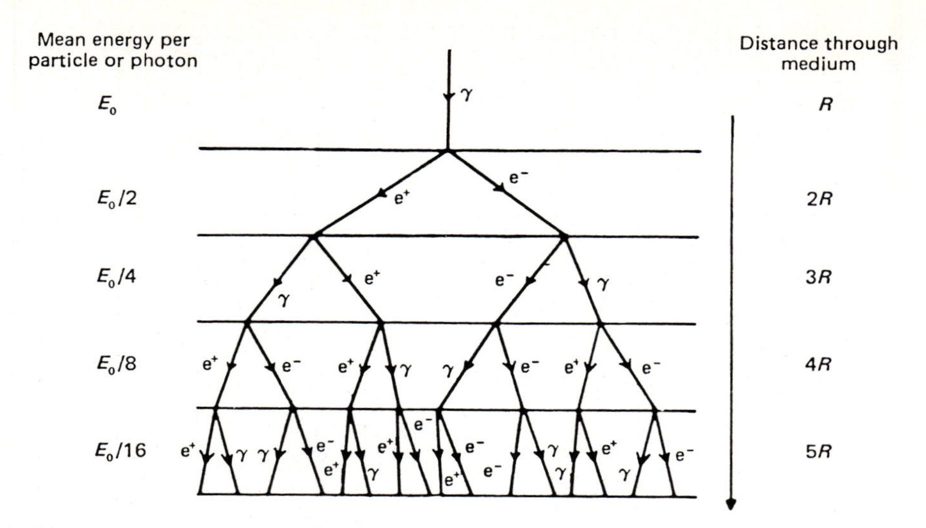

Figure 4.17. A simple model for an electromagnetic shower.

4.5 Electron–photon cascades or electromagnetic showers

We can now understand how cascades or showers initiated by high energy electrons or γ-rays can come about. A high energy photon generates an electron–positron pair, each of which in turn generates high energy photons by bremsstrahlung, each of which generates an electron–positron pair, each of which ... and so on.

Let us build a simple model of an electron–photon cascade in the following way. In the ultrarelativistic limit, the radiation lengths for pair production and bremsstrahlung are the same. Therefore the probability of these processes taking place is one-half at distance ξ, given by

$$\exp(-\xi/\xi_0) = \tfrac{1}{2}$$

i.e.

$$\xi = R = \xi_0 \ln 2$$

Therefore, if we initiate the cascade with a γ-ray of energy E_0, after a distance of, on average, R, an electron–positron pair is produced and we assume that the pair share the energy of the γ-ray, i.e. $E_0/2$ each. In the next length R, the electron and positron lose, on average, half their energy and they each radiate a photon of energy $E_0/4$. Thus, we end up with two particles and two photons, all having energy $E_0/4$ after distance $2R$. And so on ... as illustrated in Fig. 4.17.

Thus, after distance nR, the number of (photons + electrons + positrons) is 2^n and their average energy is $E_0/2^n$. It can also be seen that, on average, the shower consists of $\frac{2}{3}$ positrons and electrons and $\frac{1}{3}$ photons. The cascade eventually stops when the average energy per particle drops below the critical energy E_c. Below this energy, the dominant loss process for the electrons is ionisation losses rather than bremsstrahlung. This produces copious quantities of electron–ion pairs but they

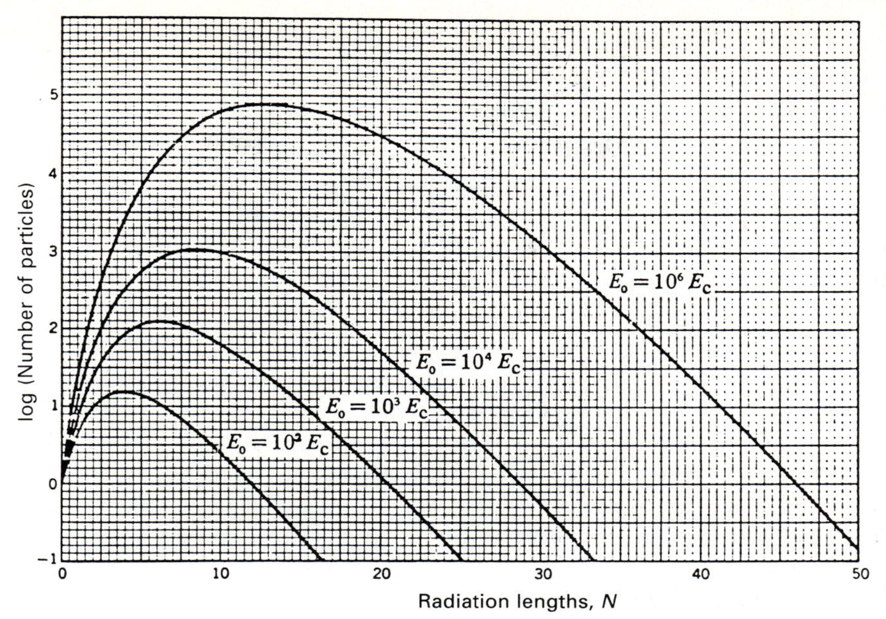

Figure 4.18. The total number of particles in a shower initiated by an electron of energy E_0 as a function of depth through the medium measured in radiation lengths N; E_c is the critical energy. (From B. Rossi and K. Greisen (1941). *Rev. Mod. Phys.*, **13**, 240.)

are all of very low energy. In addition, with decreasing energy, the production cross-section for pairs decreases until it becomes of the same order as that for Compton scattering and photoelectric absorption (see Fig. 4.16). Thus, the shower reaches its maximum development when the average energy of the cascade particles is about E_c. The number of high energy photons and particles is roughly E_0/E_c and the number of radiation lengths N is given by

$$N\xi_0 = n\xi_0 \ln 2$$
$$N = \ln (E_0/E_c)$$

At larger depths, the number of particles falls off because of ionisation losses which become catastrophic once the electrons become non-relativistic. All these calculations can be carried out properly. The appropriate cross-sections at different energies have to be used and integrations over all possible products with the relevant probability distributions have to be included. The results obtained by Rossi and Greisen (1941) are shown in Fig. 4.18. These calculations confirm the predictions of our simple model, namely, (i) the initial growth is exponential, (ii) the maximum number of particles is proportional to E_0 and (iii) beyond maximum, there is a rapid attenuation of the electron flux.

An important feature of these results is that the shower consists only of electrons, positrons and γ-rays – there are no muons, pions and other debris produced. This helps in distinguishing the arrival of very high energy γ-rays from other types of particle. These electron–photon cascades or electromagnetic

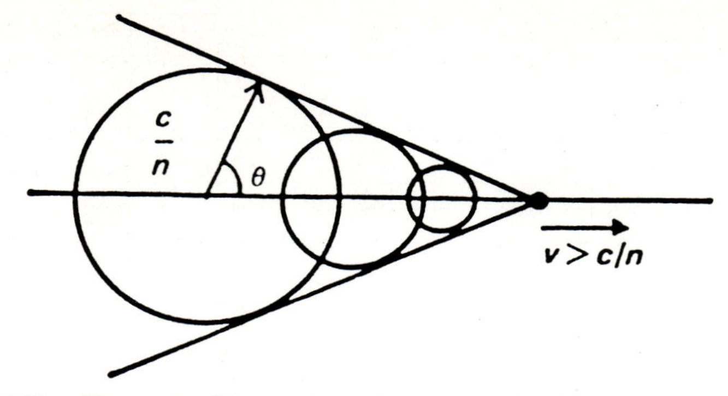

Figure 4.19. Illustrating Huygens' construction for determining the direction of propagation of the wavefront of Cherenkov radiation.

showers were among the first high interactions to be detected inside cloud chambers. They are readily recognised because of the characteristic signature of copious fluxes of electrons and positrons. These showers also accompany nuclear cascades which we shall consider in the next chapter.

4.6 Cherenkov radiation and ultrahigh energy γ-rays

When a fast particle moves through a medium at a constant velocity v which is greater than the velocity of light in that medium, it emits *Cherenkov radiation*. The simple geometric picture of this process is that, because the particle moves 'superluminally' through the medium, a 'shock wave' is created behind the particle and this results in a loss of energy by the particle. The wavefront of the radiation propagates at a fixed angle with respect to the velocity vector of the particle because only in this direction do the wavefronts add up coherently according to Huygens' construction (Fig. 4.19). The geometry of Fig. 4.19 shows that the angle of the wave vector with respect to the direction of motion of the particle is $\cos\theta = c/nv$ where n is the refractive index of the medium and c/n is the velocity of propagation of light in the medium.

Cherenkov radiation is important in two different contexts in the detection of high energy particles and photons. First, the process can be used in the construction of threshold detectors in which Cherenkov radiation is only emitted if the particle has velocity greater than c/n. Thus, if the particles pass through materials such as lucite or plexiglass, for which $n \approx 1.5$, only those with $v > 0.67c$ emit Cherenkov radiation which can be detected as an optical signal. If only particles with the most extreme relativistic energies are to be detected, gas Cherenkov detectors can be used in which the refractive index of the gas is just greater than 1.

A second application is in the detection of ultrahigh energy γ-rays when they enter the top of the atmosphere. The high energy γ-ray initiates an electron–photon cascade and, if it is of very high energy, the electron–positron pairs created acquire velocities greater than the speed of light at the top of the atmosphere. These electrons and positrons radiate optical Cherenkov radiation which can be detected by light detectors at sea-level. We will have much more to say about these types of telescopes in Section 7.4.2.

Let us derive the main features of Cherenkov radiation in a little more detail. This provides further interesting applications of the radiation formulae derived in Section 3.3. Our strategy is as follows. We consider an electron moving along the positive x-axis at a constant velocity v. This motion corresponds to a current density $\mathbf{J}$ given by

$$\mathbf{J} = ev\delta(x - vt)\,\delta(y)\,\delta(z)\,\mathbf{i}_x \tag{4.58}$$

Now let us take the Fourier transform of this current density to find the frequency components $\mathbf{J}(\omega)$ corresponding to this motion.

$$\begin{aligned} \mathbf{J}(\omega) &= \frac{1}{(2\pi)^{\frac{1}{2}}} \int \mathbf{J} \exp(\mathrm{i}\omega t)\, \mathrm{d}t \\ &= \frac{e}{(2\pi)^{\frac{1}{2}}} \delta(y)\,\delta(z) \exp(\mathrm{i}\omega z/v)\,\mathbf{i}_x \end{aligned} \tag{4.59}$$

This Fourier decomposition corresponds to representing the motion of the moving electron by a line distribution of oscillating currents. Our task is to work out the coherent emission, if any, from this distribution of oscillating currents. Since we need to know how to add together the fields from different segments of the line distribution, we write down the expressions for the *retarded* values of the current which contribute to the field at any given point $\mathbf{r}$. Now, it will not have escaped the notice of the reader that the formulae derived in Section 3.3 referred to the emission of an *accelerated* electron. How can we use these formulae to work out the radiation from an electron moving at a constant, but superluminal, velocity? The full treatments given in standard texts such as Jackson (1975) and Clemmow and Dougherty (1969) are quite complex. We adopt here an approach developed by John Peacock.

First, let us recall some of the standard results concerning the propagation of electromagnetic waves in a medium of permittivity, or dielectric constant, ε. It is a standard result of classical electrodynamics that the flow of electromagnetic energy through a surface $\mathrm{d}\mathbf{S}$ is given by the Poynting vector flux, $\mathbf{N}\cdot\mathrm{d}\mathbf{S} = (\mathbf{E}\times\mathbf{H})\cdot\mathrm{d}\mathbf{S}$. The electric and magnetic field strengths $\mathbf{E}$ and $\mathbf{H}$ are related to the electric flux density $\mathbf{D}$ and the magnetic flux density $\mathbf{B}$ by the constitutive relations

$$\mathbf{D} = \varepsilon\varepsilon_0\mathbf{E}; \quad \mathbf{B} = \mu\mu_0\mathbf{H}$$

The energy density of the electromagnetic field in the medium is given by the standard formula

$$u = \int \mathbf{E}\cdot\mathrm{d}\mathbf{D} + \int \mathbf{H}\cdot\mathrm{d}\mathbf{B} \tag{4.60}$$

A derivation of these results is given in Longair (1984). It is useful to recall the results for energy flow in the case in which the medium has a constant real permittivity ε and permeability $\mu = 1$. Then, the energy density in the medium is

$$u = \tfrac{1}{2}\varepsilon\varepsilon_0 E^2 + \tfrac{1}{2}\mu_0 H^2 \tag{4.61}$$

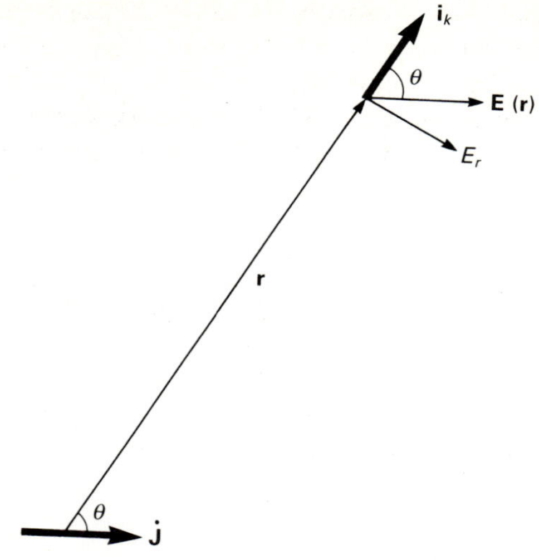

Figure 4.20.

The velocity of propagation of the electromagnetic waves is found from the dispersion relation $k^2 = \varepsilon\varepsilon_0 \mu_0 \omega^2$ i.e. $c(\varepsilon) = \omega/k = (\varepsilon\varepsilon_0 \mu_0)^{-\frac{1}{2}} = c/\varepsilon^{\frac{1}{2}}$. This demonstrates the well-known result that, in a linear medium, the refractive index n is just $\varepsilon^{\frac{1}{2}}$. Another useful result is the relation between the **E** and **B** fields in the electromagnetic wave. It is easy to show that the ratio E/B is just $c/\varepsilon^{\frac{1}{2}} = c/n$. If this result is substituted into the expression for the electric and magnetic field energies in the wave (4.61), it is found that these are the same. Thus, the total energy density in the wave is $u = \varepsilon\varepsilon_0 E^2$. Furthermore, the Poynting vector flux $\mathbf{E} \times \mathbf{H}$ is just $\varepsilon^{\frac{1}{2}}\varepsilon_0 E^2 c = n\varepsilon_0 E^2 c$. This energy flow corresponds to the energy density of radiation in the wave $\varepsilon\varepsilon_0 E^2$ propagating at the velocity of light in the medium c/n as would be expected, $N = n\varepsilon_0 E^2$. This is the result we have been seeking. It is similar to the formula we used in Sections 3.3.2 and 3.3.3 but now with the refractive index n included in the right place.

We can now write down the expression for the electric field strength **E** due to the current density **J** at distance **r**. Let us start from (3.16a).

$$\mathbf{A}(\mathbf{r}) = \frac{\mu_0}{4\pi} \int \frac{\mathbf{J}(\mathbf{r}', t - |\mathbf{r} - \mathbf{r}'|/c)}{|\mathbf{r} - \mathbf{r}'|} \mathrm{d}^3\mathbf{r}' = \frac{\mu_0}{4\pi} \int \frac{[\mathbf{J}]}{|\mathbf{r} - \mathbf{r}'|} \mathrm{d}^3\mathbf{r}' \tag{4.62}$$

Taking the time derivative

$$\mathbf{E}(\mathbf{r}) = -\frac{\partial \mathbf{A}}{\partial t} = -\frac{\mu_0}{4\pi} \int \frac{[\dot{\mathbf{J}}]}{|\mathbf{r} - \mathbf{r}'|} \mathrm{d}^3\mathbf{r}' \tag{4.63}$$

In the far field limit, the component of the radiation field is perpendicular to the radial vector **r**. As indicated in Fig. 4.20, the electric field component perpendicular to **k** is given by $E_r = \mathbf{E}(\mathbf{r}) \times \mathbf{i}_k$. i.e.

$$|\mathbf{E}_r| = \frac{\mu_0 \sin\theta}{4\pi} \left| \int \frac{[\dot{\mathbf{J}}]}{|\mathbf{r} - \mathbf{r}'|} \mathrm{d}^3\mathbf{r}' \right| \tag{4.64}$$

This formula reduces to the expression (3.6) for the radiation of a point charge by the substitution $\int[\dot{\mathbf{J}}]\,\mathrm{d}^3\mathbf{r}' = e\ddot{\mathbf{r}}$.

We now go through the same procedure described in Section 3.3.4 to evaluate the frequency spectrum of the radiation. First of all, we work out the total radiation rate by integrating the Poynting vector flux over a sphere at a large distance $\mathbf{r}$.

$$\left(\frac{\mathrm{d}E}{\mathrm{d}t}\right)_{\mathrm{rad}} = \int_S ncE_r^2\,\mathrm{d}S$$
$$= \int_\Omega \frac{n\sin^2\theta\,\mu_0^2\varepsilon_0 c}{16\pi^2}\left|\int\frac{[\dot{\mathbf{J}}]}{|\mathbf{r}-\mathbf{r}'|}\mathrm{d}^3\mathbf{r}'\right|^2 r^2\,\mathrm{d}\Omega \tag{4.65}$$

We now assume that the size of the emitting region is much smaller than the distance to the point of observation, $L \ll r$. Therefore, we can write $|\mathbf{r}-\mathbf{r}'| = r$ and then,

$$\left(\frac{\mathrm{d}E}{\mathrm{d}t}\right)_{\mathrm{rad}} = \int\frac{n\sin^2\theta}{16\pi^2\varepsilon_0 c^3}|\int[\dot{\mathbf{J}}]\,\mathrm{d}^3\mathbf{r}'|\,\mathrm{d}\Omega$$

Now, we take the time integral of the radiation rate to find the total radiated energy.

$$U = \int_{-\infty}^{\infty}\left(\frac{\mathrm{d}E}{\mathrm{d}t}\right)_{\mathrm{rad}}\mathrm{d}t = \int_{-\infty}^{\infty}\int_\Omega\frac{n\sin^2\theta}{16\pi^2\varepsilon_0 c^3}|\int[\dot{\mathbf{J}}]\,\mathrm{d}^3\mathbf{r}'|^2\,\mathrm{d}\Omega\,\mathrm{d}t \tag{4.66}$$

We now use Parseval's theorem to transform from an integral over time into one over frequency. Noting, as in Section 3.3.4, that we are only interested in positive frequencies, we find

$$U = \int_0^{\infty}\int_\Omega\frac{n\sin^2\theta}{8\pi^2\varepsilon_0 c^3}|\int[\dot{\mathbf{J}}(\omega)]\,\mathrm{d}^3\mathbf{r}'|^2\,\mathrm{d}\Omega\,\mathrm{d}\omega \tag{4.67}$$

Let us now evaluate the volume integral of the retarded potential $\int[\dot{\mathbf{J}}(\omega)\,\mathrm{d}^3\mathbf{r}'$. We take $\mathbf{R}$ to be the vector from the origin of the emitting region to the observer and $\mathbf{x}$ to be the position vector of the current element $\mathbf{J}(\omega)\,\mathrm{d}^3r'$ from the origin. Thus, $\mathbf{r}' = \mathbf{R}-\mathbf{x}$. Now the waves from the current element at $\mathbf{x}$ propagate outwards from the emitting region at velocity c/n with phase factor $\exp[\mathrm{i}(\omega t-\mathbf{k}\cdot\mathbf{r}')]$ and therefore, relative to the origin at O, the phase factor becomes

$$\exp[\mathrm{i}(\omega t-\mathbf{k}\cdot(\mathbf{R}-\mathbf{x})]$$
$$= \exp(-\mathrm{i}\mathbf{k}\cdot\mathbf{R})\exp[\mathrm{i}(\omega t+\mathbf{k}\cdot\mathbf{x})]$$

Therefore, taking the volume integral of the component $[\dot{\mathbf{J}}(\omega)]$, we find that

$$|\int[\dot{\mathbf{J}}(\omega)]\,\mathrm{d}^3\mathbf{r}'| = |\mathrm{i}\omega\int[\mathbf{J}(\omega)]\,\mathrm{d}^3\mathbf{r}'|$$
$$= |\int\omega\exp[\mathrm{i}(\omega t+\mathbf{k}\cdot\mathbf{x})]\mathbf{J}(\omega)\,\mathrm{d}^3\mathbf{r}'|$$
$$= \left|\frac{\omega e}{(2\pi)^{\frac{1}{2}}}\mathbf{i}_x\exp(\mathrm{i}\omega t)\int\exp\left[\mathrm{i}\left(\mathbf{k}\cdot\mathbf{x}+\frac{\omega x}{v}\right)\right]\mathrm{d}x\right|$$
$$= \left|\frac{\omega e}{(2\pi)^{\frac{1}{2}}}\int\exp\left[\mathrm{i}\left(\mathbf{k}\cdot\mathbf{x}+\frac{\omega x}{v}\right)\right]\mathrm{d}x\right| \tag{4.68}$$

This is the key integral in deciding whether or not the particle radiates. If the electron propagates in a vacuum, $\omega/k = c$ and we can write the exponent

$$kx(\cos\theta + \omega/kv) = kx(\cos\theta + c/v) \tag{4.69}$$

Since, in a vacuum, $c/v > 1$, this exponent is always greater than zero and hence the exponential integral is always zero. This means that a particle moving at constant velocity in a vacuum does not radiate.

If, however, the medium has refractive index n, $\omega/k = c/n$ and then the exponent is zero if $\cos\theta = -c/nv$. This is the origin of the Cherenkov radiation phenomenon. The radiation is only coherent along the angle θ corresponding to the Cherenkov cone derived from the simple Huygen's argument. We can therefore write down formally the energy spectrum by using (4.67) recalling that the radiation is only emitted at an angle $\cos\theta = -c/nv$. We therefore find

$$\begin{aligned}\frac{\mathrm{d}U}{\mathrm{d}\omega} &= \int_\Omega \frac{n\omega^2 e^2 \sin^2\theta}{16\pi^3\varepsilon_0 c^3}\left|\int \exp\left[ikx\left(\cos\theta + \frac{\omega}{kv}\right)\right]\mathrm{d}x\right|^2 \mathrm{d}\Omega \\ &= \frac{n\omega^2 e^2}{16\pi^3\varepsilon_0 c^3}\left(1 - \frac{c^2}{n^2v^2}\right)\int_\Omega \left|\int \exp\left[ikx\left(\cos\theta + \frac{\omega}{kv}\right)\right]\mathrm{d}x\right|^2 \mathrm{d}\Omega\end{aligned} \tag{4.70}$$

We now have to evaluate the integral. Let us write $k(\cos\theta + \omega/kv) = \alpha$. The integral therefore becomes

$$\int_\theta \left|\int \exp(i\alpha x)\,\mathrm{d}x\right|^2 2\pi \sin\theta\,\mathrm{d}\theta$$

Let us take the line integral along a finite path length from $-L$ to L. It should be noted that there is a problem in evaluating the integral of a function which only has finite value at a specific value of θ from $-\infty$ to $+\infty$. In fact, this is why the normal derivation involves the use of contour integration to get rid of the infinites. The integral should be taken over a small finite range of angles about $\theta = \cos^{-1}(c/nv)$ for which $(\cos\theta + \omega/kv)$ is close to zero. Therefore, we can integrate over all values of θ (or α) knowing that most of the integral is contributed by values of θ very close to $\cos^{-1}(c/nv)$. Therefore, the integral becomes

$$8\pi \int \frac{\sin^2 \alpha L}{\alpha^2}\frac{\mathrm{d}\alpha}{k}$$

Taking the integral over all values of α from $-\infty$ to $+\infty$, we find that the integral becomes $(8\pi c/n\omega)\pi^2 L$. Therefore the energy per unit bandwidth is

$$\frac{\mathrm{d}U}{\mathrm{d}\omega} = \frac{\omega e^2}{2\pi\varepsilon_0 c^3}\left(1 - \frac{c^2}{n^2v^2}\right)L \tag{4.71}$$

We now ought to take the limit $L \to \infty$. However, there is no need to do this since we obtain directly the energy loss rate per unit path length by dividing by $2L$. Therefore, the loss rate per unit path length is

$$\frac{\mathrm{d}U(\omega)}{\mathrm{d}x} = \frac{\omega e^2}{4\pi\varepsilon_0 c^3}\left(1 - \frac{c^2}{n^2v^2}\right) \tag{4.72}$$

Since the particle is moving at velocity v, the energy loss rate per unit bandwidth is

$$I(\omega) = \frac{\mathrm{d}U(\omega)}{\mathrm{d}t} = \frac{\omega e^2 v}{4\pi\varepsilon_0 c^3}\left(1 - \frac{c^2}{n^2 v^2}\right) \tag{4.73}$$

Notice that the intensity of radiation depends upon the variation of the refractive index with frequency $n(\omega)$.

In laboratory experiments, it is possible to measure the wave vector of the emitted radiation and the total emitted light provides information about the velocity and charge of the particles. In space experiments, no attempt is made to measure the wavefront but only the total optical emission. By using a number of detectors with different thresholds, for example, using different gases at different pressures, both v and z can be found for the particles. The main problem in using these detectors is that the light yield is very small and photomultipliers have to be used. One method of making most efficient use of the emitted photons is to ensure that the light is totally internally reflected – suitable treatment of the surfaces of the dielectric increases the amount of radiation detected. To optimise the sensitivity of the detectors, the variation of dielectric constant with frequency can be chosen to match the response curve of the photocathode.

4.7 Electron–positron annihilation and positron production mechanisms

Perhaps the most extreme form of energy loss mechanism for electrons is annihilation with their antiparticles, the positrons. Particle–antiparticle annihilation results in the production of high energy photons and, conversely, high energy photons can collide with ambient photons to produce particle–antiparticle pairs. The case of electron–positron annihilation is of particular interest because definite evidence for this process has been found in the central regions of our Galaxy and it is a powerful method of studying the regions in which it is expected that there are fluxes of positrons.

There are several sources of positrons in astronomical environments. Perhaps the simplest is the decay of positively charged pions π^+ described in Section 5.4. The pions are created in collisions between cosmic ray protons and nuclei and the interstellar gas, roughly equal numbers of positive, negative and neutral pions being created. Since the π^0s decay into γ-rays, the flux of interstellar positrons created by this process can be estimated from the γ-ray luminosity of the interstellar gas. A second process is the decay of long-lived radioactive isotopes created by explosive nucleosynthesis in supernova explosions. For example, the β^+ decay of ^{26}Al has a mean lifetime of 1.1×10^6 years. This element is formed in supernova explosions and so it is ejected into the interstellar gas where the decay results in a flux of interstellar positrons. A third process already mentioned is the creation of electron–positron pairs through the collision of high energy photons with the field of a nucleus (see Section 4.4).

A further process of considerable importance in certain astronomical environments is pair production in *photon–photon collisions*. The threshold for this

Table 4.2. *The energies of ultrahigh energy photons* (ε_2) *which give rise to electron–positron pairs in collision with photons of different energies* (ε_1)

	ε_1 (eV)	ε_2 (eV)
Microwave Background Radiation	6×10^{-4}	4×10^{14}
Starlight	2	10^{11}
X-ray	10^3	3×10^8

process can be worked out using similar procedures to those used in our discussion of Compton scattering. If $\mathbf{P}_1$ and $\mathbf{P}_2$ are the momentum four-vectors for the photons before the collision,

$$\mathbf{P}_1 = [(\varepsilon_1/c)\,\mathbf{i}_1, \varepsilon_1/c^2]; \quad \mathbf{P}_2 = [(\varepsilon_2/c)\,\mathbf{i}_2, \varepsilon_2/c^2]$$

then conservation of four-momentum requires

$$\mathbf{P}_1 + \mathbf{P}_2 = \mathbf{P}_3 + \mathbf{P}_4 \tag{4.74}$$

where $\mathbf{P}_3$ and $\mathbf{P}_4$ are the four-vectors of the created particles. To find the threshold for pair production, we require that the particles be created at rest and therefore

$$\mathbf{P}_3 = [0, m]; \quad \mathbf{P}_4 = [0, m]$$

Squaring both sides of (4.74) and noting that $\mathbf{P}_1 \cdot \mathbf{P}_1 = \mathbf{P}_2 \cdot \mathbf{P}_2 = 0$ and that $\mathbf{P}_3 \cdot \mathbf{P}_3 = \mathbf{P}_4 \cdot \mathbf{P}_4 = \mathbf{P}_3 \cdot \mathbf{P}_4 = m^2c^2$,

$$\mathbf{P}_1 \cdot \mathbf{P}_1 + 2\mathbf{P}_1 \cdot \mathbf{P}_2 + \mathbf{P}_2 \cdot \mathbf{P}_2 = \mathbf{P}_3 \cdot \mathbf{P}_3 + 2\mathbf{P}_3 \cdot \mathbf{P}_4 + \mathbf{P}_4 \cdot \mathbf{P}_4$$

$$2\left(\frac{\varepsilon_1 \varepsilon_2}{c^2} - \frac{\varepsilon_1 \varepsilon_2}{c^2} \cos\theta\right) = 4m^2c^2$$

$$\varepsilon_2 = \frac{2m^2c^4}{\varepsilon_1(1-\cos\theta)} \tag{4.75}$$

where θ is the angle between the incident directions of the photons. Thus, if electron–positron pairs are created, the threshold for the process occurs for head-on collisions, $\theta = \pi$, and hence,

$$\varepsilon_2 \geqslant \frac{m_e^2 c^4}{\varepsilon_1} = \frac{0.26 \times 10^{12}}{\varepsilon_1} \text{ eV} \tag{4.76}$$

where ε_1 is measured in electron volts. This process thus provides not only a means for creating electron–positron pairs but also results an important source of opacity for very-high-energy γ-rays. Table 4.2 shows the more important examples we will encounter as our story unfolds. Photons with energies greater than those in the last column are expected to suffer some degree of absorption when they traverse regions with high energy densities of photons with energies listed in the first column.

The cross-section for this process for head-on collisions in the ultrarelativistic limit is

$$\sigma = \pi r_e^2 \frac{m_e^2 c^4}{\varepsilon_1 \varepsilon_2}\left[2 \ln\left(\frac{2\omega}{m_e c^2}\right) - 1\right] \tag{4.77}$$

where $\omega = (\varepsilon_1 \varepsilon_2)^{\frac{1}{2}}$ and r_e is the classical electron radius. In the classical regime for which $\omega \approx m_e c^2$, the cross-section is

$$\sigma = \pi r_e^2 \left(1 - \frac{m^2 c^4}{\omega^2}\right)^{\frac{1}{2}} \tag{4.78}$$

(see Ramana Murthy and Wolfendale (1986)). These cross-sections enable the opacity of the interstellar and intergalactic medium to be evaluated as well as providing a mechanism by which large fluxes of positrons could be generated in the vicinity of active galactic nuclei.

Electron–positron annihilation can proceed in two ways. In the first case, the electrons and positrons annihilate at rest or in flight by the traditional interaction

$$e^+ + e^- \rightarrow 2\gamma$$

When emitted at rest, the photons both have energy 0.511 MeV. When the particles annihilate 'in flight', meaning that they suffer a fast collision, there is a dispersion in the photon energies. It is a pleasant exercise in relativity to show that, if the positron is moving with velocity v and has corresponding Lorentz factor γ, the centre of momentum frame of the collision has velocity $V = \gamma v(1+\gamma)$ and that the energies of the pair of photons ejected in the direction of the line of flight of the positron and in the backward direction are

$$E = \frac{m_e c^2(1+\gamma)}{2}\left(1 \pm \frac{V}{c}\right) \tag{4.79}$$

From this result, it can be seen that the photon which moves off in the direction of the incoming positron carries away most of the energy of the positron and that there is a lower limit to the energy of the photon ejected in the opposite direction of $m_e c^2/2$.

If the velocity of the positron is small, *positronium atoms*, i.e. bound states consisting of an electron and a positron, can form by radiative recombination; 25% of the positronium atoms form in the singlet $^1S_0$ state and 75% of them in the triplet $^3S_1$ state. The modes of decay from these states are different. The singlet $^1S_0$ state has a lifetime of 1.25×10^{-10} s and the atom decays into two γ-rays, each with energy 0.511 MeV. The majority triplet $^3S_1$ states have a mean lifetime of 1.5×10^{-7} s and three γ-rays are emitted, the maximum energy being 0.511 MeV in the centre of momentum frame. In this case, the decay of positronium results in a continuum spectrum to the low energy side of the 0.511 MeV line. If the positronium is formed from positrons and electrons with significant velocity dispersion, the line at 0.511 MeV is broadened both because of the velocities of the

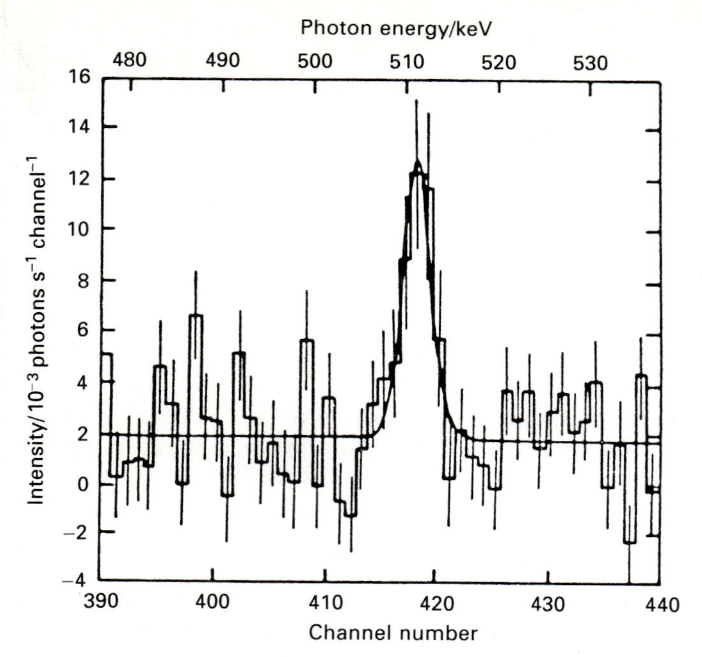

Figure 4.21. HEAO-3 observations of the 0.511 MeV electron–positron annihilation line from the general direction of the Galactic Centre. The observations were made in autumn 1979. (From G. R. Riegler, J. C. Ling, W. A. Mahoney, W. A. Wheaton, J. B. Willett, A. S. Jacobson and T. A. Prince (1981). *Astrophys. J. Lett.*, **248**, 113.)

particles and because of the low energy wing due to the continuum three photon emission. This is a useful diagnostic tool in understanding the origin of the 0.511 MeV line. If the annihilations take place in a neutral medium with particle density less than 10^{21} m^{-3}, positronium atoms are formed. On the other hand, if the positrons collide in a gas at temperature greater than about 10^6 K, the annihilation takes place directly without the formation of positronium.

The cross-section for electron–positron annihilation in the extreme relativistic limit is

$$\sigma = \frac{\pi r_e^2}{\gamma}[\ln 2\gamma - 1] \tag{4.80}$$

For thermal electrons and positrons, the cross-section becomes

$$\sigma \approx \frac{\pi r_e^2}{(v/c)} \tag{4.81}$$

One of the most exciting results of γ-ray astronomy has been the detection of the 0.511 MeV electron–positron annihilation line from the direction of the Galactic Centre (Fig. 4.21). We will have much more to say about this observation and the source of positrons in later sections.

5

Nuclear interactions

5.1 Nuclear physics and high energy astrophysics

Nuclear physics is central to many branches of astrophysics, in particular to the understanding of the processes of energy generation in stars. Perhaps surprisingly, we will treat this topic very lightly because, in these cases, the nuclear processes occur deep in the centres of stars where the products of nucleosynthesis are generally only very indirectly observable. The obvious exceptions to this statement are the observations of neutrinos from the Sun and the supernova SN1987a. We restrict attention in this chapter to nuclear processes in which the products of the nuclear activity are directly observable. We need cross-sections to study the spallation reactions of high energy particles in the interstellar medium as well as the production cross-sections and half-lives of radionuclides created in the spallation process and in sources of freshly synthesised material such as supernova remnants. We deal first with the case of nuclear interactions associated with inelastic collisions of high energy protons and nuclei.

Our treatment will be very much simpler than our analysis of extranuclear activity. Nuclear interactions are only important when the incident high energy particle makes a more or less direct hit on the nucleus. This is because the strong interaction forces which hold the nucleus together are short-range forces. Thus, the cross-section for nuclear interactions, in the sense that some form of interaction with the nucleons takes place, is just the geometric cross-section of the nucleus. A suitable expression for the radius of the nucleus is

$$R = 1.2 \times 10^{-15} A^{1/3}\ \mathrm{m} \tag{5.1}$$

where A is the mass number. In many of the cases we will deal with, the high energy particles have energies greater than 1 GeV. This introduces further simplifications in the picture since, at these energies, the de Broglie wavelength of the incident particle is small compared with the distance between nucleons in a nucleus. For

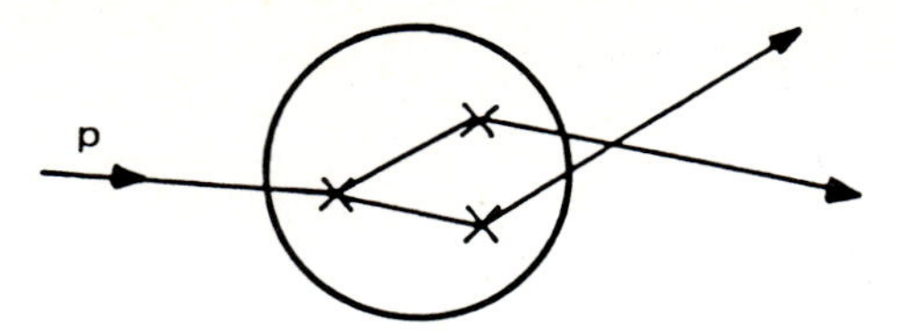

Figure 5.1. A schematic diagram showing the multiple scattering of a high energy proton inside a nucleus.

example, let us work out the effective size of an incident proton from Heisenberg's Uncertainty Principle. A proton of energy, say, 10 GeV, has 'size'

$$\Delta x \approx D = \hbar/\gamma m_{\mathrm{p}} c = 0.02 \times 10^{-15} \text{ m} \tag{5.2}$$

We can therefore think of the incident proton as being a discrete, very small particle which interacts with the individual nucleons within the nucleus. The number of particles with which it interacts is just the number of nucleons along the line of sight through the nucleus. For example, a proton passing through an oxygen or nitrogen nucleus interacts, on average, with about $15^{1/3}$ i.e. 2.5 of the nucleons. In fact, we obtain a reasonable model of the nuclear interactions if we consider the incident proton as undergoing multiple scattering inside the nucleus (Fig. 5.1).

5.1.1 *High energy protons*

The general picture of the interaction of a high energy proton with a nucleus can be described by the following empirical rules.

(i) The proton interacts violently with an individual nucleon in a nucleus and, in the collision, pions of all charges, π^+, π^- and π^0, are the principal products. Strange particles may also be produced and occasionally antinucleons as well.

(ii) In the centre of momentum frame of reference of the proton–nucleon encounter, the pions emerge mostly in the forward and backward directions but they may have lateral components of momentum of the order of $m_{\mathrm{p}} c \approx$ 100–200 MeV c^{-1}.

(iii) The nucleons and pions all possess very high forward motion through the laboratory frame of reference and so they come out of the interaction with very high energy.

(iv) Each of the secondary particles is capable of initiating another collision inside the same nucleus, provided, of course, that the initial collision occurred sufficiently close to the 'front edge' of the nucleus. Thus, a mini-cascade is initiated inside the nucleus.

(v) Only one or two nucleons participate in the nuclear interactions with the high energy particle, but they are generally removed from the nucleus leaving it in a highly excited state and there is no guarantee that the resulting nucleus is a stable species. A number of things may

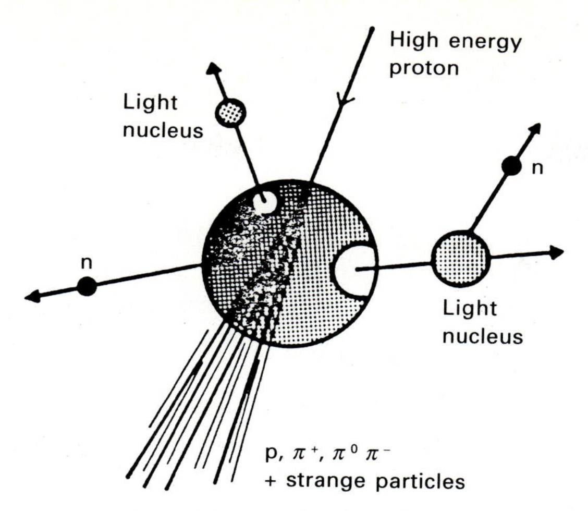

Figure 5.2. A schematic diagram showing the principal products of the collision of a high energy proton with a nucleus.

happen. What often happens is that several nuclear fragments evaporate from the nucleus. These are called *spallation fragments* and we will have a great deal to say about them in the context of the origin of the light elements. These fragments are emitted in the frame of the nucleus which is not given much forward momentum in the nuclear collision, virtually all of it going into tearing out the nucleons which interact with the high energy particle. Therefore, these spallation fragments are emitted more or less isotropically in the laboratory frame of reference. Neutrons are also evaporated from the ravished nucleus and other neutrons may be released from the spallation fragments. Remember that, for light nuclei, any imbalance between the numbers of neutrons and protons is fatal. These processes are summarised diagrammatically in Fig. 5.2. It is therefore clear that, in high energy collisions, the pions are concentrated in a rather narrow cone which is some measure of the energy of the primary high energy particle.

From the above cross-section for the interaction of high energy particles with nuclei, it can be shown in a straightforward manner that the mean free path of a high energy proton in the atmosphere is about 800 kg m^{-2}, i.e. very much less than the depth of the atmosphere which is about 10000 kg m^{-2}. In fact, because the proton often survives the interaction with some loss of energy, the flux of protons of a given energy falls off rather more slowly than this and, for particles of a given energy, the number density of protons falls off as $\exp(-x/L)$ where $L = 1200$ kg m^{-2}.

For incident protons with energy greater than 1 GeV, a useful empirical rule is that, in collisions with air nuclei, roughly $2E^{1/4}$ new, high energy, charged particles are generated in the collision, where E is measured in gigaelectron volts, although not necessarily all of them are pions. Pions of all charges are produced in almost equal numbers except at small energies at which charge conservation favours positively charged pions π^+.

5.1.2 *Cosmic ray nuclei*

The most spectacular events occur when high energy nuclei undergo collisions with other heavy nuclei, for example, with the oxygen and nitrogen nuclei of our atmosphere or with the atoms of a nuclear emulsion. Fig. 5.3 shows a rather impressive collision between a cosmic ray iron nucleus and the nucleus of a nuclear emulsion. In such collisions, several pairs of nucleons undergo pion-producing collisions and not much is left of the target nucleus. This is, of course, quite a rare occurrence; much more common are grazing encounters and in these cases only a few nucleons interact to produce a shower of pions. Again, however, the residual nuclei are left in an excited state and both eject spallation fragments as well as protons and neutrons. The only difference this time is that the incident high energy nucleus leaves with a stream of *relativistic* spallation fragments, protons and neutrons. This is very important from several points of view. First, the high energy fragments can develop into separate showers and, at the very highest cosmic ray energies, $E > 10^{17}$ eV, some of the showers which penetrate to the surface of the Earth are found to be multi-cored; these might be due to the break up of a very high energy nucleus. Second, this mechanism produces spallation products with very high energies. This will prove to be a central topic in the study of the propagation of cosmic ray nuclei in the interstellar medium. The determination of the cross-sections for the production of the various spallation products is therefore of the greatest interest.

5.2 Spallation cross-sections

The best way of determining spallation cross-sections is to undertake experiments with high energy particles in accelerators and, from these, determine directly the production cross-sections for each element as a function of energy. The problem is that for astrophysical purposes one is interested in an enormous range of species and particle energies. There has, however, been considerable progress in estimating these cross-sections for most of the important species.

There are three approaches to the determination of these cross-sections. The first is to determine purely experimentally the cross-sections for the various interactions. In these experiments, protons are fired at the target material and then the energy of the proton is the same as the energy per nucleon which the target nucleus possesses in the rest frame of the proton. Since hydrogen is by far the most common of the elements in the interstellar gas, this is the dominant process involved in the splitting up of high energy nuclei although spallation on helium nuclei also makes a significant contribution.

The results of these experiments can then be used to work out *semiempirical relations* which can be used to estimate cross-sections for rarer elements and

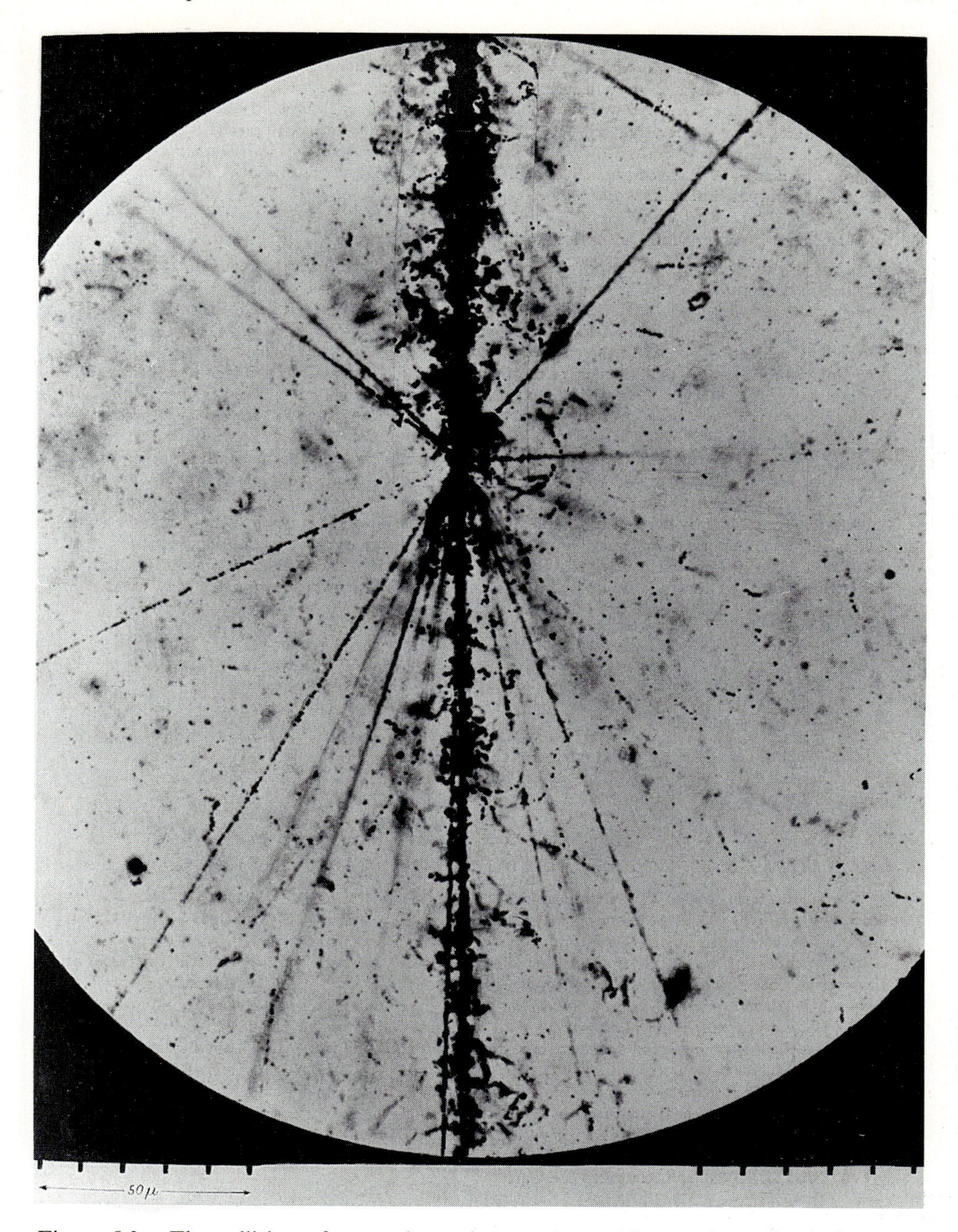

Figure 5.3. The collision of a cosmic ray iron nucleus with a nucleus of a nuclear emulsion. (From C. F. Powell, P. H. Fowler and D. H. Perkins (1959). *The study of elementary particles by the photographic method*, page 95. Oxford: Pergamon Press.)

Table 5.1. (*a*) *Partial cross-sections for inelastic collisions of selected heavy nuclei with hydrogen with $E = 2.3$ GeV per nucleon.*

Product nucleus	Z	A	Parent nucleus ^{11}B	^{12}C	^{14}N	^{16}O	^{20}Ne	^{24}Mg	^{28}Si	^{56}Fe
Lithium	3	6	12.9	12.6	12.6	12.6	12.6	12.6	12.6	17.4
		7	17.6	11.4	11.4	11.4	11.4	11.4	11.4	17.8
Beryllium	4	7	6.4	9.7	9.7	9.7	9.7	9.7	9.7	8.4
		9	7.1	4.3	4.3	4.3	4.3	4.3	4.3	5.8
		10	15.8	2.9	1.9	1.9	1.9	1.9	1.9	4.1
Boron	5	10	26.6	17.3	16.0	8.3	7.1	6.2	5.3	5.3
		11	—	31.5	15.0	13.9	12.0	10.4	9.0	8.1
Carbon	6	10	—	3.9	3.3	2.9	2.1	1.6	1.2	0.5
		11	0.6	26.9	12.4	10.6	7.9	5.9	4.5	1.3
		12	—	—	38.1	32.7	13.5	10.1	7.6	4.7
		13	—	—	10.5	14.4	10.7	8.0	6.0	3.7
		14	—	—	—	2.3	3.9	3.0	2.2	2.1
Nitrogen	7	13	—	—	10.7	3.6	2.7	2.0	1.5	0.5
		14	—	—	—	26.3	10.9	8.1	6.1	2.9
		15	—	—	—	31.5	10.0	7.5	5.7	4.3
		16	—	—	—	—	3.4	2.6	1.9	1.6
Oxygen	8	14	—	—	—	3.4	2.5	1.9	1.4	0.3
		15	—	—	—	27.8	11.8	8.9	6.7	1.0
		16	—	—	—	—	27.0	13.5	10.2	3.9
		17	—	—	—	—	15.5	11.6	8.7	4.1
		18	—	—	—	—	4.5	4.7	3.5	2.6
Fluorine	9	16	—	—	—	—	—	1.4	1.1	—
		17	—	—	—	—	8.5	6.4	4.8	—
		18	—	—	—	—	14.4	10.8	8.1	2.4
		19	—	—	—	—	21.0	10.9	8.2	4.8
		20	—	—	—	—	—	4.2	3.1	2.3
Neon	10	18	—	—	—	—	2.8	2.1	1.6	—
		19	—	—	—	—	17.3	5.3	4.0	—
		20	—	—	—	—	—	17.8	13.4	3.6
		21	—	—	—	—	—	14.0	10.6	5.4
		22	—	—	—	—	—	8.2	5.8	4.3
		23	—	—	—	—	—	—	1.3	—
Sodium	11	20	—	—	—	—	—	1.5	1.1	—
		21	—	—	—	—	—	7.7	5.6	—
		22	—	—	—	—	—	16.8	12.7	2.3
		23	—	—	—	—	—	21.0	12.0	6.4
		24	—	—	—	—	—	—	5.2	3.7
Magnesium	12	23	—	—	—	—	—	29.8	1.6	0.6
		24	—	—	—	—	—	—	17.1	3.2
		25	—	—	—	—	—	—	18.5	6.0
		26	—	—	—	—	—	—	14.4	6.8
		27	—	—	—	—	—	—	7.6	1.7

Table 5.1. (*a*) (*Continued*)

Product nucleus	*Z*	*A*	Parent nucleus ^{11}B	^{12}C	^{14}N	^{16}O	^{20}Ne	^{24}Mg	^{28}Si	^{56}Fe
Aluminium	13	25	—	—	—	—	—	—	6.3	—
		26	—	—	—	—	—	—	13.3	2.0
		27	—	—	—	—	—	—	21.0	6.7
		28	—	—	—	—	—	—	—	5.7
		29	—	—	—	—	—	—	—	2.5
Silicon	14	27	—	—	—	—	—	—	30.7	0.4
		28	—	—	—	—	—	—	—	2.7
		29	—	—	—	—	—	—	—	6.0
		30	—	—	—	—	—	—	—	10.4
		31	—	—	—	—	—	—	—	3.1
		32	—	—	—	—	—	—	—	1.2
Total inelastic cross-section			237.8	252.4	280.9	308.8	363.3	415.7	466.0	763.4

Cross-sections measured in units of millibarns = 10^{-31} m^2.
Data kindly supplied by Drs R. Silberberg and C. H. Tsao.

isotopes. This procedure is similar to that used in nuclear physics in which the semiempirical mass formula is derived based upon the liquid drop model of the nucleus. In the case of cosmic ray studies, in particular, we need the cross-sections for many of the rarer and unstable isotopes which can be created under cosmic conditions.

A third procedure is to model the spallation process by simulating the details of particle–particle collisions inside the nucleus in a high speed computer. The procedures used employ what are known as Monte Carlo techniques. The trajectory of the incoming particle inside the nucleus is followed, the initial conditions being selected at random. The proton interacts randomly with the nucleons inside the nucleus and, depending upon the particles which are knocked out of the nucleus in the interaction and the energy of the excited nucleus, the parent nucleus fragments into a number of different end products, the probability of these end products being produced being described by their partial cross-sections. In the typical Monte Carlo simulation, 100000 or more collisions are studied by computer so that good statistics can be built up even for quite rare product nuclei. This type of procedure is used widely in many aspects of physics in which enormous numbers of interactions have to be taken into account and the analytic computation of the final products would be prohibitively complicated. In the case of the spallation cross-sections, these modelling procedures are successful in accounting for the experimentally determined partial cross-sections.

Table 5.1 (*b*). *Partial cross-sections for inelastic collisions of iron (Fe) with hydrogen with $E = 2.3$ GeV per nucleon.*

Product nucleus	Z	σ
Silicon	14	24.1
Phosphorus	15	23.9
Sulphur	16	35.2
Chlorine	17	30.0
Argon	18	43.4
Potassium	19	41.6
Calcium	20	54.9
Scandium	21	55.5
Titanium	22	72.3
Vanadium	23	51.6
Chromium	24	79.6
Manganese	25	120.8
Iron	26	66.7

Cross-sections measured in units of millibarns = 10^{-31} m^2.
Data kindly supplied by Drs R. Silberberg and C. H. Tsao.

In recent years, strenuous efforts have been made to determine the spallation cross-sections for as many of the important elements and their isotopes as is practicable. Not only have the partial cross-sections for the creation of product nuclei been determined but also the variation of these partial cross-sections with energy. The results of a major programme to achieve these goals have recently been published by Webber and his colleagues. (Webber, Kish and Schrier 1990a, b, c, d.) These are the most accurate cross-sections published to date. Table 5.1 shows a compilation of partial cross-sections kindly provided by Drs R. Silberberg and C. H. Tsao who derived these from the semiempirical formulae which take into account a very wide range of additional nuclear data (see Silberberg, Tsao and Letaw (1988)). At the bottom of each column, the *total inelastic cross-section* for the break-up of the target nucleus is given. Not surprisingly, the total cross-section turns out to be similar to the geometric cross-section of the nucleus.

There is reasonable agreement between the measured cross-sections and those derived from the semiempirical formulae. Normally, the agreement is within about 25% but there are cases in which larger discrepancies are found. Note, however, that the precision of the measured partial cross-sections is about 2% for the best determinations. Whilst there are some discrepancies in the absolute values of the cross-sections, the relative cross-sections for the formation of the isotopes of a particular element from a single parent are in good agreement. Webber and his colleagues have proposed an alternative form of semiempirical formula which they find gives an improved account of the measured cross-sections. A detailed discussion of these discrepancies is far beyond the scope of this text and the interested reader should consult the papers by Webber, Silberberg and their colleagues. What is encouraging is the fact that much more experimental data is now available and that the semiempirical formulae which are still needed to account for all the possible ways in which nuclei can be fragmented are improving all the time.

There are several interesting features of Table 5.1. It can be seen that there is always a large cross-section for chipping off a single nucleon from a nucleus. This is not particularly unexpected because there are always more grazing than head-on collisions. When the product nuclei are unstable, it can be seen that the formation of pairs of nuclei with similar masses is not favoured. This is similar to what is found in nuclear fission experiments. Another interesting point is the fact that even nuclei are slightly favoured over odd nuclei as can be seen from the run of the partial cross-sections for the spallation of iron with mass number. Again this parallels the abundances of the elements as a whole which favour nuclei with even numbers of nucleons (see Section 9.2). As is well known, according to nuclear physics, this reflects the greater binding energies of nuclei with even numbers of nucleons. Finally, it will be noted that not all of the total cross-section is accounted for by the fragments listed in the table. This is largely because only the most interesting nuclei have been included in the table. It should be recalled that there are significant cross-sections for chipping protons and α-particles off nuclei. Indeed, in the spallation of ^{12}C, there is a significant cross-section for the break up of the nucleus into α-particles.

As we will find out, a wide range of energies is found among cosmic ray nuclei and it is no longer good enough to characterise the partial cross-sections by a single value independent of energy. There has been considerable progress in determining the energy dependence of these cross-sections and some examples described by Webber *et al.* (1990) are shown in Fig. 5.4. In Figs 5.4(*a*) and (*b*), the points show the experimentally determined cross-sections and the lines are the predictions of various semiempirical formulae. It can be seen that over the energy ranges shown in Figs 5.4(*a*) and (*b*), the variations of the partial cross-sections with energy are quite small. On the other hand, in the spallation of iron nuclei, there are strong variations at low energies in the partial spallation cross-sections (Fig. 5.4(*c*)). These variations are principally associated with the difference in mass number of the parent and product nuclei. A greater excitation energy is needed to break up an iron nucleus ($A = 56$) into, say, phosphorus ($A = 31$) or silicon ($A = 28$) as compared with chromium ($A = 52$) or manganese ($A = 55$) as may be appreciated from the run of nuclear binding energy with mass number. At relativistic energies, it is expected that the cross-sections should remain roughly constant and the semi-empirical formulae provided an accurate description of the partial cross-sections. Meyer (1985) has emphasised the importance of determining the partial spallation cross-sections and their variation with energy with good accuracy. It will turn out that some important astrophysical analyses will in the end be limited by the accuracy with which the experimentally determined cross-sections are known. We will return to these data in due course when we study the spallation products produced when high energy protons and nuclei interact with the interstellar gas.

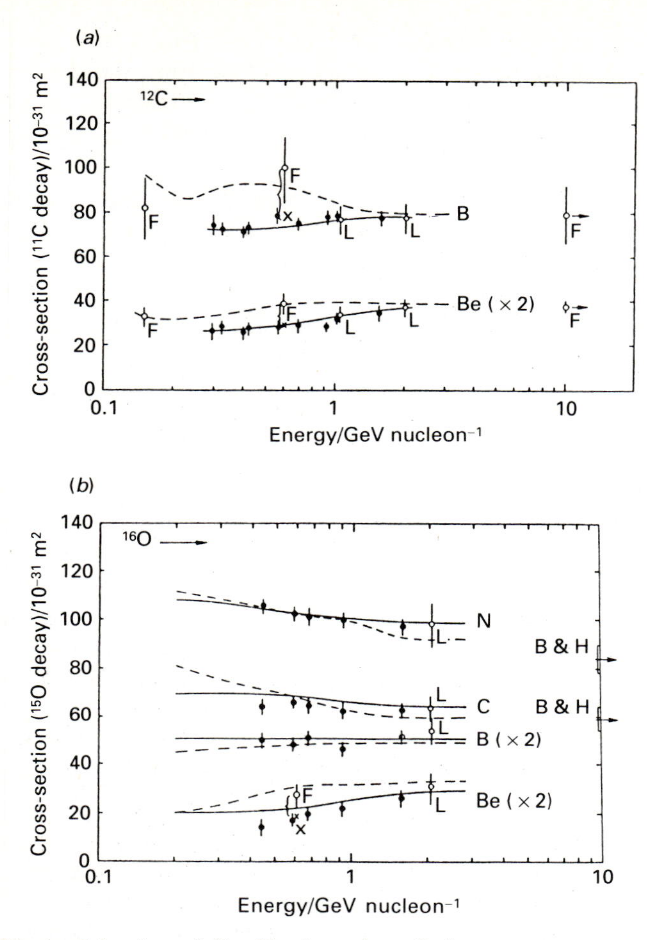

Figure 5.4. (*a* and *b*). For legend see facing page.

5.3 Nuclear emission lines

There are two types of nuclear process which are important in producing γ-ray lines in the spectra of astronomical sources. In the first of these, nuclei are excited to levels above the ground state by collisions with cosmic ray protons and nuclei. γ-rays are emitted in the subsequent decays of the nuclei to their ground states. These interactions may either take place in the diffuse interstellar gas in which case the target nuclei acquire significant velocities in the collisions, or else inside interstellar grains in which case the target nuclei emit the γ-rays essentially at rest. The physical process is similar to that of the photoexcitation of the excited electronic levels of atoms by photons and, in the same way, the cross-section for excitation of the nucleus attains a maximum value for photon energies of the same order as the energy of the excited states. Examples of the cross-sections for the photoexcitation of carbon and oxygen nuclei are shown in Fig. 5.5. It can be seen

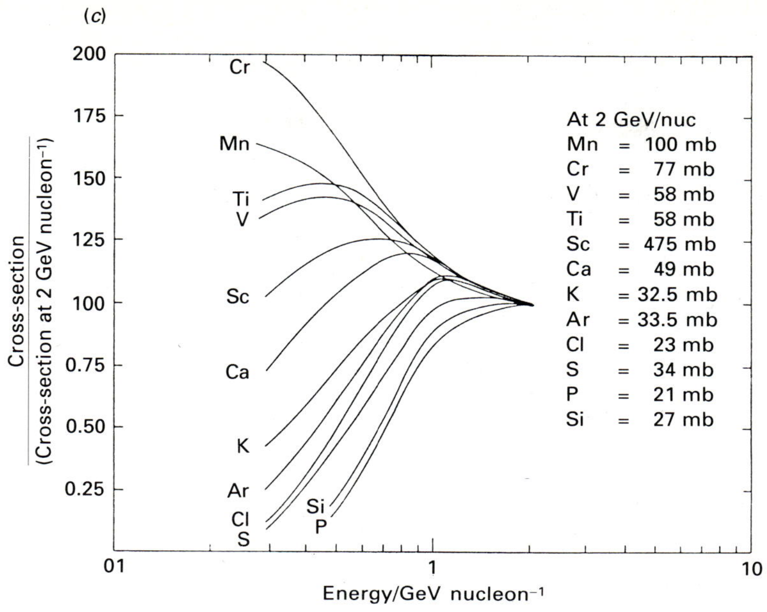

Figure 5.4. Illustrating the energy dependence of the partial cross-sections for the formation of (*a*) boron and beryllium from carbon and (*b*) nitrogen, carbon, beryllium and boron from oxygen, both in spallation interactions with protons. (From W. R. Webber, J. C. Kish and D. A. Schrier (1990). *Phys. Rev. C*, 544.) The solid lines show the expectations of the semi-empirical formulae proposed by W. R. Webber, J. C. Kish and D. A. Schrier. (*Phys. Rev. C*, 566 (1990)). The dashed lines show the expectations of the semiempirical formulae of Tsao and Silberberg (1979). The more recent forms of the semiempirical formulae provide improved fits to the experimental data (R. Silberberg, C. H. Tsao and J. R. Letaw (1985). *Astrophys. J. Suppl.*, **58**, 873.) (*c*) Relative partial cross-sections for the spallation of ^{56}Fe by protons into lighter elements as a function of energy. It can be seen that these cross-sections are strongly energy dependent at low energies. (Note: 1 mb = 1 millibarn = 10^{-31} m^2) (From W. R. Webber, J. C. Kish and D. A. Schrier (1990). *Phys. Rev. C*, 546.)

that the cross-sections for collisional excitation of these nuclei are $\approx (1\text{–}2) \times 10^{-29}$ m^2 for protons with energies $\approx$ 8–30 MeV.

Evidence for these processes occurring in astrophysical environments is provided by γ-ray spectroscopic observations of *solar flares*. Fig. 5.6 shows the spectrum of a large flare observed by the Gamma-Ray Spectrometer on board the Solar Maximum Mission which took place on 27 April 1981 (Murphy *et al.* 1985). γ-ray lines associated with many of the abundant elements are observed as well as lines associated with electron–positron annihilation ($\varepsilon = 0.511$ MeV) and the feature at 2.223 MeV which is associated with neutron capture by hydrogen nuclei – the neutrons are released in spallation interactions induced by particles accelerated in the flare. The lines of ^{12}C and ^{16}O at 4.438 and 6.129 MeV are the strongest observed in the spectrum. These observations are of the greatest interest from the point of view of the acceleration of charged particles since the particles responsible

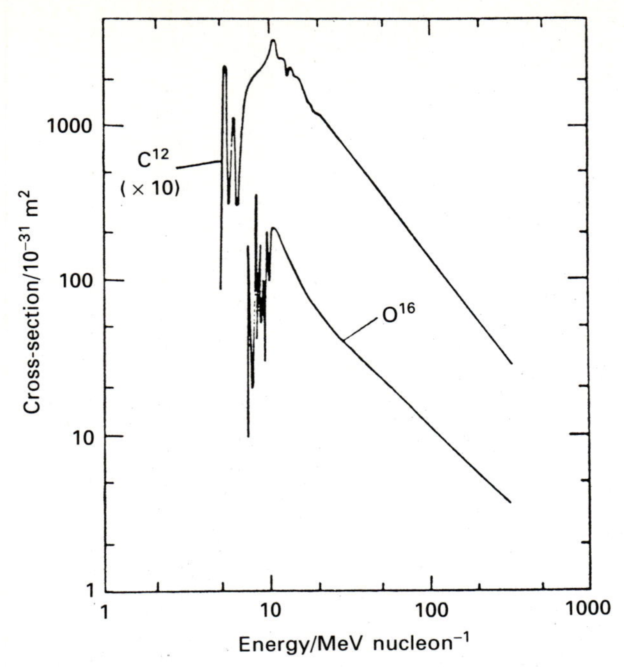

Figure 5.5. The interaction cross-sections leading to the emission of γ-ray line emission through the excitation of ^{12}C and ^{16}O by collisions with protons as a function of the kinetic energy of the incident proton. (From P. V. Ramana Murthy and A. W. Wolfendale (1986). *γ-ray astronomy*, page 4, Cambridge: Cambridge University Press).

for exciting the emission lines must be accelerated to megaelectron volt energies in the solar flare itself. We will return to this topic in Chapter 12.

Ramaty and Lingenfelter (1979) have carried out computations of the expected γ-ray spectrum of the interstellar medium due to the interaction of the interstellar flux of high energy particles with the interstellar gas. There are considerable uncertainties in these calculations, in particular, because the interstellar flux of high energy particles is poorly known in the energy range 1–100 MeV. In addition, it is not known precisely what fraction of the interstellar gas is condensed into dust grains. Nevertheless, these calculations are useful in indicating those elements which are likely to be significant emitters of γ-ray line emission, the broad lines resulting from collisions taking place in the gas phase and the narrow lines being produced from dust grains. A list of some of the more important lines is given in Table 5.2. Fig. 5.7 shows the predicted γ-ray emission spectrum in the general direction of the Galactic Centre due to these processes. It can be seen that certain lines are expected to be rather strong, the lines of ^{16}O at 6.129 MeV and of ^{12}C at 4.438 MeV being particularly prominent as in Fig. 5.6.

A second important process is the decay of radioactive isotopes created in supernova explosions. The process of explosive nucleosynthesis in supernovae creates stable and unstable nuclei and the radioactive decay of the latter are sources of γ-ray line emission. Table 5.3 is a list of some of the more important γ-ray lines

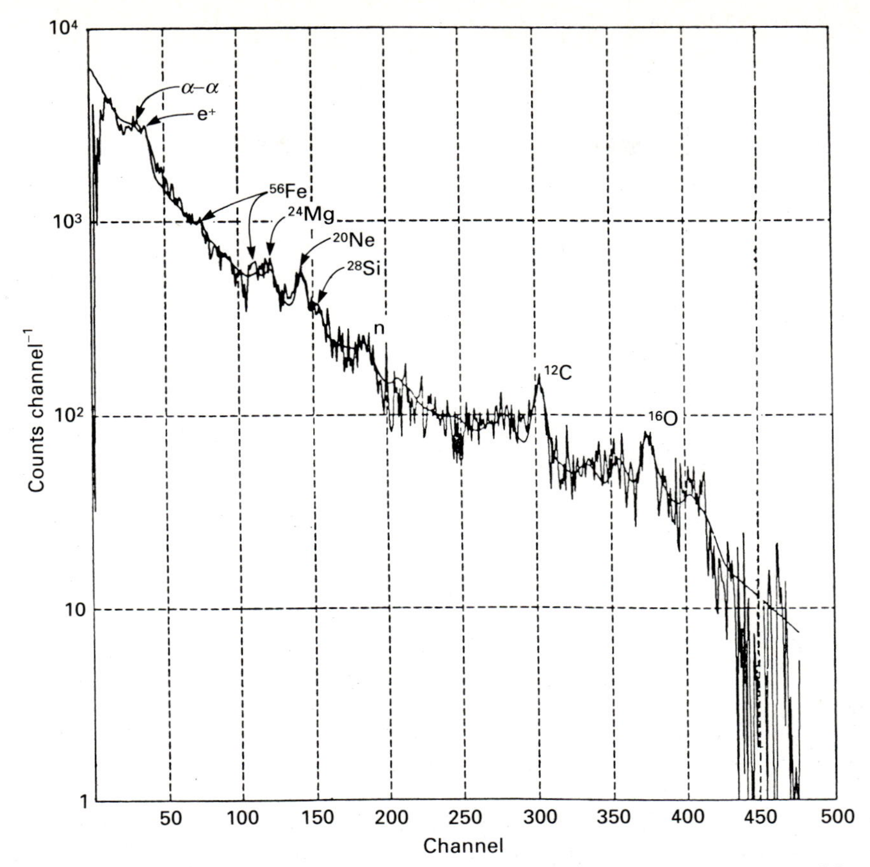

Figure 5.6. The γ-ray spectrum of the solar flare of 27 April 1981 as measured by the Gamma-Ray Spectrometer on the Solar Maximum Mission. γ-ray lines of several of the abundant elements are observed. The figure includes a prediction of the line and continuum spectrum on the basis of the excitation of the nuclei by particles accelerated in the solar flare. (From R. J. Murphy, D. J. Forrest, R. Ramaty and B. Kozlovsky (1985). *19th International Cosmic Ray Conference, La Jolla*, **4**, 253.)

which are expected to be observable as a result of *explosive nucleosynthesis* (see Section 15.2, Volume 2).

These are important diagnostic tools for the process of nucleosynthesis in supernovae and are of particular interest in the context of the recent supernova SN1987A in the Large Magellanic Cloud. Certain of the γ-ray lines are expected to be observable soon after the explosion of the supernova. The decay of ^{56}Ni to ^{56}Co and then to ^{56}Fe is of special interest because the favoured theory of the origin of the supernova's light-curve is that it is associated with the radioactive decay of ^{56}Co which has exactly the correct half-life of 77 days to explain the exponential decay of the light of the supernova.

As soon as the supernova exploded, strenuous efforts were made to detect γ-ray lines of ^{56}Co from space missions and from dedicated balloon flights. These are

Table 5.2. *Some important nuclear γ-ray lines.*

Nucleus	Energy (MeV)
^{12}C	4.438
^{14}N	2.313
	5.105
^{16}O	2.741
	6.129
	6.917
	7.117
^{20}Ne	1.634
	2.613
	3.34
^{24}Mg	1.369
	2.754
^{28}Si	1.779
	6.878
^{56}Fe	0.847
	1.238
	1.811

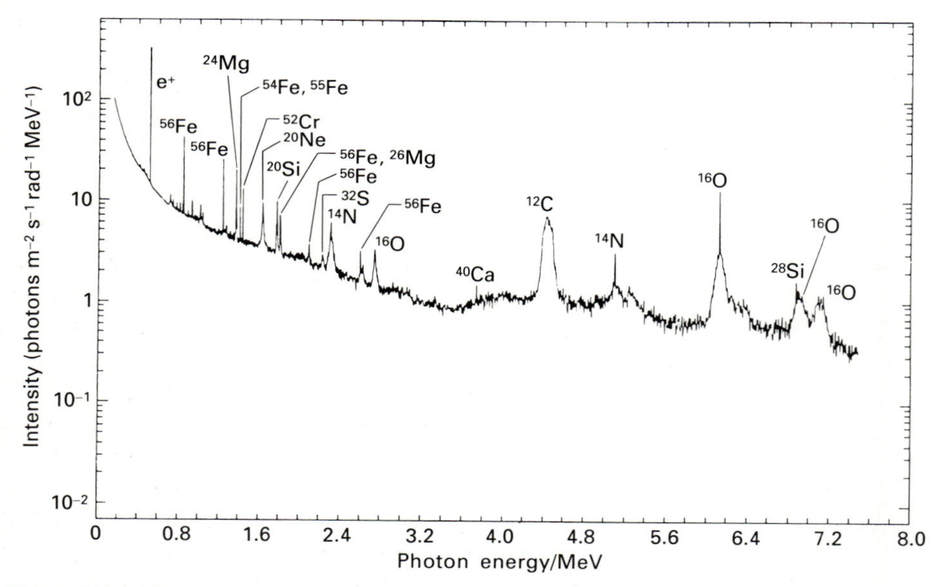

Figure 5.7. The predicted γ-ray spectrum resulting from low energy cosmic ray interactions with the interstellar gas in the general direction of the Galactic Centre. (From R. Ramaty and R. E. Lingenfelter (1979). *Nature*, **278**, 127.)

Table 5.3. *The radioactive decay chains resulting from explosive nucleosynthesis in supernovae.*

Decay chain	Mean life (years)	$Q/Q(^{56}\text{Ni})$	γ-ray energy (MeV)	Photons/positrons per disintegration
$^{56}\text{Ni} \rightarrow {}^{56}\text{Co} \rightarrow {}^{56}\text{Fe}$	0.31	1		0.2 (e^+)
			0.847	1
			1.238	0.7
$^{57}\text{Co} \rightarrow {}^{57}\text{Fe}$	1.1	2×10^{-2}	0.122	0.88
			0.014	0.88
$^{22}\text{Na} \rightarrow {}^{22}\text{Ne}$	3.8	5×10^{-3}		0.9 (e^+)
			1.275	1
$^{44}\text{Ti} \rightarrow {}^{44}\text{Sc} \rightarrow {}^{44}\text{Ca}$	68	2×10^{-3}		0.94 (e^+)
			1.156	1
			0.078	1
			0.068	1
$^{60}\text{Fe} \rightarrow {}^{60}\text{Co} \rightarrow {}^{60}\text{Ni}$	4.3×10^5	1.5×10^{-4}	1.332	1
			1.173	1
			0.059	1
$^{26}\text{Al} \rightarrow {}^{26}\text{Mg}$	1.1×10^6	1.5×10^{-4}		0.85 (e^+)
			1.809	1

$Q/Q(^{56}\text{Ni})$ is the predicted isotopic yield of each species relative to ^{56}Ni based upon Solar System abundances of the elements and the assumption that all the Solar System abundances of ^{56}Fe, ^{57}Fe and ^{44}Ca and 1%, 0.5% and 0.1% of the ^{60}Ni, ^{22}Ne and ^{26}Mg respectively are produced explosively through the above chains (Ramaty and Lingenfelter 1979).

notoriously difficult experiments because the fluxes of γ-rays are very low. Nonetheless, many observations have now been made of the 1238 and 847 keV lines of ^{56}Co and, although the detections of the lines have not been made with high signal-to-noise ratio, the evidence for their existence is convincing (for a review of these data, see Arnett, Bahcall, Kirshner and Woosley (1989)). Fig. 5.8 shows observations made by (*a*) the Solar Maximum Mission and (*b*) balloon observations carried out by the Jet Propulsion Laboratory group during a campaign in which the supernova was observed for 4.5 hours. Additional evidence for the presence of substantial quantities of cobalt and nickel in the supernova is provided by infrared spectroscopic observations in the 7–13 μm waveband in which the forbidden lines of [CoII] and [NiII] have been observed (Fig. 5.9). Analyses of these spectra indicate that the abundance of cobalt decreases with time when the envelope of the supernova becomes optically thin. The amount of radioactive nickel inferred to be created in the supernova to account for the observed mass of ^{56}Co after 370 days amounts to about 0.075 $M_\odot$. A similar figure is derived from studies of the amount of energy necessary to fuel the exponential

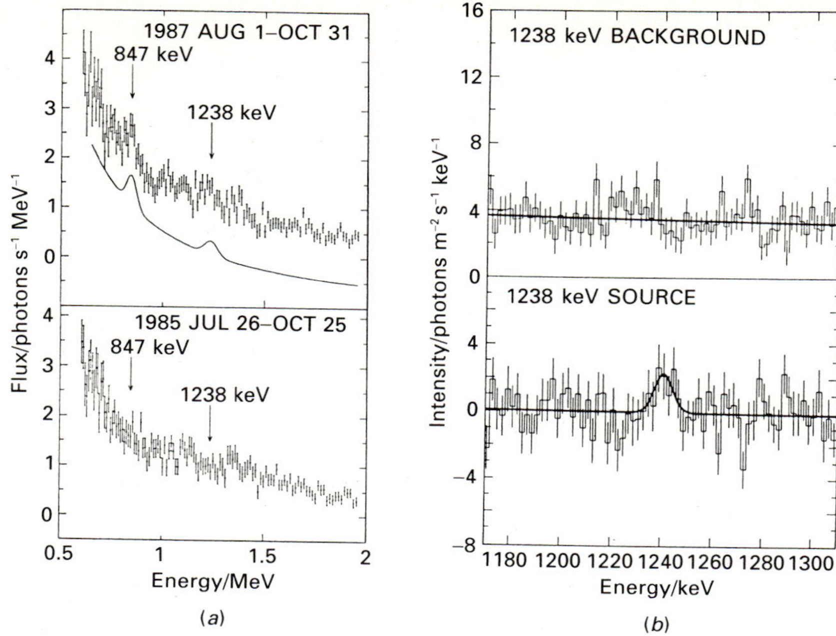

Figure 5.8. Observations of the γ-ray lines of ^{56}Co from the supernova SN1987A. (*a*) The background-subtracted spectrum accumulated between 1 August and 31 October 1987 by the Gamma-Ray Spectrometer on the Solar Maximum Mission. The expected profiles for the two ^{56}Co lines plus a power-law continuum are shown as a solid line. Also shown is an equivalent spectrum accumulated in 1985. The presence of an excess signal at the expected positions of both lines is apparent. (From S. M. Matz, G. H. Share, M. D. Leising, E. L. Chupp, W. T. Vestrand, W. R. Purcell, M. S. Strickman and C. Reppin (1988). *Nature*, **331**, 416.) (*b*) Balloon observations of the 1238 keV line of ^{56}Co made by the Jet Propulsion Laboratory group. The intrinsic width of the line is estimated to be 8.2 ± 3.4 keV. (From W. A. Mahoney, L. S. Varnell, S. A. Jacobson, J. C. Ling, R. G. Radicinski and W. A. Wheaton (1988). *Astrophys. J.*, **334**, L81.)

decay of the light of the supernova. We will have much more to say about the implications of these observations in Section 15.2, Volume 2.

The longer-lived radioactive isotopes are ejected into the general interstellar medium and should therefore be responsible for a diffuse flux of γ-ray line emission. For example, the isotopes ^{60}Co and ^{26}Al have half-lives of 4.3×10^5 and 1.1×10^6 years respectively and so they should be well mixed with the interstellar gas before a significant fraction of them have decayed into the stable isotopes listed in Table 5.3. In fact, a γ-ray line of ^{26}Al at 1.809 MeV has been observed from the interstellar gas by the HEAO-3 satellite (Fig. 5.10).

5.4 Nucleonic cascades

Just as in the case of the electron–photon cascades described in Section 4.5, we can now build models for the development of showers initiated by high

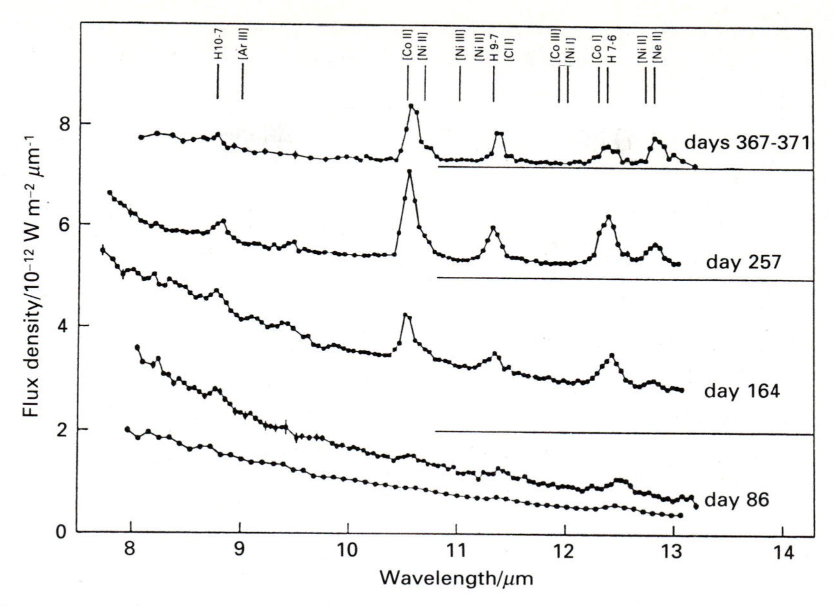

Figure 5.9. The development of the 8–13 μm spectrum of the supernova SN1987A during the first year. The positions of fine structure and hydrogenic lines are shown. The presence of strong lines of cobalt and nickel can be seen. (From D. K. Aitken, C. H. Smith, S. D. James, P. F. Roche, A. R. Hyland and P. J. McGregor (1988). *Mon. Not. R. Astron. Soc.*, **235**, 19.)

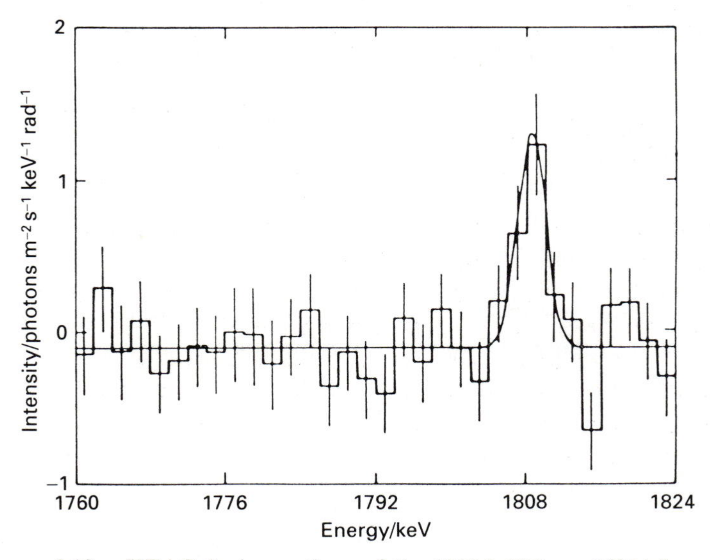

Figure 5.10. HEAO-3 observations of the 1809 keV line of ^{26}Al from the Galactic plane. The intensity is normalised to the intensity in the direction of the Galactic Centre. (From W. A. Mahoney, J. C. Ling. A. Wheaton and A. S. Jacobson (1984). *Astrophys. J.*, **286**, 578.)

energy nucleons. These are called *nucleonic cascades*. The salient features of the history of the products of nuclear collisions can be summarised as follows.

(i) The secondary nucleons and charged pions which have sufficient energy continue to multiply in successive generations of nuclear collisions until the energy per particle drops below that required for multiple pion production, i.e. about 1 GeV. This process is called a *nucleonic cascade* and in it the initial energy of the high energy particle is degraded into the energy of pions, strange particles and antinucleons. This process is sometimes referred to as *pionisation*.

(ii) The secondary protons lose energy by ionisation and most of those with energy less than 1 GeV are brought to rest.

(iii) The neutral pions π^0 are the easiest to deal with because they have very short lifetimes, 1.78×10^{-16} s, before decaying into two γ-rays, $\pi^0 \rightarrow 2\gamma$, each of which initiates an electromagnetic cascade (see Section 4.5). Many of the charged pions decay in flight into muons via the reactions

$$\left.\begin{array}{l}\pi^+ \rightarrow \mu^+ + \nu_\mu \\ \pi^- \rightarrow \mu^- + \bar{\nu}_\mu\end{array}\right\} \quad \text{mean lifetime} = 2.551 \times 10^{-8}\ \text{s}$$

The muons have virtually no nuclear interaction and are slowed down only by ionisation. The low energy muons have time to decay into positrons, electrons and muon neutrinos

$$\left.\begin{array}{l}\mu^+ \rightarrow e^+ + \nu_e + \bar{\nu}_\mu \\ \mu^- \rightarrow e^- + \bar{\nu}_e + \nu_\mu\end{array}\right\} \quad \text{mean lifetime} = 2.2001 \times 10^{-6}\ \text{s}$$

Many of the muons are produced with very high energy in the uppermost layers of the atmosphere before the pions have time to make further nuclear interactions and these muons are very penetrating indeed. Because they have virtually no nuclear interaction and their ionisation losses are small, the high energy muons arrive at the surface of the Earth intact. In their rest frames of reference, they decay with a mean lifetime of 2.2×10^{-6} s but, to the external observer, they are observed to decay with a mean lifetime of $2.2 \times 10^{-6}\gamma$ s because of relativistic time dilation, where γ is the Lorentz factor, $\gamma = (1 - v^2/c^2)^{-1/2}$. You may like to show that only muons with Lorentz factors $\gamma \geqslant 20$ survive intact to the surface of the Earth. This is the reason why observations of high energy muons at the surface of the Earth provide evidence for relativistic time dilation and length contraction. The high energy muons can penetrate quite far underground. This proves to be a very effective means of studying very high energy events and also of monitoring the average intensity and isotropy of the flux of cosmic rays arriving at the top of the atmosphere.

These interactions involving the products of nucleonic cascades are summarised in Fig. 5.11. We must remember that the development of nucleonic cascades depends

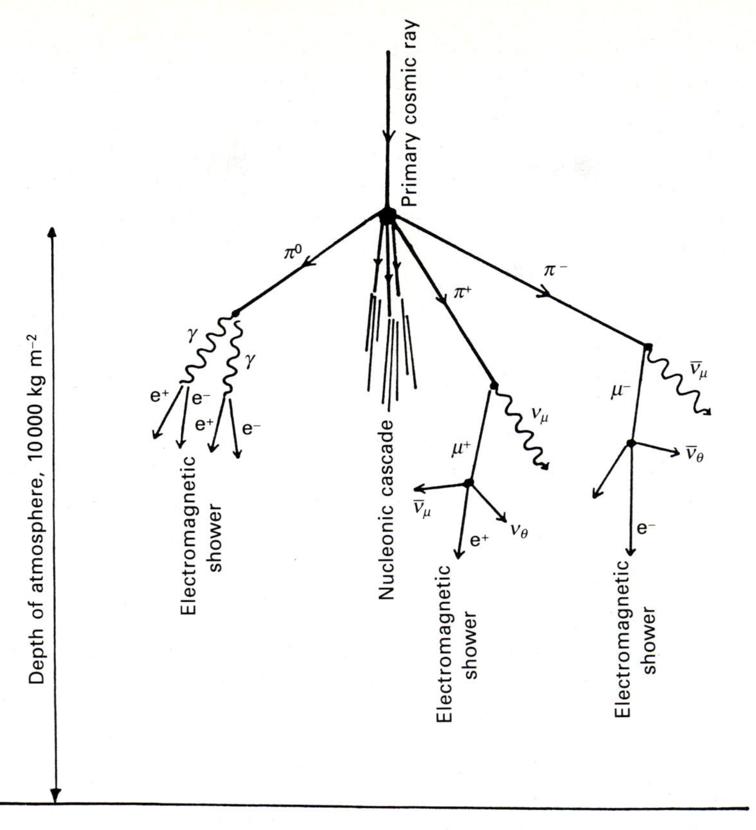

Figure 5.11. A schematic diagram showing the development of a nucleonic cascade in the atmosphere. Such cascades initiated by high energy particles develop in exactly the same fashion inside cosmic ray telescopes.

only upon the amount of matter through which the high energy particles have passed and so, if we build a detector which provides the same path length in kilograms per square metre we can ensure that these cascades develop within the detector volume. Most of the decay products are readily detectable by their ionisation losses. If there is sufficient depth to stop all the particles produced in the cascade, the total ionisation is a measure of the total energy of the primary particle. We will have much more to say about this topic later when we study cosmic ray telescopes and extensive air-showers.

5.5 Cosmic rays in the atmosphere

Let us now look at the observed distribution of high energy particles of all types in the atmosphere. This is summarised on a useful diagram presented by Hillas (1972) (Fig. 5.12). The figure shows the fluxes of different types of cosmic ray particles as observed at different heights in the atmosphere. These distributions are the result of the interactions of the flux of high energy particles incident on the

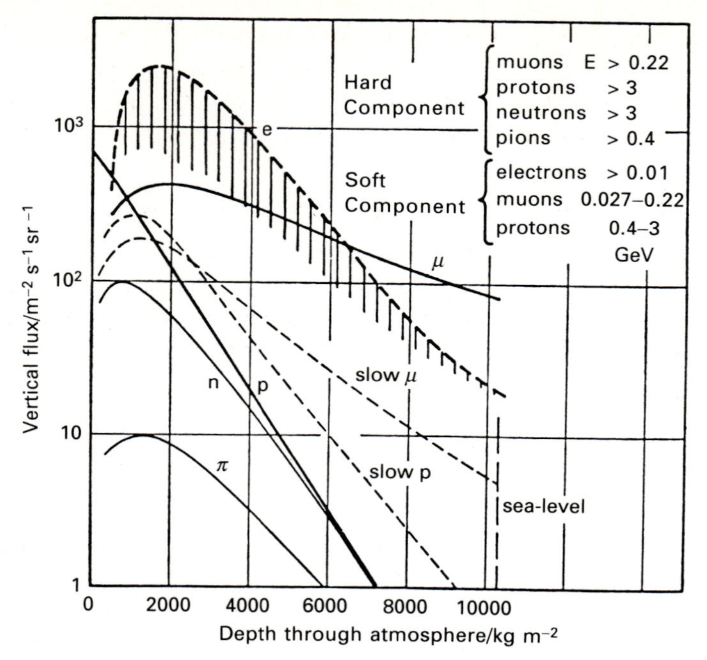

Figure 5.12. The vertical fluxes of different components of cosmic rays in the atmosphere. (From A. M. Hillas (1972). *Cosmic rays*, page 50, Oxford: Pergamon Press.)

top of the atmosphere with the molecules of the atmosphere and the subsequent nucleonic and electromagnetic cascades initiated by them. We can understand qualitatively all the features of this diagram in terms of the models of electromagnetic and nuclear cascades discussed above.

The bulk of the observed flux is caused by primary protons having energies $E \geqslant 1$ GeV. In principle, we could try to invert these distributions to derive the spectrum of the primary radiation but that effort is hardly worthwhile nowadays since the primary spectrum can be determined directly from satellite and balloon observations. Notice, in particular, the following points:

(i) The number of protons falls off exponentially and correspondingly the numbers of pions and neutrons.

(ii) The number of electrons grows exponentially to begin with – a characteristic of the electron–photon cascade – and then drops off rapidly. Notice that this means that, even at the very top of the atmosphere, there are large fluxes of secondary, relativistic, cosmic ray electrons and consequently there are problems in determining the primary spectrum of the cosmic ray electrons unless the detector is well outside the atmosphere.

(iii) The high energy muons fall off rather slowly but the low energy, or soft, muons have time to decay before reaching the surface of the Earth.

We will return to nucleonic cascades in the atmosphere when we have to deal with the most energetic cosmic rays. For the moment, we note that it is possible to develop proper models for these cascades in which the best available nuclear physics is included (see, for example Wolfendale (1973)).

5.6 Radioactive nuclei produced by cosmic rays in the atmosphere

One interesting aspect of these cosmic interactions is the production of radioactive isotopes in the upper atmosphere. We have not yet discussed what happens to the neutrons which are liberated in the spallation interactions. Most of the neutrons are eventually absorbed by ^{14}N nuclei through the reaction

$$^{14}N + n \rightarrow ^{14}C + ^{1}H$$

and about 5 % of the neutrons having energies greater than 4 MeV take part in the reaction

$$^{14}N + n \rightarrow ^{12}C + ^{3}H \text{ (endothermic)}$$

The total rate of formation of ^{14}C in the atmosphere is about 2.23×10^4 m^{-2} s^{-1} and that of tritons ^3H about 2×10^3 m^{-2} s^{-1}, this figure including the tritium which is formed directly as spallation products. These radioactive products are formed high in the atmosphere where they are rapidly oxidised to form molecules such as $^{14}CO_2$ and ^3HOH. These products are then precipitated in the normal way with CO_2 and H_2O in the atmosphere. The half-lives of ^{14}C and ^3H are 5568 years and 12.46 years respectively. The residence time of these products in the atmosphere is only about 25 years before they are absorbed in organic material or precipitated as rain and water onto the land and sea.

The abundances of ^{14}C and ^3H can then be used to date samples of material which contain residual organic matter. ^3H is used as a tracer in meteorological studies as well as being used to date agricultural products. ^{14}C is used extensively in archaeological studies and is the basis of radiocarbon dating. The success of the method depends upon the calibration of the ^{14}C ages against independently estimated ages of organic samples. The procedures for achieving this are surveyed by Damon, Lerman and Long (1978) who show that tree-ring dating (dendrochronology) provides a calibration of the ^{14}C scale back to times up to about 7500 years before the present day. The procedure is to measure the ratio of $^{14}C/^{12}C$ in the rings of very ancient trees for which reliable tree-ring ages can be determined. The best trees for this purpose are the bristlecone pines found in the Western USA. The ages of the tree rings are compared with radiocarbon ages derived assuming that the production rate of ^{14}C has been constant. The calibration curve presented by Damon *et al.* is shown in Fig. 5.13. It can be seen that the flux of cosmic rays at the top of the atmosphere must have been reasonably constant because there is an excellent correlation between the two ages. The ^{14}C ages are, however, slightly greater than they should be over the last 2500 years and then they systematically underestimate the ages at earlier times.

There is a very convincing explanation for this discrepancy. Paleo-geomagnetic studies have shown that the strength of the Earth's magnetic dipole has changed

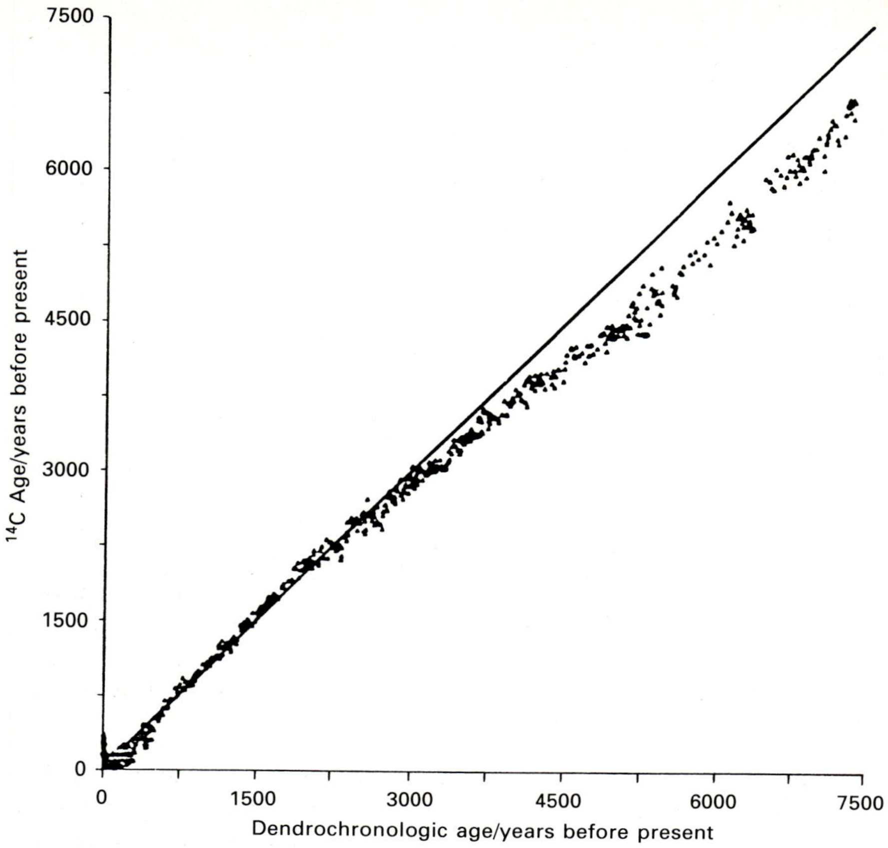

Figure 5.13. Radiocarbon dates compared with tree-ring dates. Both dates are quoted in terms of the number of decades before the present epoch. Most of these data are from bristlecone pine samples supplied by the Laboratory for Tree-ring Research, Arizona. (From P. E. Damon, J. C. Lerman and A. Long (1978). *Ann. Rev. Earth Planet. Sci.*, **6**, 457.)

significantly over the last 9000 years. Fig. 5.14 shows this variation (see Damon *et al.* (1978)). This variation of the Earth's magnetic field strength affects the flux of cosmic rays arriving at the top of the atmosphere because, as we will discuss in Chapters 10 and 11, the interstellar flux of high energy particles is impeded from arriving unmodified at the Earth by diffusion through the magnetic field in the interplanetary medium and the Earth's magnetic field. If the Earth's magnetic field strength is weaker, greater fluxes of high energy particles arrive at the top of the atmosphere where they create neutrons and consequently a greater production rate of ^{14}C is expected. Comparison of Figs 5.13 and 5.14 shows the anticorrelation between the flux of ^{14}C and the strength of the Earth's magnetic field very beautifully. When account is taken of the effect of variations in the Earth's dipole moment, the cosmic ray flux appears to have been remarkably constant over the last 10000 years.

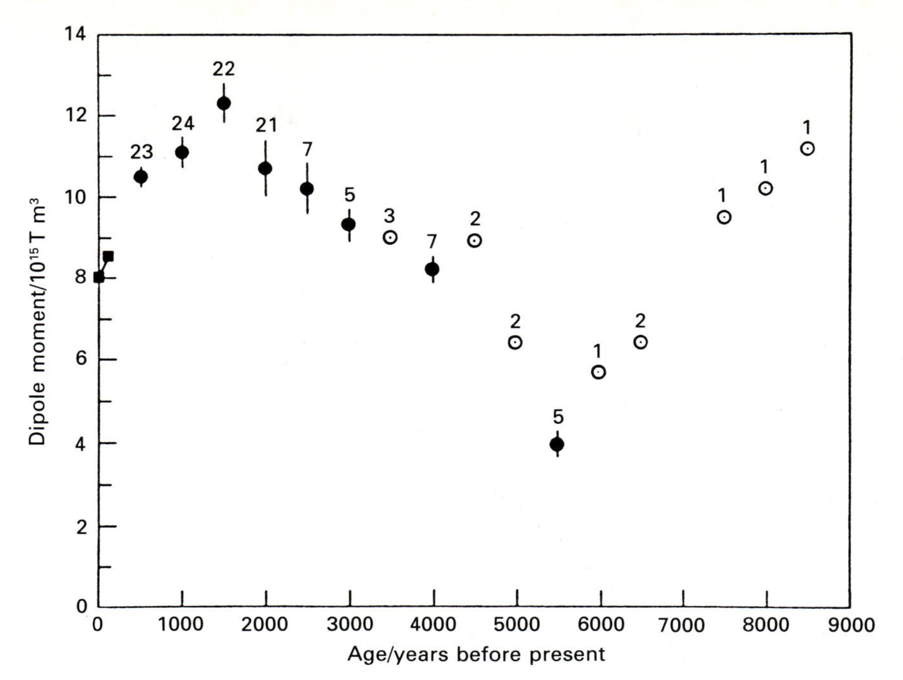

Figure 5.14. Variations in the magnitude of the Earth's magnetic dipole moment as determined by paleogeomagnetic measurements. (For discussion, see P. E. Damon, J. C. Lerman and A. Long (1978). *Ann. Rev. Earth Planet Sci.* **6**, 457.)

Even subtler effects can be found in the ^{14}C data. Fig. 5.15 shows deviations from the expected abundance of ^{14}C in tree-ring data once the correction is made for the effect of the change in incident cosmic ray flux due to the changing magnetic dipole moment of the Earth. The ordinate shows the deviations at the level of parts per thousand from the expected abundance ratio of ^{14}C. Shown along the top of the diagram are periods when there have been very low levels of solar activity as measured by low values of the sun-spot number. These periods are known as the Maunder minimum (AD 1654–1714), the Spörer minimum (AD 1416–1534) and the Wolf minimum (AD 1282–1342). It can be seen that the abundance of ^{14}C is greater during these periods. This has a natural interpretation in terms of the process known as *solar modulation* in which the irregularities in the interplanetary magnetic field which impede the passage of high energy particles to the Earth are greatest when the Sun is most active and are least when the Sun is quiet (see Chapters 10 and 11). It is remarkable that this effect shows up so clearly in the ^{14}C measurements of tree-rings.

I seem to have deviated very far from our main task which is to talk about high energy astrophysics but this is such a beautiful example of the interdisciplinary nature of much of modern research that it is irresistible to include it. There is a somewhat darker side to the story. During the period of atmospheric nuclear testing, the flux of ^{14}C increased by a factor of 2 within two years in the northern hemisphere because of the neutrons liberated in nuclear explosions. This has

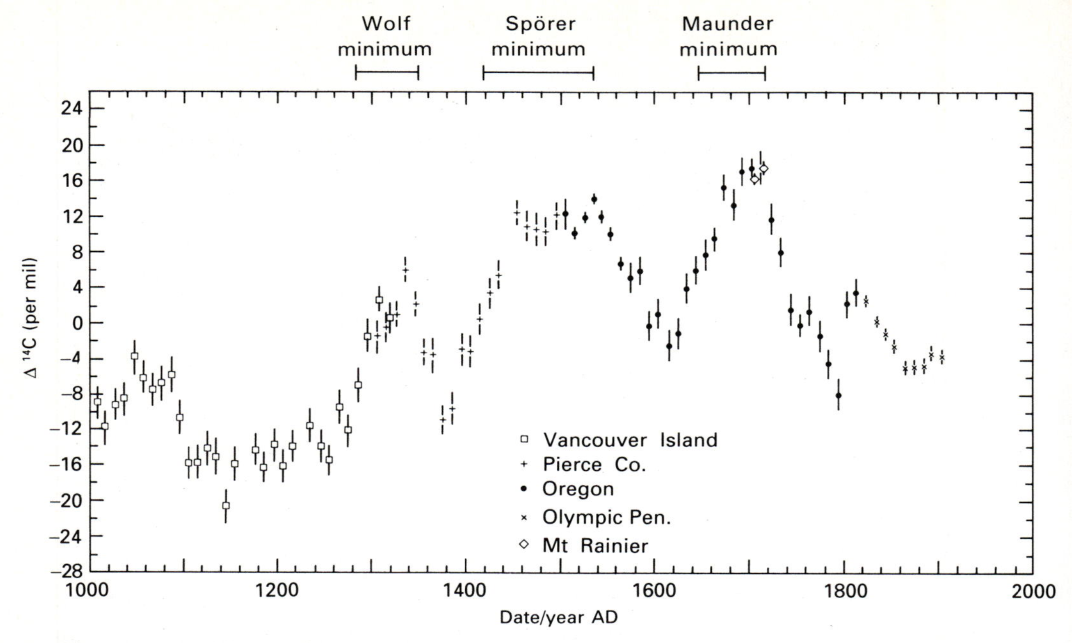

Figure 5.15. Changes in the atmospheric abundance of ^{14}C over the last 1000 years as determined by precise tree-ring measurements of the ^{14}C abundance. The ordinate scale shows deviations from the expected relative abundance of ^{14}C in parts per thousand. Long-term trends due to changes in the strength of the geomagnetic field have been removed. The periods of maximum ^{14}C abundance correspond to periods of low solar activity. (From M. Stuiver and P. D. Quay (1980). *Science*, **207**, 11.)

distorted significantly the very recent calibration curves for radioactive dating. In a somewhat lighter vein, it is an interesting calculation to estimate whether or not a nearby supernova would be detectable in the ancient tree ring data in an abrupt enhancement of the ^{14}C flux. Lingenfelter and Ramaty (1970) have estimated that a strong supernova liberating 10^{43} J would have had to have exploded within about 100 pc of the Sun to be detectable by this means. I leave this intriguing calculation to the reader.

6

Detectors for high energy particles, X-rays and γ-rays

6.1 Introduction

We can now use some of the results developed in the last four chapters to understand the practical design of detectors for high energy particles, X-rays and γ-rays. It is of particular importance to recognise the practical limitations of these devices since these determine the precision with which intensities and spectra can be measured. We will build these into complete telescopes in Chapter 7, recalling that these will often have to be flown on space vehicles. This immediately sets constraints upon the mass and volume of practicable space observatories.

6.2 Nuclear emulsions and the study of high energy particles

Nuclear emulsions are direct descendants of the photographic emulsions used by Röntgen in the discovery of X-rays and by Becquerel in the discovery of radioactivity but they are specifically designed to be sensitive to the electrons liberated by the ionisation losses of charged particles rather than to light and high energy photons. The activation energies of the emulsions are, however, similar. The nuclear emulsions consist of a high concentration of silver bromide (AgBr) crystals embedded in a matrix of gelatin. When a high energy particle enters the emulsion, its ionisation losses result in a stream of electrons along its path. The electrons activate the silver bromide crystals along the track of the particle and thus render them developable. During 'development', the activated grains are converted into grains of silver whilst the rest of the emulsion becomes transparent so that the track of the particle is revealed as a trail of developed grains. These photographic processes are complex and considered by many to be a black art – more details of them are certainly beyond the scope of this book but Appenzeller (1989) provides a useful summary of the basic physical and chemical processes involved.

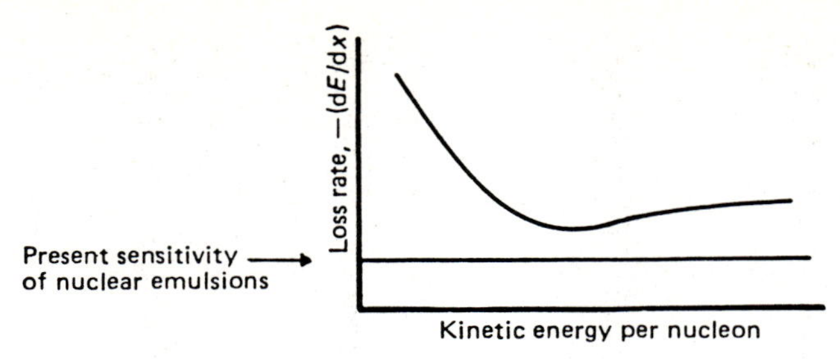

Figure 6.1. A schematic diagram showing the present sensitivity of nuclear emulsions to protons compared with the loss rate due to ionisation losses.

The sensitivities of modern emulsions are such that even a charged particle of minimum ionisation can produce a developable track (Fig. 6.1). Of course, we may not want to see every track and then less sensitive emulsions can be used as is the case, for example, if we only wish to study particles of large atomic number. One of the big advantages of nuclear emulsions is that they are dense materials, about 4000 kg m^{-3}, and therefore it is possible to obtain considerable path lengths through them. It is, however, difficult to develop thick slabs of emulsion uniformly and, in addition, the emulsions are not strong materials. It has proved possible, however, to make very thin sheets which are self-supporting and hence long path lengths through the emulsion can be obtained by constructing a large stack of these layers. These emulsion stacks are referred to as *stripped emulsions*. They are aligned by firing an X-ray through the corners of the stack and then, after exposure, the layers are peeled off and developed separately. This is the procedure which resulted in the discovery of many of the short-lived particles in the 1940s and early 1950s (see Section 1.3). Because of the large path lengths obtainable, complex nuclear interaction chains could be observed within the emulsion stacks themselves. Normally, the tracks can only be seen under a microscope and consist of rows of developed grains each about 0.5–0.8 μm in diameter. Fig. 6.2(*a*) shows sections of the track of an iron nucleus which is brought completely to rest in the emulsion. The photographs show the track at various *residual ranges* i.e. the distance the particle has yet to travel to come to rest. In Fig. 6.2(*b*), the positions along the track are compared schematically with the expected ionisation loss rate.

The problem of obtaining quantitative information from these tracks is immediately obvious. There are many delta-rays and it is clearly not a straightforward matter to determine the average amount of ionisation along the track. Empirically, it is found that the density of grains is proportional to the energy loss in the matter and hence, if we count all the exposed grains, we obtain a measure of dE/dx. If the tracks are too heavily exposed, one can use, alternatively, the width of the track but then one has to be careful about excluding delta-rays. The most accurate work involves measuring the ionisation at all points along the tracks but this is a very laborious procedure. Nowadays, automatic measuring machines are used but, even so, it remains a laborious task. In the most careful analyses, resolution of 0.2 in atomic number z has been achieved and this is as good as has been achieved by any means. One of the techniques which makes

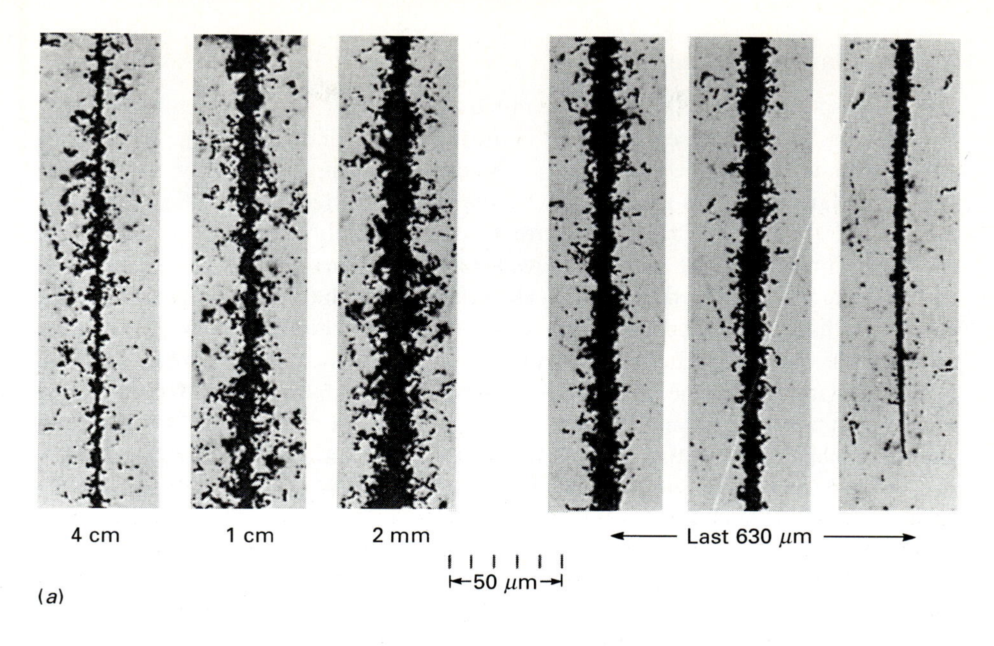

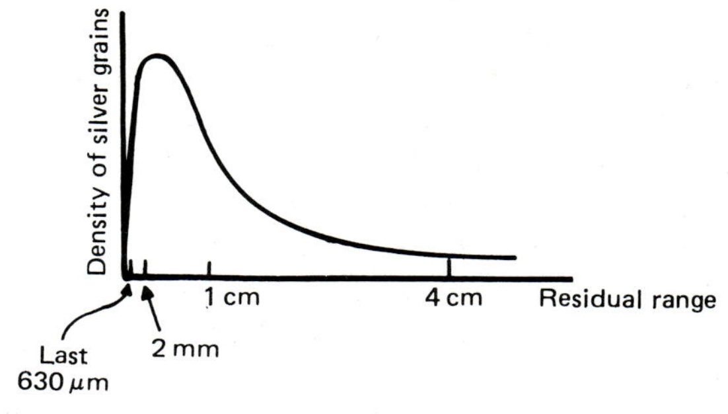

Figure 6.2. (*a*) Nuclear emulsion photographs of the track of a cosmic ray iron nucleus at various stages in its deceleration from relativistic velocities to rest. The distances are the residual ranges at which the track is observed; these positions are shown schematically in (*b*). (Photograph from M. M. Shapiro and R. Silberberg (1970). *Ann. Rev. Nucl. Sci.*, **20**, 328.)

this accuracy possible is the use of emulsions of different stopping power interleaved in the same emulsion stack. This also improves the dynamic range of the measurements.

If the composition of the primary cosmic rays is to be measured, it is essential to fly these emulsions in high altitude balloons or else to take them on manned space expeditions. It is important to remember that in this technique it is essential to retrieve one's apparatus! More details of these techniques will be found in the classic volume by Powell, Fowler and Perkins (1959) and in the review article by Shapiro (1958).

6.3 Plastics and meteorites

As long ago as Chapter 2, we noted that another way in which high energy particles make their presence known is through the radiation damage which they cause in materials. If the particles are sufficiently ionising, they leave permanent radiation damage tracks and it is found that a wide range of insulating solids are affected in this way. The key feature is that only above a certain critical or threshold ionisation rate is the damage permanent. These tracks can be revealed by using the fact that the damaged areas of the material have much higher chemical reactivity than undamaged areas and therefore, by etching, one can identify the path of the particle without dissolving away all the material. To make a good detector, a material is needed which suffers as much damage as possible by the incident particle. Experience shows that high polymers are best for this purpose, the 'theory' behind this choice being that, because they are complicated molecules, they can be disrupted and wrecked in the most interesting ways, producing displaced atoms, broken molecular chains, free radicals, etc. More details of this subject and its practical applications can be found in the reviews by Price and Fleischer (1971), Fleischer, Price and Walker (1975), by Lal (1972, 1977) and by Reedy, Arnold and Lal (1983).

6.3.1 *Radiation damage in plastics*

There is no theory of track formation but the damage can be modelled in the following way. We seek a parameter which is a measure of the density of damage as a function of the charge and velocity of the incident particle. Empirically, it is found that the radiation damage density J can be satisfactorily described by a formula similar in form to that which describes ionisation losses, with the appropriate corrections at high energies.

$$J = a\frac{Z^2}{v^2}\left[\ln(\gamma^2 v^2) - \frac{v^2}{c^2} + K - \delta\left(\frac{v}{c}\right)\right] \tag{6.1}$$

This formula is used empirically and little physical significance should be attributed to the various terms. The constants are merely parameters to be fitted to the experimental data on radiation damage. This relationship enables the sensitivity of different materials to radiation damage to be described as illustrated in Fig. 6.3, which shows the radiation damage rates for a wide range of different materials. These range from the minerals found in meteorites, through mica, Lexan polycarbonate to daicellulose nitrate which is one of the most sensitive materials yet used. A key consideration for the success of these materials as particle detectors is their uniformity and this has now been achieved for even the most sensitive materials. The choice of material restricts the range of elements and energies of particles which can be detected. For Lexan polycarbonate, for example, relativistic nuclei heavier than iodine can be detected, but only iron nuclei with velocities less than about $0.4c$ can register permanent tracks.

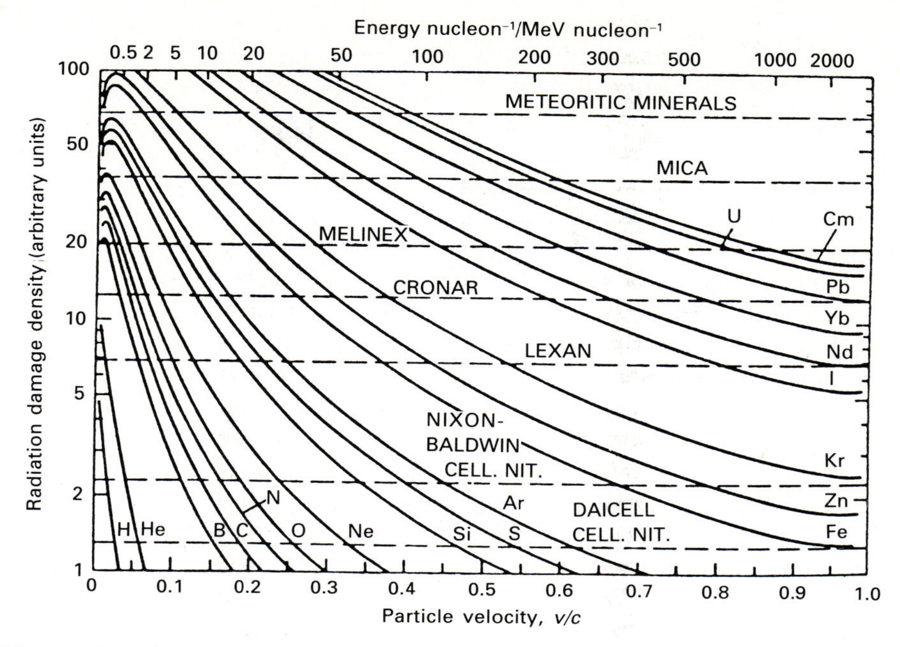

Figure 6.3. The density of radiation damage (or ionisation rate J) as a function of velocity for different incident nuclei. Approximate thresholds at which permanent tracks are formed in various materials and minerals are indicated by dashed lines. (From P. B. Price and R. L. Fleischer (1971). *Ann. Rev. Nucl. Sci.*, **21**, 295.)

How are these materials used as detectors? The technique is to measure the etching rate V_E as a function of the radiation damage density J. For example, for Lexan polycarbonate, it is found that $V_E \propto J^2$. Therefore, by measuring the etching rate as a function of position along the track, the radiation damage density can be measured directly and the atomic number of the particle can be found. Fig. 6.4(*a*) illustrates how this procedure is carried out in practice. A number of very thin sheets of the material are stacked together and, when an ionising particle enters the stack, it does not cause permanent damage until it has slowed down to the critical ionisation rate for the material. After that point, it leaves a permanent record of its passage until it is brought to rest. The radiation damage as a function of distance into the material is shown schematically in Fig. 6.4(*b*). The sheets are then separated and the surfaces are etched. Since the etch rate increases along the path, etch cones of the form shown in Fig. 6.4(*a*) are observed. For the same etching time, the depth of the etch along the track is $\int V_E \, dt$ which just measures the average etch rate along that section of track and this is related to the average radiation damage rate at that point.

Just as in the case of nuclear emulsions, the results are best plotted as a graph of etch rate against residual range, i.e. how far the particle has to travel before it is brought to rest. Just as in the case of ionisation losses, the range is a function of the mass of the particle and hence isotopes of the same element can be

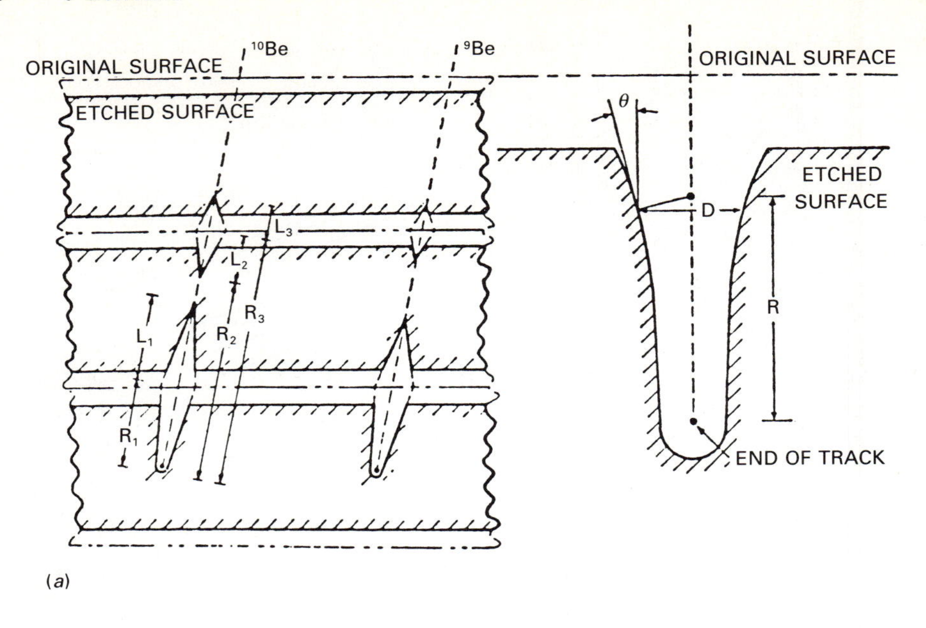

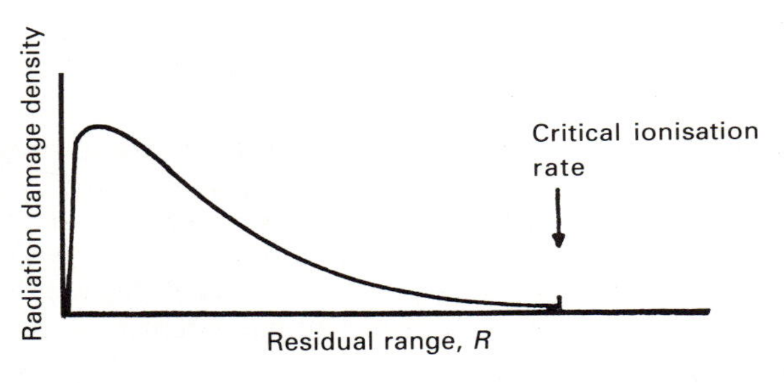

Figure 6.4. (*a*) The principles of particle identification by etch rate measurement in particle track detectors. The radiation damage density at different residual ranges R can be determined by measurements of either the etched cone length L, the taper angle θ or the diameter D of the track. (From P. B. Price and R. L. Fleischer (1971). *Ann. Rev. Nucl. Sci.*, **21**, 295.)
(*b*) A schematic diagram showing the radiation damage density as a function of residual range.

distinguished. Fig. 6.5 shows what can be achieved in laboratory experiments using daicellulose nitrate – the isotopes of boron, ^{10}B and ^{11}B, can be clearly distinguished.

The results of a balloon flight of 1969 are shown in Fig. 6.6. The experiment consisted of a large stack of plastics, emulsions, etc and was flown for 80 hours at altitude. Seven nuclei with charges greater than iron were detected. It can be seen that some very heavy elements have survived the journey through interstellar space and that one of them may well have been a uranium nucleus. This has proved to be one of the best methods of detecting heavy cosmic rays. On the Apollo space

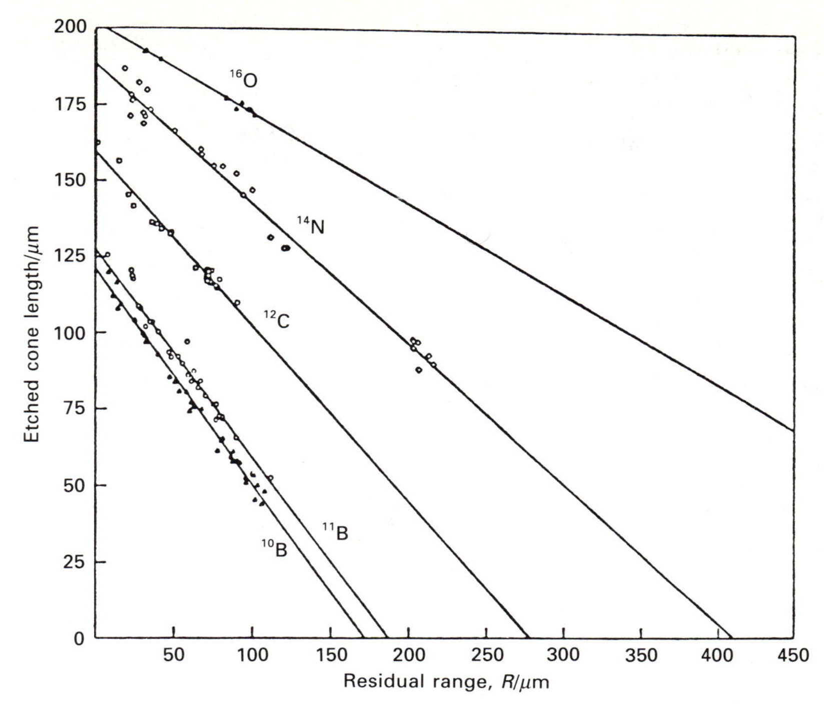

Figure 6.5. Measurements of etched cone length plotted against residual range for ions of ^{10}B, ^{11}B, ^{12}C, ^{14}N and ^{16}O in daicellulose nitrate. In these experiments ion beams were produced in an accelerator and the isotopes of individual elements were clearly resolved. In cosmic ray experiments it is difficult to achieve such resolution. (From P. B. Price and R. L. Fleischer (1971). *Ann. Rev. Nucl. Sci.*, **21**, 295.)

missions, right up to Apollo 17, plastic sheets were exposed on the Moon's surface. When the astronauts from Apollo 12 brought back the camera from the Surveyor satellite, which had landed on the Moon's surface two years earlier, etchable tracks were found in the filters of the camera.

Another useful parameter is the *total etchable length* of the cosmic ray track. Since the track only causes permanent damage below a certain velocity, the residual path length of the track is a unique function of the charge of the particle for a particular material, except for small isotopic variations. Thus, the total track length gives a quick estimate of the charge of the cosmic ray (Fig. 6.7).

6.3.2 *Meteorites*

The concept of *total etchable length* leads naturally to the idea that materials found in nature exposed to cosmic rays should contain similar tracks. Fig. 6.3 shows that meteoric materials are sensitive to cosmic rays heavier than about iron. Similar analyses can be made of lunar samples of rocks which have been exposed to the cosmic rays.

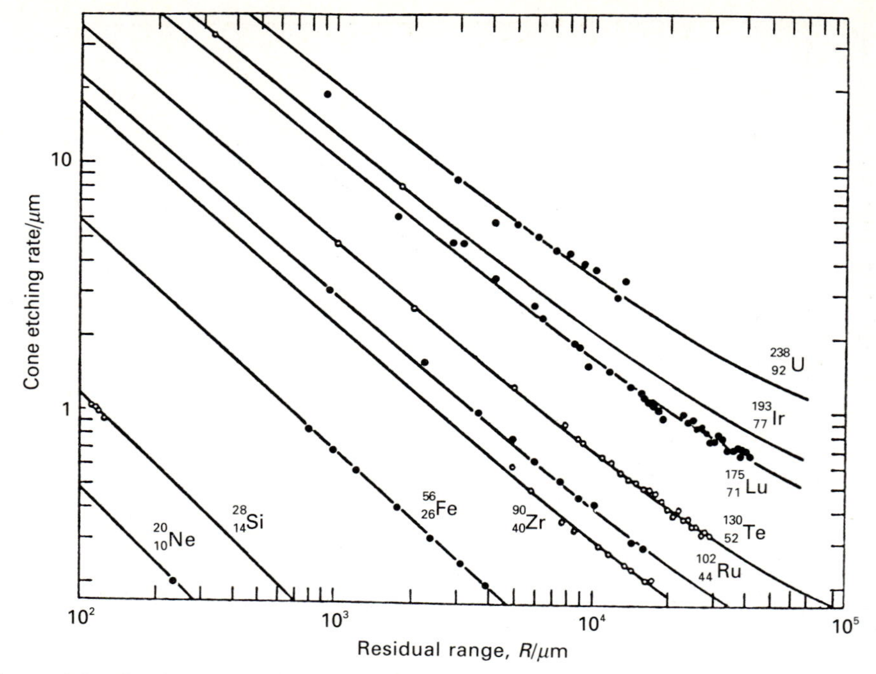

Figure 6.6. Studies of very heavy nuclei using the method of radiation damage in plastics. The neon and silicon data are averages of measurements of many tracks from accelerator calibrations. The iron data represent the spread in measurement of about 50 stopping nuclei. The data points for the six extremely heavy nuclei detected have etch rates measured at many positions along their trajectories in a large stack of Lexan polycarbonate. (From P. B. Price and R. L. Fleischer (1971). *Ann. Rev. Nucl. Sci.*, **21**, 295.)

The study of meteorites is an enormous subject and provides many important clues about the early history of the Solar System. We can only summarise very briefly some of the important points and indicate how the cosmic rays which enter the Solar System have a key role to play in determining the ages of the meteorites. In addition, they provide important constraints on variations in the interstellar flux of cosmic rays over timescales up to 10^9 years or more. There is an enormous literature on meteorites and an excellent survey of the whole field is contained in the compendium of authoritative reviews entitled *Meteorites and the early Solar System* (J. F. Kerridge and M. S. Matthews, 1988). Shorter surveys are contained in the books by Dodd (1981) and Wasson (1985).

Meteorites are interplanetary rocks which succeed in reaching the surface of the Earth without being completely vaporised by ablation in the Earth's atmosphere. The material of the meteorites is as old as the Solar System itself, i.e. about 4.6×10^9 years old. It is inferred that the parent bodies of the meteorites formed in the very early Solar System and it is probable that the asteroids which form the broad asteroid belt between Mars and Jupiter are these meteoritic parent bodies. The meteorites themselves are formed by fragmentation of these asteroids,

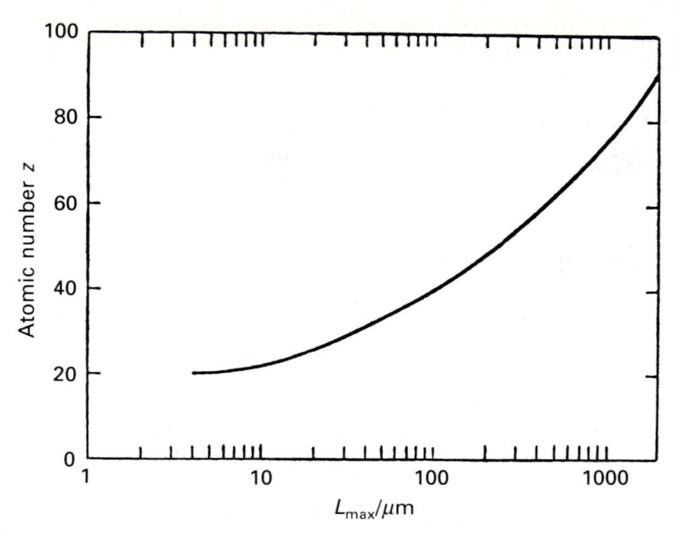

Figure 6.7. The dependence of maximum etchable path length upon the charge of the cosmic ray for tracks in the mineral pyroxene; L_{max} is given by the integral $L_{max} = \int_{E_0}^{0} 1/(dE/dx)\, dE$ where E_0 is the critical energy for permanent radiation damage. (From Price and Fleischer (1971). *Ann. Rev. Nucl. Sci.*, **21**, 295.)

probably in collisions between two asteroidal bodies. It is only when the meteorites are broken off from their parent bodies that they are exposed to the flux of cosmic rays.

The cosmic rays play a key role in two related but separate ways. First of all, the meteorites contain crystals which behave in the same way as the plastic materials described in Section 6.3.1 in that when they are bombarded with high energy particles, etchable tracks are created within the body of the crystals. Although the volume of the crystals in the meteorites is very small, the exposure times to the cosmic rays can be very long and hence they provide information about the constancy or otherwise of the cosmic ray flux over very long time periods.

Etching techniques are used to reveal the fossil tracks of cosmic rays, the etchant seeping through very fine faults in the crystals which are then rendered visible by silvering. The examples shown in the Fig. 6.8 include (*a*) a meteoritic sample and (*b*) one from a sample of lunar rock brought back by the Apollo 14 astronauts. The second picture shows many short tracks due to iron nuclei. There are, however, other much longer tracks which must be associated with elements with atomic numbers greater than that of iron. In this way the distribution of etchable path lengths provides information about the mass spectrum of cosmic ray nuclei heavier than iron.

The particles responsible for forming the tracks may be either Galactic cosmic rays or else high energy particles accelerated in solar flares. The distinction between these two types of cosmic rays is that the solar cosmic rays are generally of very much lower energy than the Galactic cosmic rays, very few indeed being observed with energies greater than 1 GeV. Consequently, they penetrate less than

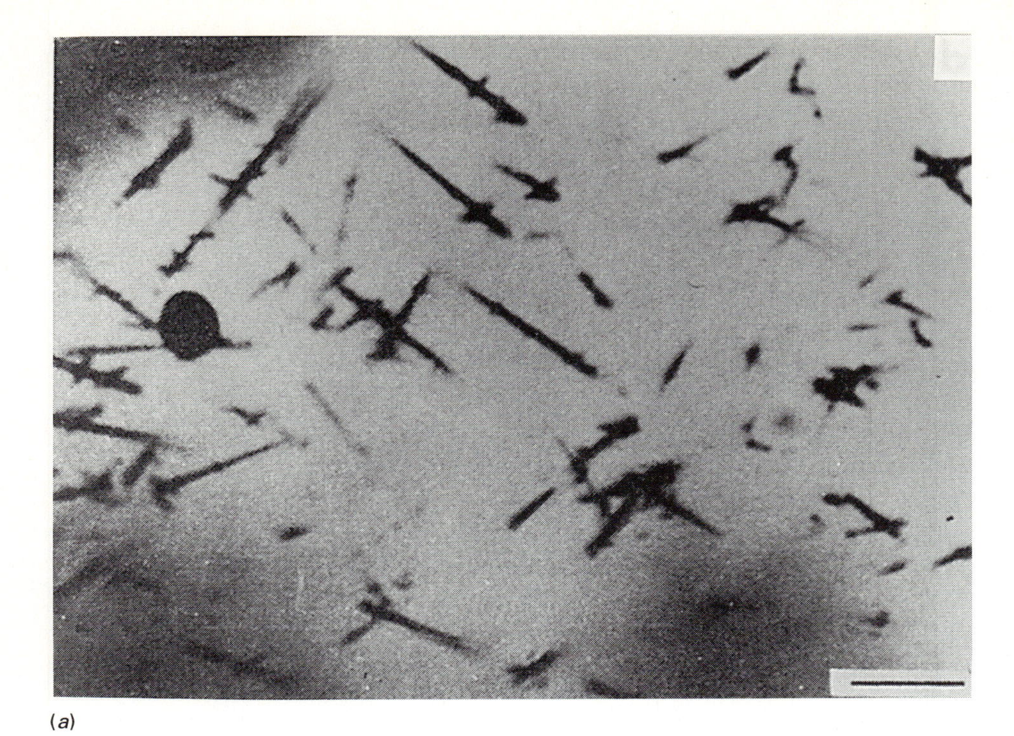

(a)

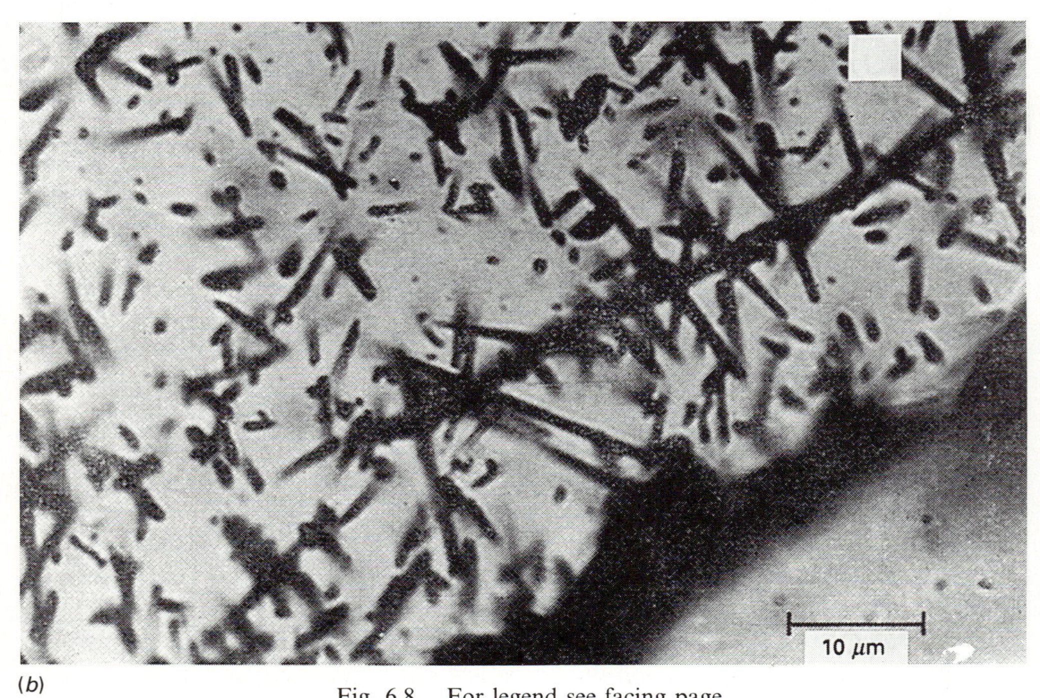

(b)

Fig. 6.8. For legend see facing page.

a few millimetres beneath the surface of the meteorite. In contrast, the Galactic cosmic rays have very much higher energies and they can penetrate much more deeply into the meteorite. The tracks detected at depths greater than 1 cm into the meteorite are certainly of Galactic origin (Fig. 6.9).

The second way in which the cosmic rays provide crucial information is through the spallation products which they induce in the material of the meteorite. In these spallation reactions, the data can be used in two quite separate ways. The spallation products produced by high energy cosmic rays entering the meteorite not only produce lighter elements as indicated in Table 5.1, they also release neutrons which can interact with the nuclei of the minerals to produce some rather rare isotopes trapped inside the meteorite. The clever thing to do is to study the abundance of those isotopes which are known to be very rare in nature and which can only have been produced by spallation inside the meteorite. Of particular value are the stable elements since they simply continue to increase in abundance with age. Good examples of stable nuclei produced as cosmogenic nuclides include rare isotopes such as ^{3}He, ^{21}Ne and ^{38}Ar. For example, Wasson (1985) quotes rates of formation of ^{3}He and ^{21}Ne of $2 \times 10^{-17}\rho$ and $3.5 \times 10^{-18}\rho$ particles per year respectively where ρ is the density of the material of the meteorite in kilograms per cubic metre, assuming the present intensity of the interstellar flux of cosmic rays.

A second way of using the same process is to study those spallation products which are radioactive. The process of spallation accounts for the persistence within the meteorites of some isotopes which have short half-lives such as tritium ^{3}H, ^{14}C and ^{10}Be, which have half-lives of 12.5, 5.6×10^{3} and 2.5×10^{6} years respectively, as well as a host of more obscure radioactivites. Table 6.1 is a list of cosmic ray induced radio nuclides which have been frequently measured in terrestrial and extraterrestrial matter (from Reedy, Arnold and Lal (1983)). This table includes the principal target nuclei as well as an indication of the source of the high energy particles which are responsible for their formation, GCR meaning Galactic cosmic rays and SCR solar cosmic rays.

These two techniques can be used to provide estimates of the exposure ages of the meteorites to the cosmic rays. The remarkable thing is that these procedures work as well as they do indicating that the intensity of the local interstellar flux of cosmic rays must have been more or less the same over very long periods of time. The exposure ages of the cosmic rays are all significantly less than the age of the Solar System. Many of the meteorites must have fragmented from their parent bodies more than about 10^{7} years ago and there is an age distribution which extends up to 10^{9} years and more.

Figure 6.8. Photomicrographs of tracks of heavy elements in meteoritic and lunar samples. (*a*) A typical example of the tracks seen in meteoritic crystals. Most of these tracks are iron nuclei. (From M. W. Caffee, J. N. Goswami, C. M. Hohenberg, K. Marti and R. C. Reedy (1988). *Meteorites and the early Solar System*, (eds J. R. Kerridge and M. S. Matthews), page 205, University of Arizona Press). (*b*) Tracks in lunar feldspar from lunar rock 14310 showing the large number of iron tracks as well as one of a much heavier nucleus. (From D. Lal (1972). *Sp. Sci. Rev.*, **14**, 3.)

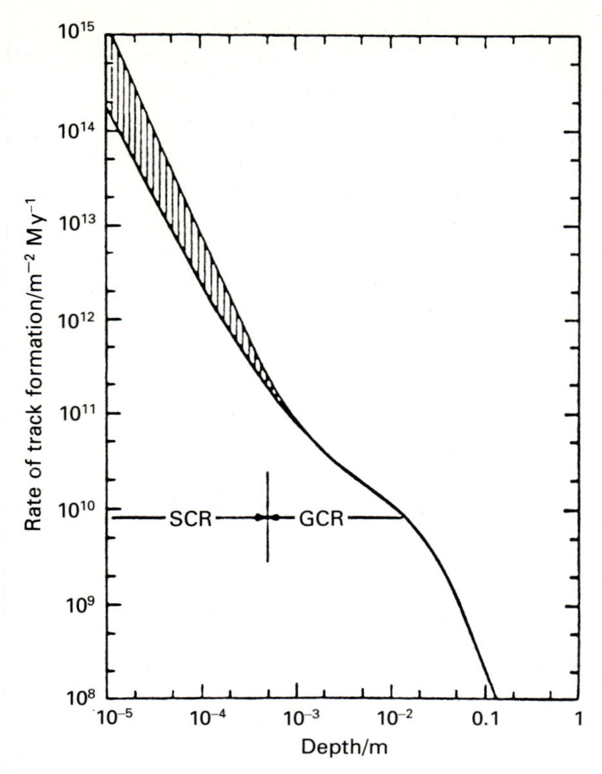

Figure 6.9. Predicted rates for the production of etchable tracks as a function of depth in a lunar rock sample directly exposed to cosmic ray particles. The etchable tracks are caused by nuclei with atomic number Z greater than 20 (see Fig. 6.3). The shaded areas reflect uncertainties in the fluxes of low energy nuclei in the solar cosmic rays. (From R. C. Reedy, J. R. Arnold and D. Lai (1983). *Ann. Rev. Nucl. Sci.*, **33**, 505.)

According to Reedy *et al.* (1983), these studies show that the cosmic ray flux must have been within about 50% of its present value over the last 10^9 years. A literal interpretation of the results suggests that over the last 10^7 years, the flux of cosmic rays has been about 50% greater than it was during the preceding 10^9 years. Thus, remarkably, it seems that our Solar System has been bombarded by roughly the same flux of cosmic rays that it experiences now for the last billion years.

In addition to studying the cosmic ray flux, many other clever things can be done. For example, the shape of the meteorite before it suffered ablation on passing through the atmosphere can be determined. Although the outline of the meteorite may now be very ragged, the equal density contours of the spallation products are often found to be smooth and these can be used to extrapolate back to the original shape of the meteorite (Fig. 6.10).

Table 6.1. *Radioactive nuclides created by spallation in meteorites*

Radionuclide	Half-life (years)	Main targets	Particles
^{3}H	12.323	O, Mg, Si	GCR, SCR
^{10}Be	1.6×10^{6}	O, Mg, Si, (N)	GCR
^{14}C	5730	O, Mg, Si, (N)	GCR, SCR
^{22}Na	2.602	Mg, Al, Si	SCR, GCR
^{26}Al	7.16×10^{5}	Al, Si, (Ar)	SCR, GCR
^{32}Si	105	(Ar)	GCR
^{36}Cl	3.0×10^{5}	Ca, Fe, (Ar)	GCR
^{37}Ar	35.0 days	Ca, Fe	GCR, SCR
^{39}Ar	269	K, Ca, Fe	GCR
^{40}K	1.28×10^{9}	Fe	GCR
^{46}Sc	83.82 days	Fe, Ti	GCR
^{48}V	15.97 days	Fe, Ti	GCR, SCR
^{53}Mn	3.7×10^{6}	Fe	SCR, GCR
^{54}Mn	312.2 days	Fe	SCR, GCR
^{55}Fe	2.7	Fe	SCR, GCR
^{56}Co	78.76 days	Fe	SCR
^{59}Ni	7.6×10^{4}	Fe, Ni	GCR, SCR
^{60}Co	5.272	Co, Ni	GCR
^{81}Kr	2.1×10^{5}	Sr, Zr	GCR, SCR
^{129}I	1.6×10^{7}	Te, Ba, La, Ce	GCR

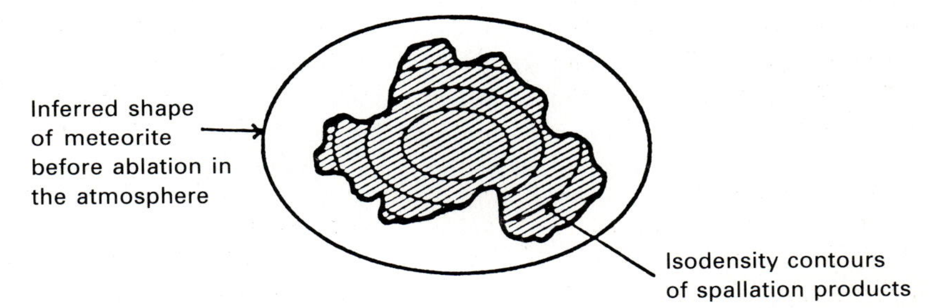

Figure 6.10. A schematic diagram showing how the original shape of a meteorite before ablation in the atmosphere can be inferred from the isodensity contours of spallation products.

6.4 Gas-filled detectors – proportional counters, Geiger counters and spark chambers

The detectors used in high energy astrophysical experiments are based upon those developed for laboratory experiments with the important difference that it is often necessary to fly them on space vehicles and therefore there are weight restrictions upon the size of the packages which can be built. This makes instruments such as cloud and bubble chambers unsuitable for space flight. *Gas-*

filled detectors have, however, been successfully developed for space experiments and have played a key role in the development of cosmic ray, X and γ-ray astronomy. These detectors may be thought of as modern versions of the Crookes tubes which played such an important role in the development of understanding of atomic structure. The typical arrangement for studying the properties of gas-filled discharge tubes is shown in Fig. 6.11 (*a*). When a high energy particle passes through the tube, it suffers ionisation losses resulting in the creation of numerous ion–electron pairs along its track whereas, when an X or γ-ray enters the tube, an energetic electron–ion pair is created at a single point by photoionisation.

The output voltage pulse from the tube varies with bias voltage V as shown in Fig. 6.11 (*b*) which compares the responses resulting from an electron and an α-particle passing through the tube. The bias voltage across the tube is arranged so that the electrons are attracted to the anode which is the thin wire running down the centre of the tube. The behaviour of the tube may be understood as follows. At very low voltages (region A), the electrons recombine and few of those liberated by the passage of the charged particle reach the anode. As the voltage increases, a steady state is set up whereby the electrons are accelerated to sufficiently high energies for recombination to be negligible (region B).

Then, in regions C and D, the electrons acquire sufficient energy to create new electron–ion pairs in collisions with the gas in the vacuum tube. The net result is an avalanche of electron–ion pairs which produces a large voltage pulse as measured by the counter circuitry. In order to obtain a good estimate of the energy loss by the high energy particle or the X or γ-ray as it passes through or is absorbed by the gas, as large an output voltage pulse as possible is needed so that an accurate estimate of the number of electron–ion pairs created is obtained. Eventually, if the voltage across the tube is too large, the response of the tube becomes non-linear and the output voltage is no longer proportional to the amount of ionisation created in the tube. The regime in which the proportionality between the amount of ionisation created by the incoming particles or photons and the strength of the output pulse is maintained is called the *proportionality regime* the devices which operate in this mode are called *proportional counters*.

At the highest voltages (regime E), the voltage is so high that a particle causing even the minimum ionisation produces a huge voltage pulse – the tube is running at saturation level and the whole tube is rapidly ionised by the passage of a single particle. Detectors operating in this regime are known as *Geiger counters* or *Geiger–Müller counters* and were used by Rutherford and Geiger in their classical experiments on the Coulomb scattering of α-particles. In practical detectors of this type, the sensitive volume contains a gas which enables the tube to recover quickly to respond to the arrival of the next ionising charge, i.e. after the voltage pulse, the ionisation is rapidly 'quenched'. These detectors find important application in *spark chambers* as we describe below.

Because proportional counters contain a gas rather than a solid, the path length through them is short and so they find their principal application as detectors for X-rays with energies $0.1 \leqslant \hbar\omega \leqslant 20$ keV. A schematic diagram of a proportional counter for use in X-ray astronomy is shown in Fig. 6.12. The X-ray enters through

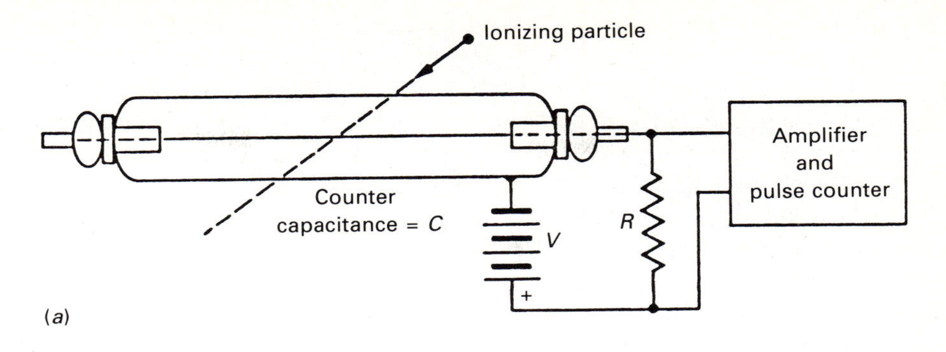

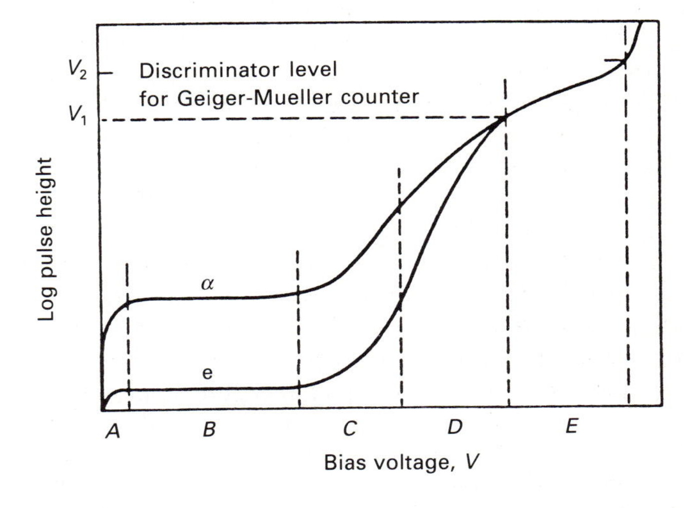

Figure 6.11. (*a*) A gas-filled discharge tube showing the arrangement used to detect pulses of current initiated by ionising particles passing through the gaseous volume. (From H. A. Enge (1966). *Introduction to nuclear physics*, page 201, London: Addison-Wesley Publishing Co.)
(*b*) The magnitude of the output pulse of the discharge tube shown in (*a*) as a function of the bias voltage V. The two curves show the response to a fast electron, e, and to a helium nucleus α. The response in the different regions A, B, C, D, and E are described in the text. (From H. A. Enge (1966). *Introduction to nuclear physics*, page 201, London: Addison-Wesley Publishing Co.)

the window of the detector and is absorbed by the gas by photoelectric absorption, ejecting an electron in the process. The excited ion then deexcites either by emitting a fluorescent X-ray or by emitting an Auger electron. The electron ejected in the initial photoionisation event is energetic enough to ionise other atoms of the gas and so, in the end, just as in the case of ionisation losses, one electron–ion pair is created for roughly every 30 eV of incident X-ray energy. These electron–ion pairs then drift to the region of the strong field where the number of pairs is amplified by a factor of between 10^3 and 10^5 and the pulse is detected. This amplification

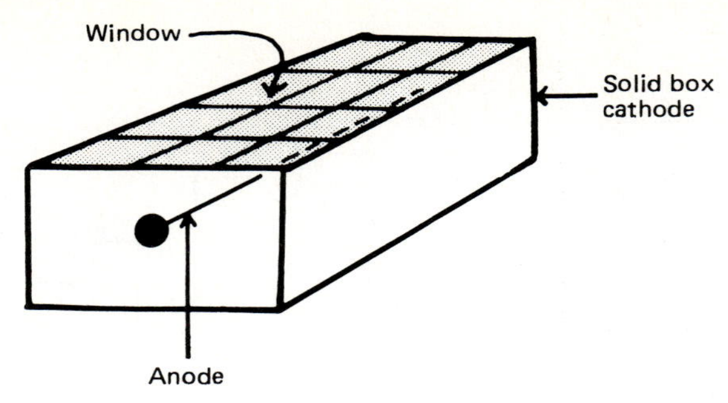

Figure 6.12. A schematic diagram of the type of proportional counter used in X-ray astronomy.

factor is essential in order to produce a signal which can be detected by the pulse counters.

Let us work out the response or quantum efficiency of such a detector for photons of different X-ray energies as used in an X-ray telescope. The probability of absorbing a photon of energy $\hbar\omega$ in the gas of the counter is

$$\eta(\hbar\omega) = \exp(-\sigma_w t_w)[1-\exp(-\sigma_g t_g)] \tag{6.2}$$

where σ_w and σ_g are the absorption coefficients and t_w and t_g the thicknesses of the window and the gas respectively. Examples of the mass absorption coefficients for different materials were described in Section 4.2 and illustrated in Fig. 4.1. Between absorption edges, the cross-section for absorption $\sigma(\hbar\omega)$ is roughly proportional to $\omega^{-3}Z^4$ where Z is the atomic number and, therefore, to maximise absorption in the detector gas and minimise that in the window, a low Z material is used for the window and a high Z substance for the detector gas. A typical proportional counter consists of a gas such as argon and a window made of a material such as mylar which is an organic plastic and therefore most of the absorption is associated with carbon atoms. The efficiency of the detector $\eta(\hbar\omega)$ as a function of the energy of the X-ray photon is illustrated in Fig. 6.13. The argon gas alone would result in the response shown in Fig. 6.13(*a*). However, this is modified by absorption in the window and therefore the probability of absorption within the gas of the detector is shown in Fig. 6.13(*b*). There is a jump in the detector efficiency at the K-edge of carbon. Thus, the response of the detector is largely determined by the materials out of which the window is fabricated and by the detector gas. Sheets of mylar as thin as 1 μm can be used as the window material which gives a path length through the window of 4400 mg m^{-2}, whereas the depth of the argon gas can be as much as 20000–40000 mg m^{-2}. There are considerable problems in the manufacture of these devices, not the least being that the detector gases can leak through such thin windows. For satellite instruments, thicker windows or stronger materials such as beryllium have to be used, which restricts the useful energy range of the instrument to greater than about 1 keV. For observations at low X-ray energies, $\hbar\omega \approx 0.25$ keV, very thin windows have to be used and then a supply of gas has to be provided to maintain the gas in the detector at constant pressure.

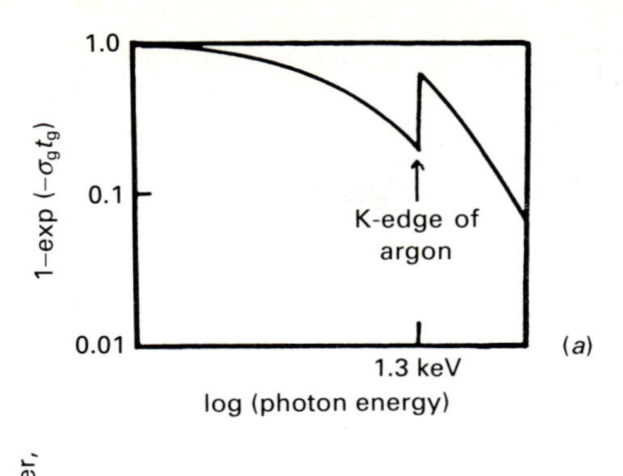

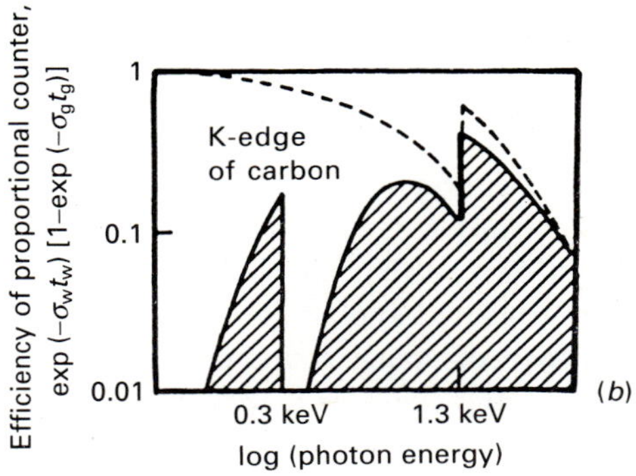

Figure 6.13. (*a*) The probability of absorbing an X-ray photon in the gaseous volume of an argon-filled proportional counter if there were no absorption by the window; σ_g is the photoelectric absorption cross-section and t_g the thickness of the gas.
(*b*) The probability of absorbing an X-ray photon in the gaseous volume of the proportional counter of (*a*) with a window of an organic substance such as mylar.

The energy resolution of the detector can be improved by using X-ray filters which utilise the fact that different materials have their K-absorption edges at different energies (Fig. 4.1). In addition, we obtain information about the energy of each arriving X-ray from the size of the voltage pulse. The accuracy with which the energy of each photon can be determined is limited by statistical fluctuations in the numbers of electron–ion pairs created by each X-ray. For example, if the efficiency of the detector were 100%, a photon of energy 10 keV would liberate only about 300 electron–ion pairs and therefore the statistical precision must be less than about $1/N^{\frac{1}{2}}$, i.e. at the very best, 5% accuracy. Typical values are somewhat greater than this. Note that inert gases are used for these devices since this means that most of the energy of the incoming X-ray goes into the kinetic energy of the electrons. If molecules were used, some of the energy would be drained into their rotational and vibrational degrees of freedom.

In proportional counters, the pulse of electrons is highly localised and hence it is possible to devise schemes whereby the location of arrival of each X-ray photon can be measured. This is achieved in *position sensitive proportional counters*. The position at which the pulse of electrons hits the anode may be estimated from the ratio of the charges collected at each end of the anode because the charge which flows in opposite directions along the wire from the location of the pulse is found to be inversely proportional to the distance from the collection point. To measure the coordinates of the pulse in the second dimension, multiple anode wires can be used and the wire along which the charge flows defines the position in the orthogonal coordinate. An alternative scheme is to use two sets of orthogonal anode wires to define the location of each event. These developments are particularly important following the development of imaging X-ray telescopes in which the X-rays are focussed at a focal plane and a two-dimensional image of the X-ray sky can be recorded.

Geiger counters are proportional counters run at saturation voltages so that any charged particle which enters the chamber causes ionisation of all the gas. This corresponds to region E of Fig. 6.11(*b*). These counters have important applications in miniaturised form as spark chambers. These have been used increasingly because of the ability to produce what are essentially Geiger counters with very small spacings between the anode and cathode. They consist of layers of these small cells, often filled with neon, with gaps a few millimetres in width. Again, these devices are run just below breakdown so that, as soon as a charged particle passes through the chamber, they are triggered and a 'spark' is produced in the element. Since they are so small, arrays of spark chambers enable the trajectory of the particle to be determined accurately and this is very important if one wants to know how much material of the telescope itself was traversed by the high energy particle. Other detectors can be used to measure the energy, charge, etc. of the particle. These instruments have proved very important in detecting γ-rays since the arrays of spark chambers can be interleaved with a material such as tungsten which 'converts' the γ-ray into an electron–positron pair, which can be followed through the spark chambers. In this way, directional information as well as information about the opening angle of the electron–positron pair can be obtained (see Fig. 6.14).

6.5 Solid state devices

When electrons are knocked out of atoms in solids, they can be swept out by an electric field just as in the case of gases. There are two main differences between gases and solids in this context. First, the high energy particles produce electron–hole pairs rather than electron–ion pairs. Second, the energy needed to produce an electron–hole pair is much smaller than that needed to produce an electron–ion pair. For example, in silicon and germanium only 3.50 eV and 2.94 eV respectively are required to produce an electron–hole pair. This means that, in solid state devices, about 10 times more electron–hole pairs are created than electron–ion pairs in a gas for which about 30 eV is needed per pair. This improvement in statistics corresponds to an increase of a factor of three in the estimation of the energy loss per unit path length.

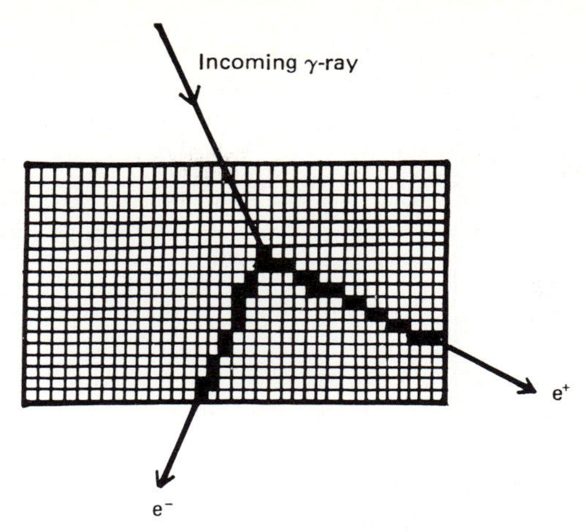

Figure 6.14. A schematic diagram showing an array of small spark chambers which enables the paths of an electron–positron pair created by an incoming γ-ray to be determined.

6.5.1 *Semiconductor devices*

To utilise the above advantages, an insulator can simply be placed between the electrodes of a detecting circuit and the pulses of current due to the release of electrons in the material measured. An insulator must be used or else leakage currents in a conducting material mask the pulse due to the charged particle. Unfortunately, ordinary insulators are too impure to be used for this purpose since many of the electrons and holes are lost in imperfections in the crystal structure. Semiconductors are used, one reason being that they are of very much greater purity.

The problem is then to produce a region of high resistance inside the semiconductor. This is done by reverse biassing a p-n junction, so that not only is the central region an insulator but also a large voltage is developed across the depletion layer (Fig. 6.15). The net effect is that, when a high energy particle causes ionisation losses in the material, the excess of holes in the p-type semiconductor is pulled into the metal and the excess electrons to the gold contact.

The big problem is to make the depletion layer as thick as possible in order to obtain a large path length through the semiconductor and also to produce a very large resistance across the depletion layer. By straightforward transistor technology, depletion layers about 1 mm thick can be made. Thicker layers can be produced by the technique known as lithium-drifting, in which one starts with not-too-pure silicon and compensates for the excess of acceptors by the drifting in of lithium atoms. Successful layers of 5 mm thickness have been achieved by this technique with the net result that path lengths of a several tens of kilograms per square metre through the device can be achieved. Because it is entirely solid state and compact, this is exactly the type of device that is best adapted to the rigours of space flight and high altitude balloons and they have been used very successfully in cosmic ray telescopes.

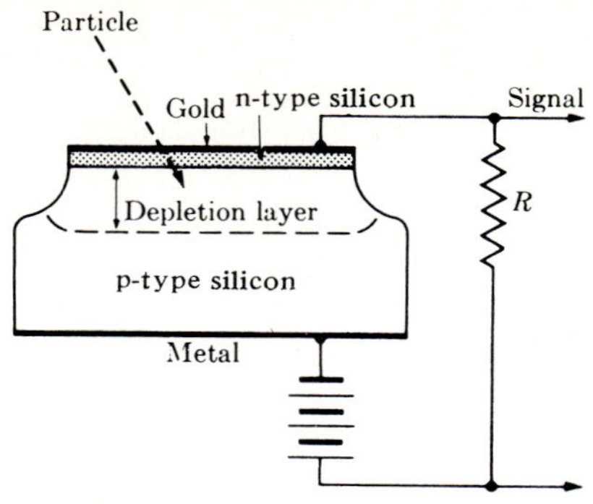

Figure 6.15. A schematic diagram of a basic semiconductor, solid state detector. (From H. A. Enge (1966). *Introduction to nuclear physics*, page 207, London: Addison-Wesley Publishing Co.)

Recent advances in semiconductor technology have also been of great importance in the detection of X-rays, specifically the use of silicon semiconductors as array detectors. The principles employed in CCDs can be used (see Section 8.6.1) in which potential wells are created within the silicon in which the electrons liberated by the X-rays are trapped. As in the case of CCDs used in optical astronomy, the charge which builds up in the potential wells is proportional to the X-ray flux density incident upon that part of the array. In the detectors available at the moment, the statistical accuracy with which the energy of the photons can be measured corresponds to an energy resolution for 3 keV X-rays of about 10%. The read-out of the device is achieved using the principles of charge transfer using charge coupling techniques. CCD detectors for X-ray energies are planned for the next generation of X-ray telescopes using X-ray imaging optics, for example, in AXAF Observatory (Section 7.3.3).

6.5.2 *Scintillation detectors*

These are historic devices. Rutherford used a scintillating zinc sulphide screen to count α-particles in the famous Rutherford scattering experiment. Nowadays, instead of viewing by eye, a photomultiplier arrangement is used to produce a large pulse of electrons. Fig. 6.16 illustrates their construction. The sequence of events is as follows: the liberated electrons produce photons inside the crystal and these photons hit the semitransparent photocathode which liberates photoelectrons. These liberated electrons are accelerated and focussed onto a multiplier arrangement of elements known as dynodes. The net result of multiplication is a large pulse arriving at the anode, the size of which is proportional to the total energy $\mathrm{d}E$ liberated in the scintillating crystal. The limitations of these devices are two-fold. First, the scintillating material converts only about 3% of the liberated electron energy into useable optical photons: second, the cathode efficiency is typically about 10–20% so that for every 5–10 photons arriving at the photocathode, only one electron is liberated which will initiate a cascade in the dynodes. To give an example of the implication of these

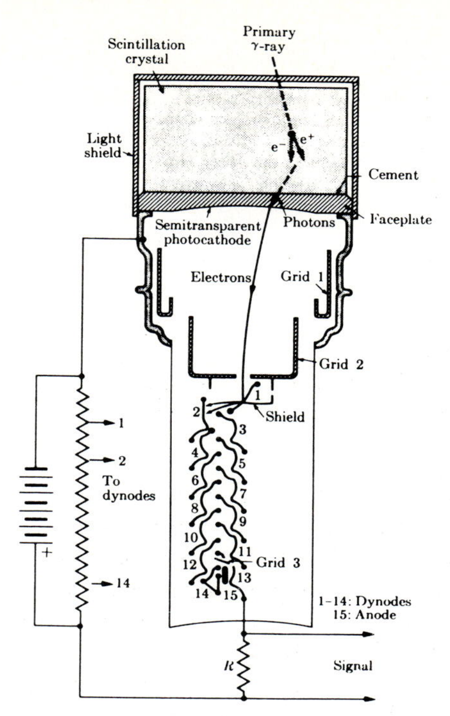

Figure 6.16. A scintillation detector showing the large scintillating crystal and the photomultiplier tube (RCA 7046). (From H. A. Enge (1966). *Introduction to nuclear physics*, page 210, London: Addison-Wesley Publishing Co.)

figures, consider the liberation of 5 MeV of particle energy within the scintillator. This energy loss liberates about 150 keV of photon energy, corresponding to about 60 000 photons. A cathode efficiency of 10 % means that we obtain only about 6000 useful photoelectrons from the cathode and hence a statistical accuracy of at best about 2 %.

These devices are often used in cosmic ray telescopes because they can be made compact and rugged. They are also popular as hard X-ray detectors at energies greater than about 20 keV because the efficiencies of proportional counters decrease as ω^{-3} with increasing photon energy $\hbar\omega$. The materials used in scintillation detectors are sodium iodide, which has maximum photon emission at −188 °C, sodium iodide doped with thallium, which has higher efficiency at room temperature, or caesium iodide. Organic scintillators such as *p*-terphenyl, which is a liquid, or plastic scintillators such as anthracene and *trans*-stiblene are also used. Liquids are particularly useful since anticoincidence enclosures are often needed and one simple way of achieving this is to surround the detector with a jacket of the scintillating fluid and to note all events recorded by that scintillator.

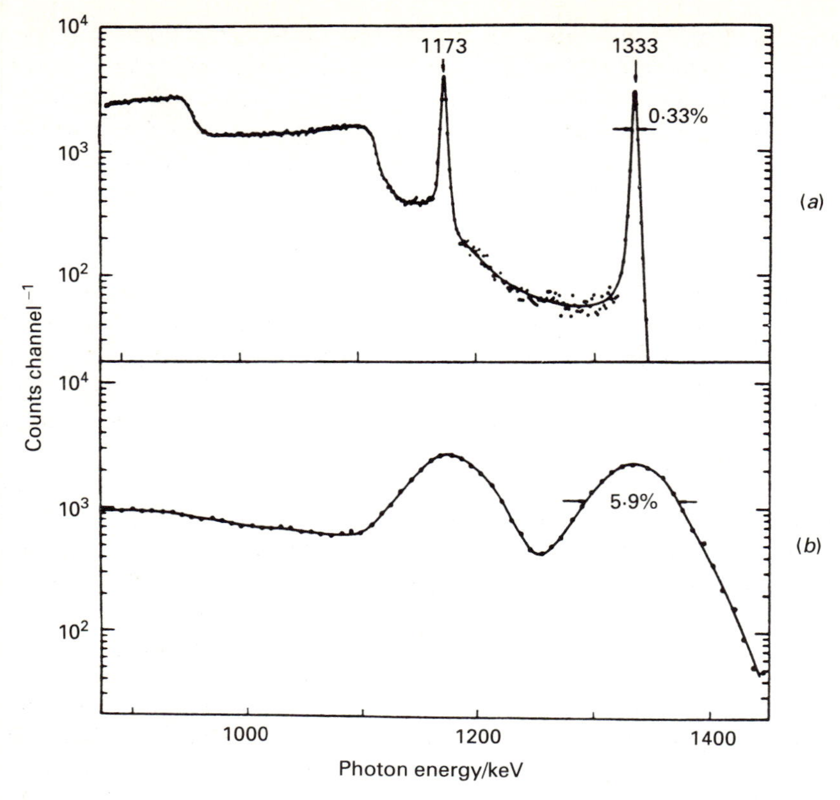

Figure 6.17. Comparison of the energy loss spectra observed by (*a*) a 3.5 mm deep lithium-drifted germanium crystal detector and (*b*) by a scintillating crystal of sodium iodide of the nuclear γ-ray lines of ^{60}Co. (From R. Hillier (1984). *Gamma ray astronomy*, page 55, Oxford: Clarendon Press. The data are from G. T. Ewan and A. J. Tavendale (1964). *Can. J. Phys.*, **42**, 2286.)

6.5.3 *Crystal detectors*

Another way of detecting high energy X and γ-rays is to make use of the photoelectric effect but now in a pure semiconductor material. The trick in this case is to use a very pure semiconductor crystal with a high atomic number so that there is a strong absorption edge at a high energy. Crystals which have proved to be very effective as hard X and γ-ray detectors for high resolution spectroscopy are single germanium crystals. As illustrated in Fig. 4.16, at energies greater than 100 keV, Compton scattering and photoelectric absorption are competing processes for the attenuation of the flux of high energy photons. The advantage of the crystal detector is that the energies of the electrons liberated as photoelectrons are large and their total energy loss can be determined with high accuracy by measuring the numbers of electron–hole pairs in the usual way. Fig. 6.17 compares the energy spectrum of the radioactive lines of cobalt ^{60}Co at 1.173 and 1.333 MeV as measured by a single germanium crystal and by a scintillation detector in a laboratory experiment (Hillier 1981). It can be seen that the crystal provides high spectral resolution ($\Delta\varepsilon/\varepsilon \approx 0.33\%$) but there is in addition a background due to the Compton scattering of the lines in the crystal at lower energies. In contrast, the spectrum measured by the scintillation detector does not provide such high resolution, about 6%, but more of the detected photons lie in the two peaks.

6.5.4 *Cherenkov detectors*

When a fast particle moves through a medium at a velocity greater than the velocity of light in that medium, it emits Cherenkov radiation as demonstrated in Section 4.6. The wavefronts only add up to produce coherent radiation in a particular direction θ with respect to the velocity vector of the electron. This angle is given by $\cos\theta = c/vn$ where n is the refractive index of the material. The intensity of the radiation is given by the expression (4.73).

We can see immediately that this phenomenon can be used to construct *threshold detectors*. Only if the velocity of the particle is large enough, will it emit Cherenkov radiation and hence a detectable signal. For charged particles with velocities $v \geqslant 0.6c$, solid dielectric materials can be used, for example, lucite or plexiglass for which $n \approx 1.5$. If one is interested in only the most extreme ultrarelativistic energies, gas Cherenkov detectors can be used. The refractive indices of gases differ only very slightly from 1 and hence only ultrarelativistic electrons are detected.

In practice, one does not attempt to measure the wavefront of the emitted radiation in cosmic ray experiments, although this can be achieved in laboratory experiments. The total emitted light is measured and this provides information about the charge and velocity of the particles passing through the material. By placing two detectors in series with different thresholds for Cherenkov radiation, for example by using different gases or the same gas at different pressures, both z and v can be found for the particles.

The main problem with these detectors is that the light yield is very small and photomultipliers have to be used to produce a measurable signal. One method is to use the fact that light is totally internally reflected within the crystal and suitable treatment of the surfaces of the enclosure increases the intensity of light which reaches the photomultiplier. These devices have proved to be very successful in achieving high charge resolution. This is realised by making the refractive index a suitable function of energy in the region in which the photomultipliers are most sensitive.

7

Cosmic ray, X-ray, γ-ray and neutrino telescopes

7.1 Introduction

The next step is to build the detectors described in Chapter 6 into telescopes for the study of cosmic rays, X-rays and γ-rays. It is interesting to contrast these telescopes with those which operate at optical, infrared and radio wavelengths which are the subject of Chapter 8. The main difference is that, in most cases, the photons and particles to be detected have such high energies that they can be considered to move in straight lines – in other words, we can use Newton's (or Einstein's!) corpuscular picture of electromagnetic radiation without worrying about wave optics and diffraction effects. From the point of view of explaining the principles of these telescopes, the early satellite instruments illustrate most clearly the features of successful telescopes for these wavebands. One common feature of the telescopes considered in this chapter is the role of anticoincidence techniques.

7.2 Anticoincidence techniques and cosmic ray telescopes

The principle is straightforward in that photons and cosmic rays are allowed to pass through a number of different detectors which are sensitive to different properties of photons or high energy particles. Particles or photons of a particular type are identified according to whether or not a signal in one detector is accompanied by responses in the other detectors. These procedures require high speed electronic circuitry designed so that an event is only registered if it possesses the *correct signature*. In the case of telescopes flown in orbit, these data have to be transmitted back to Earth.

Let us take as an example one of the early cosmic ray telescopes on board the IMP-III satellite, the acronym meaning the third Interplanetary Monitoring Platform. The layout of the detectors labelled D_1, D_2, D_3 and D_4 is shown in Fig. 7.1. D_1 and D_2 are lithium-drifted semiconductor detectors and are thin so that

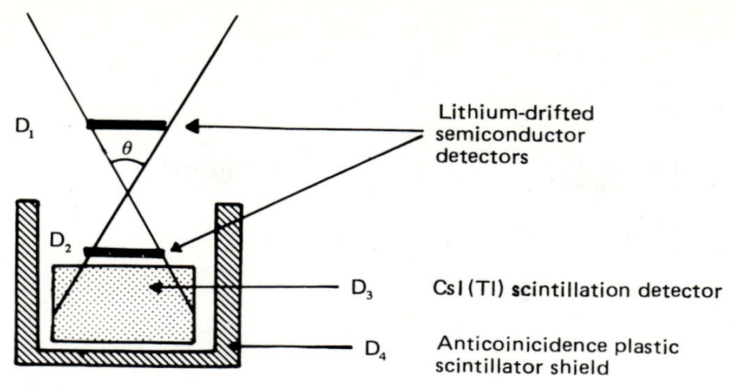

Figure 7.1. A simplified diagram showing the layout of the detectors of the cosmic ray telescope flown on board the IMP-III space probe. (From M. M. Shapiro and R. Silberberg (1970). *Ann. Rev. Nucl. Sci.*, **20**, 323.)

most cosmic ray particles pass through them but the ionisation losses in these detectors register the fact that a particle has passed through. D_3 is a scintillation detector which consists of a thallium-doped, caesium iodide crystal. D_4 is an anticoincidence shield made of a plastic scintillator. To use this combination of detectors as a telescope, we select events which give a positive signal in detectors D_1, D_2 and D_3 but give no signal in D_4. This signature is written $D_1 D_2 D_3 \bar{D}_4$ and defines the properties of the telescope.

(i) The beam of the telescope is defined by detectors D_1 and D_2 as illustrated geometrically in Fig. 7.1. Thus, only particles arriving within the cone of apex angle θ can trigger both D_1 and D_2. The angle θ is the beam-width of the telescope.

(ii) The range of energies to which the telescope is sensitive is determined by the requirement that the particles must be energetic enough to pass through D_1 and D_2 and into D_3 but not so energetic that they can pass all the way through to D_4.

(iii) The properties of the particles which have signature $D_1 D_2 D_3 \bar{D}_4$ can be inferred from the fact that the signals registered in D_1 and D_2 give a measure of $\mathrm{d}E/\mathrm{d}x$, the loss per unit path length, and the total signal $D_1 + D_2 + D_3$ is a measure of the total kinetic energy of the particle. Thus, for each particle, we obtain one point on the energy loss rate versus total kinetic energy relation.

In practice, considerable redundancy is built into space experiments so that, if one of the detectors fails, the telescope can still be operated in some other mode. For example, if either D_1 or D_2 failed, $\mathrm{d}E/\mathrm{d}x$ can still be measured by the other detector. Another important point is that we need to know the direction in which the telescope is pointing on the celestial sphere. This means that either a continuous record must be kept of the orientation of the telescope in space or else the spacecraft must be stabilised to point in one direction for a long period.

We therefore expect that, if we plot the response $D_1 + D_2$ against $D_1 + D_2 + D_3$, we should obtain the ionisation loss relations for particles of different species.

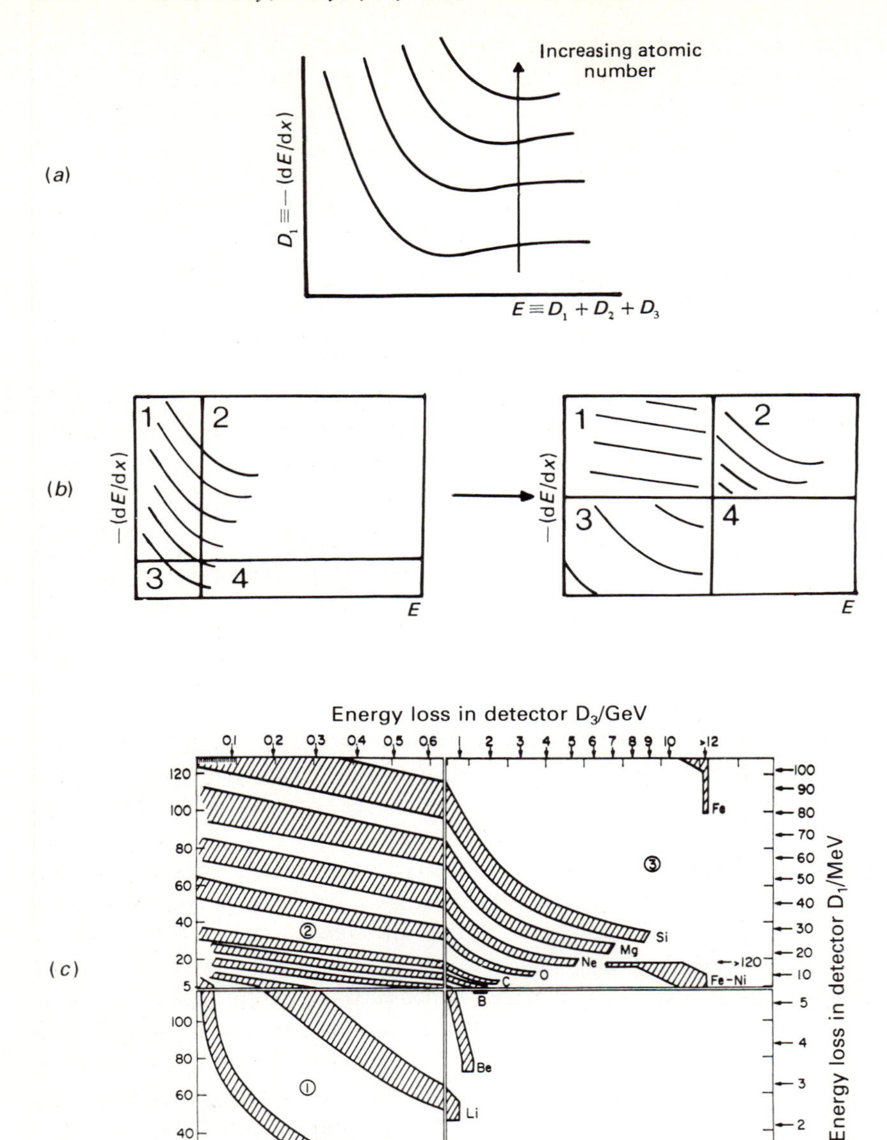

Figure 7.2. (*a*) A schematic diagram showing where cosmic rays of different energies and atomic numbers which satisfy the condition $D_1 D_2 D_3 \bar{D}_4$ are expected to lie in a plot of D_1 against $(D_1 + D_2 + D_3)$.
(*b*) A schematic diagram illustrating the different sensitivity ranges which were available on the cosmic ray telescope of the OGO-1 satellite.
(*c*) Detailed predictions of the expected location of cosmic rays in the D_1–D_3 plane when the OGO-1 telescope was operated in different sensitivity modes. The widths of the shaded regions are primarily defined by the cone angle within which particles enter

Notice that, when we evaluate $D_1 + D_2 + D_3$, we determine the total kinetic energy of the particle rather than its kinetic energy per nucleon. Thus, for a cosmic ray of atomic number z and mass number A, the standard ionisation loss curve of Fig. 2.7 is shifted along the abscissa by a factor A and along the ordinate by a factor z^2. Thus, schematically, we expect particles of different elements to lie along the loci illustrated in Fig. 7.2(*a*). There is considerable scatter about these predicted relations because the cosmic rays arrive at different angles within the cone of acceptance θ and the detectors themselves have finite energy resolution.

A good example of how this works in practice is provided by the results of the Chicago group whose experiment was carried on board the OGO-1 satellite (the first Orbiting Geophysical Observatory). In this early experiment, it was only possible to use a finite number of channels into which to sort the energies of the particles and so the telescope was operated in what were referred to as the high and low gain modes, meaning that the energy scales could be stretched to span different ranges of kinetic energy and energy loss rate. The predicted responses are shown in Fig. 7.2(*b*) and (*c*). Fig. 7.3 shows what was actually observed, the solid lines indicating the predicted loci for different elements. By counting the number of particles in the appropriate energy ranges, the relative abundances of different elements can be found and by counting the relative numbers of particles along each locus, the energy spectrum of each element can be found. The separation between the tracks in Fig. 7.3 shows how it is possible to discriminate between cosmic rays of different atomic number.

In the discussion of the astrophysics of the propagation and origin of cosmic rays, *isotopic abundances* are of the greatest interest. The problem in determining these abundances is that the mass of the cosmic ray does not appear explicitly in the expression for the ionisation loss rate of the charged particle. The differences between isotopes of the same species only become apparent when the total kinetic energies of the particles are measured. The loss rate $-\mathrm{d}E/\mathrm{d}x$ is independent of M but the kinetic energy of the particle $E_{\mathrm{kin}} = (\gamma - 1)Mc^2$ does depend upon M. As a result, isotopes of the same element can be distinguished as illustrated in Fig. 7.4 – the ionisation loss curves are slightly displaced with respect to one another. This indicates how important high precision is in measuring both the charge and total kinetic energy of the high energy particles.

There have been several important advances in determining the charges and masses of cosmic ray particles, the aim being to determine both $\mathrm{d}E/\mathrm{d}x$ and E_{kin} with as high precision as possible. Simpson (1983) describes some of the developments which have been incorporated into recent experiments. A key aspect of measuring $\mathrm{d}E/\mathrm{d}x$ and E_{kin} as accurately as possible is the determination of the precise trajectory, and hence path-length, of each cosmic ray through the various elements of the detectors. One way of obtaining an immediate improvement in the determination of the path-length is to use curved rather than flat detectors D_1 and

---

the telescope. The high energy limits of each locus are determined by the particle range limits in the telescope. (From G. M. Comstock, C. Y. Fan and J. A. Simpson (1966). *Astrophys. J.*, **146**, 57.)

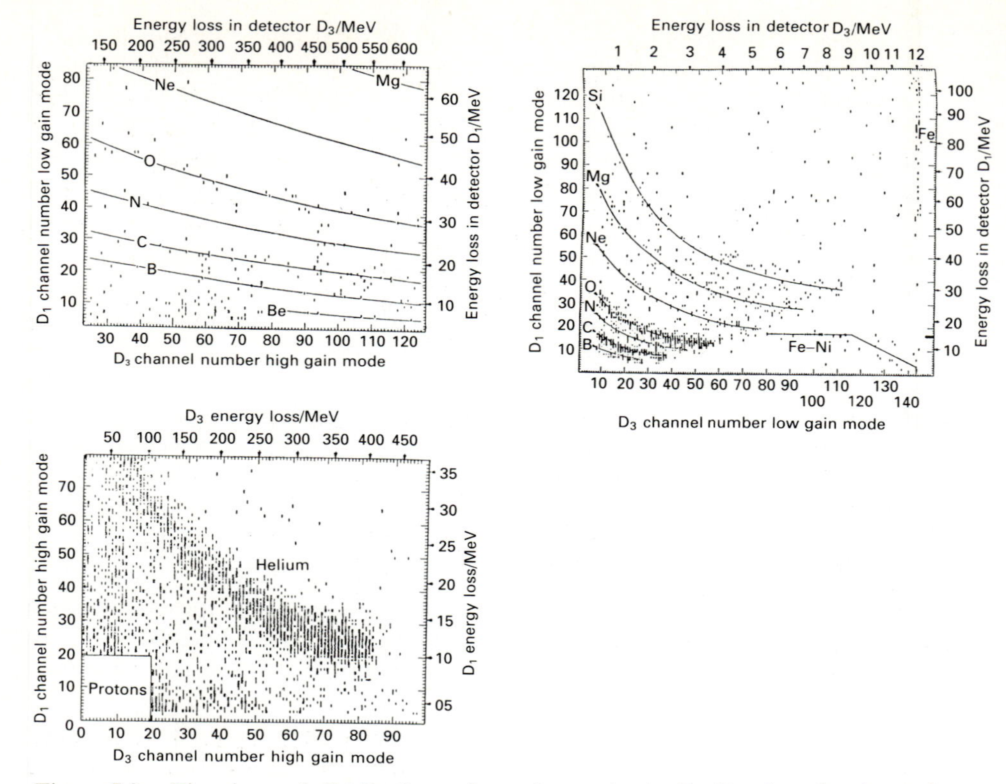

Figure 7.3. The observed distributions of cosmic rays in the D_1–D_3 plane for three of the data formats (or modes) shown in Fig. 7.2. The region populated by protons has been deleted. The solid lines indicate the predicted loci of the different elements. (From G. M. Comstock, C. Y. Fan and J. A. Simpson (1966). *Astrophys. J.* **146**, 60–1.)

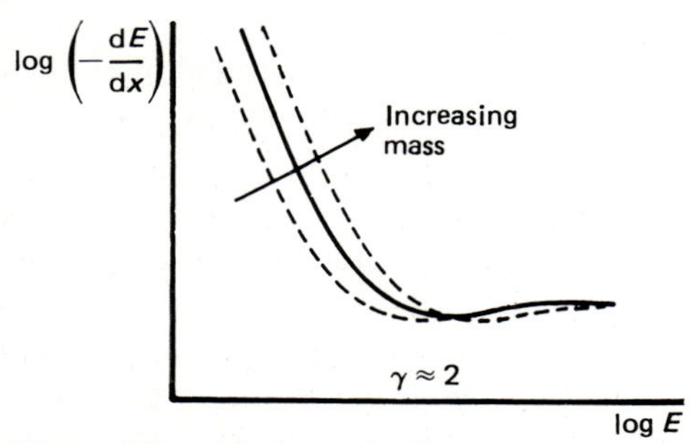

Figure 7.4. A schematic diagram illustrating the problem of distinguishing isotopes of a particular element.

D_2 to measure dE/dx. This much reduces the scatter in path lengths in the type of experiment shown in Fig. 7.1.

Examples of three satellite packages are shown in Fig. 7.5. The instrument shown in Fig. 7.5(*a*) is the Heavy Isotope Spectrometer (HIST) which was flown on the ISEE-C spacecraft. Two similar cosmic ray telescopes were flown on this

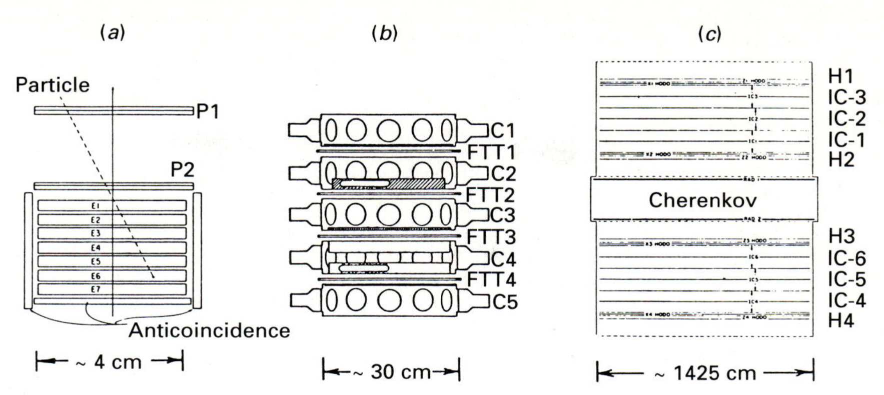

Figure 7.5. Examples of three important cosmic ray satellite telescopes. (From J. A. Simpson (1983). *Ann. Rev. Nucl. Part. Sci.*, **33**, 351.)
(*a*) A schematic diagram of an ISEE class telescope for the measurement of cosmic ray isotopes. (W. E. Althouse, A. C. Cummings, T. L. Garrard, R. A. Meawaldt, E. C. Stone and R. E. Vogt (1978). *IEEE Trans. Geosci. Electron.*, **GE15**, 204.)
(*b*) The HEAO-C2 experiment of the Danish–French collaboration. (Copenhagen–Saclay Collaboration (1981). *Adv. Space Res.*, **1**, 173.)
(*c*) The HEAO-C3 ultraheavy nuclei telescope for the study of cosmic rays with $Z \geqslant 30$. (W. R. Binns, M. H. Israel, J. Klarmann, W. R. Scarlett, E. C. Stone and C. J. Waddington (1981). *Nucl. Instrum. Methods*, **185**, 415.)

mission. In the example shown, the detectors P1 and P2 are position sensitive detectors. These detectors form what is known as a *hodoscope* from the Greek word *hodos* meaning 'way' – in space astronomy, the word means a device for measuring the direction of arrival of the cosmic ray. Two different types of position sensitive detector were employed on the spacecraft. In one case, each of the detectors consisted of a matrix of thin strips of detector material. One set of 24 strips formed the 'x' coordinate and an orthogonal set of 24 the 'y' coordinate. By measuring the (x, y) coordinates of the passage of a particle through both detectors, its trajectory through P1 and P2 and into the main body of the detector could be found. In the other telescope, thin electron drift chambers performed the same function. The detectors E1–E7 are lithium-drifted, solid state detectors designed to measure the total energy loss of the particles. By measuring precisely the path length of the particles through the detector and the energy loss, a mass resolution of better than 0.2 amu was achieved (Fig. 7.6). The telescope system was sensitive to particles with energies in the range 30–130 MeV per nucleon.

Fig. 7.5(*b*) is a plan of the telescope flown on the highly successful HEAO-C2 experiment which was constructed by a French–Danish collaboration. The telescope consisted of five closely spaced Cherenkov detectors (*C*1–*C*5) which contained materials of different refractive indices so that they were sensitive to particles of different velocities. The total mass transversed by particles passing through the telescope was 167 kg m^{-2} so that the abundances of elements between beryllium and iron could be determined up to a kinetic energy of 25 GeV per nucleon. Detectors *C*1 and *C*5 had the highest refractive indices and are made of

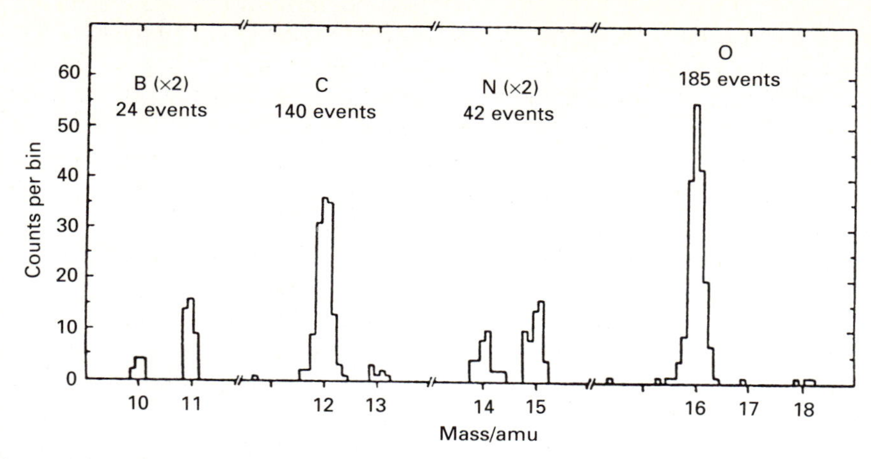

Figure 7.6. Observations made with the ISEE-3 HIST. The mass histograms of boron, carbon, nitrogen and oxygen nuclei are observed in the energy range roughly 30–120 MeV nucleon^{-1}. Note that the boron and nitrogen isotope distributions have been scaled up by a factor of 2. (From R. A. Mewaldt, J. D. Spalding, E. C. Stone and R. E. Vogt (1981). *Astrophys. J.*, **251**, L27.)

lead glass; the detectors *C*2, *C*3 and *C*4 were materials of lower refractive index, the detector *C*4 consisting of aerogel sand which has refractive index 1.015 and a corresponding threshold energy for the production of Cherenkov radiation of about 4.5 GeV per nucleon. The lower energy limit for the detection of particles was set by the requirement that events were only registered if detectors *C*1, *C*3 and *C*5 were simultaneously triggered. Since the detector *C*3 consisted of Teflon, which has refractive index 1.33, particles of energy less than 0.5 GeV per nucleon were not detected. Between the Cherenkov detectors, there were placed what are called flash tube trays (FTT). These act as a hodoscope and consist of two perpendicular layers of 128 tubes each so that when a cosmic ray passes through them, the coordinates of the event are determined. There are four flash tube trays (FTT1–FTT4) between the Cherenkov detectors and these enable the trajectories of the cosmic rays through the telescope to be determined accurately. The momenta of the particles and their charges are determined from the strengths of the Cherenkov light signals registered in the detectors *C*1–*C*5. It can be seen from equation (4.73) that both the charge z and the velocity of the particle can be evaluated if the trajectory of the particle is known since it produces a flux of Cherenkov radiation in four different dielectrics. The arrangement was successful in producing charge resolution of better than 0.2.

An ingenious technique was used to determine isotopic abundances of different elements. We show in the Appendix to Chapter 10 that, for particles detected in near Earth orbit, there are permitted and forbidden zones for particles propagating from interstellar space through the Earth's dipole magnetic field. Detailed analyses can define these zones for particles coming into the Earth at different geomagnetic latitudes and at different angles to the vertical. In particular, at certain geomagnetic latitudes, there are forbidden bands in which particles of a given rigidity should not be observed and this is reflected as an absorption feature in the momentum spectrum of these particles. Since the isotopes of the same element have different masses, the absorption features should appear at different momenta and thus the ratio of the depths of the absorption features provides a measure of the ratio of abundances of these isotopes. Applying this method in practice is complex because the forbidden zones vary along the orbit of the spacecraft but the technique was successfully demonstrated in the HEAO-C2 experiment.

The third example shown in Fig. 7.5(*c*) is the HEAO-C3 experiment which was designed to measure the abundances of the elements in the cosmic rays beyond the iron peak. The first evidence for such particles was obtained from nuclear emulsion experiments (Fig. 6.6) and the importance of establishing whether or not there are radioactive elements in the cosmic radiation is readily appreciated. The components of such a telescope are now familiar. The basic instrument consists of two sets of three ionisation chambers, (IC-1, IC-2, IC-3) and (IC-4, IC-5, IC-6), located above and below a Cherenkov counter. There is a pair of hodoscopes associated with the two sets of ionisation counters which determine the trajectories of the cosmic rays through the detectors. Events are only registered if at least two of the hodoscopes and at least two ionisation chambers and the Cherenkov detector are triggered. The charges of the incoming particles can be found from the fact that, although the signals in the ionisation chambers and the Cherenkov detectors depend upon the square of the charge of the particle z^2, they have different dependences upon the velocity of the particles as can be seen from the comparison of equations (2.19) and (4.73). Because the abundances of the elements beyond the iron peak are very low, the instrument had to be large to detect a significant number of particles and in the case of the HEAO-C3 experiment, it was about 1.5 m in length, width and height. Examples of the quality of the charge discrimination achieved by this instrument are shown in Fig. 7.7. The experiment was launched on board the HEAO-C satellite in 1979 and the exposure time of the experiment was 20 months. At about the same time, a similar experiment called BUGS-4, meaning the fourth instrument in the series of Bristol University Gas Scintillators was carried on the UK Ariel-VI satellite. Using a novel design, a single set of phototubes viewed the emission from both a spherical volume of scintillating gas and from a Cherenkov detector in the form of a concentric spherical shell. It remained in orbit for about two years.

To measure the properties of higher energy particles than those detected in the HEAO-C2 experiment, very much larger telescopes have to be constructed because there are far fewer of them and therefore, to measure significant numbers of them, we need both a large collecting area and long exposure times. In addition, we have to increase the path length through the detector so that the higher energy particles are brought completely to rest within the detector. The implication of these requirements is that, to conduct abundance and isotope studies at high energies, very massive payloads are needed which can be exposed for a long period in space. Such a mission has not yet been undertaken.

The problems with *electron detectors* are quite different from those of protons and nuclei. The first difference arises because the rest masses of the electron and

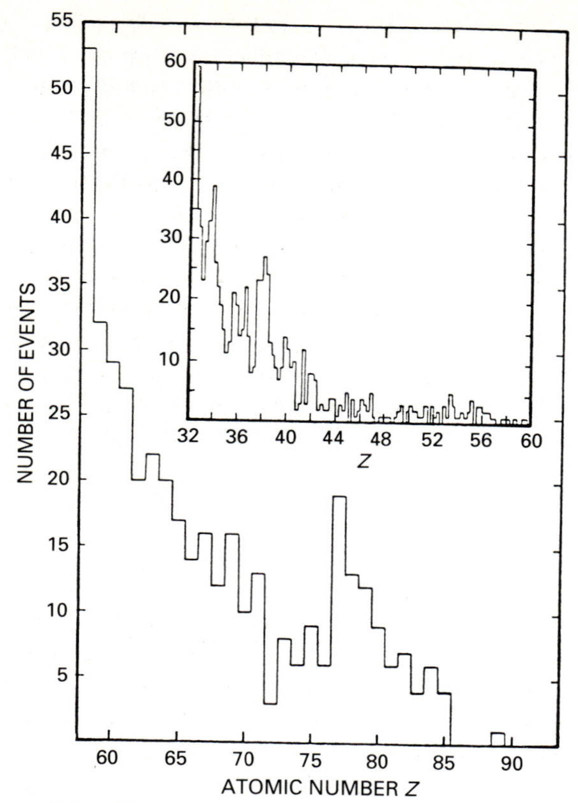

Figure 7.7. Observations of ultraheavy cosmic ray nuclei by the HEAO-C3 experiment. The histogram shows the observed number of events plotted against their assigned charge z. The main part of the figure shows all events from 450 days of data for which charge assignments could be made. The charge resolution of this set does not allow separation of the abundances of individual elements. The inset shows a selected subset for which individual element abundances can be determined for even-z elements at least to $z = 42$. (From W. R. Binns, R. K. Fickle, T. L. Garrard, M. H. Israel, J. Klarmann, E. C. Stone and C. J. Waddington (1982). *Astrophys. J.*, **261**, L117.)

the proton differ by a factor of 1840. The effects of a magnetic field upon a charged particle depend only upon the ratio of its momentum to its charge and therefore protons and electrons of the same momentum, and hence rigidity, have similar dynamics in the Solar Wind and the Earth's magnetic field, apart from their opposite charges. However, whereas the Lorentz factors of the protons are of the order unity, those of the electrons are about 2000 i.e. the 1 GeV electrons detected at the top of the atmosphere are ultrarelativistic. Because the effects of solar modulation become large at a kinetic energy of about 1 GeV and lower (see Section 9.1), only the most energetic electrons can reach the upper layers of the atmosphere where they can be detected from high flying balloons. Electrons of these energies are still very difficult to study at the top of the atmosphere because of the copious generation of relativistic secondary electrons and positrons from nuclear interactions of primary cosmic ray protons and nuclei with atoms and molecules of the atmosphere. The combination of the effects of solar modulation and secondary electron and positron production means that it is difficult to establish by direct measurement the interstellar electron spectrum at energies less than about 10 GeV. There is now, however, reasonable agreement between the various measurements throughout the energy range from about 1 to 10^3 GeV. The electron (and positron) component of cosmic rays is best studied from space vehicles and gas-filled Cherenkov counters and threshold detectors are ideal for discriminating between electrons and protons. It will be noted, however, that there is a problem similar to that of measuring the energy spectra of high energy cosmic rays. If 10^3 GeV electrons are to be brought to rest within the detector volume, the cosmic ray electron telescope has to be massive. The most successful observations have been carried out from high flying balloons carrying large payloads. We will have more to say about observations of cosmic ray electrons when we study the origin of the Galactic radio emission.

7.3 X-ray telescopes

We discussed in Chapter 6 the principal detectors used in X-ray astronomy – proportional counters below about 20 keV and scintillation detectors for energies up to about 1 MeV. More recently, microchannel plate and CCD detectors have been developed successfully for X-ray observations in space. One of the major problems is to reject cosmic rays which also cause ionisation losses inside the counters. Three techniques have been used.

(i) The first involves the use of *anticoincidence detectors*. In this case, the counters are surrounded by a scintillating material, either a plastic scintillator or a scintillating liquid, and any events which trigger both the counter and the scintillating material can be safely rejected as cosmic rays (Fig. 7.8(*a*)).

(ii) The second technique involves determining the shape of the pulse of electrons as a function of time. A fast particle, either a lower energy cosmic ray or a fast electron ejected from the walls of the detector by a cosmic ray, produces a trail of ionisation and consequently results in a broad pulse. On the other hand, an X-ray photon results in only local ionisation and the detected pulse is much sharper than that produced by the continuous losses of a charged particle. It is found that the X-ray pulses are very sharp, particularly at their leading edges. This technique of discriminating between cosmic rays and X-rays is called *rise-time* or *pulse-shape discrimination* (see Figs 7.8(*b*) and (*c*)).

(iii) A third technique involves what is known as a *phoswich detector* and is used at hard X-ray and γ-ray energies. The detector consists of alternate layers of materials which have different responses as scintillation detectors for photons and charged particles. Of the two materials, one is made of a substance such as caesium iodide which

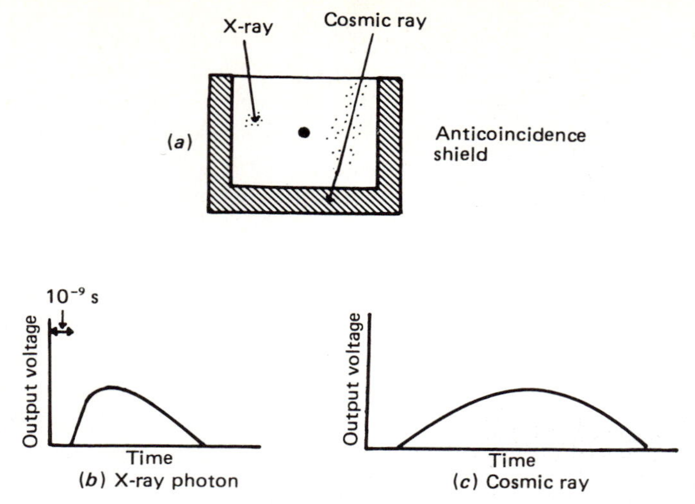

Figure 7.8. Diagrams illustrating the principles of rise-time or pulse-shape discrimination between X-rays and cosmic rays.

is sensitive to photons, as in the standard scintillation detector, whereas the other, which may be made of a plastic scintillator, is insensitive to photons. Thus, photons only produce a signal in the first detector whereas charged particles produce light signals in both materials. A clever aspect of the phoswich detector is that the materials can be arranged to have different response times so that a charged particle passing through the device results in two light pulses separated in time by about 10 μs. A photon, on the other hand, produces only a single light pulse. The light signals can therefore be collected by a single photomultiplier, associated with which there is logic circuitry which can recognise the characteristic signature of cosmic rays and reject them. The intensity of the light signals in the case of photon events is a measure of the photon energy and an energy resolution of about 10% or better can be achieved at γ-ray energies.

The beam of the X-ray telescope must also be defined and in many experiments this is achieved using a mechanical *collimator*. In the simplest case, this consists of a set of hollow rectangular tubes or pipes. The response of the collimator is thus just a triangular function since, at X-ray energies, the photons may be considered to propagate in straight lines according to geometric optics. An important exception to this rule is at very large angles of incidence at the surface of highly conducting materials such as copper, when the phenomenon of *grazing incidence reflection* can take place. For photons of energy less than about 1 keV, reflection takes place for angles between the direction of the beam and the surface of the material of less than a few degrees. The reflection process is similar to that of the deflection of radio waves in an ionised plasma, in which the plasma frequency increases with depth into the plasma. This type of reflection takes place when low frequency

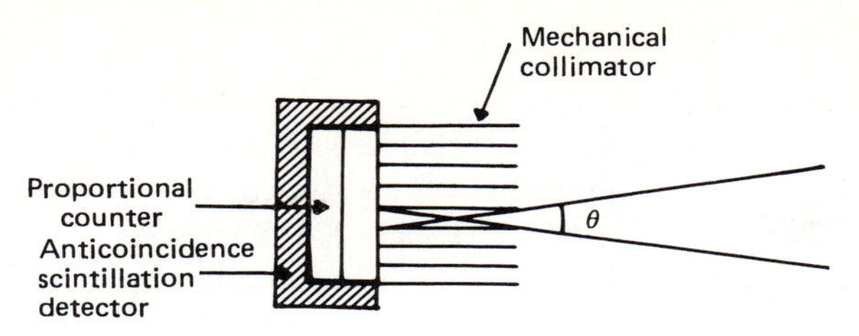

Figure 7.9. The layout of a simple X-ray telescope of the type flown on the UHURU and Ariel-V satellites.

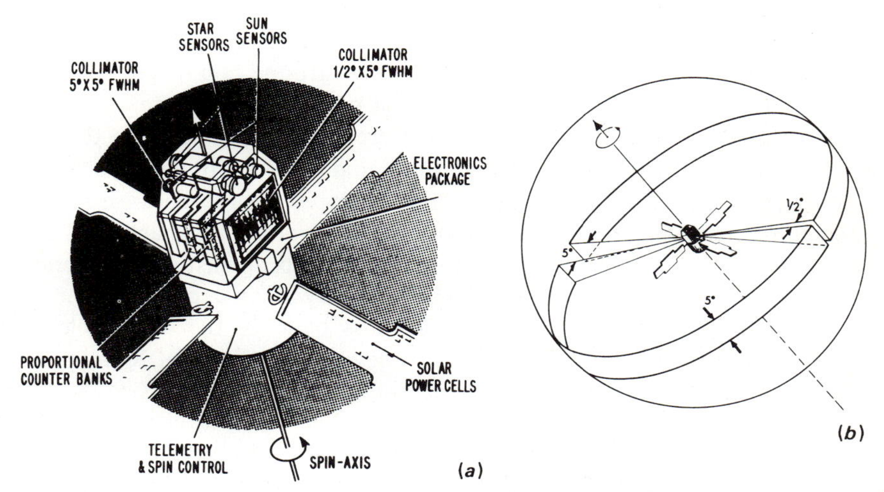

Figure 7.10. The UHURU X-ray satellite showing (*a*) the disposition of the components and (*b*) the beam pattern of the telescopes on the sky showing how the sky was scanned by the two telescope beams. (From R. Giacconi *et al.* (1971). *Astrophys. J.*, **230**, 540.)

electromagnetic waves are reflected by the ionosphere. Although the reflection only occurs at very small angles, this is adequate to design grazing incidence telescope mirrors which by careful design can produce an image of the sky in the focal plane (see Section 7.3.2).

We can now assemble a simple X-ray telescope as indicated in Fig. 7.9. Notice again the essential role played by advanced electronic circuitry in the pulse height discrimination systems and in the anticoincidence circuits. This type of telescope has been flown with outstanding success on a number of small X-ray satellites, among which the UHURU satellite is historically perhaps the most important example.

7.3.1 *The UHURU X-ray satellite*

The UHURU X-ray satellite (Fig. 7.10(*a*)) was launched from the coast of Kenya in December 1970. The satellite package consisted of two beryllium window proportional counters, each of area 10^{-3} m^2. They pointed in opposite directions perpendicular to the rotation axis and the telescope employed mechanical collimators which gave beam-widths (in fact, 'full-width half maxima') of $\frac{1}{2}^\circ \times 5^\circ$ and $5^\circ \times 5^\circ$ as illustrated in Fig. 7.10(*b*). The satellite completed one 360° rotation every ten minutes. The proportional counters were sensitive to X-ray photons in the energy range 2–10 keV. The UHURU X-ray telescope provides an excellent simple example of the way in which the observing capability of the telescope is related to its technical specification. Let us take the various items in turn.

First, we can work out the *limiting sensitivity* of the telescope. In X-ray astronomy, the limiting sensitivity is determined by the unwanted background events which cause signals to be registered by the detector. There are two types of unwanted background. First, there is the background counts per second B_1 in the detector itself due to inadequate rejection of γ-rays and cosmic rays. This is an intrinsic property of the satellite and telescope configuration; in the case of the UHURU satellite, this background amounted to about 10 counts s^{-1}. The second contribution to the background is the isotropic cosmic background of X-rays which is remarkably bright. The astrophysical origin of this background is not yet established but it results in a background X-ray flux against which discrete sources have to be detected. This background intensity B_2 is expressed in terms of the number of photons $m^{-2}\,s^{-1}\,sr^{-1}$ in the pass band of the X-ray telescope. The limiting sensitivity is thus determined by counting statistics of X-ray photons. We can derive an expression for the signal-to-noise ratio for faint X-ray sources in the following way. If the solid angle subtended by the beam of the telescope on the sky is Ω and t is the integration time on the source, the number of photons detected from the source itself is SAt where S is the flux density of the source in counts $m^{-2}\,s^{-1}$ and A is the area of the detector. The signal due to the cosmic rays and γ-rays in the detector is $B_1 t$ while the number of photons due to the cosmic X-ray background is $B_2 A\Omega t$. The total background signal is the sum of these. The source itself, which is normally very much fainter than these background signals, has to be distinguished above the noise in the background which is just the statistical fluctuations in the number of background counts. According to the standard statistical result, the fluctuations about the mean value N is just $N^{\frac{1}{2}}$. Therefore the rms fluctuations in the background signal are

$$\Delta N \approx [(B_1 + \Omega A B_2)t]^{\frac{1}{2}} \tag{7.1}$$

Therefore, the signal-to-noise ratio of an observation of a source of flux density S is

$$\frac{S(At)^{\frac{1}{2}}}{[(B_1/A) + \Omega B_2]^{\frac{1}{2}}}$$

If we require a source to be, say five times above the noise level to be reliably identified as a real source, the limiting flux density of the telescope for an integration time t is

$$S_{\min} = 5\left[\frac{(B_1/A) + \Omega B_2}{At}\right]^{\frac{1}{2}} \tag{7.2}$$

It is interesting to insert the observed values of B_1 and B_2 into this expression. The X-ray background in the energy range 2–10 keV is 10^{-11} W m^{-2} sr^{-1} and has an intensity spectrum which can be roughly described by $I(\omega)\,\mathrm{d}\omega \propto \omega^{-1}\,\mathrm{d}\omega$. It is a useful exercise to show that for the 5° collimator, the two contributions to the background signal are more or less equal but that, for the smaller field of view, only the background due to particle events is important. As a general rule, the cosmic X-ray background is unimportant as a source of noise for fields of view less than a few degrees.

In its normal mode of operation, the satellite scanned a single strip of sky for many revolutions. It is interesting to work out the faintest detectable source in a single day's observation and to compare this with the limiting flux density quoted in the UHURU catalogue of X-ray sources which is $S_{\mathrm{min}} = 1$ UHURU unit $= 1.7 \times 10^{-14}$ W m^{-2} in the 2–6 keV waveband. It is instructive to work out how long it would take to survey the whole sky to this level of sensitivity. These are left as exercises for the reader.

Another interesting exercise concerns perhaps the most spectacular discovery made by the UHURU satellite, the discovery of the pulsating X-ray sources. The narrow-beam telescope, with beam area $\frac{1}{2}° \times 5°$, recorded the X-ray flux detected by the telescope every 0.096 s and transmitted this information to the ground. The average flux density of the X-ray source Hercules X-1 (Her X-1) is 10^3 photons m^{-2} s^{-1} and its period of pulsation is 1.24 s. How far was the source above the noise when it was discovered to be pulsating? You will discover that the source was not very far above the noise and this is beautifully illustrated by Fig. 7.11 which shows the discovery record of the pulses from Her X-1. Obviously, it was not very far above the noise in a single pulsation period but, of course, Fourier or power spectrum analysis techniques could have been used to extract pulsations of very much lower amplitude from these types of data. These sources turn out to be rotating, magnetised, accreting neutron stars about which we will have much more to say in Volume 2, Section 15.5.

7.3.2 *The Einstein X-ray Observatory*

One of the most important X-ray astronomy missions following the great breakthroughs made by the UHURU Observatory was the HEAO-B X-ray satellite, which was named the Einstein X-ray Observatory. This was the first complete satellite observatory for X-ray astronomy and consisted of a suite of five scientific instruments for photometry, spectroscopy and imaging of celestial X-ray sources. The instrument package included an Imaging Proportional Counter (IPC), which contained a position sensitive proportional counter, a Focal Plane Crystal Spectrometer (FPCS) which was a variant of a Bragg crystal spectrometer, a Solid State Spectrometer, a Monitor Proportional Counter and, perhaps most important of all, a High Resolution Imaging Telescope (HRI). The observatory was launched into orbit on 13 November 1978 and the mission lasted $2\frac{1}{2}$ years.

The HRI is of special interest since it led to many of the most important discoveries of the mission. The key element in its design was the use of grazing

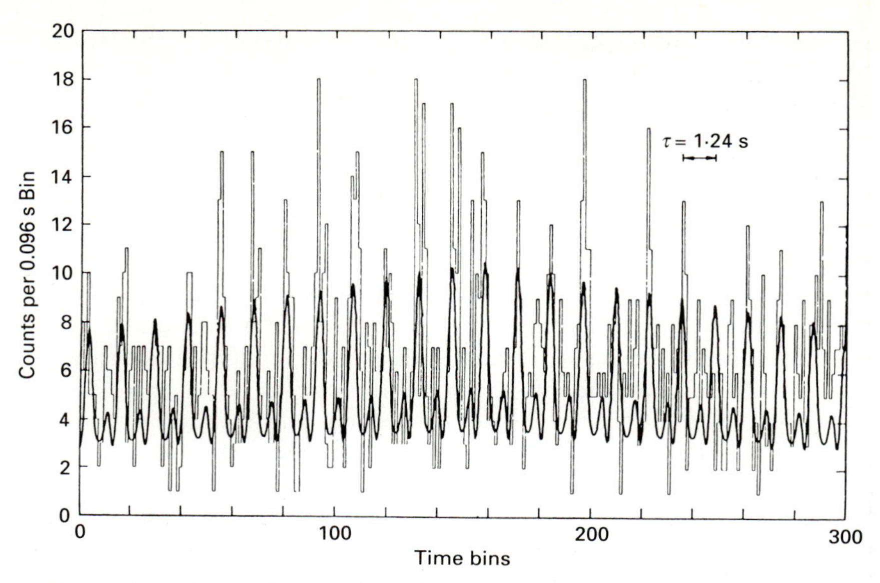

Figure 7.11. A section of the records of the intensity of the X-ray source Her X-1 as a function of time. The histogram shows the number of counts observed in successive 0.096 s bins. The continuous line shows a best-fitting harmonic curve to the observations, taking account of the varying sensitivity of the telescope as it swept over the source. (From H. Tananbaum, H. Gursky, E. M. Kellogg, R. Levinson, E. Schreier and R. Giacconi (1972). *Astrophys. J.*, **174**, L144.)

incidence optics to produce high angular resolution images of the fields of the X-ray sources. X-rays are reflected from the surfaces of conducting materials if the angle of incidence is large. Typically, at energies $h\nu \sim 1$ keV, reflection only occurs for angles between the surface and the direction of the incident photons of a few degrees and this angle decreases rapidly with increasing energy. Therefore, to focus the X-rays from a celestial source, a parabolic reflector would have to be very long indeed and the central portions of the reflector could not be used. The focal length of the telescope can be shortened if a second reflecting mirror is included, the preferred configuration consisting of a paraboloid–hyperboloid combination (Fig. 7.12). This arrangement only focuses X-rays incident on the annular area corresponding to the projected area of the paraboloid–hyperboloid surfaces. To increase the collecting area, a number of these combination mirrors can be nested one inside the other. This was the configuration used in the HRI of the Einstein X-ray Observatory. It enabled the telescope to take images over angular fields 25 arcmin in diameter with angular resolution better than 4 arcsec within the central 5 arcmin of the field.

The two-dimensional detector in the focal plane matched the angular resolution of the telescope and consisted of two microchannel plate detectors run in series. These detectors consist of an array of very narrow tubes along which a high potential difference is maintained. When an electron enters one end of a tube, it is accelerated and, on colliding with the walls, ejects further electrons which are, in

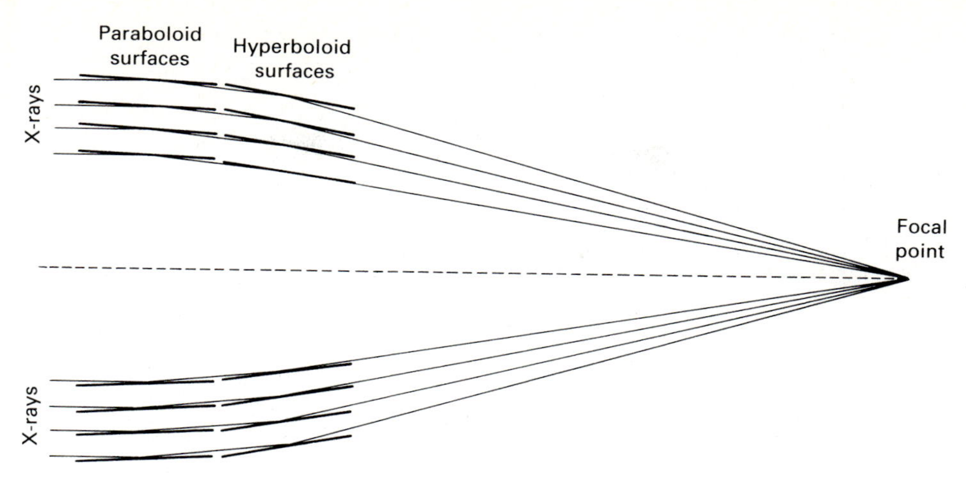

Figure 7.12. Illustrating the focussing of X-rays by a nested set of paraboloid–hyperboloid mirrors. This is the arrangement which was used in the Einstein X-ray Observatory. (From W. Tucker and R. Giacconi (1985). *The X-ray universe*, page 105, Cambridge: Harvard University Press.)

turn, accelerated down the tube and so on. As in the case of a proportional counter, the aim is to convert what starts off as a single photon into an intense burst of electrons. In the case of the HRI, there was a coating of magnesium fluoride on the front face of the first microchannel plate and when the two microchannel plates were run in series, an incident X-ray photon resulted in a burst of about 5×10^7 electrons at the output of the second plate. This burst of electrons was detected by a cross-grid charge detector so that the coordinates of the incident X-ray photon could be determined precisely.

The only other information we need is the effective area of the telescope and the background signal in the detector. Because the phenomenon of grazing incidence reflection is a strong function of photon energy and because of the presence of absorption edges in the material of the windows of the detector, the effective area is a strong function of energy (Fig. 7.13). The maximum effective area is found to occur at energies $\varepsilon \lesssim 1$ keV as expected and was typically about 0.04 m^2. The response of the detectors could be modified by inserting X-ray filters in the beam of the telescope, thus providing crude spectral information.

The background radiation in the detector, which was mostly due to charged particles, amounted to about 5×10^{-3} counts arcmin^{-2} s^{-1}. To give a rough estimate of what this means in terms of sensitivity, a point source of flux density 10^{-3} UHURU units, i.e. a source of flux density of about 1.7×10^{-17} W m^{-2} in the 2–6 keV waveband, could be detected at the 5σ level in an integration time of 50000 s.

To match the high performance of the imaging telescope, it was necessary that the pointing stability of the telescope also be very good, i.e. about 1 arcsec stability, but this was not attempted. Rather, the absolute pointing of the telescope was known with somewhat poorer accuracy but, at any instant, the pointing

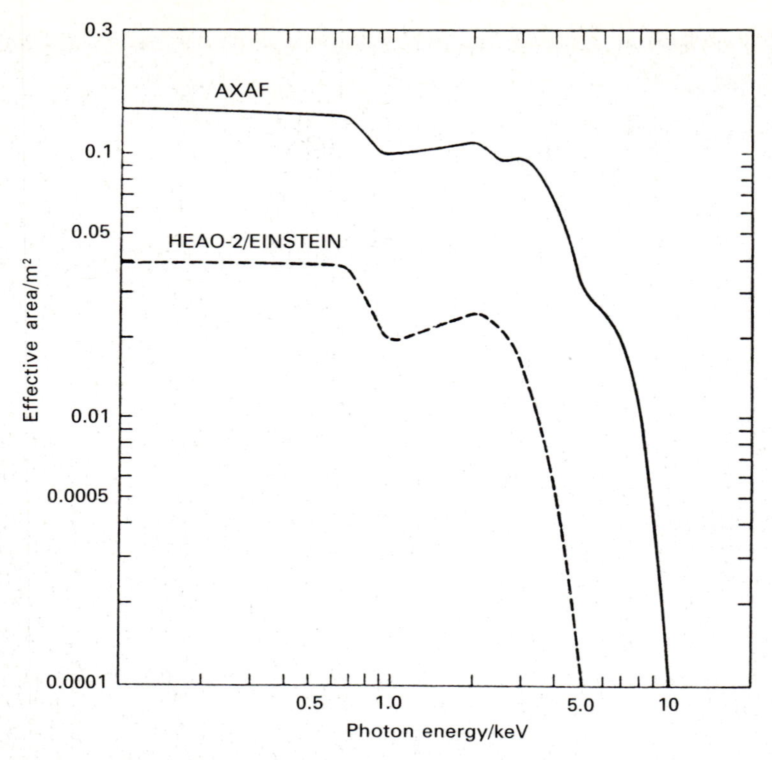

Figure 7.13. The effective areas of the HEAO-2/Einstein X-ray telescope and the AXAF telescope as a function of the energy of the X-rays. These relations refer to imaging on the axis of the telescope system. (From M. C. Weisskopf (1987). *Astrophys. Lett. and Commun.*, **26**, 3.)

direction of the telescope relative to bright stars was known precisely. Therefore, once the observation was completed, a map of the sky could be reconstructed with the full angular resolution of the high resolution imager. An example of the quality of the images obtained with the HRI is shown in Fig. 7.14.

A brief list of the types of science which were opened up by the Einstein observations would include the following: the detection of X-ray emission from all classes of star including the whole of the main sequence, supergiants and white dwarfs; the discovery of about 80 sources of the Andromeda Nebula (M31) and a similar number of sources in the Magellanic Clouds; high resolution images of the X-ray emission from clusters of galaxies; the detection of X-ray emission from all known classes of active galactic nuclei including the most extreme quasars known; deep X-ray source counts extending to flux densities about 10^3 times fainter than the limits of the UHURU catalogue. All branches of astronomy have been significantly influenced by observations made with the Einstein Observatory and a selection of scientific highlights is given by Tucker and Giacconi in their book *The X-ray Universe*.

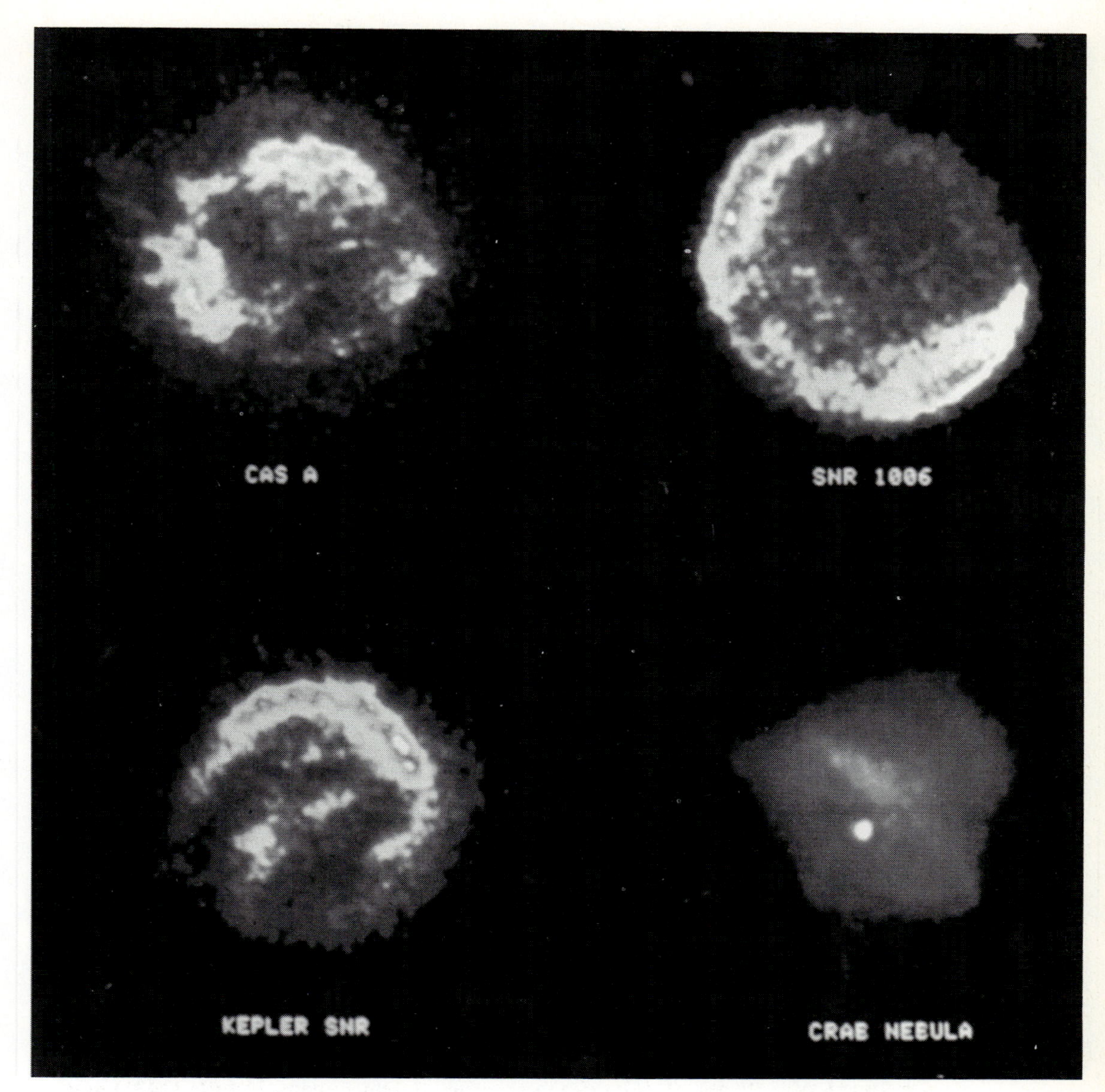

Figure 7.14. X-ray images of four supernova remnants observed by the HRI of the Einstein X-ray Observatory. Cas A, the supernova of 1006 and Kepler's supernova are shell-like X-ray sources. In the case of the Crab Nebula, the intense central X-ray source pulsates at the frequency of the pulsar which is the energy source for the high energy particles in the Nebula. (From W. Tucker and R. Giacconi, (1985). *The X-ray universe*, pages 116–17, Cambridge: Harvard University Press.)

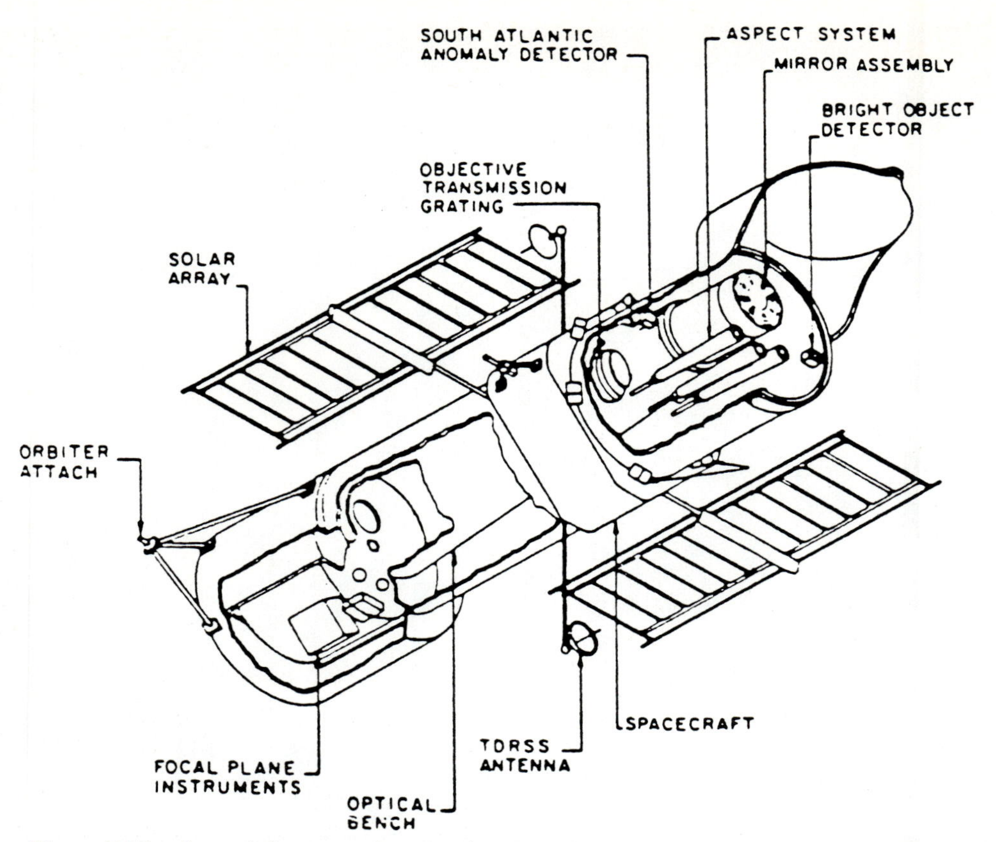

Figure 7.15. An artist's conception showing the principal components of the AXAF X-ray observatory. (From M. C. Weisskopf (1987). *Astrophys. Lett. and Comm.*, **26**, 2.)

7.3.3 *The Advanced X-ray Astronomy Facility (AXAF) and other future X-ray missions*

There is a natural progression from facilities such as the Einstein Observatory to large dedicated observatories which are planned to have long lifetimes in space and which are designed to be used by the astronomical community at large. In the NASA astronomy programme, the concept has developed of a series of 'Great Observatories' which will be long-lived large astronomical observatories in space. Four of these are planned, the Hubble Space Telescope (HST), the Gamma-Ray Observatory (GRO), the Advanced X-ray Astronomy Facility (AXAF) and the Space Infrared Telescope Facility (SIRTF), and we will meet them all in this and the following chapter. They are all characterised as being large telescopes which will be launched by the Space Shuttle – they all weigh about 10 tonnes or more and each will completely fill the cargo bay of the Shuttle.

AXAF is the first of these we meet and it is designed to be a complete observatory in space. We will simply highlight the distinctive features of the

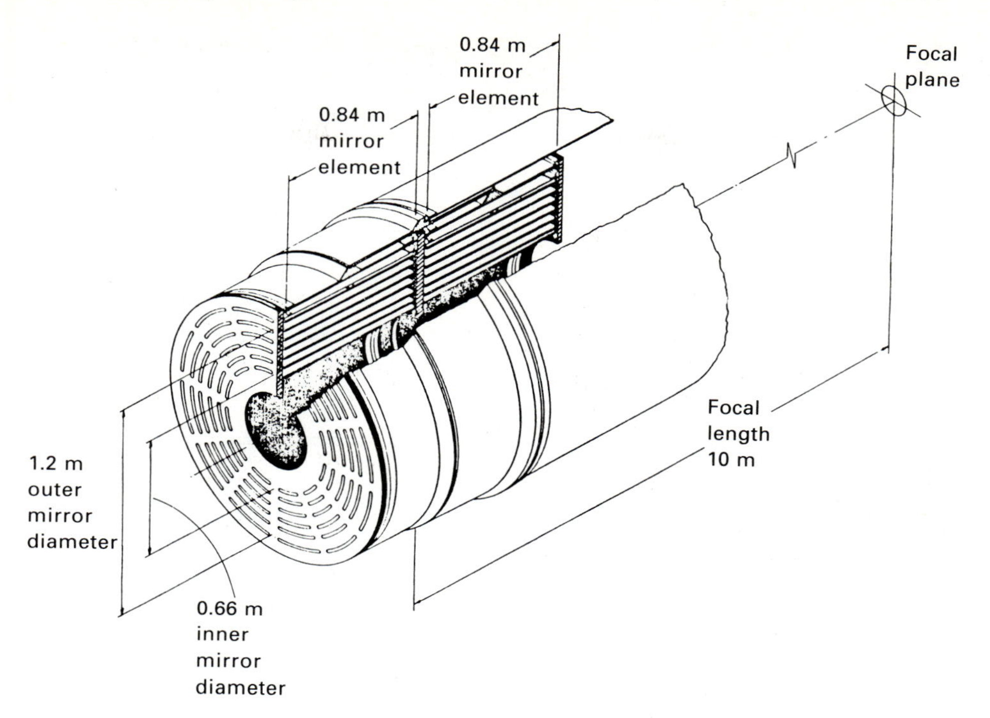

Figure 7.16. The AXAF High Resolution Mirror Assembly baseline design which consists of six nested paraboloid–hyperboloid pairs. (From L. P. van Speybroeck (1987). *Astrophys. Lett. and Commun.*, **26**, 129.)

mission. The overall layout of the telescope is shown in Fig. 7.15. The configuration is driven by the requirement that the telescope have outstanding image quality. This means that the telescope will use grazing incidence optics and, to attain high sensitivity, six large parabola–hyperbola mirrors will be nested to form the mirror assembly (Fig. 7.16). The focal length of the mirror assembly is 10 m and so the X-ray images are formed at the focal plane at the other end of the spacecraft. The scientific instruments are located within the aft shroud behind the focal plane. The telescope parameters are compared with those of the Einstein Observatory in Table 7.1 (*a*) which also lists the proposed package of scientific instruments. The effective area of the telescope is compared with that of the Einstein Observatory in Fig. 7.13. The major gains in sensitivity result, not only from the increased collecting area of the telescope but also from the very high angular resolution of the telescope which is an order of magnitude better than that of the Einstein Observatory. In addition the telescope will have good sensitivity at energies up to 8–10 keV. The goal is to launch AXAF in about 1997. Current thinking is that following the launch by the Shuttle, it will be serviceable by the Space Station which is planned to be permanently manned.

The scientific instruments listed in Table 7.1 (*b*) will by now be familiar to the reader. The table lists the instruments, their spectral resolutions and the type of detector used. These capabilities represent an enormous advance over the present

Table 7.1 (*a*) *Comparison of the Einstein Observatory and AXAF*

	HEAO-2/Einstein	AXAF
Number of mirror elements	4 nested pairs	6 nested pairs
Outer diameter (m)	0.58	1.2
Focal length (m)	3.44	10.0
Geometric area (m^2)	0.046	0.170
Resolution (arcsec)	4.0	0.5
Field of view (degrees)	1	1

Table 7.1 (*b*). *AXAF scientific instruments*

Instrument	Waveband (keV)	Resolution ($\lambda/\Delta\lambda$) (or $\Delta\varepsilon$)	Comments
Low energy transmission gratings	0.1–4.0	2000 at low energies 100 at 2 keV	
High energy transmission gratings	0.5–9.0	1000 at low energies 100 at 9 keV	
Bragg crystal spectrometer	A variety of selected energies within AXAF waveband	50–70 $\varepsilon < 0.5$ keV 200 $0.5 < 0.8$ keV 1000–2000 $0.8 < \varepsilon < 2$ keV 500–1000 $2 < \varepsilon < 8$ keV	
CCD imagers	An array of CCD devices 0.1–10 keV	150 eV	Quantum efficiencies $\geqslant 30\%$ over most 0.1–10 keV waveband
X-ray calorimeter	0.1–10 keV	10 eV	
Microchannel plate imager	0.1–8 keV	1 at 1 keV	Field of view 32×32 $(\text{arcmin})^2$ Quantum efficiencies 20–50 % 0.1–3 keV 10–20 % 3–8 keV

generation of X-ray facilities. It will not, however, be the only advanced X-ray telescope to fly during the 1990s. Other key missions will include the ROSAT observatory (standing for ROentgen SATellit) which is a joint Germany–UK–USA mission and which will perform a complete deep survey of the sky to roughly the depth of the deep Einstein surveys. It will provide the first large X-ray survey of the whole sky and the resulting catalogues of X-ray sources will contain

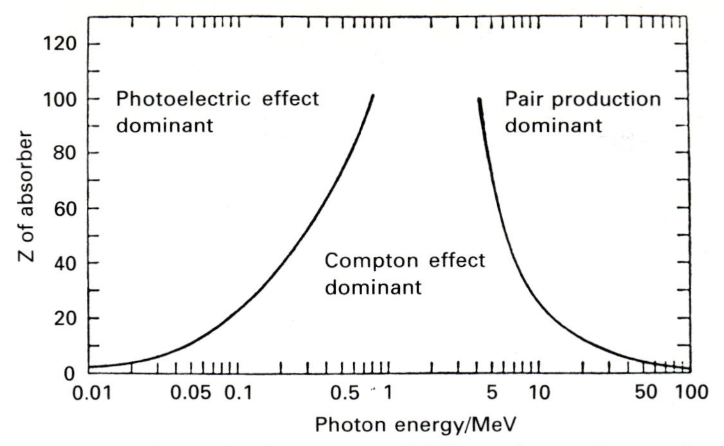

Figure 7.17. The relative importance of different forms of energy loss mechanisms for γ-rays as a function of photon energy and the atomic number of the material. (From R. Hillier (1984). *Gamma-ray astronomy*, page 52, Oxford: Clarendon Press.)

hundreds of thousands of objects. It will provide an essential finding list for sources which will be followed up by the AXAF Observatory. ROSAT was launched successfully in the first week of June 1990.

Another large mission is the XMM mission of the European Space Agency (the acronym originally meant X-ray Multi-Mirror Telescope). Unlike the AXAF Observatory which includes imaging as well as spectroscopic instruments, the XMM mission is primarily dedicated to X-ray spectroscopy, the goal being to provide the largest collecting area possible and to maximise the throughput of the X-ray spectrometers (Peacock, Taylor and Ellwood, 1990). The baseline design incorporates 58 nested paraboloid–hyperboloid mirrors and high sensitivity CCD X-ray detectors. The effective area of the telescopes at 2 keV will be about 0.6 m^2, complementing the capabilities of AXAF. The telescope will be launched in 1998.

7.4 γ-ray telescopes

Our task is now to build the detectors for γ-rays described in Section 6.5 into γ-ray telescopes. It is convenient to split up the energy range into three separate regions depending upon which of the three processes of energy loss for the photons is dominant. Fig. 7.17 shows the relative importance of the processes of photoelectric absorption, Compton scattering and electron–positron pair production as a function of the atomic number Z of the absorber. In the low energy range $\varepsilon \lesssim 0.1$–0.3 Mev, *photo-electric absorption* is dominant and the types of detector and telescope already described in the context of X-ray astronomy can be used up to energies of about 0.5 MeV if materials with high atomic number are used as detector materials. The scintillation counters and solid state detectors described in Section 6.5 are commonly used in these telescopes.

At high energies, $\varepsilon \gtrsim 30$ MeV, *electron–positron pair production* is the dominant loss process and we have already described how arrays of spark chambers can be

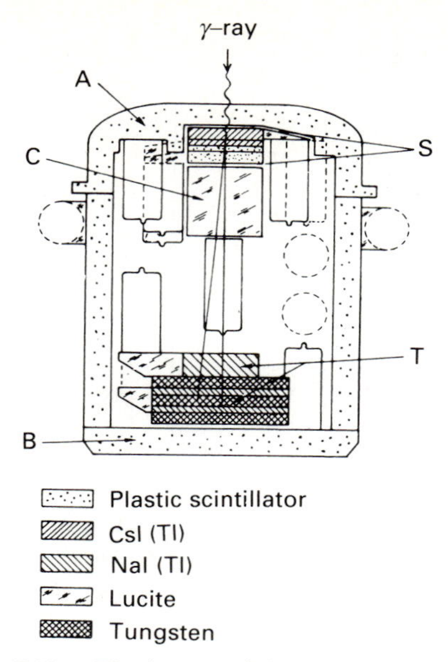

Figure 7.18. The layout of the components of the OSO-III γ-ray telescope. The main components are: A and B, anticoincidence shields; S, the converter consisting of about one radiation length of caesium iodide; C, the lucite Cherenkov counter system; T, sodium iodide scintillation counters interleaved with tungsten. The diameter of the instrument was less than 30 cm. (From W. L. Kratuschaar, G. W. Clark, G. P. Garmire, R. Borken, P. Higbie, C. Leong and T. Thorsos (1972). *Astrophys. J.*, **177**, 341.)

used to register the conversion of a high energy γ-ray into an electron–positron pair. At extremely high energies, $\varepsilon \gtrsim 300$ GeV, a completely different approach is possible through the detection of the Cherenkov radiation emitted by the electron–photon cascades initiated by the γ-ray in the upper atmosphere (Section 7.4.4).

The difficult region for observation is the intermediate range 0.5–10 MeV in which Compton scattering is dominant but the other processes cannot be wholly neglected. Let us deal separately with each of these energy ranges.

7.4.1 *γ-ray telescopes at energies $\varepsilon \gtrsim 30$ MeV*

In this waveband, electron–positron pair production by γ-rays is the principal loss mechanism. A *converter* is needed which converts the γ-ray into an electron–positron pair which can then be detected by conventional means. The first successful γ-ray telescope was flown on the OSO-III mission (Orbiting Solar Observatory) in 1967–8, although evidence for cosmic γ-rays had previously been found by the Explorer II satellite. The layout of the detectors is shown in Fig. 7.18 which includes all the important features of a successful γ-ray telescope. The γ-ray is converted into an electron–positron pair in the caesium iodide crystal and the energy loss rates of these are measured in the lucite Cherenkov counters which provide an estimate of dE/dx. The electron and positron are then brought to rest

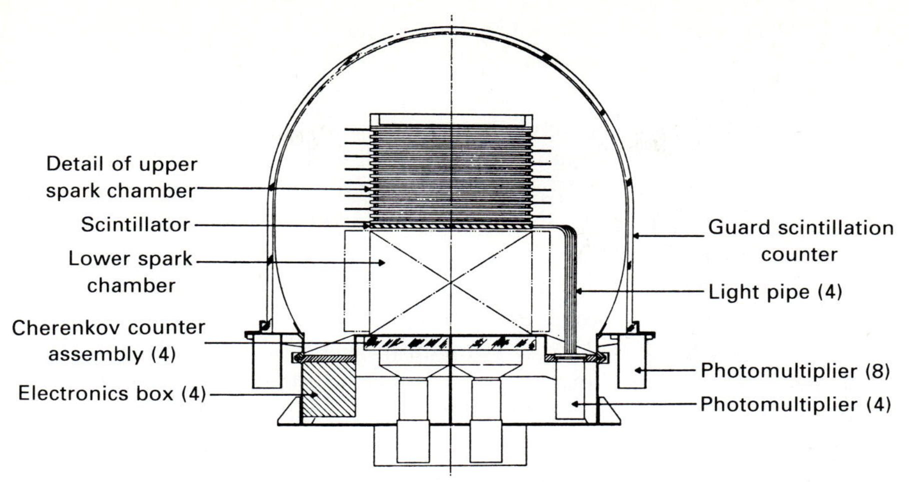

Figure 7.19. The SAS-II γ-ray telescope showing the arrays of spark chambers within the spherical anticoincidence shield. (From C. E. Fichtel (1974). *Phil. Trans. Roy. Soc. Lond.*, **277**, 367.)

in the scintillation counters and ionisation chambers below the Cherenkov detectors. The major problems with these telescopes are as follows.

(i) The anticoincidence rejection techniques for cosmic ray events must be very good indeed. The cosmic γ-rays in this energy range constitute only about 1 in 100 000 of the events recorded by the instrument, most of which are due to cosmic rays.

(ii) Below about 30 MeV, Compton scattering becomes important and therefore, at low energies, γ-rays are scattered into the telescope from large angles of incidence and hence the beam of the telescope becomes poorly defined at these energies. For this reason, this type of detector is only used to detect γ-rays with energies significantly greater than 30 MeV and the energies of these photons are determined from the energy losses in the ionisation chambers. Therefore, at energies greater than about 100 MeV, the beam of the telescope is determined by the geometry of the detector.

It was from observations with this telescope on the OSO-III satellite that the γ-radiation from our own Galaxy was discovered as well as the existence of a cosmic γ-ray background established. The resolution of this telescope was rather poor, about 20°, and much more sophisticated devices have been flown since that time, in particular through the use of spark chambers which enable the arrival directions of the γ-rays to be determined much more accurately.

Spark chambers were used in the SAS-II and COS-B satellites with great success. The layout of the detectors was similar in the two experiments, the configuration of the SAS-II experiment being shown in Fig. 7.19. The key advance is the ability to measure the paths of the electron and positron through the spark chamber array (see Fig. 6.16) and hence to work out the initial direction of the γ-ray. The limitation upon the accuracy with which the direction of arrival of the γ-ray can be measured is set by Coulomb scattering of the paths of the electron and positron. For γ-rays of energy 100 MeV and greater, the uncertainty in the directions of the electrons and positrons, and hence of the incident γ-ray, is of the order of 1–2°. Therefore, if we know the orientation of the satellite, relatively high resolution studies of the distribution of γ-rays on the celestial sphere can be made. The satellite packages included a considerable amount of sophisticated electronic logic so that the details of all the relevant γ-ray events could be transmitted to the Earth.

The SAS-II spacecraft was operational for seven months before an electronic failure resulted in premature termination of the experiment. In contrast, the COS-B satellite remained operational for $6\frac{1}{2}$ years during which time it collected about 100000 γ-rays. The beautiful γ-ray map of the Galactic plane made by the COS-B satellite is displayed in Fig. 1.8(*f*). In addition to the diffuse emission from the plane of our Galaxy, the COS-B satellite detected 25 discrete γ-ray sources including the quasar 3C 273 and pulsed γ-ray emission from the Crab and Vela pulsars at the period of these pulsars. We will have a great deal more to say about all these observations when we study the astrophysics of the interstellar gas, X and γ-ray pulsars and active galactic nuclei.

7.4.2 *Compton telescopes*

In the intermediate energy band 0.3–30 MeV, Compton scattering is the dominant loss process. The principles by which the energy and direction of arrival of these γ-rays can be determined is best illustrated by the Compton telescope shown in Fig. 7.20 which was developed by the group at the Max Planck Institute for Extraterrestrial Physics at Garching, West Germany. The telescope consists of two arrays of detectors separated typically by about 1 m. The upper array of detectors acts as the 'converter' and consists of 16 cells of liquid scintillator. The incoming γ-ray undergoes a Compton scattering interaction within one of the cells of the detector, liberating an electron. This electron deposits its energy within the scintillating liquid and, in the usual manner, produces an optical signal proportional to the energy loss which is measured by the photomultiplier tubes attached to each cell (see Section 6.5.2). The scattered γ-ray leaves the converter with reduced energy and at an angle ϕ with respect to its arrival direction. It is then absorbed in one of the elements of the sodium iodide scintillation detector array which is separated by about a metre from the converter and which consists of 32 blocks, each with its own photomultiplier. By noting the detectors which are involved in the two detector arrays, the trajectory of the scattered γ-ray between them can be found.

From our analysis of the Compton scattering process (Section 4.3.2), we know that the frequencies of the photon before and after the scattering and the scattering angle ϕ are related by

$$\frac{\omega}{\omega'} = 1 + \frac{\hbar\omega}{m_e c^2}(1 - \cos\phi) \qquad (7.3)$$

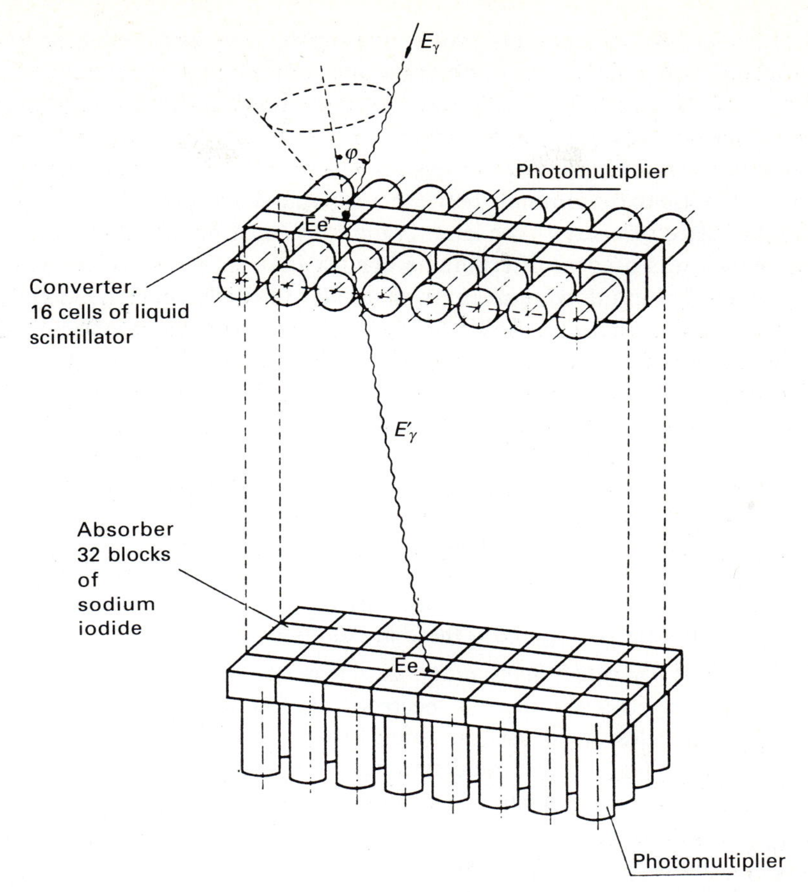

Figure 7.20. A schematic drawing of the MPIfEP Compton telescope for γ-rays, illustrating the principle of measuring the scattering angle ϕ. The distance between the converter and the absorber is 1.2 m. (From V. Schonfelder, F. Graml and F.-P. Penningsfeld (1980). *Astrophys. J.*, **240**, 350.)

Converting this relation into energies, we can write the energy loss in the converter as $E_c = \hbar\omega - \hbar\omega'$ and the remaining energy loss in the absorber as $E_a = \hbar\omega'$ and hence

$$\cos\phi = 1 - \frac{E_c m_e c^2}{E_a(E_c + E_a)} \tag{7.4}$$

Thus, the angle ϕ can be found from the energies deposited in the converter and the absorber, assuming that all the energy of the scattered photon is absorbed in the second array. Therefore, we now know the angle of the incoming γ-ray with respect to its path between the two detector arrays. It could have arrived anywhere along a cone of half-angle ϕ as illustrated in Fig. 7.20. The trick is to plot these cones for all the events incident upon the telescope and then sources can be identified by the clustering of the intersections of the cones at the positions of sources. In the MPI experiment, 3 MeV photons were located within an annulus 1° wide. To achieve this quality of observation a number of clever procedures were developed. For example, in addition to the anticoincidence shield and its associated electronics, the arrival times of events in the two detectors were measured very precisely as well as the pulse shapes of the events. The timing was sufficiently precise to discriminate between γ-rays passing downwards and upwards through the detector arrays – the latter could thus be eliminated. The pulse shape discrimination was used to discriminate against low energy neutrons created by cosmic ray spallation interactions in the material of the telescope. The significance of this technique is that knowledge of the direction of arrival of the γ-rays with much increased precision means that the effective angular resolution of the detector is much improved which lowers the effective background signal.

7.4.3 *The Gamma-Ray Observatory* (*GRO*)

The second of the Great Observatories to be launched will be the GRO. An artist's impression of this large space observatory is shown in Fig. 7.21. There are four scientific instruments on board, each of which should have a familiar ring by now. On the main body of the spacecraft there are three instruments. Going from left to right across Fig. 7.21, these are as follows.

(i) The *Oriented Scintillation Spectroscopy Experiment* (OSSE). This device is a scintillation spectrometer with a collimated field of view. The telescope consists of four identical phoswich detectors of NaI(Tl) and CsI(Na). The detectors can be rotated about a single axis and they are used in pairs. Whilst one of a pair is observing the source, the other is observing a blank piece of sky. After a period of roughly two minutes, the roles of the detectors are reversed.

(ii) The *Compton Telescope* (COMPTEL) is of exactly the same design as that described in Section 7.4.2 and is sensitive in the energy range 1–30 MeV.

(iii) The *Energetic Gamma-Ray Experiment* (EGRET) is a classical spark chamber experiment for γ-ray photons with energies greater than about 20 MeV. It is of exactly the same type as that described in Section 7.4.1 but is very much larger. Photons with energies up to 30 GeV should be stopped within the detector volume.

Finally, at the four corners of the satellite are the four pairs of detectors of the fourth instrument, the *Burst And Transient Source Experiment* (BATSE). These detectors form an all-sky monitor for transient and burst sources of γ-rays. These are uncollimated scintillation detectors made of NaI(Tl). One of each pair is a large area detector with an anticoincidence shield of plastic scintillator. The location of burst sources can be estimated from the strength of the signal detected by the four detectors. The second of each pair is a smaller NaI(Tl) crystal optimised for good energy resolution so that the spectra of burst and transient sources can be measured. Whenever BATSE detects a burst source, it triggers the other three GRO instruments.

Figure 7.21. A schematic view of the GRO showing the locations of the four instruments. OSSE is on the left, EGRET is on the right and COMPTEL is in the middle. The eight BATSE detectors are arranged in pairs at the four corners of the spacecraft. (From V. Schonfelder (1990). *Adv. Space. Res.*, **10**, No. 2, 243.)

The main characteristics of these instruments are summarised in Table 7.2. In general terms, the four instruments on the GRO are more than ten times more sensitive than any other γ-ray telescopes previously flown. The GRO was successfully launched by the Space Shuttle in April 1991. The significance of these observations for high energy astrophysics will become apparent as our story unfolds.

7.4.4 *Ultrahigh energy γ-ray telescopes*

We described in Section 4.6 how ultrahigh energy γ-rays entering the top of the atmosphere give rise to ultrahigh energy electron–positron pairs which emit Cherenkov radiation because they travel at velocities greater than the local speed of light in the atmosphere, $v > c/n$ where n is the refractive index of the medium. The prediction that Cherenkov light pulses from cosmic ray showers should be detectable from the surface of the Earth was made by Blackett in 1948 and confirmed by observation several years later by Galbraith and Jelley. More recently, it has become possible to use the technique to study the flux of ultrahigh energy γ-rays incident on the top of the atmosphere using the same technique. An excellent review of the development of ultrahigh energy γ-ray astronomy is given by Ramana Murthy and Wolfendale (1986).

The threshold for the emission of Cherenkov radiation depends upon the

Table 7.2. *Summary of some of the characteristics of the four instruments on board the GRO*

	OSSE	COMPTEL	EGRET	BATSE
Energy range	0.1–10 MeV	1–30 MeV	20–30 MeV	50 keV–1 MeV (large area detector) 20 keV–30 MeV (spectroscopy detector)
Field of view	Collimator with 3.8 × 11.4° full-width half maximum	1 sr	Opening angle 45°	Full unocculted sky
Angular resolution		0.75–2.2°	0.4–2°	1–10° location accuracy
Energy resolution	8 % at 662 keV 3.2% at 6.13 MeV	5–8 %	15 %	6 energy bands (large area detector) 7 % at 662 keV (spectroscopy detector)

refractive index n of the atmosphere as described in Section 4.6. The three key properties of Cherenkov radiation which enable the properties of the γ-rays to be determined are

(i) The Cherenkov condition $v > c/n$;
(ii) The angle of the wavefront is given by $\cos\theta = c/nv$;
(iii) The intensity of emission per unit path length is

$$\frac{dU(\omega)}{dx} = \frac{\omega e^2}{4\pi\varepsilon_0 c^3}\left(1 - \frac{c^2}{n^2 v^2}\right) \tag{4.72}$$

These relations can be used to work out the properties of γ-rays which can be detected by their Cherenkov light emission in the atmosphere (Fig. 7.22).

To understand this diagram, we need a model for the variation of the refractive index of the atmosphere with height. The variation of the density of the atmosphere can be represented by an exponential function

$$\rho = \rho_0 \exp(-x/x_0): \quad \rho_0 = 1.35 \text{ kg m}^{-3}, \quad x_0 = 7.25 \text{ km} \tag{7.5}$$

where x_0 is the scale height of the density distribution and ρ_0 is the density of the atmosphere at sea-level. The path-length l in kg m^{-2} through the atmosphere can be written

$$l = \int_{\infty}^{x} \rho \, dx = 10000 \exp(-x/x_0) \text{ kg m}^{-2} \tag{7.6}$$

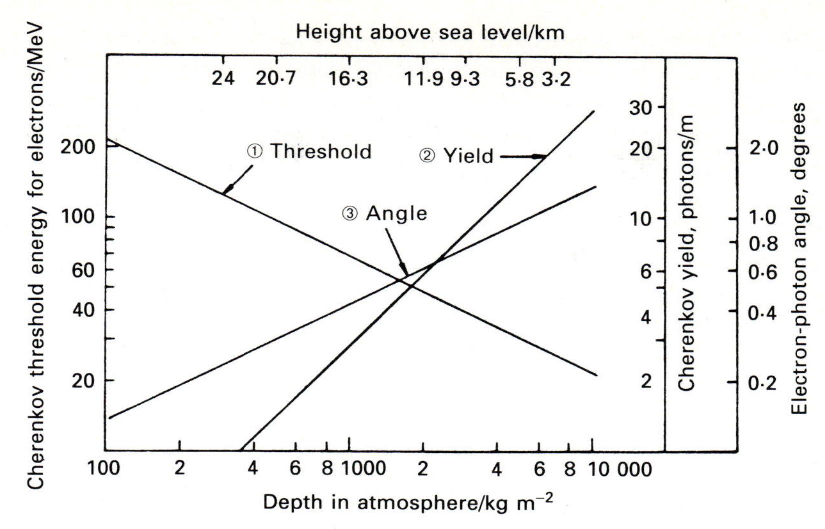

Figure 7.22. The variation with depth in the atmosphere of different properties of the Cherenkov radiation associated with electron–positron pairs created by very high energy γ-rays. (1) is the threshold energy for the electrons to emit Cherenkov radiation; (2) is the intensity of Cherenkov radiation; (3) is the angle between the direction of propagation of the wave and the direction of motion of the electron. (From P. V. Ramana Murthy and A. W. Wolfendale (1986). *Gamma-ray astronomy*, page 182, Cambridge: Cambridge University Press.)

Thus, the total path-length through the atmosphere is $l_0 = 10\,000$ kg m^{-2}. For gases, for which the refractive index is just slightly greater than unity, we can write $n = 1 + \alpha\rho$ where α is a constant and $\alpha\rho$ is always very much less than 1. Let us work out the threshold for the emission of Cherenkov radiation using the relation $v/c = 1/n$. Expressing this condition in terms of the Lorentz factor of the electron $\gamma = (1 - v^2/c^2)^{-\frac{1}{2}}$, we find

$$\gamma_t = (1 - v^2/c^2)^{-\frac{1}{2}} = (1 - 1/n^2)^{-\frac{1}{2}} \approx 1/(2\alpha\rho)^{\frac{1}{2}} \tag{7.7}$$

Inserting the refractive index of the atmosphere at sea level $n = 1 + 2.763 \times 10^{-4}$ into this relation, we find $\gamma_t = 40$; i.e. $E_t \approx 20$ MeV, in agreement with the value shown in Fig. 7.22. The variation of γ_t with pathlength through the atmosphere can be written

$$\gamma_t \propto \rho^{-\frac{1}{2}} \propto \exp(x/2x_0) \propto l^{-\frac{1}{2}} \tag{7.8}$$

in agreement with Fig. 7.22. In the same approximation, we can use equation (4.72) to work out the intensity of emission as a function of path length

$$I(\omega) \propto (1 - c^2/n^2v^2) \propto \rho \propto l \tag{7.9}$$

Similarly, the angle of the wavefront to the direction of the fast particle is

$$\cos\theta \approx 1 - \theta^2/2 = c/nv \approx 1 - \alpha\rho$$
$$\theta \propto \rho^{\frac{1}{2}} \propto l^{\frac{1}{2}} \tag{7.10}$$

The significance of this diagram is that it shows that the threshold energy for Cherenkov emission decreases as the particles penetrate further through the atmosphere and that the intensity of the radiation increases as well. Thus, the intensity of Cherenkov radiation depends upon the development of the electron–photon cascade through the atmosphere. For example, consider a 300 GeV γ-ray. The critical energy of air is 83 MeV and hence $E \approx 3.5 \times 10^3 E_c$. Referring to Fig. 4.18, we see that the maximum development of the cascade corresponds to about seven or eight radiation lengths from the top of the atmosphere. Since the radiation length for air is 365 kg m^{-2}, the maximum development occurs at a depth of about 3000 kg m^{-2}, i.e. about a height of 10 km above sea-level.

The electrons and positrons responsible for the Cherenkov light are absorbed in the upper atmosphere but their Cherenkov light propagates to sea-level where it can be detected as a very short pulse of photons. For a 300 GeV γ-ray, for example, a few million photons are produced and the light flash as detected at sea-level lasts only about 10 ns. These photons are typically emitted within a cone of about 1–2°, this being partly due to the fact that the Cherenkov radiation is emitted within a cone with half-angle about 1–2° and partly to the fact that the electrons and positrons in the electron–photon shower suffer Coulomb scattering and hence the emitting particles do not all emit along the same axis. The net result is that the Cherenkov radiation is not focussed at a single point on the ground but is diffused over an area of up to about 100 m from the axis of the shower. For the example given above, the Cherenkov photon density near the central axis of the shower is about 7 m^{-2}.

The ultrahigh energy γ-ray telescope therefore consists of an array of optical telescopes which act as light flux collectors (Fig. 7.23). The signal has to be detected against the background of night-sky photons which amounts to about 7×10^{11} photons m^{-2} s^{-1} sr^{-1}. In addition, the Cherenkov light from the γ-rays has to be distinguished from that due to cosmic ray showers. The night-sky background can be eliminated by very fast accurate timing of the arrival times of the light pulses as detected by the separate elements of the telescope array. It is more difficult to discriminate against cosmic ray showers but in general their Cherenkov radiation is more diffuse because of the production of pions and other particles at larger angles to the core of the shower. At high energies, $E > 10^{13}$ eV at which the muons produced in cosmic ray showers reach the surface of the Earth, a muon detector can be used as an anticoincidence detector so that only the muon-free showers, which should be γ-rays, are detected. The observing technique for steady sources is to nod on and off the source location, or, in the case of pulsating sources, to observe the source continuously and search for correlated signals at the pulsation frequency.

This is a new and difficult type of astronomy because the source intensities are very weak. In addition, there is the possibility that the sources of the very high energy photons are variable. For example, on occasion, the Crab Nebula has been detected as a source of ultrahigh energy γ-rays at about the 5σ level of significance and on other occasions, it is not detected. Similar results have been found for the

Figure 7.23. A photograph of the 10 m reflecting night sky Cherenkov telescope of the Fred Lawrence Whipple Observatory at Mount Hopkins, USA. (Courtesy of the Smithsonian Institution.)

active galaxy Centaurus A and possibly SS433. The pulsating sources such as the Crab and Vela pulsars have shown significant fluxes of ultrahigh energy γ-rays at the expected periods. These results are of the greatest interest because these photons are the most energetic yet detected and provoke a number of intriguing astrophysical problems which we will deal with in due course.

7.5 Neutrino telescopes

Having developed the concepts necessary for understanding the construction of detectors and telescopes for high energy photons, it is intriguing that many of the same ideas are found in the systems designed to detect neutrinos of astronomical origin. Neutrinos are generated in very large numbers in astronomical objects. For example, stars like the Sun generate a neutrino luminosity of about 9×10^{24} W (Bahcall 1989) and when a neutron star forms during a supernova explosion, a significant fraction of the binding energy of the neutron star is released in the form of neutrinos. It is confidently expected that there is a universal background of neutrinos with temperature about 1.7 K which originated in the early Universe when the neutrinos were in thermal equilibrium with the matter. The problem is that the neutrino has an extremely small cross-section for interaction with matter, for example the electron–neutrino scattering cross-section is only about 10^{-49} m^2, and so enormous detectors are required to register even a few neutrino events from astronomical sources. Remarkably, neutrinos from both the Sun and a supernova have now been detected. Solar neutrinos originating in the nuclear reactions responsible for the luminosity of the Sun have been detected in a 20 year campaign by Dr Raymond Davis and his colleagues at the Homestake gold-mine in the USA and neutrinos were detected from the supernova SN1987A which exploded in the Large Magellanic Cloud on 23 February 1987 by detectors primarily designed to detect the products of proton decay. These remarkable stories are elegantly described by Bahcall (1989).

Let us deal first of all with the detectors which are closest in spirit to those which we have described already. The large proton (or nucleon) decay experiments are designed on the principle that the proton might decay in many ways but these all end up producing a fast electron or positron. Some examples of possible decay modes of the proton might be

$$
\begin{aligned}
&p \rightarrow e^+ + \pi^0 \\
&p \rightarrow \mu^+ + K^0; \quad \mu^+ \rightarrow e^+ + \nu_e + \bar{\nu}_\mu \\
&p \rightarrow e^+ + \gamma \\
&p \rightarrow e^+ + e^+ + e^- \\
&\quad \vdots \\
&\quad \text{etc.}
\end{aligned}
$$

It is the fast electrons resulting from such possible decay modes that the Kamioka Nucleon Decay Experiment (Kamiokande) in Japan and the Irvine–Michigan–Brookhaven (IMB) experiment in the USA were designed to detect.

The requirements of these experiments are not dissimilar from those of γ-ray telescopes. There is a clear requirement to shield the detector from cosmic ray particles which would cause unwanted high energy secondary electrons in the detector volume. In the case of the Kamiokande experiment, the apparatus is located underground at a depth of about 1 km in the Kamioka metal mine of the Mitsui Mining and Smelting Company which is located in the Japanese Alps about

Figure 7.24. The installation of the phototubes in the Kamiokande II experiment. The picture shows workers installing the phototubes on the walls of the water Cherenkov detector in 1983. (From J. N. Bahcall (1989). *Neutrino astrophysics*, page 387, Cambridge: Cambridge University Press.)

300 km west of Tokyo (Fig. 7.24). The active volume of the detector consists of a total of 3000 tonnes of very pure water and is enclosed in a large steel tank. In the proton decay experiments, only the inner 680 tonnes are used because of the need to eliminate background events. The active volume is viewed by a large number of photomultipliers tubes which cover about 20% of the inner surface of the tank. The objective is to detect the Cherenkov radiation of any fast electrons released in the active volume. When used as a neutrino detector, this experiment has the advantage that the arrival directions of the neutrinos can be estimated from the directionality of the Cherenkov radiation emitted by the fast electron. In addition, the time of arrival of the neutrinos can be accurately measured as well as spectral information from the intensity of the Cherenkov radiation. Anticoincidence detectors surround the detector and great care has to be taken to reduce the background due to natural radioactivity. By taking these precautions and many others, the threshold energy of the Kamiokande experiment for the detection of fast electrons has been reduced to 7.5 MeV. The IMB experiment has a larger active volume but is sensitive only to higher energy electrons, $E > 20$ MeV. So far,

no evidence has been found for fast electrons resulting from the decay of the proton, the corresponding half-life of the proton being estimated to be greater than about 10^{33} years. This result is of considerable concern to particle theorists because some of the preferred versions of Grand Unified Theories of elementary particles predict a lifetime for the proton which is significantly less than this figure.

The process of the greatest interest for the detection of neutrinos is the scattering of neutrinos by electrons

$$\nu + e \rightarrow \nu' + e'$$

The high energy neutrino transfers energy to the electron and since the latter acquires a relativistic velocity in the scattering process, the electron emits Cherenkov radiation (Section 4.6) which is detected by the photomultiplier tubes. One of the most remarkable results of modern astrophysics has been the simultaneous detection of neutrinos by the Kamiokande and IMB experiments from the supernova SN1987A. Within a period of 12 s at about 07 h 36 m UT on 23 February 1987, the Kamiokande experiment registered 12 neutrino events with electron energies in the range 6.3–35.4 MeV and the IMB experiment registered 8 events with energies in the range 19–39 MeV. In both cases, these signals were far above the background signal. The supernova was first observed optically the following day. The astrophysical implications of these observations for our understanding of stellar evolution are profound and are discussed in Volume 2, Section 15.2.2.

The most famous astronomical neutrino experiment uses a different technique to measure the rate of generation of neutrinos from the nuclear burning regions of the Sun. Most of the neutrinos produced by the Sun result from the basic proton–proton reaction which initiates the pp chain

$$p + p \rightarrow {}^{2}H + e^{+} + \underset{\substack{\uparrow \\ \varepsilon_\nu \leqslant 0.420\ \mathrm{MeV}}}{\nu_e} : {}^{2}H + p \rightarrow {}^{3}He + \gamma;\ {}^{3}He + {}^{3}He \rightarrow {}^{4}He + 2p \tag{7.11}$$

These neutrinos are of relatively low energy with a maximum energy of 0.420 MeV. Higher energy electron neutrinos with energies up to 15 MeV are generated in a side chain of the main pp chain.

$$\left.\begin{array}{l} {}^{3}He + {}^{4}He \rightarrow {}^{7}Be + \gamma :\ {}^{7}Be + p \rightarrow {}^{8}B + \gamma \\ {}^{8}B \rightarrow {}^{8}Be^{*} + e^{+} + \underset{\substack{\uparrow \\ \varepsilon_\nu \leqslant 15\ \mathrm{MeV}}}{\nu_e} :\ {}^{8}Be^{*} \rightarrow {}^{4}He + {}^{4}He \end{array}\right\} \tag{7.12}$$

The predicted energy spectra of the neutrinos from these and other processes are displayed in Fig. 7.25.

In the famous ^{37}Cl experiment, (Fig. 7.26) these neutrinos interact with ^{37}Cl nuclei to produce the radioactive species ^{37}Ar through the reaction

$${}^{37}Cl + \nu_e \rightarrow {}^{37}Ar + e^{-} \tag{7.13}$$

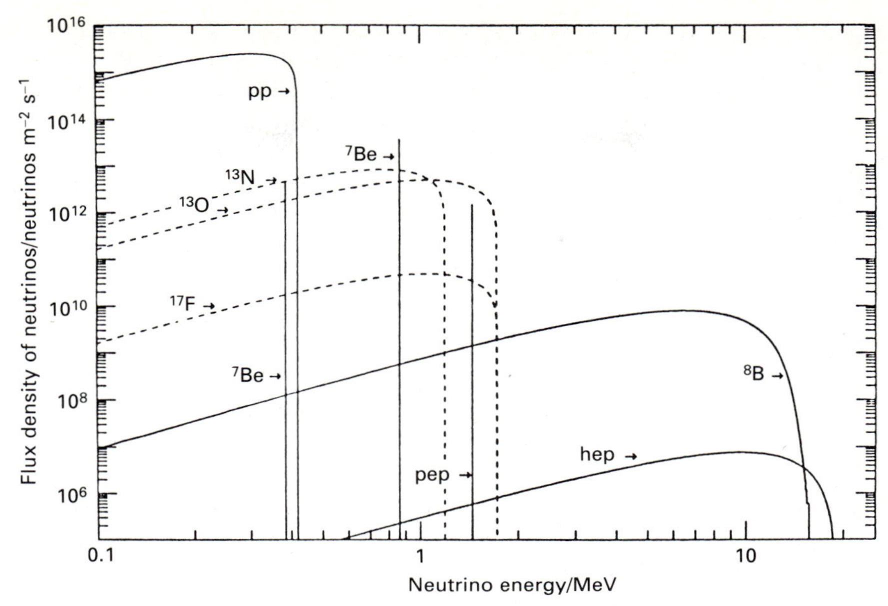

Figure 7.25. The predicted neutrino spectrum of the Sun for the standard solar model. All the fluxes refer to the intensity observed at a distance of 1 astronomical unit (au). The units for continuum spectra are numbers m^{-2} s^{-1} MeV^{-1}. The line fluxes are numbers m^{-2} s^{-1}. The spectra originating from the pp chain are drawn as solid lines; those from the CNO cycle are shown as dashed lines. (From J. N. Bahcall (1989). *Neutrino astrophysics*, page 13, Cambridge: Cambridge University Press.)

The threshold energy for this reaction is 0.814 MeV. This is greater than the maximum energy of the electron neutrinos generated in the main pp chain but is much less than that of those produced in decay of ^{8}B.

In the experiment conducted by Dr Raymond Davis in the deep mine of the Homestake Gold Mining Company located at Lead, South Dakota, the amount of radioactive ^{37}Ar generated within a large volume of perchloroethylene, C_2Cl_4, is measured. The incoming neutrinos have sufficient energy to release the argon nucleus from the perchloroethylene molecule and the argon atoms are then flushed out of the fluid by passing helium through the volume. The half-life of ^{37}Ar is 35 days and so the flushing out of the argon gas does not have to be undertaken too regularly. The amount of radioactive argon present is estimated from its radioactive decay products, specifically the 2.82 keV Auger electrons produced in the decay of ^{37}Ar by electron capture. These electrons are identified by placing the gas in a sensitive proportional counter and looking for the signature of a 2.82 keV electron.

It is this experiment which has led to the solar neutrino problem in which only about one-third the flux of neutrinos from the Sun has been detected as compared with the expectations of the standard model of the solar interior. A summary of the results of this experiment up to 1988 is shown in Fig. 7.27. For many years this was the only experiment to report a deficit in the numbers of solar neutrinos.

Figure 7.26. The ^{37}Cl neutrino detector developed by R. Davis Jr. and his colleagues. The photograph shows the tank which contains 400 000 litres of perchloroethylene in a cavity 1500 m below ground at the Homestake Mine. (From J. N. Bahcall (1989). *Neutrino astrophysics*, page 309, Cambridge: Cambridge University Press.)

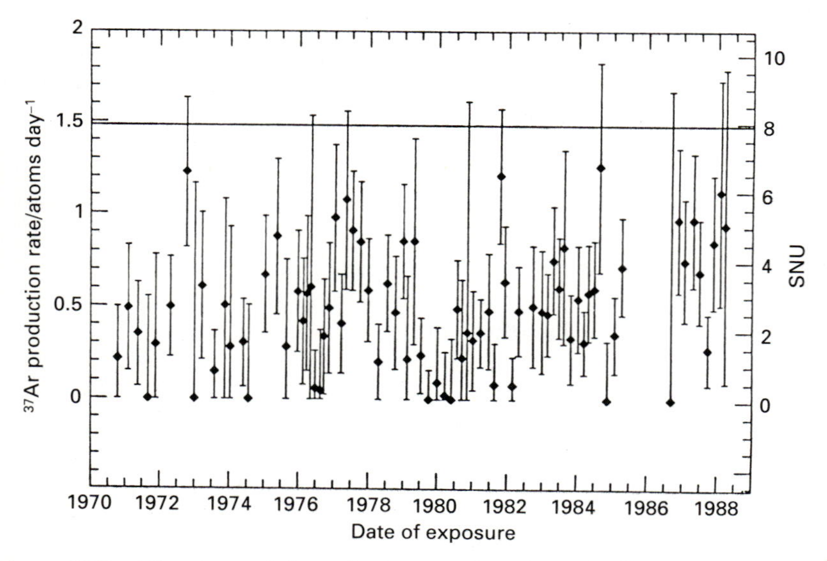

Figure 7.27. The observed flux of solar neutrinos from the ^{37}Cl experiment for the period 1970–88. The solid line at 8 solar neutrino units (SNUs) is the expectation of the standard solar model. (From J. N. Bahcall (1989). *Neutrino astrophysics*, page 319, Cambridge: Cambridge University Press.)

Recently, however, the Kamiokande experiment with the many refinements described above has confirmed that there are fewer neutrinos than predicted by the standard model of the Sun. That experiment provides an upper limit to the solar neutrino flux of less than 0.55 of the expected value.

The next important experiments will be those which attempt to measure the much more plentiful low energy neutrinos from the main pp chain. Among the most important of these are the GALLEX (GALLium EXperiment) and SAGE (Soviet-American Gallium Experiment) projects which exploit the interaction of the low energy neutrinos with gallium, ^{71}Ga.

$$\nu_e + {}^{71}Ga \rightarrow e^- + {}^{71}Ge \tag{7.14}$$

The threshold energy for this reaction is 0.2332 MeV.

The principles of the experiment are similar to those of the ^{37}Cl experiment. In the GALLEX experiment, which is a European collaboration with US and Israeli participation, 30 tonnes of gallium are used in an aqueous solution of gallium chloride and hydrochloric acid. When ^{71}Ge nuclei are created, the germanium atoms form volatile germanium tetrachloride $GeCl_4$ molecules. These are extracted from the volume and the amount of ^{71}Ge present is estimated from the number of radioactive decays of the ^{71}Ge. This decay, which is the inverse of the reaction (7.14), has a half-life of 11.43 days. The GALLEX experiment is being conducted in the Gran Sasso Underground Laboratory in the Italian Alps. The SAGE experiment is a similar Soviet experiment with US participation which will use 60 tonnes of gallium and which will be located in an underground laboratory in the Baksan Valley in the Caucasus mountains. The main difference between these two experiments is that the USSR experiment uses solid gallium as a detector. Gallium melts at about 30 K and so it can be heated and flushed with hydrochloric acid to produce the volatile germanium tetrachloride, $GeCl_4$. These experiments are now in progress.

8

Detectors and telescopes for optical, infrared, ultraviolet and radio astronomy

8.1 Introduction

Most textbooks on astrophysics are written the other way round – they begin with the classical techniques of optical astronomy and branch out from there to the new astronomies. In our case we have begun with high energy astronomies and only now, eight chapters into our story, do we introduce 'low energy' astronomy. There are two crucial differences between the detectors and telescopes for high energy photons and for the astronomies listed at the head of this chapter. First of all, up till now, we have been able to treat the radiation as consisting of individual photons and the problem has been to register as many of them as possible. In most cases, we have neglected the wave properties of electromagnetic radiation. In the low energy astronomies, we can no longer neglect the wave properties of radiation and diffraction effects play a crucial role in determining the angular resolving power of telescopes. The second crucial difference is that, in the high energy astronomies, the photon energies normally far exceed the typical thermal energies of background radiation from the sky, the telescope and the detectors, i.e. $\varepsilon = h\nu \gg kT$ where T is the temperature of the sky, telescope or detector. Perhaps the most extreme examples of this second difference are the telescopes and detectors which operate at $\lambda = 10\ \mu$m in which case, $T \approx h\nu/k \approx 300$ K, i.e. roughly room temperature. Thus, the astronomical sources have to be detected against a huge thermal background originating from the sky and the telescope itself. As Gareth Wynn-Williams put it to me, it is as if optical astronomers had to observe the sky in the daytime with telescopes constructed out of neon tubes!

From the point of view of basic physics, these differences originate in the different properties of radiation in the Rayleigh–Jeans and Wien regions of the Planck spectrum. As Einstein demonstrated in his remarkable papers of 1905 and 1909, in the Wien region of the spectrum $h\nu \gg kT$, the radiation may be considered to consist of photons and their statistics are similar to those of particles

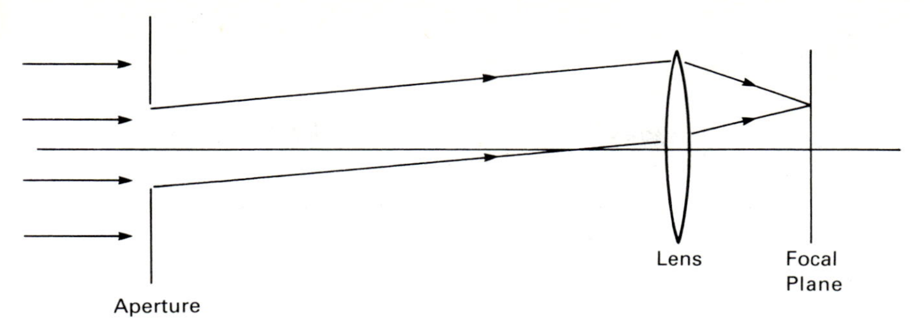

Figure 8.1.

confined in a box. In the Rayleigh–Jeans region, $h\nu \ll kT$, the radiation behaves like electromagnetic waves with all the characteristic properties of waves such as diffraction, refraction etc. The statistical properties of the radiation are those of the superposition of waves. I have discussed the story of these great discoveries in Longair (1984), Chapters 7–11. Of course, it is an oversimplification to make this distinction because the radiation possesses residual wave-like properties in the Wien region and particle-like properties in the Rayleigh–Jeans region but it helps in understanding the differences in approach to the construction of telescopes and detectors in different wavebands.

These considerations are important in understanding the two basic properties of the telescope and detector systems used in the low energy astronomies – these are their *angular resolving powers* (or *angular resolution*) and their *sensitivities*, i.e. their ability to detect faint objects. As we showed in Chapters 6 and 7, for high energy photons, the angular resolution is normally determined by geometric optics and the sensitivity by the ability to detect the individual photons against unwanted background radiation, normally resulting from other high energy processes occurring in the detector volumes. In contrast, in the radio, infrared, optical and ultraviolet wavebands, the fundamental limitation to the angular resolving power is determined by diffraction optics.

8.2 Diffraction-limited telescopes

The theory needed to understand the imaging properties of telescope systems for the low energy astronomies is *Fourier Transform optics*. The angular resolving power of a telescope is defined in terms of the angular distribution of radiation received from a point source at infinity. In the case of high energy telescopes, the angular resolving power is determined by geometric optics but for ultraviolet, optical, infrared and radio telescopes, the spread in the distribution of radiation is determined by diffraction effects. Let us build up the necessary theoretical apparatus in simple stages.

We represent the telescope by an aperture and a lens (Fig. 8.1). We consider the radiation to originate from a point source at infinity and so the radiation incident on the aperture can be considered to be plane parallel light. On passing through

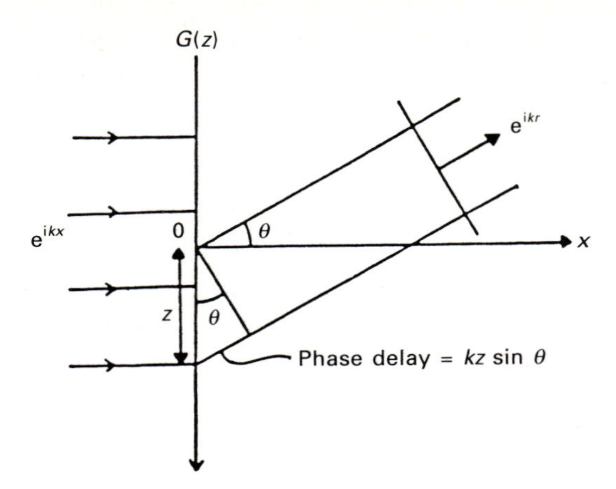

Figure 8.2. Illustrating the phase delay associated with the constructive interference of waves at an angle θ with respect to the direction perpendicular to the aperture.

the aperture, we have to take account of diffraction effects which means that the radiation is no longer plane-parallel but that there is an angular distribution of radiation. The function of the lens is to focus the radiation onto a detector located in the focal plane of the lens so that the image of the point source is produced at the detector rather than at infinity.

Let us consider first a one-dimensional aperture and suppose that the transmission function of the aperture is $G(z)$, i.e. the fraction of the electromagnetic radiation incident in the incremental distance dz at the point z on the aperture which is transmitted by the aperture is $G(z)$. We now use Huygen's principle to replace the wavefront of the radiation by a superposition of point sources of spherical waves. Thus, in our present problem, we can represent the light transmitted at z by a point source of spherical waves of amplitude (or field strength) $G(z)\,dz$. It is apparent that there is constructive interference of the sources of radiation in the direction of the x-axis but the waves interfere more and more destructively as the angle θ from the axis increases. We work out the effects of interference by evaluating the phase delays with respect to transmission through the centre of the aperture. If $\mathbf{k}$ is the wave vector of the radiation transmitted by the aperture, the field strength of the radiation in the direction θ from the increment dz of the aperture is proportional to $G(z)\,dz\,\exp(-ikz\sin\theta)$ where the exponential factor is no more than the phase delay of the radiation propagating at angle θ (Fig. 8.2). The total field strength is therefore

$$U(\theta) = \int_{-\infty}^{\infty} G(z)\exp(-ikz\sin\theta)\,d\theta \tag{8.1}$$

In the limit of small angles, $\sin\theta \approx \theta$ and hence

$$U(\theta) = \int_{-\infty}^{\infty} G(z)\exp(-ik\theta z)\,d\theta \tag{8.2}$$

We notice that $U(\theta)$ is just the Fourier transform of the aperture distribution $G(z)$. We can easily extend this analysis to two dimensions, in which case we can consider a rectangular aperture in the y–z plane with the corresponding angles being ϕ and θ. By exactly the same procedure, we find

$$U(\theta, \phi) = \int_{-\infty}^{\infty} \int_{-\infty}^{\infty} G(y, z) \exp[-ik(\phi y + \theta z)] \, dy \, dz \tag{8.3}$$

Let us look at the cases of uniformly illuminated one and two-dimensional apertures of length $2a$ and area $2a \times 2b$ respectively. In the one-dimensional case, $G(z) = 1$ if $-a \leqslant z \leqslant a$ and is zero otherwise. Therefore

$$U(\theta) = \int_{-a}^{a} \exp(-ik\theta z) \, dz = 2\frac{\sin(k\theta a)}{k\theta}$$

To find the intensity of radiation we take the square of the field strength

$$I(\theta) \propto U^2(\theta) \propto \frac{\sin^2(k\theta a)}{(k\theta)^2}$$

We normalise this intensity distribution to the central intensity I_0 at $\theta = 0$ and then

$$I(\theta) = I_0 \left[\frac{\sin(k\theta a)}{k\theta a}\right]^2 \tag{8.4}$$

By exactly the same procedure, the rectangular aperture has diffraction pattern

$$I(\theta, \phi) = I_0 \left[\frac{\sin(k\theta a)}{k\theta a}\right]^2 \left[\frac{\sin(k\phi b)}{k\phi b}\right]^2 \tag{8.5}$$

Expression (8.4) is the well-known *Fraunhofer diffraction pattern* for a single slit and the expression (8.5) is the corresponding pattern for a rectangular slit, the function $[\sin(k\theta a)/k\theta a]^2$ being displayed in Fig. 8.3. Notice that the term Fraunhofer diffraction is used when the approximation $\sin\theta \approx \theta$ can be applied and higher off-axis terms in θ^2, etc. are neglected. Technically, the condition for Fraunhofer diffraction is that the input and output beams from the aperture have the same curvature, the simplest case being when both beams are parallel at infinity, as in our treatment.

It can be seen that the first zero of the diffraction pattern occurs at $k\theta a = \pi$ and that the intensity of radiation falls to half its central intensity at $k\theta a = 1.39$. Let us rewrite this result in terms of the wavelength λ

$$\theta_{\frac{1}{2}} = 0.44\lambda/2a \tag{8.6}$$

What this result means is that only within this angle $\theta_{\frac{1}{2}}$ do the phases of the waves add constructively to produce a strong signal. Outside this angle, the phases of the waves become progressively more and more out of step and they do not interfere constructively. The same considerations apply to the rectangular aperture.

Let us repeat the calculation for a circular aperture which is the analogue of a circular mirror or lens of an optical telescope. We transform the expression (8.3)

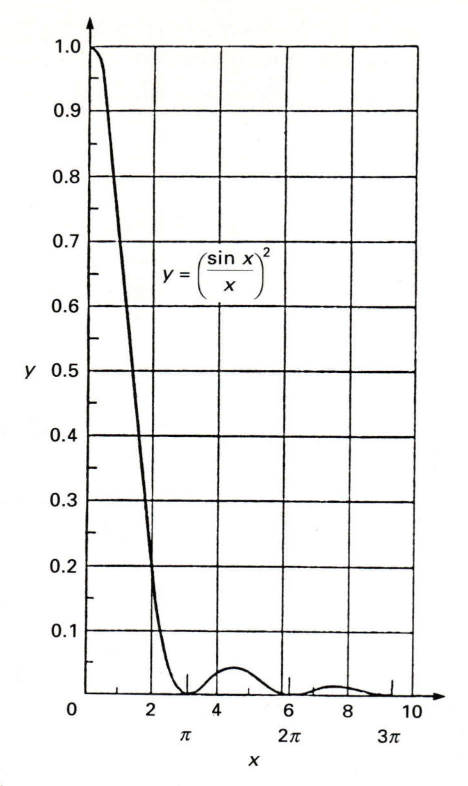

Figure 8.3. The diffraction pattern of rectangular one-dimensional aperture. (From M. Born and E. Wolf (1980). *Principles of optics*, 6th edition, page 393, Oxford: Pergamon Press.)

into polar coordinates (r, ψ) such that the point (y, z) on the aperture is given by $y = r \sin \psi, z = r \cos \psi$ and the element of area becomes $r\,\mathrm{d}r\,\mathrm{d}\psi$. Correspondingly, the angles on the sky are transformed from the 'rectangular' coordinates (θ, ϕ) to corresponding 'angular' coordinates (ρ, σ) such that $\theta = \rho \cos \sigma, \phi = \rho \sin \sigma$. Thus, ρ measures the angular distance from the axis of the circular aperture. The expression (8.3) therefore becomes

$$U(\rho, \sigma) = \int_0^a \int_0^{2\pi} G(r, \psi) \exp[-ik\rho r(\cos \psi \cos \sigma + \sin \psi \sin \sigma)]\, \mathrm{d}\psi\, r \mathrm{d}r$$

$$= \int_0^a \int_0^{2\pi} G(r, \psi) \exp[-ik\rho r \cos(\psi - \sigma)]\, \mathrm{d}\psi r\, \mathrm{d}r \tag{8.7}$$

where the radius of the aperture is taken to be a. We assume a uniformly illuminated aperture so that $G(r, \psi) = 1$ if $r \leqslant a$ and $G(r, \psi) = 0, r > a$. The ψ integral is evaluated using the integral representation of the Bessel function $J_0(z)$

$$J_0(z) = \frac{1}{2\pi} \int_0^{2\pi} \exp(\mathrm{i}z \cos \alpha)\, \mathrm{d}\alpha$$

Then

$$U(\rho, \sigma) = \frac{2\pi}{(k\rho)^2} \int_0^a J_0(x) x \, dx$$

where $x = k\rho r$.

We use the further Bessel function relation

$$\frac{d}{dx}[x^{n+1} J_{n+1}(x)] = x^{n+1} J_n(x)$$

so that, setting $n = 0$, we find

$$U(\rho) = \frac{2\pi a}{k\rho} J_1(k\rho a) = 2A\left[\frac{J_1(k\rho a)}{k\rho a}\right] \tag{8.8}$$

where $A = \pi a^2$ is the area of the aperture. As before the intensity is found by taking the square of the field strength and, normalising to the central intensity I_0, we find

$$I(\rho) = I_0 \left[\frac{2J_1(k\rho a)}{k\rho a}\right]^2 \tag{8.9}$$

Note that in the limit of small values of x, $J_1(x)/x \approx \frac{1}{2}$. The diffraction pattern (8.9) is shown in Fig. 8.4 and represents the well-known Airy pattern of a circular aperture. Outside the central bright region, which is known as the Airy disc, there are concentric rings which contain a small but significant fraction of the energy transmitted by the aperture. This is illustrated in Fig. 8.4 in which the total encircled energy is displayed as a function of angular distance from the axis of the aperture, i.e.

$$f(\rho) = \frac{\int_0^\rho I(\rho)\rho \, d\rho}{\int_0^\infty I(\rho)\rho \, d\rho}$$

It can be seen that 84% of the light lies within the first dark ring and more than 90% within the second. This function, the encircled energy fraction, is an important measure of the performance of a telescope system, particularly when high precision photometry is required.

The first dark ring of the Airy diffraction pattern occurs at $k\rho a = 1.220\pi = 3.883$. Expressing the angle ρ in terms of the wavelength λ and a, the radius of the aperture, we find

$$\rho = 1.22\lambda/D$$

where $D = 2a$ is the diameter of the aperture. This corresponds to the *Rayleigh criterion* for resolving two nearby point sources. The criterion corresponds to one object being located on the first dark ring of the Airy diffraction pattern.

The above discussion describes the simplest results of the diffraction theory of telescope systems. There are a number of points which we should note.

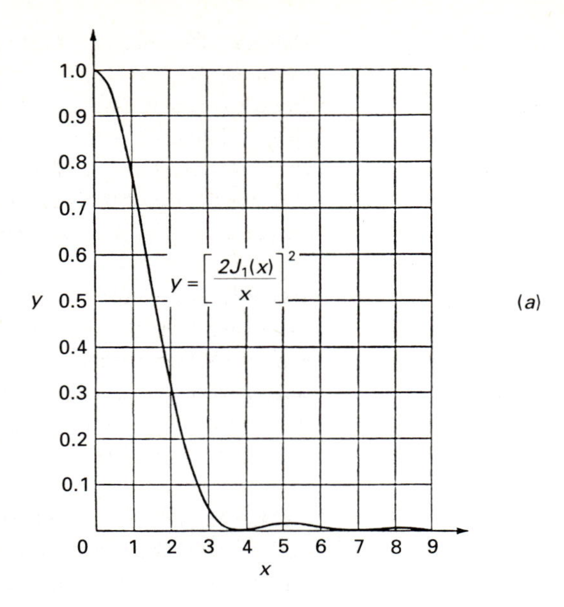

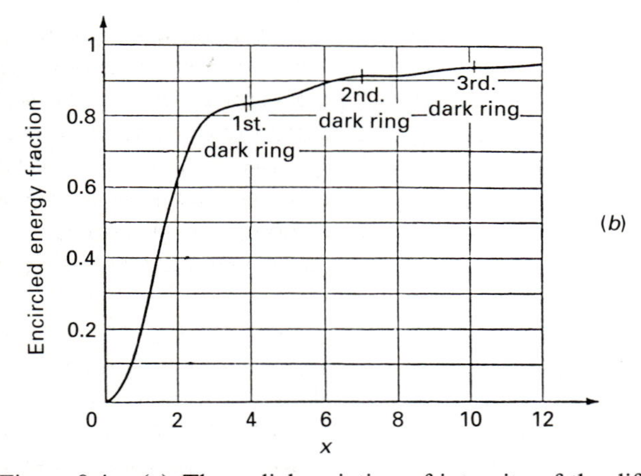

Figure 8.4. (*a*) The radial variation of intensity of the diffraction pattern of a uniform circular aperture. (From M. Born and E. Wolf (1980). *Principles of optics*, 6th edition, page 396, Oxford: Pergamon Press.) (*b*) The encircled energy fraction as a function of angular distance from the axis of a uniform circular aperture. (From M. Born and E. Wolf (1980). *Principles of optics*, 6th edition, page 398, Oxford: Pergamon Press.)

(i) In a perfect imaging system, the images of point sources should display the Airy diffraction pattern with a succession of dark rings surrounding the Airy disc.

(ii) Even in a perfect imaging system, only 84% of the radiation is contained within the first dark ring and so inevitably radiation is scattered outside an angular radius $\rho \approx \lambda/D$. This is an important factor in the design of imaging and spectroscopic instruments.

(iii) The point-spread function of the aperture depends upon the function $G(z)$ and so can be modified by adopting different forms of the transmission function $G(z)$. For example, if $G(z)$ is taken to be a Gaussian function $G(z) \propto \exp(-\rho^2/2\rho_0^2)$, the point spread function $I(\rho)$ will also be of Gaussian form because the Fourier transform of a Gaussian function is just another Gaussian. The effect of a Gaussian transmission function is to broaden the point spread function but it has the advantage of eliminating the dark rings around the Airy disc. Thus, for an aperture of a given size, there is a trade off between angular resolution and eliminating the dark rings in the Airy pattern. In radio astronomical terms, the rings external to the Airy disc are called side lobes of the main beam and can be eliminated or strongly attenuated by 'grading the aperture'. In optical parlance, the selection of a particular function $G(z)$ is called *apodisation*.

(iv) The success of an imaging system is crucially dependent upon the quality of the optical system, in particular, how accurately the phases of the waves originating at different points in the aperture are preserved. This is normally quoted in terms of the *wavefront error* of the optical system. If there are random phase errors imposed on the wavefront, the amplitudes of the waves add together slightly out of phase. This has two immediate consequences – the central intensity is reduced and the point spread function is broadened. In particular, power is removed from the central maximum and scattered into the wings of the point-spread function. The effect of random phase errors on the point spread function is illustrated in Fig. 8.5 in which it can be seen that there is a significant decrease in the central intensity if the random phase errors are even as small as $\lambda/10$. The usual criteria for a diffraction-limited system is that the random phase errors should amount to less than $\lambda/20$, i.e. less than a twentieth of the wavelength of radiation. This is the primary limitation to the construction of very large telescopes. All errors due to the atmosphere, the telescope, its optics and its instruments should cumulatively add up to an error of less than $\lambda/20$ wherever the telescope is pointed on the sky if diffraction-limited imaging is to be achieved.

Table 8.1 lists the diffraction limits of various radio, millimetre, infrared, optical and ultraviolet telescopes. I have given in this list only fully steerable single dishes since we will look at interferometry and aperture synthesis in the next section. The angular resolutions quoted in Table 8.1 represent the theoretical maximum resolution which can be obtained. In practice, these values are not always obtained in ground-based observations because of the effect of the atmosphere. Generally speaking, diffraction-limited performance is obtained at radio, millimetre and far-

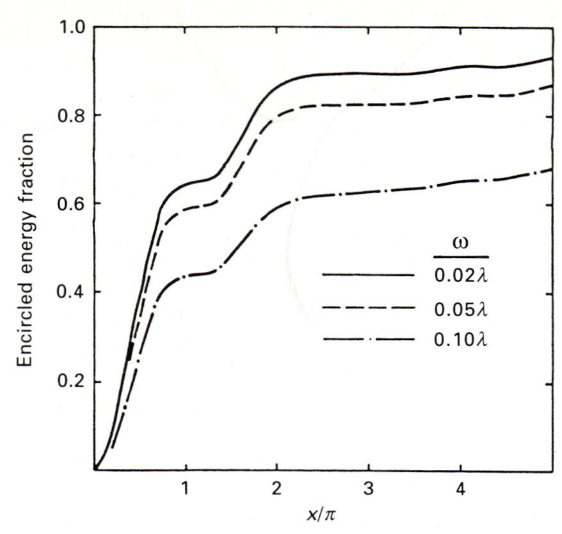

Figure 8.5. The encircled energy fraction as a function of angle from the axis of a Cassegrain telescope and its dependence upon the rms surface errors of the surface of the primary mirror. The radius of the central hole is 33% of that of the primary mirror. The solid line shows essentially the encircled energy fraction for a perfect mirror (see also Fig. 8.8). The dashed line shows the response when the rms surface errors on small scales is $\lambda/20$. The dot-dashed line shows the response if the rms error is $\lambda/10$. It can be seen that small scale irregularities in the surface of the mirror result in energy being removed from the main beam and scattered into the wings of the diffraction pattern. These calculations have been carried out for the mirror of the Hubble Space Telescope by D. J. Schroeder. (From D. J. Schroeder (1987). *Astronomical optics*, page 211, San Diego: Academic Press, Inc.)

infrared wavelengths and the various telescopes listed in Table 8.1 have achieved this at the wavelengths quoted. An interesting example is that of the James Clerk Maxwell Telescope which was designed to be the largest sub-millimetre telescope in the world. At the time of writing the rms surface accuracy of the telescope is less than 30 μm. Using the criterion that the telescope should be diffraction-limited at a wavelength 20 times the rms wavefront errors, it can be seen that the telescope is diffraction-limited at a wavelength of 600 μm and that it still has good performance but with somewhat less power in the main beam at a wavelength of 350 μm, the shortest wavelength which is accessible from the summit of Mauna Kea.

Going to shorter wavelengths, infrared space telescopes such as the Infrared Astronomical Satellite (IRAS) which operated in space from 1981 to 1983 and the Infrared Space Observatory of ESA which will be launched in 1993 are both small telescopes with cooled primary mirrors of diameters 60 cm. They are diffraction-limited at thermal infrared wavelengths. However, large ground-based optical–infrared telescopes begin to lose diffraction-limited performance at wavelengths shorter than about 5–10 μm. The reason for this is the phenomenon known as *astronomical seeing*. The atmosphere above the telescope is not totally

Table 8.1. *Examples of the diffraction-limited performance of radio, infrared, optical and ultraviolet telescopes*

There are several different ways of expressing the angular resolution of telescope systems. For the Airy pattern, the Rayleigh criterion corresponds to $1.22\,(\lambda/D)$. Another useful measure is the 'full-width half maximum' which corresponds to the diameter of the image of a point source at which the intensity falls to half the central value. For the Airy pattern, this corresponds to $\theta_{FWHM} = 1.02\,(\lambda/D)$. Notice that, for radio astronomical antennae, the illumination of the dish is normally chosen to be Gaussian rather than uniform in order to reduce the amplitude of the side lobes which pick up stray radiation. If a Gaussian illumination function is chosen which falls to -10 dB (i.e. to 10% of the central intensity) at the edge of the dish, $\theta_{FWHM} = 1.2\,(\lambda/D)$. Notice that another way of expressing the angular resolution of the telescope system is in terms of the *encircled energy fraction*. For example, for the Airy function, 90% of the encircled energy lies within an angular radius of $1.97\,(\lambda/D)$ which is just inside the second dark ring. The specification of the Hubble Space Telescope, for example, is that 90% of the encircled energy should lie within an angular radius of 0.1 arcsec. In general, the best way of describing the imaging properties of the telescope is in terms of the *modulation transfer function* (see, e.g. Lena (1988)). Note that, in column 4, these are theoretical maximum resolutions and may, in practice, be degraded by other aspects of the optical system.

Telescope	Diameter of aperture	Wavelength	Angular resolution λ/D (see above note)
MPIfRA Effelsberg, West Germany	100 m	74 cm	25 arcmin
		6 cm	2 arcmin
James Clerk Maxwell Telescope, Mauna Kea	15 m	1 mm	14 arcsec
		0.35 mm	5 arcsec
Infrared Astronomical Satellite (IRAS)	60 cm	100 μm	34 arcsec
ESA Infrared Space Observatory (ISO)	60 cm	10 μm	4 arcsec
UK Infrared Telescope (UKIRT)	3.9 m	10 μm	0.5 arcsec
		2.2 μm	0.1 arcsec[a]
Palomar 5 m Telescope	5 m	1 μm	0.04 arcsec[a]
		500 μm	0.02 arcsec[a]
NASA-ESA Hubble Space Telescope (HST)	2.4m	1 μm	0.09 arcsec
		500 nm	0.043 arcsec
		200 nm	0.017 arcsec
		120 nm	0.010 arcsec

[a] Not achieved because of the effects of astronomical seeing (see text).

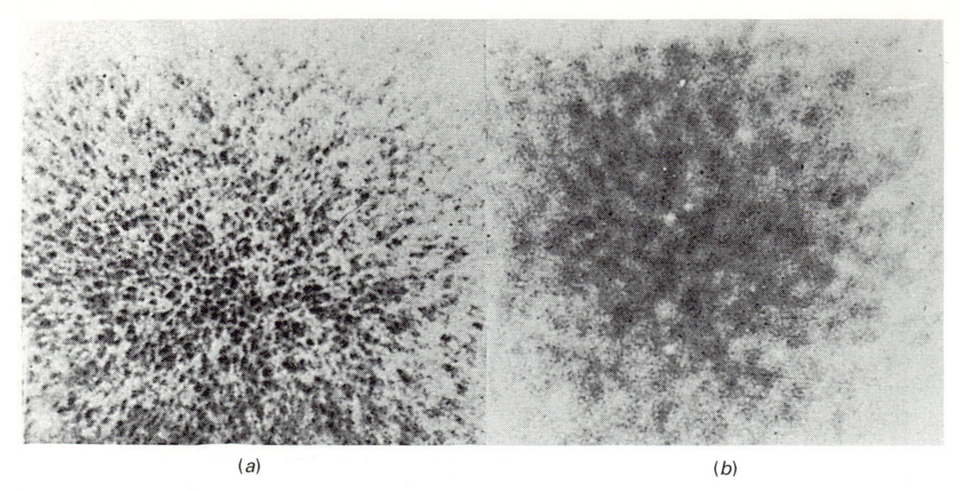

Figure 8.6. Illustrating the phenomenon of astronomical seeing. (*a*) A very short exposure (0.02 s) of an unresolved star showing the splitting up of the image into a large number of 'speckles'. Roughly speaking, the angular size of each of the speckles corresponds to the diffraction limit of the aperture. (*b*) A similar very short exposure of the giant star Betelgeuse illustrating the difference in speckle pattern as compared with (*a*) when the star has an angular size greater than the diffraction limit of the telescope. In both images (*a*) and (*b*), the overall angular size of the image is about 2 arcsec. When observations are made with longer integration times, the speckles overlap and the image is blurred with a typical full-width half maximum of about 1 arcsec. (From A. Labeyrie (1978). *Ann. Rev. Astr. Astrophys.*, **16**, 81.)

smooth – there is small scale turbulence which gives rise to fluctuations in the refractive index of the atmosphere and these turbulent cells cause the image of a point source to be scattered in angle. This problem is best illustrated by very short exposure images of bright stars (Fig. 8.6). If a star is photographed with an exposure of about 10^{-2} s, its image appears as a random superposition of speckles, each speckle corresponding to the image formed by the light passing coherently through one of the turbulent cells. This phenomenon of astronomical seeing is the prime cause of degradation of the diffraction limited performance of all large optical–infrared telescopes. On a good site, the size of the 'seeing disc' can be as small as 0.5 arcsec and typically, under good observing conditions, it is about 1 arcsec. Thus, for a 5 m class telescope, the degradation of the image quality corresponds to about a factor of 20–40 as compared with the theoretical angular resolution.

If the seeing disc is characterised by an angular scale θ_s, it is found that this characteristic size decreases slowly with increasing wavelength as $\theta_s \propto \lambda^{-0.2}$ so that if the seeing disc is typically about 1 arcsec at 500 nm, it is expected to be only about 0.6 arcsec at 5 μm and 0.5 arcsec at 10 μm. In fact, it can be seen from Table 8.1 that a 4 m infrared telescope such as the UK Infrared Telescope has a diffraction limit of 0.5 arcsec at 10 μm so that diffraction-limited performance is obtained at these mid-infrared wavelengths.

Astronomical seeing limits the sensitivity as well as the angular resolution of ground-based telescopes in many of the most important types of observation and in recent years very strenuous efforts have been made to minimise the problem. The most important consideration is to build the telescope on a very good site which means locating it on a dark site far from population centres and above the inversion layer in the atmosphere at which the thick cloud layer forms. High altitude island sites are an advantage because, in addition to the above considerations, the flow of air over the telescope is smoother than if the air flow has the opportunity to become strongly turbulent on passing over continental landmasses. Sites such as Mauna Kea in Hawaii at 4200 m and the Roche de los Muchachos in the island of La Palma in the Canary Islands at an altitude of 2800 m are typical of very good sites. Much of the seeing can originate in the vicinity of the telescope dome itself and nowadays strenuous precautions are taken to ensure that the air around the telescope and within the dome is as stable and smooth as possible. For example, in the Canada–France–Hawaii Telescope on Mauna Kea, which has probably achieved the best imaging of any ground-based telescope, the floor of the observing floor is refrigerated in order to prevent thermal convection currents developing within the dome.

In order to improve the imaging properties of telescopes, there is now considerable interest in *adaptive optics* which employ techniques by which the effects of seeing can be minimised by real time image stabilisation. The idea is to use flexible mirrors which can compensate in real time for the distortions imposed on the wavefronts of the incoming waves and thus provide diffraction limit performance. It is likely that the next generation of large optical–infrared telescopes will be designed to be operated in an adaptive optics mode once this technology becomes available.

One way of avoiding the problems of astronomical seeing is to place the telescope above the Earth's atmosphere and this will be achieved by the NASA–ESA Hubble Space Telescope. Being located above the Earth's atmosphere, the problem of astronomical seeing is eliminated and the whole of the ultraviolet waveband is opened up for observation. It also means, however, that the optics of the telescope have to be of very high quality to take advantage of the space environment. In fact, the 2.4 m primary mirror of the telescope has been polished to an accuracy of $\lambda/60$ at a wavelength of 533 nm. This means that the telescope will certainly be diffraction-limited at optical wavelengths and that it will still provide diffraction limited imaging at wavelengths as short as 200 nm. By image reconstruction techniques, it will be possible to achieve even higher angular resolution at wavelengths as short as 120 nm, the short wavelength limit of the telescope. Unfortunately the primary mirror was polished to the wrong figure and the full capability of the telescope will only be realised once correction optics are included in the optical path for each scientific instrument (see Section 8.7.1).

8.3 A pedagogical interlude

In Section 8.2, we determined the diffraction pattern of an aperture by shining upon it coherent, plane-parallel light of a definite angular frequency ω. In reality, we observe the radiation from sources which consist of the random superposition of the emissions of vast numbers of independent radiators which normally emit broad-band emission. How can we use the results of Section 8.2 in this case?

Let us perform some simple sums to indicate what happens. Our aim is to work out the amplitude, intensity and fluctuations in the radiation from a source consisting of the superposition of a large number of radiators with angular frequency ω but random phases. The amplitude of the field is the vector sum of the fields from each radiating particle and can be evaluated using an Argand diagram which shows the summation of the amplitudes and phases of the individual waves (Fig. 8.7). Thus, we can write

$$\mathbf{E}(\mathbf{r}) = \sum_i \mathbf{E}_i \exp[-\mathrm{i}(\mathbf{k}_i \cdot \mathbf{r} - \omega_i t + \phi_i)] \tag{8.10}$$

We make the observations at a fixed point $\mathbf{r}$ and so we can write

$$\mathbf{E}(\mathbf{r}) = \sum_i \mathbf{E}_i \exp[\mathrm{i}(\omega_i t - \phi_i)]$$

To find the intensity of the field, we take the average of the product of the amplitude of the radiation and its complex conjugate over some finite time T

$$\langle E^2 \rangle = \langle EE^* \rangle = \frac{1}{T}\int_0^T \sum_i \sum_j E_i E_j \exp[\mathrm{i}(\omega_i t - \phi_i - \omega_j t + \phi_j)]\,\mathrm{d}t$$

Obviously, over any finite period of time, the only terms which survive the averaging process are those for which $\omega_i = \omega_j$ and $\phi_i = \phi_j$ i.e.

$$\langle E^2 \rangle = \sum_i E_i^2 \tag{8.11}$$

This is a well-known and reassuring result – it states that the total intensity of radiation is the sum of the energies in the individual waves. This has a very natural interpretation in terms of the random emission of photons from a source. In the simple case of the random superposition of the fields of n sources, each of field strength E but with random phases, the total intensity is nE^2 while the field strength grows only as $n^{\frac{1}{2}}E$. This makes sense in terms of the vectors performing a random walk on the Argand diagram in which, in n steps, the magnitude of the resultant E vectors reaches an average distance of $n^{\frac{1}{2}}E$ from the origin. Notice that these results are not correct if the sources themselves are coherent i.e. the phases are not random but, as in a laser, are arranged to emit with a fixed phase relation. In this case, the intensity is approximately equal to nE^2.

The key point is that the resultant $\mathbf{E}$ vector has a definite magnitude E and phase ϕ relative to some arbitrary phase reference. How long does this pattern persist? Classically, if the oscillators continue to radiate indefinitely and all have precisely

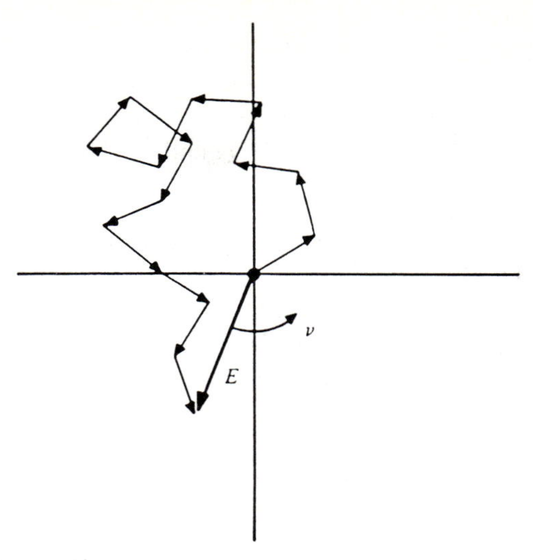

Figure 8.7. Illustrating the addition of a random superposition of electric field vectors on an Argand diagram.

the same frequency, the resultant magnitude and phase would remain fixed for all time. In practice, however, the observations are made within some finite frequency bandwidth $\Delta\nu$ and so the individual **E** vectors rotate on the Argand diagram with slightly different frequencies. This means that, in a time $t \approx 1/\Delta\nu$, the waves shift out of phase with respect to one another. The resultant amplitude is still E on average but the resultant phase has changed. This may be expressed by stating that the field remains *phase-coherent* only for a time $\sim \Delta\nu^{-1}$ which is often called the *coherence-time* τ. Thus, in imaging systems and interferometers, it is essential that the radiation which follows different routes to the detector be combined within a coherence time or the phase information and consequently constructive interference is lost.

There is an important distinction between those detector systems which respond to the total power $\langle E^2 \rangle$ of the radiation and those which measure the amplitude and phase of the incoming signal. The former are referred to as *incoherent* or *bolometric* detectors whereas the latter are known as *coherent* detectors. Generally speaking, the detectors used in radio astronomy are coherent detectors whereas those used at infrared, optical and higher frequencies employ incoherent systems.

Let us now consider the amplitude of the fluctuations in the intensity of the signal. I have treated this problem in Section 12.2.2 of Longair (1984) where it is shown that, superimposing electromagnetic waves of the same amplitude but random phase results in a fluctuating electric field intensity, the rms deviation of which is the same as the intensity of the waves themselves i.e.

$$\langle \Delta E^2 \rangle = NE^2 \tag{8.12}$$

This is a very important result and, as shown in Longair (1984), corresponds to considering the interference of all pairs of waves contributing to the intensity. It is this simple but crucial fact which makes interferometry and aperture synthesis possible. Because the fluctuations have amplitude $\Delta E \sim E$, it is possible to measure the phase difference of the signals arriving at separate detectors by cross-correlating the signals.

Let us have a brief look at what can be achieved using interferometry and aperture synthesis.

8.4 Interferometry and synthesis imaging

The simplest way of extending the above ideas on image formation to interferometers, aperture synthesis, speckle interferometry and so on is to use the basic theorem derived in expression (8.3) that the amplitude of the response to a point source is just the two-dimensional Fourier transform of the aperture distribution. We can determine the response to a point source at infinity by using the many useful theorems of Fourier analysis. A complete description will be found in Bracewell's monograph *The Fourier Transform and Its Applications*. First of all, it is useful to introduce a number of mathematical tools used in interferometry and aperture synthesis.

In general, the functions are complex, having both amplitude and phase, and so we have to consider the Fourier properties of complex functions $f(x)$ and their complex conjugates $f^*(x)$. We use the definition of Fourier transforms given in Chapter 3, i.e.

$$F(\omega) = \frac{1}{(2\pi)^{\frac{1}{2}}}\int_{-\infty}^{\infty} f(x)\exp(\mathrm{i}\omega x)\,\mathrm{d}x$$

The *convolution* of two function $f(x)$ and $g(x)$ is defined to be

$$f(x) * g(x) = \int_{-\infty}^{\infty} f(u)\,g^*(x-u)\,\mathrm{d}u \tag{8.13}$$

The *cross-correlation* of two functions $f(x)$ and $g(x)$ is defined to be

$$f(x) \bigstar g(x) = \int_{-\infty}^{\infty} f(u-x)\,g^*(u)\,\mathrm{d}u \tag{8.14}$$

Note carefully the distinction between the star and pentagram symbols for convolution and cross-correlation respectively. It follows that the *autocorrelation function* of a function $f(x)$ with itself is

$$f(x) * f(x) = \int_{-\infty}^{\infty} f(u-x)\,f^*(u)\,\mathrm{d}u \tag{8.15}$$

Often, the autocorrelation function is normalised to its value at $x = 0$ in which case the autocorrelation function is

$$\gamma(x) = \frac{\displaystyle\int_{-\infty}^{\infty} f^*(u)f(u-x)\,\mathrm{d}u}{\displaystyle\int_{-\infty}^{\infty} f^*(u)f(u)\,\mathrm{d}u} \tag{8.16}$$

with $\gamma = 1$ at $x = 0$.

The following theorems simplify many of our calculations.

> The *Similarity Theorem*: If $f(x)$ has Fourier transform $F(s)$, then $f(ax)$ has Fourier transform $|a|^{-1} F(s/a)$.
>
> The *Addition Theorem*: If $f(x)$ and $g(x)$ have Fourier transforms $F(s)$ and $G(s)$, then the Fourier transform of $f(x)+g(s)$ is $F(s)+G(s)$.
>
> The *Shift Theorem*: If $f(x)$ has Fourier transform $F(s)$, then $f(x-a)$ has Fourier transform $\exp(-2\pi i a s) F(s)$.
>
> The *Modulation Theorem*: If $f(x)$ has Fourier transform $F(s)$, then $f(x) \cos \omega x$ has Fourier transform $\frac{1}{2}F(s-\omega/2\pi)+\frac{1}{2}F(s+\omega/2\pi)$.
>
> The *Convolution Theorem*: The Fourier transform of the convolution of two functions $f(x)*g(x)$ is the product of their Fourier transforms $F(s)G(s)$. There are many different forms of this theorem.
>
> *Parseval's Theorem* (referred to as *Rayleigh's Theorem* by Bracewell) states that, if $F(s)$ is the Fourier transform of $f(x)$, then

$$\int_{-\infty}^{\infty} |f(x)|^2 \, dx = \int_{-\infty}^{\infty} |F(s)|^2 \, ds \tag{8.17}$$

> The *Power Theorem* states that if $f(x)$ and $g(x)$ have Fourier transforms $F(s)$ and $G(s)$, then

$$\int_{-\infty}^{\infty} f(x)\, g^*(x) \, dx = \int_{-\infty}^{\infty} F(s)G^*(s) \, ds \tag{8.18}$$

> Finally, the *Autocorrelation Theorem* states that the Fourier transform of the autocorrelation function $f(x) \bigstar f(x)$ is $|F(s)|^2$.

These theorems are all simply derived from the definitions of Fourier transforms and are very clearly described by Bracewell.

These theorems find immediate application in the theory of antennae and interferometers. For example, the shift theorem shows us directly that if a point source is located at an angle α with respect to the axis of the telescope beam, there is a phase shift across the aperture of the antenna and the position of the source may be found by measuring this change of phase. This is the principle of measuring very accurate positions by radio interferometry, i.e. radio astrometry.

Another pleasant example is to work out the point-spread function for a mirror with a circular hole in the centre, such as is adopted in the Cassegrain configuration found in most modern telescopes, including the Hubble Space Telescope. The amplitude of the response to a point source can be found using the addition theorem by subtracting the Fourier transform of the central hole from that of the full mirror (Fig. 8.8). Because of the very high quality of the optics of the Hubble Space Telescope, it is expected that this aperture distribution will eventually be achieved once the correction optics are installed. Notice that there are many ways of specifying the angular resolving power of such a mirror. The

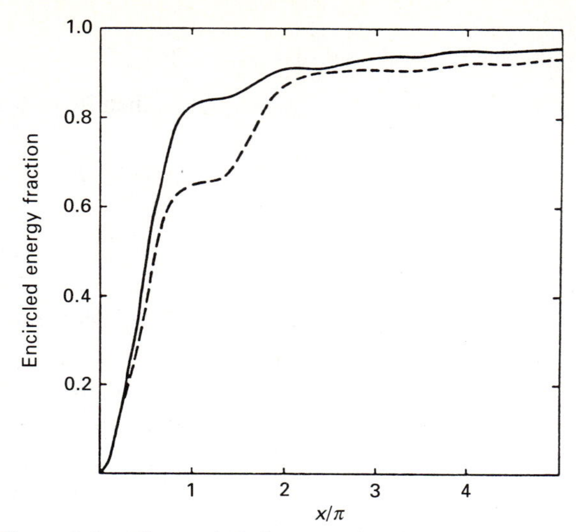

Figure 8.8. The encircled energy fraction as a function of angle from the axis of a circular aperture (solid line) compared with that of the same aperture but with a circular hole cut in the middle of the primary with radius 33% of that of the primary mirror (dashed line). This encircled energy distribution is similar to that expected for the Hubble Space Telescope. (From D. J. Schroeder (1987). *Astronomical optics*, page 136, San Diego: Academic Press.)

different prescriptions include the 90% encircled energy contour, the full-width at half maximum, the fraction of the power within the central diffraction spike of the unobscured aperture, etc. The modulation transfer function includes all this information.

Let us consider next the simple case of a two element interferometer in one dimension which we represent by two uniform apertures separated by a distance D (Fig. 8.9(*a*)). To describe this aperture distribution, we can convolve the aperture distribution of a single aperture with two delta-functions separated by distance D. i.e.

$$T(z) = f(z) * g(z)$$

in which $f(z) = 1$ if $-a \leqslant z \leqslant a$ and $f(z) = 0$ otherwise; $g(z)$ is represented by two delta-functions $\delta(z)$ separated by distance D.

$$g(z) = \delta(-D/2) + \delta(D/2)$$

Then, the amplitude of the diffracted image is given by the Fourier transform of $T(x)$ which, according to the convolution theorem, is just proportional to the product of the Fourier transforms of $f(z)$ and $g(z)$. For the case of two uniform apertures the amplitude is therefore

$$\begin{aligned} F(\theta) &= I_0 \sin(k\theta a)/k\theta a \\ G(\theta) &= (2\pi)^{-\frac{1}{2}} \int g(z) \exp(\iota k\theta z)\mathrm{d}z \\ &= [\exp(\iota k\theta D/2) + \exp(-\iota k\theta D/2)]/(2\pi)^{\frac{1}{2}} \end{aligned}$$

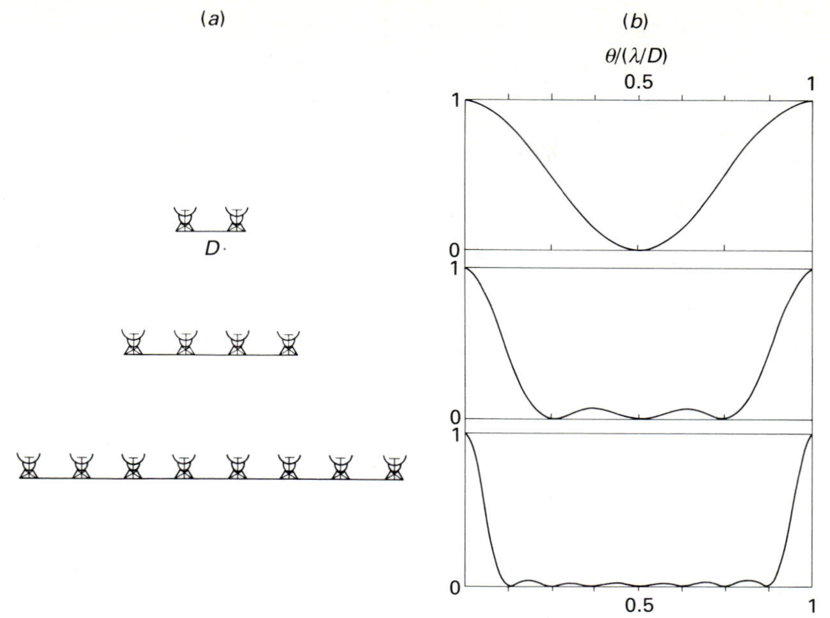

Figure 8.9. (*a*) A schematic diagram showing grating arrays with increasing numbers of antennae. (*b*) The angular response of the grating arrays shown in (*a*). These diagrams show only the part of the beam associated with the interferometer pattern and should be multiplied by the response of an individual antenna to find the overall response of the grating array, i.e., in the case of the rectangular apertures used in the example in the text, the response should be multiplied by the function $[\sin(k\theta a)/k\theta a]^2$ (see expression (8.19)). It can be seen that the more antennae added to the array, the narrower the beam becomes relative to the separation between the grating lobes. This is the principle of grating array telescopes.

and so

$$U(\theta) = F(\theta)G(\theta) = \frac{2I_0}{(2\pi)^{\frac{1}{2}}}\frac{\sin(k\theta a)}{k\theta a}\cos\left(\frac{k\theta D}{2}\right)$$

The power polar diagram of the interferometer is proportional to the square of the amplitude $I(\theta) \propto U^2(\theta)$

$$I(\theta) \propto \left[\frac{\sin(k\theta a)}{k\theta a}\right]^2 \cos^2\left(\frac{k\theta D}{2}\right) \tag{8.19}$$

This response is shown in Fig. 8.9(*b*). It is straightforward to extend this analysis to an array of telescopes, each separated by distance D to produce what is known as a *grating telescope*. For example, for a four-element interferometer, the function $g(z)$ becomes

$$g(z) = \delta\left(-\frac{3D}{2}\right) + \delta\left(-\frac{D}{2}\right) + \delta\left(\frac{D}{2}\right) + \delta\left(\frac{3D}{2}\right)$$

and so on for more antennae. Figs. 8.9(*a*) and (*b*) show the distribution of antennae and the resulting power polar diagrams. It can be seen that, as the number of antennae increases, the polar diagram becomes sharper and the first grating side lobe occurs several beam-widths away from the central maximum.

The extension to two dimensions is straightforward and this technique is employed to map the distribution of radio emission from extended sources. The strategy is to arrange for the first grating side lobe to lie outside the angular distribution of radiation on the sky. In the case of *aperture synthesis*, the above procedure is used to fill in all the baselines which would be needed to reconstruct a fully-filled aperture of diameter equal to the longest baseline used in the interferometric array.

What was realised by the radio astronomers was that, since, in general, the sky does not change with time, there is no need to obtain all the information at one time. Provided the amplitude and relative phases of the signals arriving at two separate antennae can be measured, the information appropriate to that separation of elements of a large antenna is obtained. In the most elegant realisation of this scheme, the rotation of the Earth is used to move one antenna with respect to another and this technique is often referred to as Earth-rotation aperture synthesis. The Very Large Antenna (VLA) in the USA is the ultimate realisation of this principle which was pioneered in Cambridge by Martin Ryle and for which he was, with Anthony Hewish, the first astronomer to receive the Nobel Prize for Physics.

The principles of aperture synthesis are succinctly described by Barry Clark (1986) in the introductory chapter of the excellent monograph *Synthesis imaging* which includes many excellent chapters on the problems of realising aperture synthesis in practice. I especially recommend Clark's article which proves the key theorem, understandable from our above discussion, that an interferometer is a device which measures the *complex spatial coherence function* of the sky as measured by antennae separated by the vector distance $\mathbf{r}_1 - \mathbf{r}_2$. The complex spatial coherence function is defined to be

$$V_\nu(\mathbf{r}_1, \mathbf{r}_2) = \langle E_\nu(\mathbf{r}_1)\, E_\nu^*(\mathbf{r}_2)\rangle$$

and is the complex Fourier transform of the distribution of radiation $I_\nu(\mathbf{s})$ on the sky

$$V_\nu(\mathbf{r}_1, \mathbf{r}_2) = \int I_\nu(\mathbf{s}) \exp[-2\pi i \nu \mathbf{s}\cdot(\mathbf{r}_1 - \mathbf{r}_2)/c]\, d\Omega$$

where $\mathbf{s}$ is the unit vector in the direction in which the intensity of radiation is $I_\nu(s)$ and $d\Omega$ is the element of solid angle. If only a small region of sky is considered and measurements are made relative to a tracking centre in the direction $\mathbf{s}_0$, the complex spatial coherence function can be written

$$V_\nu(\mathrm{u}, \mathrm{v}) = \int I_\nu(l, m) \exp[-2\pi i(ul + vm)]\, dl\, dm$$

where u and v are distances measured in wavelengths in an orthogonal system of coordinates and l and m are angles measured from the phase tracking centre $\mathbf{s}_0$. The performance of an interferometer system can therefore be described in terms of the completeness of the information in the u–v plane, meaning simply how successfully the plane has been covered for observations of a source of a particular angular size. Two examples of the results of these types of observation are shown in Figs. 8.10 and 8.11.

The first example shows synthesis imaging of the radio source Cygnus A which is the brightest extragalactic radio source in the northern sky. In Fig. 8.10(*a*), the typical configuration of the 27 antennae of the US VLA which is located in the New Mexico desert near Soccorro is shown. The antennae are spaced logarithmically along each arm to provide maximum coverage of the u–v plane and can be moved to fixed locations along these arms which are about 20 km in length. The array can be used in a number of configurations; in the most compact configuration, and hence the lowest angular resolution mode, the lengths of the arms are about 0.6 km whilst, in the highest resolution mode, the lengths of the arms are about 20 km. The coverage of the u–v plane in an 8 h observation of a source at declination 45° is shown in Fig. 8.10(*b*). A map of Cygnus A created from a number of observations of the source using the VLA in a variety of different configurations is shown in Fig. 8.10(*c*). The incredible amount of detail and large dynamic range present in this image are apparent..

In the second example, recent Very Long Baseline Interferometry (VLBI) observations of the strong radio source 3C 273 are shown in Fig. 8.11. The global distribution of the telescopes used in the observations is shown in Fig. 8.11(*a*) and their geographical locations included in the figure caption. The resulting coverage of the u–v plane is much more sparse than it is in the case of observations with the VLA. The maps created from a programme of VLBI observations of 3C 273 over the period 1984–5 are shown in Fig. 8.11(*b*). The angular scale on both axes is milliarcseconds. It can be seen that the source components are moving apart and, at the distance of the source, the separation of the components takes place at a velocity of about eight times the speed of light. We will have a great deal to say about these remarkable observations in Volume 2, Chapter 24.

There is an important difference in approach between these two techniques. In the case of the VLA, the signals from each antenna are taken along waveguides to a central location where the coherence function is measured. One of the very important tricks of aperture synthesis is to compensate for the rotation of the Earth by continuously adding delay lines into one arm of any pair of antennae so that the baseline always appears to be in a fixed plane perpendicular to the direction of the source. This is no more than applying the Shift Theorem to aperture synthesis. In the case of VLBI, the antennae are not connected. Instead, very accurate clocks are placed at each antennae and the electric field intensity $I(t)$ measured as a function of time on very high speed tape recorders. The signals are then correlated off-line to find the complex spatial coherence functions.

How are these ideas extended to higher frequencies, in particular to infrared and optical wavelengths? At wavelengths shorter than about 1 cm, the atmosphere becomes progressively more unstable and introduces phase variations which become increasingly difficult to eliminate. At infrared and optical wavelengths, the problem is compounded by the problems of atmospheric seeing. For a typical large telescope, say 4 m diameter, the seeing cells which split up the wavefront are typically $\approx$ 10–20 cm in size and so the image is split into many 'speckles' so that the phase of the signals is very difficult to measure. Various ingenious procedures have been devised to overcome these problems and this is an important developing field.

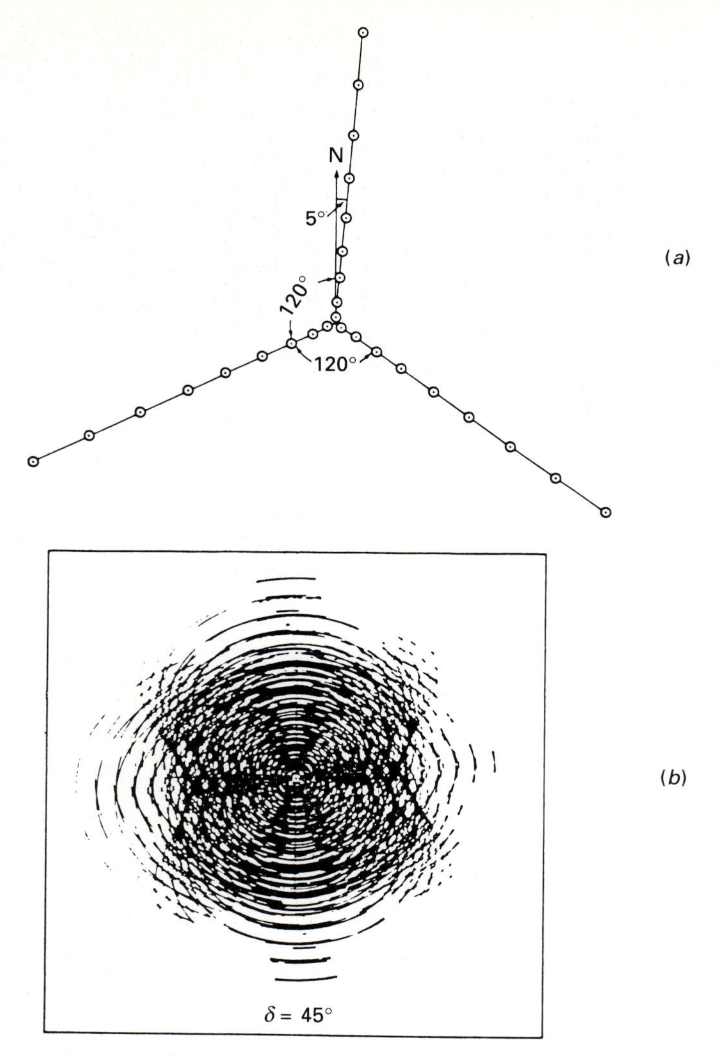

Figure 8.10. (*a* and *b*) For legend see facing page.

In one approach, small aperture telescopes are used which can be matched to the size of the seeing cells. Then, there should normally be only a single seeing cell over the aperture of each telescope. The clever thing to do is to introduce path compensation for the phase errors over each telescope very rapidly and then combine the signals to produce the complex spatial coherence function. This approach is being adopted by John Baldwin and his colleagues in Cambridge.

An alternative is to use the full aperture of the large telescope and attempt to deconvolve the two-dimensional image seen in the focal plane into a diffraction limited image. Very rapid photographs are taken of the object so that the speckles are 'frozen'. Typically, the cameras have to be run at a rate of 50 Hz to achieve this. The objective is then to use Fourier transform optics to derive the Fourier

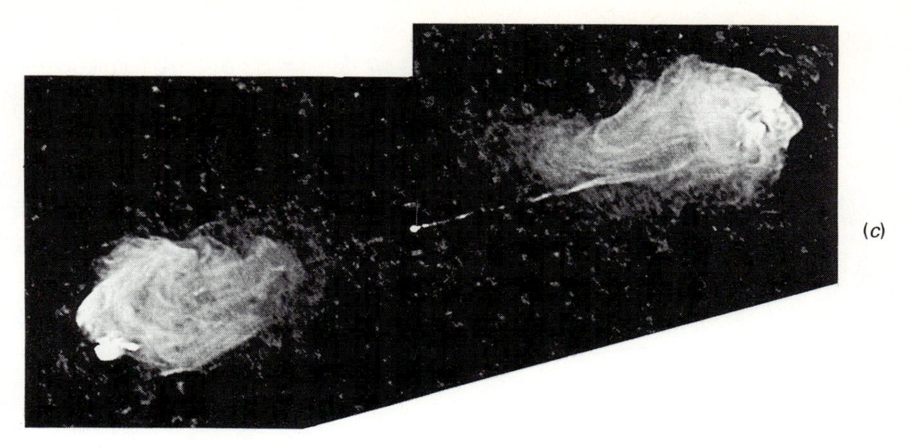

Figure 8.10. (*a*) A typical configuration of the 27 antennae of the US VLA. (*b*) The typical coverage of the aperture (or *u–v*) plane by the VLA in an 8 h observation of a source at declination 45°. (From A. R. Thompson (1986). *Synthesis imaging*, eds R. A. Perley, F. R. Schwab and A. H. Bridle, page 25, NRAO Publications. (*c*) A map of the radio source Cygnus A made using fully sampled observations by the VLA. (From R. A. Perley, J. W. Dreher and J. J. Cowan (1984). *Astrophys. J.*, **285**, L35.)

transform of the object with the diffraction limited resolution of the full aperture. Weigelt (1989) describes a number of ingenious ways of performing these deconvolutions, all of which are very demanding in terms of image processing.

8.5 The sensitivities of astronomical detectors

All signals in astronomy can only be measured with a certain statistical precision. It is again useful to make the distinction between those wavebands in which it is most convenient to think of the intensity of radiation as a flux of photons and those in which it can be considered a superposition of electromagnetic waves. In both cases, there are limitations in terms of the number of photons counted per second or the strength of the signal and in terms of the background against which the signal has to be detected. Let us deal with the two cases separately.

8.5.1 *Optical and infrared detectors*

If $h\nu \gg kT$, it is convenient to consider the flux of radiation to be composed of photons of energy $h\nu$. The first limit to the precision with which this flux of radiation can be measured is simply the statistics of the number of photons counted. In the absence of noise, Poisson statistics tells us that the uncertainty in this number is roughly $n \pm n^{\frac{1}{2}}$. Normally, however, this signal is observed against a noise background, for example, thermal fluctuations in the receiver, fluctuations in the intensity of the sky background and so on. Very often the background signal will far exceed the strength of the signal we are trying to detect. We can however

(*a*)

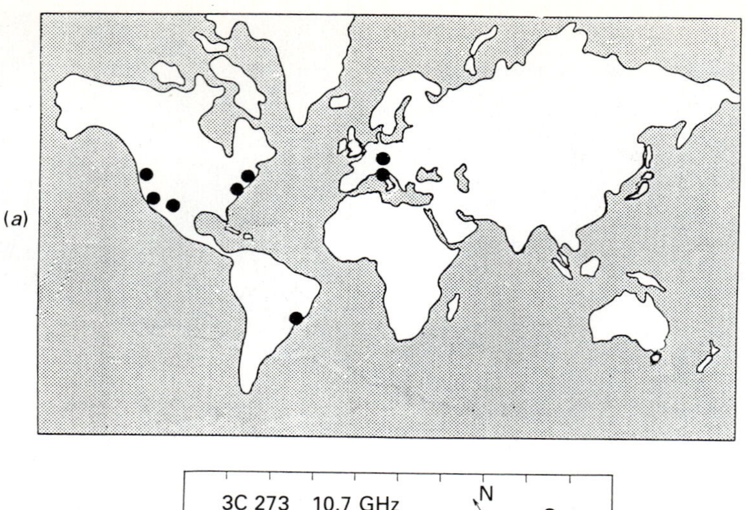

(*b*)

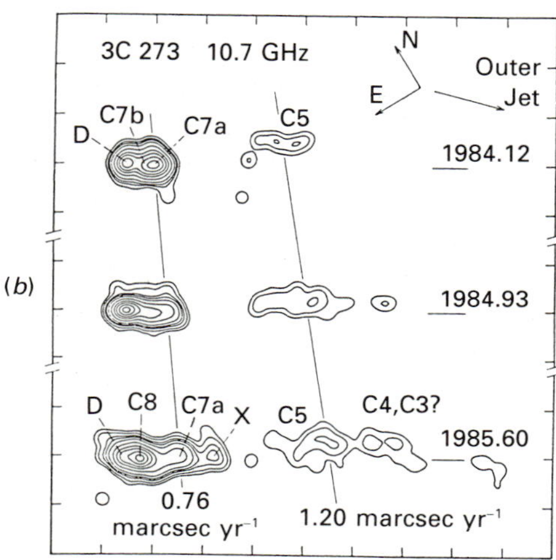

Figure 8.11. (*a*) The distribution of radio telescopes used in VLBI observations of the radio source 3C 273. The telescopes in the USA are located at the Haystack Observatory, the National Radio Astronomy Observatory at Green Bank, the GR Aggasiz Station, the Owens Valley Radio Observatory and the Hat Creek Radio Observatory. Those outside the USA are located at the MPIfRA Observatory at Effelsberg, West Germany, at the Istituto di Radioastronomia, Medicina, Italy and at the Instituto de Pesquisas Espaciais, Atibaia, Brazil. (After R. T. Schillizzi (1989). *Very Long Baseline Interferometry – Techniques and Applications*, eds M. Felli and R. E. Spencer, page 397, Dordrecht, D. Reidel Publishing Co.) (*b*) The very high resolution radio maps of 3C 273 at 10.7 GHz reconstructed from VLBI observations made at the above eight station VLBI network. The expansion of the source at a rate of about 1.20 marcsec per year can be seen. (From M. H. Cohen, J. A. Zensus, J. A. Biretta, G. Comoretto, P. Kaufmann and Z. Abraham (1987). *Astrophys. J.*, **315**, L89.)

use the *Central Limit Theorem* to show how the background intensity can be reduced statistically. The theorem may be expressed in the following way:

> If N estimates of a quantity x_i are made which are randomly selected from an arbitrary probability density function $p(x)$, the best estimate of the mean value of x is
>
> $$\bar{x} = \frac{1}{N}\sum_{i=1}^{N} x_i$$
>
> and the probability distribution of this value $\bar{x}$ about the true mean is a Gaussian distribution with standard deviation $\sigma_0/N^{\frac{1}{2}}$ where σ_0 is the standard deviation of the probability density function $p(x)$.

This theorem is proved in all the standard text books on statistics and is very important. Notice that the probability density function can be of any form, in particular of non-Gaussian form, and we need only know σ_0 for this non-Gaussian distribution. The other theorem we need concerns the variance of the sum of two quantities x and y each of which have separately variances σ_x^2 and σ_y^2, σ_x and σ_y being their standard deviations. The variance of $x+y$ is $\sigma_{x+y}^2 = \sigma_x^2 + \sigma_y^2$.

The fundamental point is that, every time we make an astronomical observation, we average over a large number of independent estimates of the random noise signal, x_i. Consequently, whatever the characteristics of the noise described by $p(x)$, it is safe to assume that the probability distribution of $\bar{x}$ is of Gaussian form. Furthermore, as N increases, the mean value of the noise is determined with greater precision and, specifically, the standard deviation about the mean decreases as $N^{-\frac{1}{2}}$ so that, provided the backgrounds do not change, and this is an important assumption, the noise can be determined with very high accuracy by making a sufficiently large number of observations.

Let us apply these concepts to a photon counting detector, such as an image photon counting system (IPCS) or a CCD which are now in common use on all classes of optical telescope and, in particular, on the Hubble Space Telescope. The problem is as follows. The flux density of the source we wish to detect is $N(\nu)$ photons $m^{-2}\ s^{-1}\ Hz^{-1}$; the source is observed for a time t in a waveband $\Delta\nu$. We let the effective aperture of the telescope be A_{eff} in the sense that this is the mirror area which corresponds to the number of photons registered by the detector in the focal plane for a source of known flux density; in other words, A_{eff} is the effective area of the telescope once account is taken of all the losses between the flux of radiation entering the aperture and the flux being detected as photons. A_{eff} clearly includes, for example, the quantum efficiency of the detector and losses in the optical system. Therefore, the number of photons detected from the source is

$$S = N(\nu) A_{eff} \Delta\nu t \text{ photons}$$

The statistical uncertainty in this number of photons is just $\pm n^{\frac{1}{2}}$ so that the signal-to-noise ratio, in the absence of all other sources of noise is just

$$\frac{S}{N} = (N(\nu) A_{eff} \Delta\nu t)^{\frac{1}{2}}$$

To work out the noise in the presence of other sources of noise we have to add the variances of all the sources of noise in the detector system and those due to the environment. Let us list some of these.

(i) First there is the variance associated with the total number of photons detected, S. According to Poisson statistics, the standard deviation of S is $S^{\frac{1}{2}}$ and hence the variance is $\sigma_1^2 = S$.

(ii) There is unwanted background radiation from both the sky and the telescope. In both cases, the radiation enters the detector just as if it consisted of photons from the source. In the optical and near-infrared wavebands, $\lambda \lesssim 2\ \mu m$, the background from the sky is the primary source of unwanted background radiation. If the intensity of the night sky is $B(\nu)$ photons $m^{-2}\ s^{-1}\ Hz^{-1}\ sr^{-1}$, we need to know the solid angle Ω subtended by the detector on the sky. The smallest this angle could be is the diffraction limit of the telescope but this is not normally attained at optical and near-infrared wavelengths because of the effects of astronomical seeing. In addition, it may be desirable to work with larger apertures than the seeing disc in order to ensure that the observations are of high photometric accuracy. Therefore, the signal from the night sky is $A_{\text{eff}}\Omega B(\nu)\Delta\nu t$, which by the same reasoning as in (i) is also the variance σ_2^2 about the mean value of the background. In the case of the thermal infrared wavebands, $\lambda \gtrsim 2\ \mu m$, the telescope itself is a strong source of background radiation, particularly at wavelengths $\lambda \sim 10\ \mu m$ at which a Planck function at 300 K has maximum intensity. The aim of the design of infrared telescopes is to minimise the emissivity of the telescope and, in particular, to ensure that the detector sees as little of the telescope structure as possible. For example, the primary and secondary mirrors must be of very high reflectivity and the secondary mirror should underilluminate the primary mirror so that as little as possible of the thermal background from the telescope enters the detector. The vanes holding the secondary mirror in place must have as small a projected area as possible as seen by the detector. We include the thermal background of the telescope in the same formula for the background given above.

(iii) There will be a certain 'dark current' in the detector, by which is meant noise electrons produced within the detector volume due to a variety of causes, for example, the thermal excitation of electrons into the potential wells of a CCD detector. If C is the number of electrons generated per second, the total variance during the observation is $\sigma_3^2 = Ct$, following the same reasoning as above.

(iv) There may also be a noise contribution when the signal is read out through the output amplifier. If this amounts to R electrons rms, the variance is $\sigma_4^2 = R^2$ for a single read-out. If the detector has to be read out many times in the course of an observation, this source of noise can amount to an important contribution to the noise signal against which faint objects have to be detected.

We can now find the total noise signal by summing the variances of these individual noise contributions

$$\sigma^2 = \sigma_1^2 + \sigma_2^2 + \sigma_3^2 + \sigma_4^2 \tag{8.20}$$

Thus, provided all the contributions to the total noise signal remain constant during the integration, the standard deviation of the fluctuations about the mean noise level is σ. The signal-to-noise ratio is therefore

$$S/N = N(\nu)A_{\text{eff}}\,\Delta\nu t/\sigma \tag{8.21}$$

Let us look at some simple applications of these formulae. First of all we can recover our previous result when the primary source of noise is simply photon statistics. In this case, $\sigma = \sigma_1 = (N(\nu)A_{\text{eff}}\,\Delta\nu t)^{\frac{1}{2}}$ and

$$S/N = [N(\nu)A_{\text{eff}}\,\Delta\nu t]^{\frac{1}{2}} \tag{8.22}$$

If the observations are *detector-noise-limited*, $\sigma = \sigma_3 = (Ct)^{\frac{1}{2}}$ and

$$\frac{S}{N} = \frac{N(\nu)A_{\text{eff}}\,\Delta\nu}{C^{\frac{1}{2}}}t^{\frac{1}{2}} \tag{8.23}$$

For very faint objects, particularly in the infrared waveband, the observations are often *background-limited* in which case $\sigma = \sigma_2$ and

$$S/N = N(\nu)\,(A_{\text{eff}}\,t/\Omega B(\nu))^{\frac{1}{2}} \tag{8.24}$$

The importance of these results is in showing how the signal-to-noise ratio depends upon the size of the telescope, A_{eff}, and the integration time necessary to achieve this signal-to-noise ratio. One of the more interesting ways of writing these relations is in terms of the time needed to achieve a given signal-to-noise ratio. If the diameter of the primary mirror of the telescope is D, $A_{\text{eff}} \propto D^2$ and then the relations (8.22), (8.23) and (8.24) become

$$\text{Photon-noise-limited} \quad t \propto A_{\text{eff}}^{-1} \propto D^{-2} \tag{8.25}$$
$$\text{Detector-noise-limited} \quad t \propto A_{\text{eff}}^{-2} \propto D^{-4} \tag{8.26}$$
$$\text{Background-limited} \quad t \propto \Omega A_{\text{eff}}^{-1} \propto \Omega D^{-2} \tag{8.27}$$

Another way of expressing these results is in terms of the limiting flux density which can be observed with a given signal-to-noise ratio in a given time. In this case we find

$$\text{Photon-noise-limited} \quad S \propto A_{\text{eff}}^{-1} \propto D^{-2} \tag{8.28}$$
$$\text{Detector-noise-limited} \quad S \propto A_{\text{eff}}^{-1} \propto D^{-2} \tag{8.29}$$
$$\text{Background-limited} \quad S \propto \Omega^{\frac{1}{2}} A_{\text{eff}}^{-\frac{1}{2}} \propto \Omega^{\frac{1}{2}} D^{-1} \tag{8.30}$$

In many programmes, the observations are made at the very limit of what is possible technologically and generally speaking these observations are background-limited. It can be seen from the relations (8.27) and (8.30) that the limiting flux density is then strongly dependent upon Ω, the solid angle subtended by the detector on the sky. Two examples are of special interest.

In the optical waveband, the angular resolution is limited by the effects of astronomical seeing and this determines the minimum pixel size which it is sensible to use in the observations. This means that, for ground-based observations, the solid angle Ω is limited to about 1 arcsec^2 and is more or less independent of wavelength. What this calculation shows is how sensitive the quality of the observations is to good seeing. On those occasions when the astronomical seeing is a factor of 2 better than the above figure, the limiting flux density is smaller by a factor of 2. This same calculation also shows the dramatic improvement in limiting flux density which can be achieved by the Hubble Space Telescope. With an angular resolution about 10 times better than that normally achievable in ground-based observations, the limiting flux density for the detection of point-like sources is about a factor of 10 fainter than for the identical telescope observing from the surface of the Earth or, equivalently, the same source can be observed with the same signal-to-noise in only 1/100 of the time needed on the ground.

In the second example, we consider the advantages which are gained when the telescope is diffraction-limited as occurs in the thermal infrared region of the spectrum. As shown above, a 4 m telescope becomes diffraction-limited at about 5–10 μm because of the increase in the size of the seeing cells with increasing wavelength. In this case, $\theta \approx \lambda/D$ and hence $\Omega \propto \lambda^2/D^2$. As a result, the relations (8.27) and (8.30) become

$$t \propto D^{-4}; \quad S_{\text{min}} \propto D^{-2}$$

Thus, the gains in using a large telescope in the thermal infrared region of the spectrum are large, provided the detectors are matched to the diffraction pattern of the telescope.

These simple calculations are indicative of those necessary to evaluate the performance of telescopes at optical, infrared and ultraviolet wavelengths. More detailed calculations are necessary to assess the performance of telescopes for spectroscopy since the spectrograph has to be matched to the optics of the telescope. In this type of calculation, the sensitivity depends strongly upon whether the object lies completely within the slit of the spectrograph or whether only a portion of the image passes through the jaws of the spectrograph. A detailed discussion of the signal-to-noise ratios expected for different classes of observation is very clearly explained by Schroeder (1987), who makes reference to earlier works by Code (1973) and Bowen (1964).

Observations of faint objects from the surface of the Earth are background-limited and, for rough calculations, it may be assumed that the background in the V waveband corresponds to 22 magnitudes arcsec^{-2}. This intensity correponds to $B(\nu) = 6 \times 10^{-32}$ W m^{-2} Hz^{-1} sr^{-1} at $\lambda = 550$ nm. Some other useful background intensities are given in Table 8.2. It should be noted that these figures should be used with caution because the background intensity is strongly wavelength dependent, particularly due to the presence of strong atmospheric emission lines at certain wavelengths.

Table 8.2. *The intensity of the sky background in the optical and infrared wavebands*

(1)	(2)	(3)	(4)	(5)	(6)	(7)	(8)	(9)
U	0.365	8.2×10^{14}	0.068	4.2×10^{-8}	1.88×10^{-23}	22	150	10
B	0.44	6.8×10^{14}	0.098	7.2×10^{-8}	4.64×10^{-23}	23	100	10
V	0.55	5.5×10^{14}	0.089	4.0×10^{-8}	3.95×10^{-23}	22	170	15
R	0.70	4.3×10^{14}	0.22	1.8×10^{-8}	2.87×10^{-23}	21	250	55
I	0.90	3.3×10^{14}	0.24	8.3×10^{-9}	2.24×10^{-23}	18.5	1.5×10^{3} [b]	370
J	1.25	2.4×10^{14}	0.28	3.07×10^{-9}	1.60×10^{-23}	16	1.0×10^{4} [b]	2.8×10^{3}
H	1.65	1.8×10^{14}	0.30	1.12×10^{-9}	1.02×10^{-23}	13	5.6×10^{4} [b]	1.7×10^{4}
K	2.2	1.4×10^{14}	0.42	4.07×10^{-10}	6.57×10^{-24}	12.5	4.4×10^{4}	1.8×10^{4}
L	3.45	8.7×10^{13}	0.60	7.30×10^{-11}	2.90×10^{-24}	5.5	8.0×10^{6}	5.0×10^{6}
M	4.7	6.4×10^{13}	0.67	2.12×10^{-11}	1.63×10^{-24}	2	1.0×10^{8}	7.0×10^{7}
N	10.2	2.9×10^{13}	5.2[a]	1.10×10^{-12}	3.90×10^{-25}	−3	1.0×10^{9}	5.0×10^{9}
Q	20.0	1.5×10^{13}	5.2[a]	7.80×10^{-14}	1.04×10^{-25}	−5	6.0×10^{8}	3.0×10^{9}

[a] These bands may be observed with narrower filters to reduce the background intensity.
[b] In these wavebands the background is dominated by hydroxyl (OH) airglow emission lines.

Column Labels

(1) Name of waveband
(2) Effective wavelength of the waveband ($\lambda_{eff}/\mu m$).
(3) Effective frequency of the waveband (ν_{eff}/Hz).
(4) Effective width of the band ($\Delta\lambda_{eff}/\mu m$).
(5) Flux density of a zero magnitude star per unit wavelength ($S_\lambda(0)$/W m^{-2} μm^{-1}).
(6) Flux density of a zero magnitude star per unit frequency range ($S_\nu(0)$/W m^{-2} Hz^{-1}).
(7) Background intensity in magnitudes (arcsec^{-2}).
(8) Background photon intensity per unit waveband ($I(\lambda)$/photons m^{-2} arcsec^{-2} s^{-1} μm^{-1}).
(9) Background photon intensity in standard waveband given in column (4) (I /photons m^{-2} arcsec^{-2} s^{-1}).

The names of the standard astronomical wavebands are given in column (1) with their mean wavelengths and frequencies in columns (2) and (3). Column (4) gives the effective width of the standard wavebands. In the optical waveband ($\lambda \leqslant 1$ μm), these are determined by the properties of the standard filters. In the infrared waveband ($\lambda \geqslant 1$ μm), the widths of the bands are primarily determined by the widths of the atmospheric windows. Columns (5) and (6) give the flux densities of a zero magnitude star per unit wavelength and per unit frequency respectively. Column (7) gives the background sky brightness in magnitudes (arcsec^{-2}) and in terms of the intensity per unit bandwidth in column (8). In column (9), the background intensities in the standard wavebands of column (4) are given.

The background intensities should only be used as a rough guide in making estimates of the observing time necessary to reach a given limiting magnitude. In the above table, it is assumed that the observations are made from a good dark site in the absence of

8.5.2 *Radio and millimetre-wave receivers*

The basic theorem which describes the amplitude of thermal fluctuations in an electrical circuit is *Nyquist's theorem.* I have shown in the Appendix to Chapter 12 of Longair (1984) how this theorem can be rigorously derived from Einstein's prescription that, in thermal equilibrium, the average energy per mode is given by the relation

$$\bar{E} = \frac{h\nu}{\exp(h\nu/kT) - 1} \tag{8.31}$$

where ν is the frequency of that mode. By considering the modes of a transmission line of wave impedance Z_0 terminated by matched resistors R, it can be readily shown that the resistors each deliver a noise power

$$P = \frac{h\nu}{\exp(h\nu/kT) - 1} \text{ W Hz}^{-1} \tag{8.32}$$

if they are maintained at temperature T. At low frequencies, $h\nu \ll kT$, which is normally the case for radio and microwave receivers, this expression reduces to

$$P = kT, \tag{8.33}$$

which is the familiar form of Nyquist's theorem. Thus, at radio wavelengths, the noise power delivered by the resistor in the frequency range ν to $\nu + \Delta\nu$ is $P\mathrm{d}\nu = kT\mathrm{d}\nu$. In the opposite limit, $h\nu \gg kT$, the noise power decreases exponentially as

$$P = h\nu \exp(-h\nu/kT)$$

For radio receivers, the expression (8.33) provides a convenient way of describing the performance of a receiver which delivers a certain noise power P_n. We can define an *equivalent noise temperature* T_n for the performance of the receiver at frequency ν by the relation

$$T_n = P_n/k \tag{8.34}$$

There is another important feature of the fluctuations in this noise signal which we have already noted. We note that, per unit frequency range per second, the electrical noise power is kT which corresponds exactly to the energy of a single wave mode in thermodynamic equilibrium. Since this energy is in the form of the superposition of electromagnetic waves, the fluctuations in this wave mode

---

scattered moonlight. In the infrared waveband at wavelengths $\lambda \geq 2\ \mu$m, it is assumed that the telescope and the atmosphere are at a temperature of 288 K and that the emissivity of the telescope is 8%. Lower background fluxes are obtained if narrow filters are used. In addition, at those wavelengths at which air glow is the dominant source of noise, lower background fluxes can be obtained by observing between the strong OH emission lines. This is now possible with the latest generation of infrared spectrometers.

correspond to $\Delta E/E = 1$. Therefore, we expect the amplitude of the noise fluctuations per unit frequency interval to be kT.

We have already described in Section 8.3 how a wave packet consisting of waves with frequencies in the interval ν to $\nu + \Delta\nu$ remains coherent for a time $\tau \approx \Delta\nu^{-1}$ which is called the coherence time. To make independent estimates of the field strength, the samples should not be taken more than about once per coherence time. Thus, whereas for photons, we obtain an independent piece of information every time a photon arrives, in the case of waves, we obtain independent estimates only once per coherence time $\tau \approx \Delta\nu^{-1}$. If we observe a source for time t, we obtain $t/\tau = t\Delta\nu$ independent estimates of the intensity of the source.

Normally, we are interested in detecting very weak signals in the presence of a much greater noise signal generated by the receiver. As discussed above, this noise power fluctuates with amplitude $\Delta E/E = 1$ per mode per second. We can therefore reduce the amplitude of the fluctuations by increasing the length of the integration and by increasing the bandwidth of the observations. In both cases, we increase the number of independent estimates of the strength of the signal by a factor $\Delta\nu t$ and hence the amplitude of the power fluctuations is reduced to

$$\Delta P = kT/(\Delta\nu t)^{\frac{1}{2}} \tag{8.35}$$

Thus, if we are able to integrate long enough and use large enough bandwidths, very weak sources can be observed.

Radio astronomers extend the use of temperatures to define other aspects of the performance of the telescope, its receivers and the objects to be observed. If the intensity of radiation from a region of sky is I_ν, an equivalent *brightness temperature* T_b can be defined using the expression for the intensity of black-body radiation

$$I_\nu = \frac{2h\nu^3}{c^2}\frac{1}{\exp(h\nu/kT_b) - 1} \tag{8.36}$$

This brightness temperature T_b is a lower limit to the temperature of the region itself because thermodynamically no region can emit radiation with intensity greater than that of a black-body at its thermodynamic temperature, unless the population of the levels is inverted as in the case of masers and lasers – the latter are scarcely systems with a thermodynamic distribution of particle energies. The concept of brightness temperature is thus a very useful way of setting a lower limit to the temperature of the region for sources in which the radiation mechanism is incoherent. Very often in radio astronomy it is safe to use the Rayleigh–Jeans approximation for the brightness temperature because $h\nu \ll kT$ and consequently the expression for brightness temperature reduces to

$$I_\nu = 2kT_b/\lambda^2 \tag{8.37}$$

We have already shown that there is a certain amount of noise power available at the terminals of a resistance R at temperature T. We can replace the resistance R by a matched antenna of radiation resistance R and, if the antenna is then placed in a black-body cavity at temperature T, the same noise power will be available at the output of the antenna because the system is in thermodynamic equilibrium. If instead, the antenna is looking at the sky and the brightness temperature distribution is $T_b(\theta, \phi)$, the mean power available at the output of the antenna is

$$W = \tfrac{1}{2}A_e \int T_b(\theta, \phi)\, P(\theta, \phi)\, d\Omega = kT_A \tag{8.38}$$

where A_e is the effective area of the antenna and is related to the beam area of the antenna Ω_A by $A_e \Omega_A = \lambda^2$ (see Section 8.2), $P(\theta, \phi)$ is the polar diagram of the antenna and T_A is defined to be the *antenna temperature* due to the radiation incident upon it. If the source is a point object or small compared with the half-power beam width, the flux density of the source is

$$S = (2kT_A/\lambda^2)\,\Omega_A \tag{8.39}$$

This antenna temperature has to be detected in the presence of noise and the main contributions are from the radio background emission from our Galaxy T_{gal}, the emission from the Earth's atmosphere T_{atm} and the noise temperature of the complete receiving system T_{sys}. The latter term contains all components which contribute to the noise power present in the receiver at the point where the signal is detected. This includes the contributions from the antenna itself, the antenna noise temperature T_{na}, from the transmission lines T_m and the noise temperature of the receiver T_n. The sum of these sources of noise can be expressed

$$T_{tot} = \sum_i T_i$$

In an observation of integration time t and with bandwidth $\Delta\nu$, the minimum detectable antenna temperature is therefore

$$\Delta T_{min} = T_{tot}/(\Delta\nu t)^{\frac{1}{2}}$$

as shown above.

Various potential contributions to T_{tot} are illustrated in Fig. 8.12. In many cases the main source of noise is the noise temperature of the receiver and the emphasis has been upon the construction of low noise amplifiers of the very weak astronomical signals. The main types of amplifiers used are transistors, tunnel diodes, parametric amplifiers and masers. At low frequencies, $\nu \lesssim 1$ GHz, the dominant source of noise is the radio background emission from the Galaxy and consequently there is no point in using the ultimate in low noise amplifiers at these frequencies. Room temperature FET amplifiers are adequate for many purposes. At millimetre wavelengths, the atmospheric and cosmic background noise increase and rather different techniques are used as described below.

8.6 Detectors

At the heart of any astronomical instrument there is a detector which converts the astronomical signal into some analogue or digital form which is used for analysis by the astronomer. Many of the principles of detection are similar to

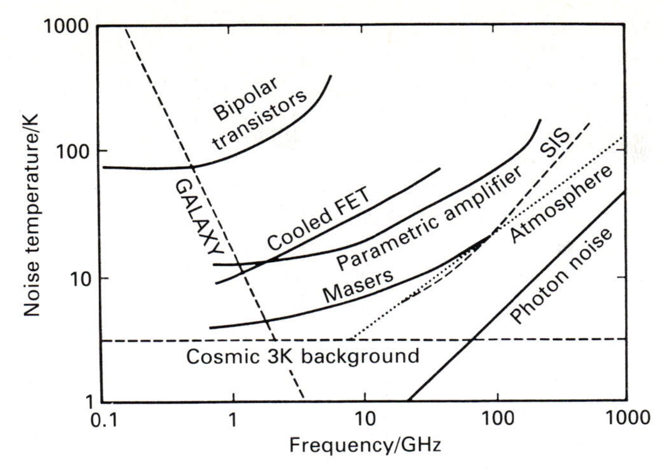

Figure 8.12. A comparison of the different contributions to the noise performance of receivers for radio and millimetre astronomy as a function of frequency. The noise temperatures of the receivers shown are among the best achieved to date. Care should be exercised in using the data in this diagram for detailed calculations. For example, there are strong variations in the intensity of the Galactic radio emission as a function of direction and the atmospheric noise is strongly site-dependent. (From I. Appenzeller (1989). *Evolution of galaxies: astronomical observations*, eds I. Appenzeller, H. J. Habing and P. Lena, page 355, Berlin: Springer-Verlag.)

those already described in the context of the detectors which are used in the high energy astronomies and which were described in Chapter 6. In this section, we highlight the important differences between these detectors and those employed in the low energy astronomies. There are several excellent books which describe in detail the operation of astronomical detectors and I can recommend those by Eccles, Sim and Tritton (1983), Kitchin (1984), Walker (1987), Lena (1988) and McLean (1989) as well as the excellent introductory survey by Appenzeller (1989).

8.6.1 *Optical and ultraviolet detectors*

Until the mid-1960s, most astronomical signals were recorded on *photographic plates*, either in the form of direct images of the sky or in the form of photographic spectra. The principles of the use of photographic plates are exactly the same as those described in Section 6.2 in the context of the nuclear emulsions used in cosmic ray studies except that now photons rather than electrons excite the silver atoms. The silver bromide crystals are semiconducting materials with band-gaps of about 2 eV and so they should only be sensitive to blue light. With the addition of impurities, however, the sensitivities of photographic plates can be extended into the red region of the optical spectrum. The problem with the photographic plates used in optical astronomy is that they have low quantum efficiencies, typically about 1–2% for the best emulsions such as the IIIaJ emulsions produced by Kodak. This means that most of the light incident upon the photographic plate does not result in activated grains which can be revealed by the process of development. In addition, the emulsions have a somewhat limited dynamic range in that they can only accommodate a relatively small range of intensities before the brightest images are saturated.

They have, however, the enormous advantage that they are able to store a huge amount of information. For example, in a large wide-field Schmidt Telescope, such as the UK Schmidt Telescope at the Siding Spring Observatory in New South Wales, the 14 in plates correspond to a field of view on the sky of about $6° \times 6°$ so that, if the typical *picture-element* or '*pixel*' at the focal plane is 1 arcsec^2, the total number of pixels observed in a single exposure is about 5×10^8. When we take account of the intensity information contained in each pixel, it is apparent that each Schmidt Telescope plate contains several gigabits of information. This is a huge amount of information and large photographic plates such as these remain the most important detectors when very wide fields of view are required – they have a unique role to play in sky-survey work. In addition, photographic plates have been available for astronomical use since the beginning of the present century and so the photographic record remains an important source of information about astronomical objects which change their positions or intensities over periods up to about 100 years.

In order to take advantage of the enormous amounts of information present on Schmidt plates, *ultrahigh speed measuring machines* have been developed to convert the photographic data into digital data which can be used, for example, to distinguish between stars and galaxies in an automatic fashion and also to provide catalogues of the order of the 10^5–10^6 stars and galaxies present on each plate (Fig. 8.13). It is studies using Schmidt Telescope plates in conjunction with the high speed measuring machines which have led to the discovery of many of the most distant quasars and of candidates for the elusive brown dwarfs. In both of these examples, only one or two objects of each class are expected from among the 100 000 images which are also present on the plate. Another excellent example of the use of these facilities is in the determination of the large scale distribution of faint galaxies which is of crucial importance for cosmological studies (Fig. 8.14).

Photoelectric detectors were introduced into astronomy in the 1950s and they continue to be used for the precise photometry of stars and galaxies. In these devices, the light from an object within a given angular aperture and filter band is incident upon a photocathode which liberates electrons by the photoelectric effect. As in the case of the scintillation detectors described in Section 6.5.2, the liberated electrons are accelerated down a photomultiplier tube resulting in an exponentially increasing cascade of electrons. Thus, for each detected photon, a large pulse is registered at the output and so these devices can be used in a photon counting mode. Photomultipliers of this type are used in the High Speed Photometer/Polarimeter on board the Hubble Space Telescope.

A major advance occurred with the introduction of *image intensifier tubes* into optical astronomy. In a typical arrangement, the images are focussed onto the photocathode of an image intensifier. The resulting photoelectrons are accelerated to several kiloelectron volts and then they encounter a phosphor screen which

Figure 8.13. The COSMOS high speed measuring machine at the Royal Observatory, Edinburgh. This machine is used to scan wide angle astronomical plates very rapidly and has particular application in the scanning of large astronomical plates taken with large Schmidt telescopes. (Courtesy of the Royal Observatory, Edinburgh.)

liberates typically about 100 photons for each incident electron. One possibility is then to place a photographic plate behind the phosphor and thus obtain a much higher quantum efficiency, this factor being determined by the quantum efficiency of the first photocathode which is typically about 20%. Alternatively, the flux of

Figure 8.14. An example of the type of large scale survey work which can be undertaken using the combination of Schmidt Telescope plates and high speed measuring machines. This diagram shows the distribution of faint galaxies in the Edinburgh–Durham Southern Galaxy Catalogue to a limit of $b_j = 20$ over an area of 1700 square degrees of the Southern sky centred on the galactic south pole. (b_j is a photographic magnitude corresponding to the sensitivity of the IIIaJ emulsions used in the Sky Survey.) Carefully calibrated surveys such as these are crucial in defining the large scale structure of the Universe. (Courtesy of Drs C. Collins and H. MacGillvray, Royal Observatory, Edinburgh.)

photons from the first stage of amplification may strike another photocathode. Then, magnetic focussing is needed to focus the accelerated pulse of electrons onto the next phosphor screen. In the devices used in astronomical detectors, three or four stages of image intensification may take place.

There are different ways of dealing with the bunches of electrons, each of which corresponds to the arrival of a single photon at the entrance window to the photocathode. One way of dealing with the output signal is to collect the charge on an array of detectors which accumulate the total charge as a measure of the total number of photons incident on each detector element. This is the principle of the *digicon detector* and these devices are used in the Faint Object Spectrograph and the Goddard High Resolution Spectrograph of the Hubble Space Telescope.

Another very successful technique has been developed by Boksenberg and his colleagues – this mode of operation is known as an *image photon counting system* (IPCS). In this approach, the intense burst of output photons associated with each detected photon is viewed by a television scanner. The centroid of each burst is found by electronic logic circuits and its position and time of arrival is stored in a computer. For each photon detected by the first photocathode, a single photon is labelled in the computer memory. The first application of this technique was in astronomical spectroscopy but it has been extended to two-dimensional imaging so that 'long-slit' spectra can be taken as well as direct imaging. Detectors of this type are present on most large telescopes and the Faint Object Camera of the Hubble Space Telescope contains a detector of this type. The great advantage of this type of detector is that it is linear so that the fluxes of photons measured are directly proportional to the intensities observed at the focus of the telescope.

In the case of digicon detectors, in which individual photons are not detected, the dynamic range of the individual detectors elements must be very large. In the case of the IPCS detectors, the dynamic range of the instrument is limited by the ability of the centroiding logic to distinguish between successive events arriving at the same point on the output phosphor of the image tube. Typical arrays consist of, say, 512 digicons in a linear array or about 500×500 picture elements available in an IPCS detector working in imaging mode. Thus, the fields of view accessible to these devices are very much smaller than those obtainable with photographic plates. The quantum efficiencies of these detectors reach about 20–25%, the best figures being achieved in the ultraviolet region of the spectrum. Generally, the quantum efficiency of these detectors falls off rapidly at wavelengths $\lambda \gtrsim 600$ nm.

The most important advance in optical astronomy over the last ten years has been in the field of solid-state detectors, particularly those based upon silicon technology. The most important of the new devices are those known as *charge coupled devices* (CCDs). The band-gap of silicon is 1.12 eV which means that photons with wavelengths shorter than 1.1 μm can excite electrons into the conduction band. The design of the CCD chip is shown schematically in Fig. 8.15(*a*). By selective doping, the chip is divided into a number of sensitive strips separated by narrow channel stops. Down each channel, there is a very thin insulating layer and on top of that there are the read-out electrodes, in the typical three-phase CCD, there being three read-out electrodes per pixel of the array. In the case of a three-phase CCD, the central electrode is maintained at a positive voltage relative to the other two so that, when an electron–hole pair is created by an incident photon, the electrons accumulate beneath the central electrode whilst the holes are swept out into the substrate of the chip. Thus, each pixel contains a small electrostatic potential well in which the electrons accumulate.

The term *charge coupling* refers to the way in which the charges are read out from each pixel at the end of the exposure. In the typical three-phase CCD, each pixel is read out by changing the positive voltages on the read-out electrodes in sequence so that the accumulated charge is shifted down each channel (Fig. 8.15(*b*). The read-out of the chip is arranged so that each channel is read out sequentially through a single amplifier which converts the charge in each pixel into an amplified digital signal which is stored in the computer memory. The charge transfer efficiency of the CCD is crucial for its success as a detector. The great importance of these devices for astronomy is that silicon has a very high quantum efficiency for the detection of photons in the wavelength range 400–1000 nm, typical values being about 60–70%, corresponding to an improvement by a factor of about 50 over photographic plates. In addition, they have a wide dynamic range. Up to about 10^5 electrons can be trapped in an individual well before overflow into neighbouring wells becomes important. The main source of noise is

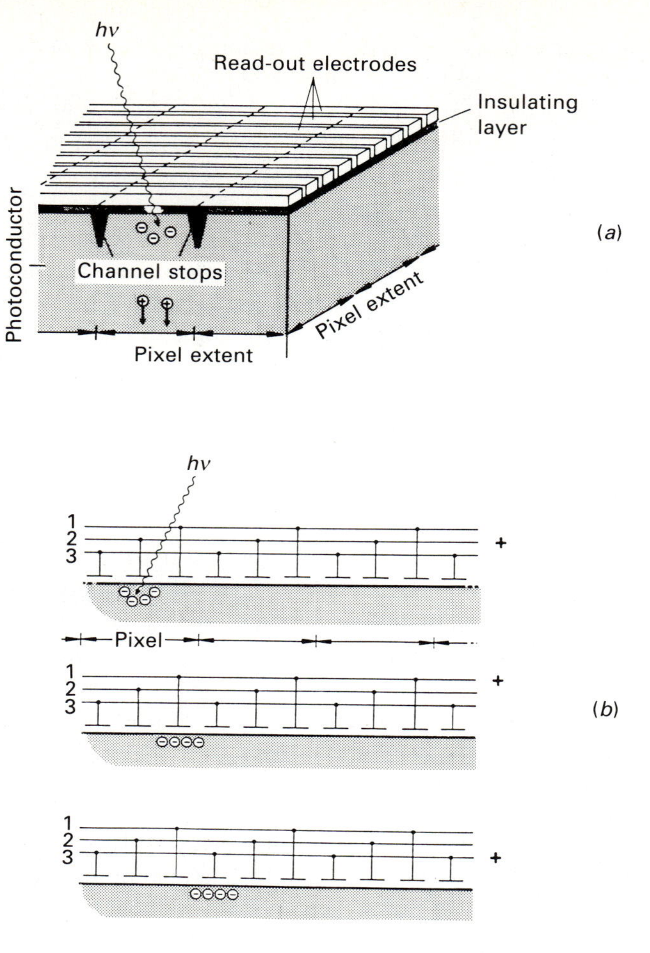

Figure 8.15. (*a*) Illustrating the construction of a CCD on a chip of pure silicon. The channel stops are created by heavily doping the material and then a very thin insulating layer is deposited on top of the chip. On top of the insulator are the thin conducting read-out electrodes. Each pixel has three read-out electrodes and the central one is maintained at a positive potential to form a potential well in which the electrons are trapped.
(*b*) Illustrating how charge is transferred along a row of pixels by changing the positive voltage on the read-out electrodes sequentially. (From I. Appenzeller (1989). *Evolution of galaxies: Astronomical observations*, page 316, Berlin: Springer-Verlag.)

the dark current associated with the thermal excitation of electrons into the conduction band. For this reason, the detectors are cooled to about −100 °C which reduces the dark current to a low level. An important source of noise is then the read-out noise which is added to the signal when it is read out through the amplifier. As an example of the performance of a modern CCD detector, the expected instrumental performance of the Wide Field Planetary Camera of the Hubble Space Telescope is listed in Table 8.3.

Table 8.3. *The wide field/planetary camera*

The camera can operate with two different focal ratios, $f/12.9$ or $f/30$, the former being known as the wide field mode and the latter as the planetary mode. In each mode the detector consists of an array of four thinned, back-illuminated CCDs, each consisting of 800×800 pixels

	Wide field mode	Planetary mode
Focal ratio	$f/12.9$	$f/30$
Picture format	1600×1600 pixels	1600×1600 pixels
Field of view	2.57×2.47 (arcmin)^2	66.7×66.7 (arcsec)^2
Pixel size	0.1 arcsec	0.043 arcsec
Wavelength range	115–1100 nm	115–1100 nm
Dynamic range (per pixel, single exposure)	460[a]	460[a]
Maximum S/N (per pixel, single exposure)	170	170
Overall dynamic range for stars (V)	about 9 to 29	about 8 to 29
Exposure times	Minimum exposure time is 0.11 seconds. Maximum length of target intergrations is 100000 seconds. In practice, the exposure times will be limited by saturation effects and cosmic ray events	

The camera contains twelve filter wheels, each with four filters and a clear 'home' position. The filters include a wide range of narrow and broad-line filters as well as polarisation filters and transmission gratings.

[a] The full-well capacity of the CCD detectors is 30000 electrons and the rms read-out noise per pixel is 13 ± 2 electrons. The quoted dynamic range is the full-well capacity divided by five times the read-out noise.

A major problem has been the production of high quality CCD chips needed in astronomical applications. The performance of the chips is very sensitive to chip defects. Arrays of 500×500 pixels are now routinely available and larger chips are being developed. The increased power of optical telescopes which has resulted from the introduction of these devices on large telescopes has been spectacular. It should be noted that, since the quantum efficiencies of these detectors are now approaching 100%, further gains in sensitivity or speed of observation can only be obtained by making the aperture of the telescopes larger.

8.6.2 *Infrared detectors*

As has already been described in Section 1.4.3, atmospheric absorption and emission become progressively more important as observations are made further into the infrared waveband. A larger scale diagram showing the transmission of the atmosphere as a function of wavelength in the 1–30 μm waveband is displayed in Fig. 1.11. It can be seen that there are a number of atmospheric windows in this wavelength region in which the atmospheric absorption is small. At wavelengths longer than 30 μm however, the atmosphere becomes opaque until wavelengths of about 350 μm at which the atmosphere again becomes more or less transparent. On a high, dry site, such as Mauna Kea, the near-infrared windows at 1.1 (J), 1.65 (H) and 2.2 (K) μm are essentially 100% transparent but there is also very good transparency throughout the 1–5 μm region with the exception of a few wavelengths at which there is very strong atmospheric absorption. With modern infrared detectors, observations are nearly always background-limited in the infrared waveband. One of the problems with making observations in the presence of a significant background component is that there are fluctuations in the atmospheric absorption and emissivity. The standard procedure for dealing with this problem is to 'chop' between the source and a nearby blank piece of sky. Provided the chopping takes place at a frequency which is greater than that over which the sky changes, the difference in these signals provides a measure of the source intensity.

Despite the availability of infrared windows, there are several important regions of the spectrum which cannot be observed because of atmospheric absorption and the whole of the wavelength region from 30 to 300 μm is essentially completely opaque as can be seen from Fig. 1.11. To obtain a clear view of the sky, it is essential to make observations from above the Earth's atmosphere. Ideally, the observations are made from satellite observatories as was the case for the *Infrared Astronomy Satellite* (IRAS) and will be for the next generation of space missions such as the *Infrared Space Observatory* (ISO) of the ESA.

One particular problem which is unique to these wavelengths is the fact that the telescopes and the instruments themselves are strong sources of thermal infrared background radiation. For the instruments, the solution is to cool them to a low temperature so that the thermal background is reduced to a negligible level. For example, for observations in the 1–5 μm window, it is customary to cool the components in the optical train to temperatures of about 50–70 K to eliminate the thermal background. For space experiments such as ISO which are designed to operate in space throughout the 3–100 μm wavelength range, the telescope and the elements in the optical train have to be cooled to about 4 K to minimise the thermal background radiation. In the case of ground-based telescopes, it is impractical to cool the whole telescope to a low temperature and so every precaution is taken to reduce the thermal emissivity of the telescope to a minimum.

Modern detectors of infrared radiation operate in a similar mode to that of silicon detectors in the optical waveband but there are some important differences. First, it is apparent that the band-gaps in the semiconductor materials must be much smaller than in the case of silicon. There are various ways of achieving this. One approach is to develop semiconductor materials which have much smaller band-gaps than silicon – these are known as *intrinsic* semiconductors. Materials such as indium antimonide (InSb) and mercury cadium telluride ($Hg_{1-x}Cd_xTe$) have been extensively developed as suitable materials for infrared detectors. Another approach is to 'dope' the semiconductor so that an impurity band is created just below the conduction band – these are known as the *extrinsic*

Table 8.4. *The band-gaps and maximum wavelengths for different semiconductor materials*

Composition	E_G (eV):	λ_{max} (μm):
Intrinsic photoconductors		
GaAs	1.35	0.92
Si	1.12	1.10
Ge	0.68	1.82
InAs	0.33	3.80
InSb	0.18	6.95
HgCdTe	adjustable	adjustable
Impurity band photoconductors		
Si:In	0.16	8
Si:Ga	0.07	18
Si:Bi	0.06	18
Si:As	0.05	23
Si:P	0.05	28
Ge:As	0.01	95
Ge:Ga	0.01	120
Ge:Ga stressed	< 0.01	240

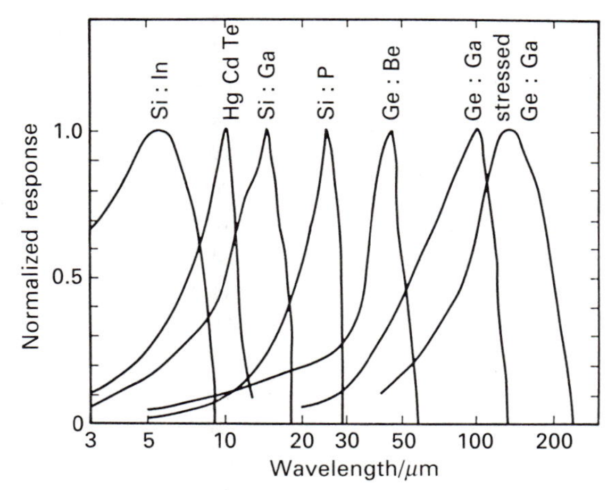

Figure 8.16. The wavelength response of different types of material used as detectors of infrared radiation (see Table 8.3). (From I. Appenzeller (1989). *Evolution of galaxies: Astronomical observations*, pages 312–13, Berlin: Springer-Verlag.)

semiconductors. Some examples of commonly-used materials of both kinds are listed in Table 8.4 and their wavelength responses are shown in Fig. 8.16. For infrared detectors, the sensitivites are quoted in terms of their *noise equivalent powers*. This is the power received by the detector such that for a bandwidth of 1 Hz, the signal-to-noise ratio is 1. Since the noise power decreases as $(\Delta\nu)^{-\frac{1}{2}}$, it follows that the noise equivalent power (or NEP) is quoted in terms of units of

W Hz$^{-\frac{1}{2}}$. For the best modern detectors NEPs of about 10^{-17} W Hz$^{-\frac{1}{2}}$ or less are found.

A second problem is that the fluxes of photons are much greater in the infrared as compared with the optical waveband. There are two reasons for this – first, the energies of the photons are smaller and consequently more photons are obtained for a given flux density. Second, often the signals have to be detected against a very much larger background of unwanted radiation. This is particularly the case for observations made at about 10 μm at which the background flux of radiation may be about 10^6 times greater than the signal. The consequence is that the detectors must have the capacity for accepting very large fluxes of photons, meaning that they must have a linear response over a wide dynamic range. In the case of observations in the thermal infrared region of the spectrum, it is often necessary to read the detector out at a high rate, for example at 50 Hz or more, to prevent saturating the detector. With the development of high speed electronic circuitry and the availability of large computer storage capacity, these problems can now be solved.

Until about five years ago, infrared astronomy was severely handicapped by the lack of availability of infrared arrays similar to the CCD and the digicon. One of the most exciting developments in astronomy over the last few years has been the introduction of infrared array cameras on instruments such as the UK Infrared Telescope in Hawaii. Perhaps the most successful of these devices have been the indium antimonide arrays produced by the Santa Barbara Research Center (SBRC) which operate in the 1–5 μm waveband. These arrays are what are known as *hybrid* arrays. In the case of the silicon CCD, the detector material and the means of reading out the array are incorporated into a single silicon chip. In contrast, the infrared materials are used in the same manner for the detection of the radiation but the array is then read out through some other type of array bonded onto the infrared array. This read-out array may be a CCD array or some other form of direct read-out device. In the SBRC arrays, an array of MOSFET amplifiers are bonded directly onto the indium antimonide array and each of these elements of the array can be addressed electronically. (MOSFET stands for Metal Oxide Semiconductor Field Effect Transistor which has a very high input impedance which results in a very low noise figure.) The devices which have been used most successfully so far have array sizes of 58 × 62 elements but much larger arrays are now under development. These developments have led to a completely new generation of cameras and spectrographs for the infrared waveband. At last, the scientific grasp of the infrared waveband is approaching that of the optical waveband. The first of these infrared cameras are producing spectacular new science which has not been possible up till now (Fig. 8.17). In the near future similar arrays will be available for the longer infrared wavelengths as well.

8.6.3 *Radio and millimetre receivers*

Unlike the detectors described in the previous subsection, those used in radio and millimetre astronomy are coherent detectors in which the receiver

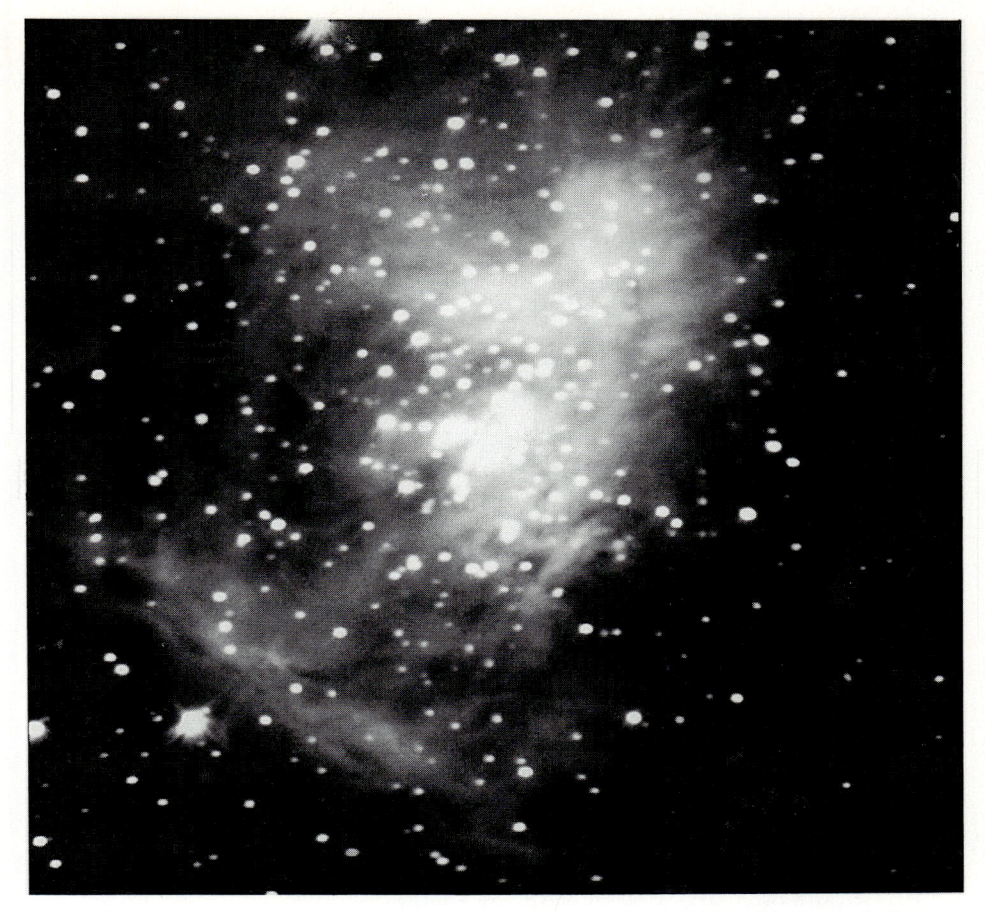

Figure 8.17. A composite three-colour infrared image of the Orion Nebula made from observations at 1.1, 1.65 and 2.2 μm using the infrared camera IRCAM on the UK Infrared Telescope. (Courtesy of Drs M. McCaughren and I. S. McLean.)

responds to the instantaneous field strength of the signal rather than to its intensity. Generally, in radio and millimetre astronomy, the signal strengths are very weak compared to other sources of noise and therefore long integration times are needed. It is a characteristic of radio receivers that very large gains are needed to make the electric field strengths measurable and this is the source of much of their complexity. Excellent introductions to many aspects of receiver systems and their performance can be found in Kraus (1986) and Rohlfs (1986).

The limiting sensitivity of a radio telescope system can be worked out using the relations given in Section 8.5.2. A point source of flux density S has a certain antenna temperature T_1 given by the expression (8.39). As an example, let us work out the antenna temperature of a bright radio source which has flux density $1\ \mathrm{Jy} = 10^{-26}\ \mathrm{W\ m^{-2}\ Hz^{-1}}$ at a wavelength of 6 cm as observed by a 25 m radio telescope. The antenna temperature of the source is then roughly $T_A \approx 0.2$ K. This

Table 8.5(*a*). *The specification and performance goals of the Hubble Space Telescope*

Telescope aperture	2.4 m
System focal ratio	$f/24$
Optical design	Ritchey–Chrétien Cassegrain
Total obscured area	13.8% aperture area
Total field of view	28 arcmin diameter
Spectral range	115 nm–1 mm
Optical performance: system wave front error	$\lambda/13.5$ at 632.8 nm
Radius of 70% encircled energy from a point source at 633 nm	0.1 arcsec
Faint object sensitivity subject to viewing in directions greater than (1) 50° from Sun, (2) 70° from Earth's limb, and (3) 15° from the Moon	Point objects having $m = 27$ or brighter; extended sources to a limit of 23 mag arcsec^{-2} in a visual waveband. Both observations to be made in 10 hours of observation
Pointing stability	0.007 arcsec on a guide star having $m_v = 14$–14.5 in the wavelength range 400–800 nm
Minimum throughout to focal plane	60% at 632.8 nm 40% at 120 nm
The satellite	
Launch weight	11600 kg (25500 lb or 12.75 tons)
Satellite length	13.1 m (43.5 ft)
Satellite diameter	4.27 m (14 ft)
Size of solar panels	Each 11.8 × 2.3 m (39.4 × 7.8 ft)
Power supplied by solar array	Minimum of 2400 W
Satellite orbit	
Circular earth orbit serviceable by the Space Shuttle.	
Altitude	About 600 km
Inclination	28.5°

is only a rough estimate because an accurate calculation would have to use the correct expression for the effective area of the telescope and also we have neglected any losses in the system. This figure should be compared with the system temperature which for a modern radio telescope should be of the order of 50–100 K. To observe this source in the presence of system noise, it is apparent that the product $(\Delta\nu t)^{\frac{1}{2}}$ must be at least 10^3 and preferably much larger.

Since the signal power may often be only one-millionth of the system noise, the receivers must be very stable and they must be very carefully calibrated. The calibration can be undertaken by switching the receiver between the object and a stable reference source. Stable amplifiers are available at low radio frequencies but large amplification factors and stability become problems at high frequencies. To overcome this problem extensive use is made in radio astronomy of *heterodyne* techniques. In this procedure, the signal is mixed with a reference signal at a nearby

Table 8.5(*b*) *Hubble Space Telescope scientific instruments*

All five scientific instruments can be operated in a very large number of modes and the instrument handbooks for each of them should be consulted for a complete description of their capabilities

Instrument	Field of view	Picture format	Wavelength region (nm)	Spectral resolution ($\lambda/\Delta\lambda$)
(1) Wide Field-Planetary Camera (WF-PC)	WF 2.6×2.6 arcmin^2	1600×1600 pixels	115–1100	Many narrow and broad-band filters
	PC 67×67 arcsec^2	1600×1600 pixels	115–1100	
(2) Faint Object Camera (FOC)	*f*/48 44×44 arcsec^2	512×512 pixels	115–700	Many narrow and broad-band filters
	f/96 22×22 arcsec^2	512×512 pixels	115–700	
	f/288 7.6×7.6 arcsec^2	512×512 pixels	115–700	
(3) Faint Object Spectrograph (FOS)	Apertures 0.1–1 arcsec diameter	512 linear diode array	110–850	1300
				250
(4) High Resolution Spectrograph (HRS)	0.25×0.25 arcsec^2	512 linear diode array	105–180	2000
	2×2 arcsec^2		105–320	2×10^4
			105–320	10^5
(5) High Speed Photometer (HSP)	0.4–1.0 arcsec diameter	4 imager dissector tubes and photomultiplier tube	120–700	Many ultraviolet and visual filters

In addition, astrometric observations can be carried out using the Fine Guidance Sensors (FGS). The relative separations of stars within the field of view of the sensors are expected to be measured with an accuracy of about 0.0016 arcsec.

frequency produced by a stable *local oscillator*. The signals are passed through a non-linear *mixer* which has the effect of producing signals corresponding to the sum and difference frequencies of the two signals. This can be demonstrated by a simple calculation. For example, if the signal is $V_i = V_1 \sin(\omega_1 t + \phi_1)$ and the local oscillator signal is $V = V_{Lo} \sin(\omega_{Lo} t + \phi_2)$, when these signals are passed through a mixer which gives an output voltage proportional to the square of the input voltage, we find

$$\begin{aligned} V_{out} &= AV_{in}^2 \\ &= A[V_1 \sin(\omega_1 t + \phi_1) + V_{Lo} \sin(\omega_{Lo} t + \phi_2)]^2 \\ &= AV_1^2 \sin^2(\omega_1 t + \phi_1) + AV_{Lo}^2 \sin^2(\omega_{Lo} t + \phi_2) \\ &\quad + 2AV_1 V_{Lo} \sin(\omega_1 t + \phi_1) \sin(\omega_{Lo} t + \phi_2) \\ &= AV_1^2 \sin^2(\omega_1 t + \phi_1) + AV_{Lo}^2 \sin^2(\omega_{Lo} t + \phi_2) \\ &\quad + AV_1 V_{Lo} \sin[(\omega_1 + \omega_0) t + \phi] + AV_1 V_{Lo} \sin[(\omega_1 - \omega_0) t + \phi'] \end{aligned}$$

By inserting filters, the difference frequency, $\omega_1 - \omega_0$, which is known as the intermediate frequency ω_{IF}, can be selected and the frequency of the signal reduced to a much lower value at which it can be amplified by standard radio frequency techniques. Notice that the output signal V_{out} at the intermediate frequency ω_{IF} is proportional to the input signal V_i and so, although the device is non-linear, it is linear so far as the transformation of the field strength is concerned between the frequency ω_1 and ω_{IF}.

The noise temperatures obtained with various types of device have already been displayed in Fig. 8.12. The principles of transistor amplifiers, tunnel diode mixers, parametric amplifiers and masers are well known and a clear description of their modes of operation and noise figures can be found in Kraus (1986). Perhaps less familiar are the problems of the receivers for use at millimetre and sub-millimetre wavelengths. Besides the problem that it is difficult to obtain large amplification factors, the problems with mixers at these very high frequencies is in enabling them to respond sufficiently rapidly to the high frequency signal. Schottky diodes and SIS (superconductor–insulator–superconductor) diodes are highly non-linear devices with the ability to respond rapidly enough to enable mixer action to take place. They produce strong signals at the intermediate frequency and the SIS devices, in particular, have good noise figures. The principles of their operation is described in simple terms by Appenzeller (1989). At millimetre and sub-millimetre wavelengths, the signal is normally mixed directly with the local oscillator before any amplification takes place.

8.7 Two more Great Observatories

8.7.1 *The Hubble Space Telescope*

On 24 April 1990, more than twenty five years after it was first conceived and thirteen years after its formal approval by the US Congress, the NASA–ESA Hubble Space Telescope was launched into orbit by the Space Shuttle *Discovery*. Unquestionably, it represents one of the greatest achievements of modern

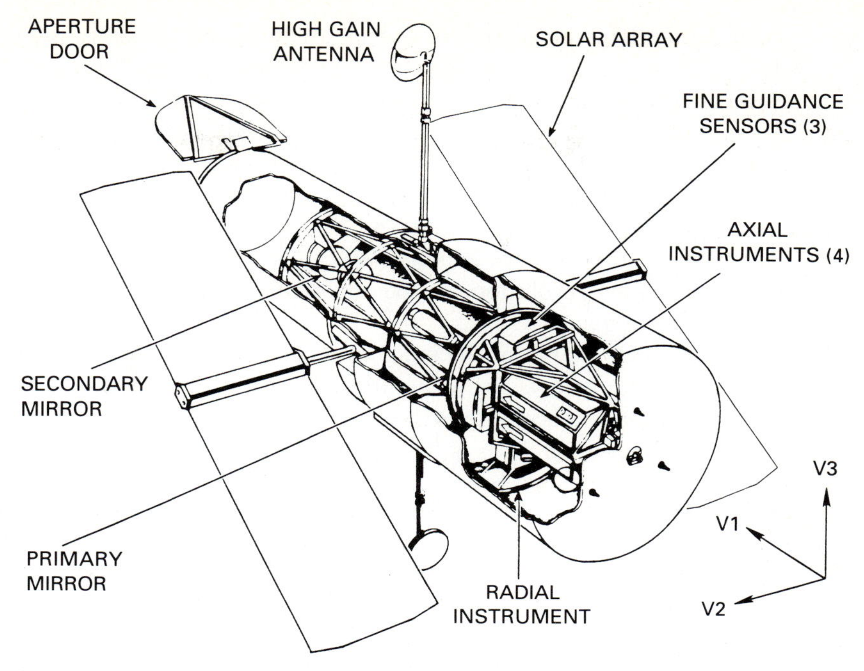

Figure 8.18. Illustrating the overall layout of the components of the Hubble Space Telescope. (Courtesy of the Space Telescope Science Institute.)

technology in that a large optical–ultraviolet observatory was placed in Earth orbit, eliminating at a stroke all the problems which arise because we have to view the heavens through the distorting screen of the Earth's atmosphere. Tragically, it has turned out that the primary mirror was polished to the wrong figure resulting in very severe spherical aberration at the focal plane. As a result, the full capabilities of the telescope have not been realised. Plans are afoot, however, to include additional optical elements in the optical path to each scientific instrument so as to eliminate largely this problem. In Table 8.5, I give the formal specifications and performance goals of the Hubble Space Telescope and its first generation of scientific instruments. I will restrict the discussion to some comments about key aspects of the performance of the telescope and its instrumentation.

The layout of the components of the Hubble Space Telescope are shown in Fig. 8.18. A Cassegrain configuration with Ritchey–Chrétien optics was adopted with the focal plane being located behind the primary mirror. This optical design provides a wide field of view over which the optical aberrations are within the specifications of Table 8.5. The scientific instruments are located in the aft shroud and also surround the focal plane. The major technical problems arise from the requirement that the 2.4 m telescope be diffraction-limited in the optical waveband. The key requirement of Table 8.5 is that 70% of the encircled energy of a point source should lie within a radius of 0.1 arcsec which places stringent demands upon the surface accuracy of the primary mirror. This topic was discussed in section 8.2 in which it was shown that the surface irregularities should be less than $\lambda/20$ to ensure a minimum degradation of the optical quality of the telescope system (see Fig. 8.5). In fact, this specification has been far exceeded by the optical engineers of the Perkin–Elmer Corporation and the primary mirror is known to have surface smoothness better than $\lambda/50$ at a wavelength of 633 nm although the overall figure of the telescope was in error. As a result, the telescope is expected to be diffraction-limited at a wavelength at least 2.5 times shorter than this reference wavelength with the implication that the 70% encircled energy radius will be 0.04 arcsec at a wavelength of about 250 nm. With this quality of imaging, it is very likely that even higher resolution can be achieved at shorter wavelengths using deconvolution techniques.

The extremely high quality of the imaging performance of the telescope brings with it stringent requirements for its pointing and stability. Clearly the telescope must be able to point very accurately indeed or else the value of the high resolution of the optics will be wasted. This is why there is a requirement that the pointing stability of the telescope be about 0.01 arcsec. This is an extremely important requirement and places enormous demands upon the quality of the *Fine Guidance Sensors* which keep the telescope pointing in precisely the same direction. This very high pointing accuracy is achieved by locking the Fine Guidance Sensors onto bright stars which lie within an annulus between 10 and 14 arcmin of the field of interest. As a result there is a further requirement to know all the bright stars which are available as guide stars and the Space Telescope Science Institute has produced an enormous catalogue of all the bright stars in the sky down to about 15 magnitude which are suitable as guide stars for the telescope.

A second major feature of the telescope is the fact that it is an ultraviolet as well as an optical telescope. For the first time, deep imaging will be possible in the ultraviolet waveband, especially with the Faint Object Camera which is designed to exploit the ultimate in angular resolution in the ultraviolet region of the spectrum. The two spectrographs have high efficiencies in the ultraviolet waveband and indeed the High Resolution Spectrograph only operates in the waveband $115 < \lambda < 320$ nm (see Table 8.5).

A third major advantage of the telescope is that it can be operated in essentially perfect observing conditions whenever it is able to observe the skies. I express this feature in this somewhat roundabout way because the telescope cannot observe the sky all the time. There are constraints upon the directions in which the telescope can be pointed because the light from the Sun and Moon cannot be allowed to enter the telescope or else the sensitive detectors could suffer permanent damage. In fact, all the highest quality observations will be made when the telescope is on the dark side of the Earth and so shaded from direct sunlight. During these times of optimum observing conditions, the telescope should operate as a perfect diffraction-limited 2.4 m telescope.

A fourth important point is that the Hubble Space Telescope is planned to be a long-lived observatory in space. The nominal lifetime of the Hubble Space Telescope is 15 years in space. This can be achieved because the telescope has been placed in a relatively low Earth orbit which is serviceable by the Space Shuttle. This means that the telescope can be visited at regular intervals for refurbishment and maintenance by space-suited astronauts. Among the most important aspects of these maintenance missions is the requirement to boost the telescope to a high enough orbit so that it does not suffer orbital decay and reenter the atmosphere. It is very important scientifically that the telescope have this long lifetime in space because, despite the enormous power of the telescope, all observations take a finite amount of time and many of the most important projects will require large numbers of lengthy observations which will take many years to complete.

The instrumentation package listed in summary form in Table 8.5 should be familiar to the reader by now. It is salutory to remember that this package of instruments was approved in 1977 and yet only now are they beginning their operational lifetimes in space. A second generation of instruments is now being planned and among these are a powerful two-dimensional optical–ultraviolet spectrograph as well as near-infrared instruments for observations at wavelengths as long as about 2.4 μm.

The Hubble Space Telescope is operated on behalf of NASA and the scientific community by the Space Telescope Science Institute (STScI) which is located on the campus of the Johns Hopkins University in Baltimore. The operations are extremely complex because of the enormous number of modes in which the telescope can be operated. All the observations will be pre-planned well in advance in order to make optimum use of the observing time. Thus, there is little sense in which the astronomer 'makes an observation' with the Hubble Space Telescope. Rather, the data are acquired by the STScI and then provided to the observer who has sole rights to the data for a period of normally one year.

Because the primary mirror has the wrong shape, the capabilities described above are not yet available to astronomers. The fact that the mirror is the wrong shape has resulted in severe spherical aberration which means that the light from the outer regions of the mirror are focussed at a different point from the inner regions. There is therefore no position in which the telescope can be focussed. In best focus position, only about 15–20 % of the encircled energy from a point source lies within a radius of 0.1 arcsec, the rest of the energy being scattered to form a complex diffraction pattern. The result is that the primary images are blurred. This problem can be partially remedied by deconvolving the blurred images if the point-spread function of the telescope is known well enough. Using these techniques, it has been possible to recover the full angular resolution of the telescope. However, these deconvolution techniques are not effective for very faint objects or for diffuse structures. This problem is most severe for the two cameras of the Hubble Space Telescope. Spectroscopic observations can be carried out with larger slit-widths than originally intended and so many of these programmes can be carried out, particularly in the ultraviolet wavebands which are inaccessible from the ground.

The proposed solution to this problem involves placing pairs of correction mirrors in front of each of the scientific instruments which compensate for the effect of spherical aberration and which should largely recover the lost capability of the telescope. It is proposed that these changes be made during the first maintenance and refurbishment mission which is scheduled for about two years from the time of amending this text (February 1991).

We will have more to say about the economics of a project such as the Hubble Space Telescope in the final section of this chapter. A gentle introduction to many of its aspects will be found in my book *Alice and the Space Telescope* (Longair 1989).

8.7.2 *The Space InfraRed Telescope Facility (SIRTF)*

The fourth of the Great Observatories is the Space Infrared Telescope Facility (SIRTF). Its role can be most simply summarised by stating that it provides for the infrared waveband from 3 to 700 μm the same type of capability as those provided by the Hubble Space Telescope, the AXAF and the GRO for their respective wavebands. It is in the 3–700 μm waveband that observations from the ground are severely hampered by the background radiation from the Earth's atmosphere and from the telescope itself. The *IRAS mission* was the first space telescope to survey the whole sky in the difficult waveband from 12 to 100 μm from space. The IRAS instruments were state-of-the-art when the telescope was launched in 1981 but they have now been superseded by array detectors with very high sensitivites (see Section 8.6.2). In addition, as was pointed out in Section 8.5.1, there are enormous gains in sensitivity as well as angular resolution to be gained from a telescope which is diffraction limited in the thermal infrared waveband.

In 1993, the ESA will launch the *Infrared Space Observatory* which will be the first cooled infrared telescope to be flown in space with modern infrared array detectors. Then, towards the end of the decade, it is planned that the SIRTF observatory will be launched. The specifications and the scientific instruments are summarised in Tables 8.6(*a*) and (*b*) and a plan view of the telescope is shown in Fig. 8.19. The key feature of the telescope is the cooling of the complete telescope assembly to cryogenic temperatures in order to eliminate the thermal background radiation of the telescope throughout these wavebands. For this reason, the telescope assembly is housed inside a superfluid liquid helium tank and the baffling of the telescope from external radiation has to be of the very highest quality. Thus, the telescope may be thought of as being confined within a cryostat which maintains the telescope structure at about 4 K. The configuration of the telescope is again of Cassegrain type with the scientific instruments being located behind the primary mirror. The instruments themselves have to operate at very low temperatures so that they do not contribute more background than the residual background incident at the focal plane from the telescope.

It can be seen from the table of instruments that it is planned to make full use of recent advances in infrared detector array technology. The current package consists of a complete set of photometric, imaging and spectroscopic instruments using large format infrared arrays. To simplify the operations and increase the efficiency of the observational programme, it is planned to place the telescope in a 100000 km circular orbit. The great advantage of this is that the telescope is not continually moving in and out of the Earth's shadow as is the case for low Earth orbit. The disadvantage is that the telescope will not be serviceable by the Space

Table 8.6(*a*). *Space Infrared Telescope Facility* (*SIRTF*)

Mirror diameter	> 90 cm
Wavelength coverage	2–700 μm
Diffraction-limited wavelength	3 μm
Angular resolution	(λ/4 μm) arcsec
Pointing stability accuracy	0.15/0.25 (arcsec)
Field of view	7 arcmin
Sensitivity	
10 μm	6 μJy
60 μm	100 μJy
Spectral resolving power	> 2000
Lifetime	5 yr
Orbit	100000 km circular

Shuttle and consequently the telescope will only remain a cryogenically cooled telescope until the time when all the cryogens have boiled off. It is planned that SIRTF will have sufficient cryogens to guarantee a five-year lifetime in orbit. A new start for this project is proposed for the US Fiscal year 1993.

8.8 Economic and political considerations

In the first edition of this book, the section with a similar title to this was the subject of considerable debate among my colleagues. There can be no argument about the fact that many of the most important facilities described in this and the previous chapters are very expensive and represent enormous investments by the communities and countries involved. Whilst the pioneering studies were undertaken with modest instruments and satellite packages often built by small groups of dedicated scientists, that epoch of exploration is largely over for most fields of astronomy. Once the exploratory phase is over, observatories are needed to convert the pioneering observations into real astrophysics and this means building large observatories with much increased sensitivity and angular resolution. High energy astrophysics and astronomy are seen as part of *Big Science*, by which is meant that they are expensive. There is also the implication that the science is pure science and hence of no immediate practical application.

An interesting part of the argument concerns what the costs of a large project really are and whether or not some of the large space projects are really vastly more expensive than what are conceived of as being less expensive ground-based facilities. Let us sharpen up the argument by listing some of the important scientific projects we have just described, their capital costs and their expected lifetimes. In Table 8.7, the figures are approximate but they are good enough to make the essential point. First of all, there is the important distinction between the *captial costs* of building a facility and the *operating costs* needed to make sure that the facility remains operating with good efficiency and that the instrumentation remains up-to-date. The usual working rule which applies to order of magnitude

Table 8.6(*b*). *SIRTF scientific instruments*

Instrument	Characteristics	Array format		Performance (1σ 500 s)
Infrared Array Camera (IRAC)	Wide field and diffraction limited imaging, 1.8–30 μm, using arrays with up to 256 × 256 pixels. Polarimetric capability. Broad and narrow filters	2–5 μm	256 × 256, InSb	0.3 μJy at 3 μm
		5–15 μm	128 × 128, SiAs IBC	6 μJy at 10 μm
		15–28 μm	128 × 128, SiAs IBC	20 μJy at 20 μm
Infrared Spectrograph (IRS)	Long slit and echelle spectrographs, 2.5–200 μm using detector arrays up to 256 × 256 pixels. Resolving power from 100 to > 2000	2.5–4 μm	256 × 256 InSb	Line flux ~ 1.5×10^{-19} W m^{-2}
		4–28 μm	128 × 128 Si: As IBC	Continuum ~ 0.03–3 mJy
		28–120 μm	2 × 16 Ge:Be	
		120–200 μm	2 × 16 Ge:Ga	
			Stressed Ge:Ga	
Multi-band Imaging Photometer (MIPS)	Background-limited imaging and photometry, 15–200 μm, with pixels sized for complete sampling of Airy disk. Array sizes up to 32 × 32 pixels. Broadband photometry, 200 to > 700 μm. Polarimetric capability	15–30 μm	10 × 50 Si:AS IBC	150 μJy at 60 μm
		30–55 μm	16 × 32 Ge:Be	280 μJy at 100 μm
		50–120 μm	32 × 32 Ge:Ga	9 mJy at 300 μm
		120–200 μm	2 × 8 Stressed Ge:Ga	
		200–1200 μm	5 Ge Bolometers	

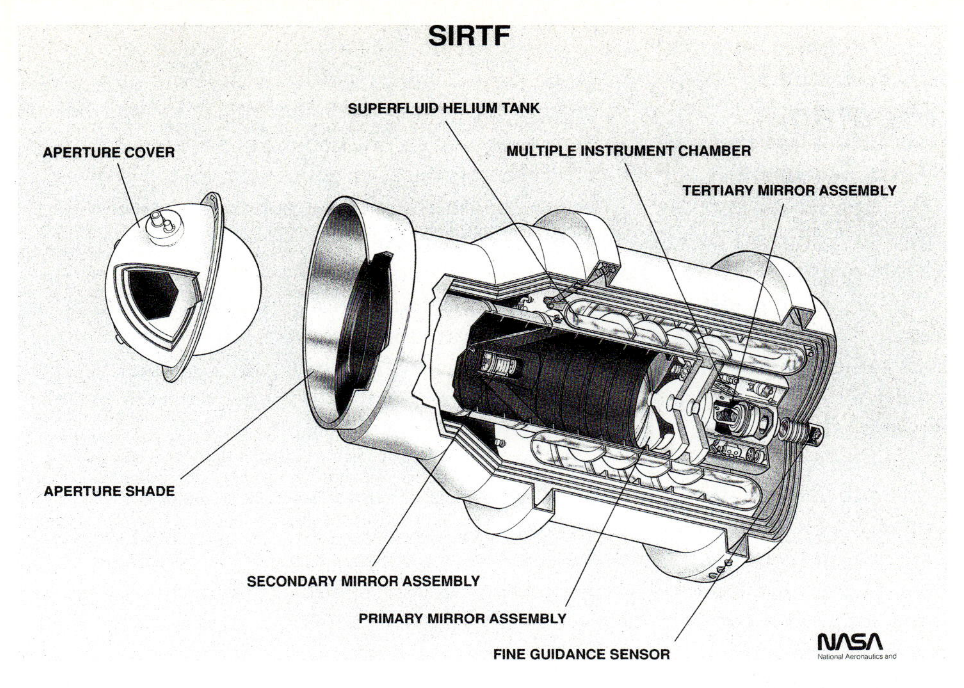

Figure 8.19. Illustrating the overall layout of the components of the SIRTF. (Courtesy of NASA.)

Table 8.7. *Typical costs and lifetimes of large facilities for astronomy and high energy astrophysics*

Facility	Capital cost (10^6\$) x	Lifetime (years) y	Total cost (10^6\$) $x+0.1xy$
100 m fully steerable radio telescope (GB)	100	50	600
Synthesis array e.g. VLA (GB)	150	30	600
Infrared – SIRTF (SB)	800	5	1200
Optical–infrared 8–10 m (GB)	100	50	600
Optical–ultraviolet – HST (SB)	1500	15	3750
X-ray – AXAF (SB)	1000	15	2500
X-ray survey – ROSAT (SB)	250	3	325
Gamma-ray – GRO (SB)	750	10	1500

GB = ground-based facility; SB = space-based facility.

to all projects is that the operating costs per year are about 10 % of the capital cost. The nature of the game can now be appreciated.

Is space astronomy more expensive than ground-based astronomy? Let us compare the overall costs of a space mission which lasts five years with that of a ground-based observatory which has a lifetime of fifty years. According to the above prescription of the total costs, these facilities will have the same total costs if the capital cost of the space experiment is four times that of the ground-based experiment! For example, if a 10 m optical telescope project has a capital cost of 100 M$ and operates for 50 years, the total cost is equivalent to the cost of a 400 M$ space project lasting five years which is the order of magnitude of a cornerstone project of ESA, for example the High Throughput Spectrometer (or XMM) project (Section 7.3.3).

This example is just playing games but there are serious questions behind it. In the end, the total costs of long-term projects are totally dominated by the operating costs rather than by the capital cost of the facility. Any director of a ground-based observatory, including the present author in his previous position, will confirm that getting the capital costs of a facility is trivial compared with securing a proper level of funding to ensure that the facility can be operated properly in the long term.

The other side of the coin is the maintenance of the essential expertise to enable the facilities to perform at a high level of effectiveness. This is where the argument is turned on its head so far as the space community is concerned. Whilst it is an advantage from the programmatic point of view if the projects have a short rather than a long duration, what do you do with your staff when they are not in the thick of a major project. In fact, you have to keep the team together if you want to have a realistic hope of participating in the next big project. Another way of expressing this is that these large projects need an ongoing scientific, engineering and managerial infrastructure to make them feasible and in addition an ongoing programme of missions has to be planned.

These points may seem to be rather far from the mainstream of a text on high energy astrophysics but they are absolutely vital for the health and well-being of what are in the end observational and instrumental sciences. My conclusions which the reader may like to ponder are the following.

(i) It is inevitable that the facilities needed for innovative research at the frontiers of present capabilities will require large and expensive facilities. These have grown to a scale where they are becoming beyond the capabilities of even the richest countries and international collaboration is crucial for many of the next generations of facilities. There is a real danger that facilities will not be built because they are just too expensive even for sympathetic governments. It will be appreciated that a strong political element enters the argument in all these expensive projects.

(ii) In the end, the overall cost of a programme may well be dominated by the operations costs and not the capital cost of the facility. The costs are roughly equal when the lifetime of the project is ten years. The success of projects is largely determined by how well the operating phase is supported.

(iii) There must be scope for small as well as large programmes within the scientific programme. In my view, the most beautiful example of this is the Japanese space programme which has not gone in for very large programmes but which has launched modest packages to study specific goals in X-ray astronomy. Thus, provided one is prepared to limit the scope of one's ambitions, the small dedicated project can be very successful. The dichotomy here is that a specialised programme may not gain the necessary community support if it becomes too expensive.

(iv) Every discipline needs an intellectual and managerial infrastructure before any of these projects can be realised. In the end, none of these projects is feasible without the dedicated scientists and engineers prepared to put an enormous amount of life-blood into them. In addition, it is not just good enough to have inspired science and engineering. The management has to be of the highest order or the project will simply fail.

It is inevitable that the scientists, engineers and astrophysicists involved in these disciplines become involved in precisely these arguments. It is not surprising that they are required to be powerful advocates for their disciplines. If they are not, there is a real danger that progress will grind to a halt.

9

The cosmic ray flux at the top of the atmosphere

It is now time to summarise the data which have been accumulated over the last 20 years on the properties of the fluxes of high energy particles arriving at the top of the atmosphere. In terms of numbers, if we simply count the numbers of high energy particles of different types, about 98% of the particles are protons and nuclei whilst about 2% are electrons. Of the protons and nuclei, about 87% are protons, 12% are helium nuclei and the remaining 1% are heavier nuclei. For particles with energies up to about 1000 GeV per nucleon, the best data have been obtained from space experiments. At energies greater than about 10^{14} eV, the data are derived from studies of extensive air-showers and we discuss these techniques in Sections 9.4 and 9.5.

9.1 The energy spectra of cosmic rays

The most striking feature of the cosmic rays is the fact that their energy spectra span a very wide range of energies indeed. The energy spectra are not of Maxwellian form but can be well represented by power-law energy distributions as illustrated in Fig. 9.1, which shows the energy spectra for protons, helium, carbon and iron nuclei as a function of the kinetic energy per nucleon of the particles. Let us note some of the features of these energy spectra.

(1) First of all, at energies less than about 1 GeV nucleon^{-1}, the energy spectra of all four species show a pronounced attenuation relative to the power-law observed at high energies. The energy and shape of the cut-off vary with the phase of the solar cycle so that the fluxes of low energy particles decrease during periods of high solar activity and are at a maximum during phases of low solar activity. In fact, the cut-off is entirely an artefact of the fact that the cosmic ray particles have to diffuse in towards the Earth from interstellar space through the outflowing Solar Wind. This phenomenon is known as *solar modulation* of the flux of cosmic rays. It appears that the greater the solar activity, the greater the disturbances in

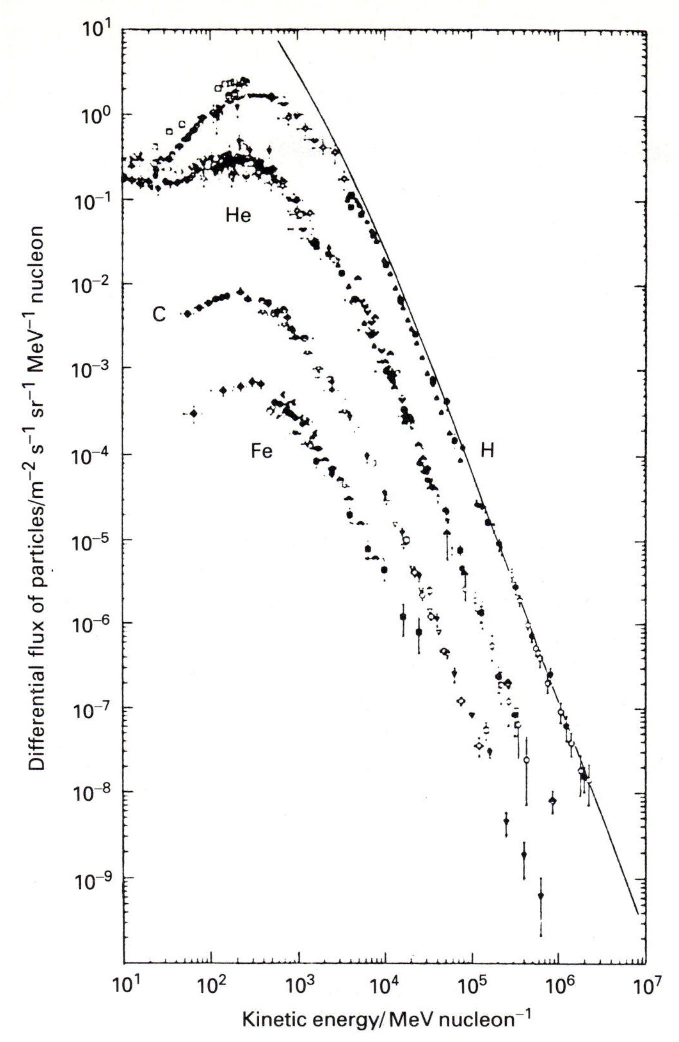

Figure 9.1. The differential energy spectra of cosmic rays as measured at the Earth from observations made from above the Earth's atmosphere. The spectra for hydrogen, helium, carbon and iron are shown. The solid line shows the unmodulated spectrum for hydrogen, i.e. the effects of propagation through the interplanetary medium upon the energy spectra of the particles have been eliminated using a model for the modulation process. The flux of helium nuclei below about 60 MeV nucleon^{-1} is due to an additional flux of these particles which is known as the anomalous ^{4}He component. (From J. A. Simpson (1983). *Ann. Rev. Nucl. Part. Sci.*, **33**, 330).

the interplanetary magnetic field which impede the propagation of particles with energies less than about 1 GeV nucleon^{-1} to the Earth. This is a topic which we take up in much more detail in Chapter 11.

For present purposes, two points should be noted. First of all, an example is given in Fig. 9.1 of what the shape of the energy spectrum of cosmic ray protons is expected to be in local interstellar space once the effects of solar modulation have been removed. It can be seen that the cut-off more or less disappears. The second point follows from one of the important results established in Section 11.1 and that is that the dynamics of high energy particles in any magnetic field configuration depends upon what is known as the *rigidity* or *magnetic rigidity* of the particles. This is defined to be the quantity $R = pc/ze$ where p is the relativistic three-momentum of the particle and z is its electric charge. Particles of different charges and masses but with the same rigidities have the same dynamics in any magnetic field configuration. It might appear that this makes the comparison of the relative abundances of the different elements in the cosmic rays more difficult to interpret but in fact this is not so. In Table 11.1, we compare the various ways of describing the energies, momenta and rigidities of particles. The normal units for expressing the energies of the particles are in terms of their kinetic energies per nucleon and, as was shown in Section 2.4.2, this is a measure of the velocity, or Lorentz factor γ, of the particles. Therefore, if the particles have the same velocity, their rigidities $R = (A/z)(m_p \gamma v c/e)$ depend only upon the mass to charge ratio A/z of the particles where A is the mass number and z the atomic number. Since this number is always close to 2 for elements up to about iron, it is expected that different elements are influenced in the same way, provided the energies are expressed in terms of the kinetic energy per nucleon. More detailed calculations show that, even in energy ranges where it is known that the effects of solar modulation are strong, the relative abundances of the different elements are well represented by the ratio of particle fluxes shown in Fig. 9.1.

(2) The *differential energy spectra* of the various cosmic ray species can be well represented by power-law distributions over the energy range $E > 1$ GeV nucleon^{-1} as can be seen in Fig. 9.1. The spectra are conventionally written

$$N(E)\,\mathrm{d}E = KE^{-x}\,\mathrm{d}E \tag{9.1}$$

The energy E is expressed in terms of the kinetic energy per nucleon. The exponent x lies in the range 2.5–2.7. The spectra shown in Fig. 9.1 extend to the highest energies which can be attained with clear discrimination of the charge of the species. As can be seen in Fig. 9.1, these spectra extend to about 10^3 GeV nucleon^{-1}. At higher energies, it is normally possible only to measure the total energy of the particles.

This power-law distribution of particle energies is only the first of many occasions upon which we will encounter power-law energy distributions and continuum spectra in high energy astrophysics. We will demonstrate in due course that this power-law distribution of particle energies represents the spectrum of high energy particles in the interstellar medium.

(3) Although the spectra shown in Fig. 9.1 look remarkably similar, there are significant differences between the energy spectra of different elements. An example of this is given in Fig. 9.2 which shows the observed spectra of certain elements relative to the differential spectrum of iron. Some elements have energy spectra similar to iron, for example nickel, but others show significantly steeper spectra, some of them such as boron, titanium, vanadium and potassium markedly

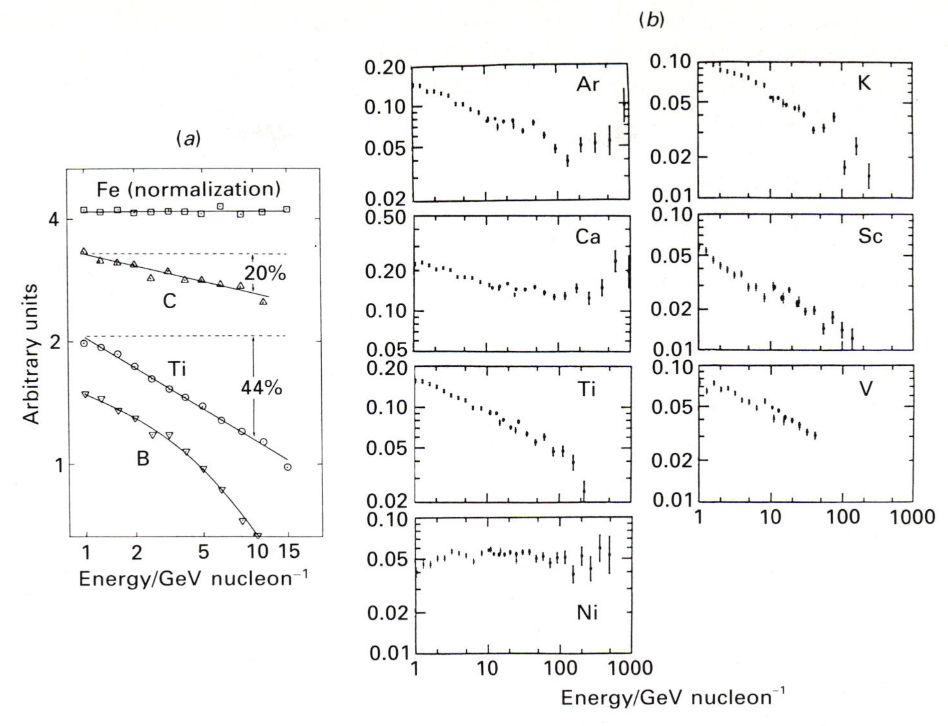

Figure 9.2. Examples of the variation of the differential energy spectra of cosmic rays of different elements. (*a*) Differential spectra of carbon, titanium and boron relative to iron. (From N. Lund. *Cosmic radiation in contemporary astrophysics*, ed. M. M. Shapiro, page 1, D. Reidel Publishing Co., Dordrecht, 1984.) (*b*) The differential spectra of argon, calcium, titanium, nickel, potassium, scandium and vanadium relative to iron. (From M. D. Jones, J. Klarmann, E. C. Stone, C. J. Waddington, W. R. Binns, T. L. Garrard and M. H. Israel (1985). *19th Intl. cosmic ray conference*, La Jolla, USA, Volume 2, page 28.)

so. As we will discuss in Volume 2, Chapter 20, there is an important distinction between those cosmic ray species which are accelerated in sources of high energy particles such as supernovae and those species which are created by nuclear interactions of these species with the nuclei of atoms and molecules of the interstellar gas – this process is known as *spallation*. Those elements which are produced in large abundances in the sources are called primary elements while those which are principally produced by spallation in the interstellar gas are called secondary elements. It will be demonstrated in Chapter 20 that carbon and iron are primary species while titanium, vanadium, potassium and boron are secondary products. This result suggests that part of the difference between these energy spectra can be attributed to differences in the path lengths for spallation as a function of particle energy.

(4) It will be noted that, at the very lowest energies below about 60 MeV nucleon^{-1}, there is a turn-up in the energy spectrum of helium nuclei. This

component is referred to as the *anomalous* ^4He *component*. The spectral turn-up only appeared in 1972 and its intensity began to decrease again in 1981. Thus, it was at its maximum intensity during the period of minimum Solar activity. Apparently there was no evidence for this component in 1965 during the previous Solar minimum. The nature of this component is far from clear. The flux of these particles increases with increasing distance from the Sun suggesting that they are accelerated in the outer regions of the heliosphere. The variability of the component and the fact that it does not conform to any simple model of Solar modulation is consistent with the hypothesis that the particles are accelerated in the outer heliosphere.

The energy spectra of *cosmic ray electrons* have been determined both from high flying balloon experiments and from satellites. The problem with the balloon experiments is that large fluxes of high energy electrons are generated as secondary particles in collisions between cosmic rays and the nuclei of the atoms and molecules in the upper layers of the atmosphere. The best data have been summarised by Webber (1983) who has shown that they are all consistent with a differental energy spectrum which follows the hatched area in Fig. 9.3(*b*). In this figure, the differential energy spectrum is displayed in the form $E^3N(E)$ so that a power-law distribution of electron energies $N(E) \propto E^{-3}$ would correspond to a straight line parallel to the x-axis across the diagram. It is immediately apparent that the spectrum of the electrons is strongly influenced by the effects of solar modulation at energies $E \lesssim 1$ GeV and that it is probable that only at energies greater than about 10 GeV are the observations free of these effects. In the high energy range, $E \geqslant 10$ GeV, the observations can be described by a spectrum of power-law form

$$N(E)\,\mathrm{d}E = 700\, E^{-3.3}\,\mathrm{d}E \text{ particles m}^{-2}\text{ s}^{-1}\text{ sr}^{-1} \tag{9.2}$$

where the energies of the particles are measured in gigaelectron volts. This spectrum is somewhat steeper than that of cosmic ray protons and nuclei. We will show in Volume 2, Chapters 18 and 19 that this spectrum has been significantly influenced by the effects of synchrotron radiation losses and hence it probably does not reflect the injection spectrum of the electrons.

Fortunately, there is complementary information about the spectrum of ultrarelativistic electrons in the interstellar medium from observations of their synchrotron radio emission and also from the low energy γ-ray emission of the Galaxy. These data enable reasonable estimates of the demodulated electron spectrum to be made and we take this story up in much more detail in Chapters 11 and 18.

9.2 The abundances of the elements in the cosmic rays

The chemical abundances of the cosmic rays provide important clues to their origin and to the processes of propagation from their sources to the Earth. If we can deduce from these data the abundances of the cosmic rays at their

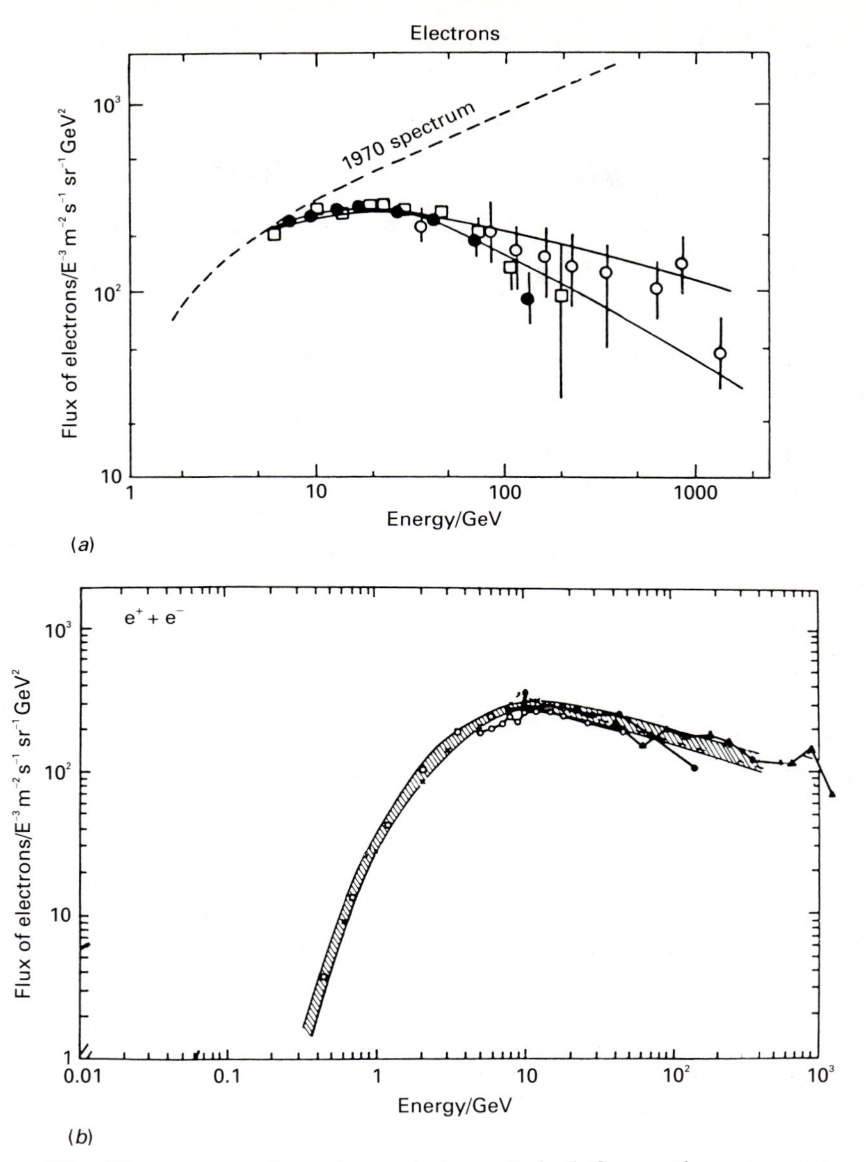

Figure 9.3. The spectrum of cosmic ray electrons. In both figures, the spectra are plotted as $(\mathrm{d}N/\mathrm{d}E)E^3$. (*a*) The differential spectrum as presented by Lund (1984) showing how estimates of the spectral shape have changed since the earliest estimates. (From N. Lund (1984) *Cosmic radiation in contemporary astrophysics*, ed. M. M. Shapiro, page 1, D. Reidel Publishing Co., Dordrecht.) (*b*) The differential energy spectrum as presented by Webber (1983). According to Webber, all the data are consistent with the spectrum indicated by the shaded region in (*b*). (From W. R. Webber (1983). *Composition and origin of cosmic rays*, ed. M. M. Shapiro, page 83, Dordrecht: D. Reidel Publishing Co.)

Table 9.1. *Galactic cosmic ray elemental abundances at 1 AU, normalised to Si = 100 compared with the Solar System and local interstellar abundances. (energy intervals given in MeV nucleon^{-1})*

	Cosmic rays				
Element	70–280	600–1000	Average at 1000–2000	Solar System	Local Galactic
He	41700 ± 3000	27030 ± 580			$(0.27 \pm 0.06) \times 10^6$
Li	100 ± 6	136 ± 3		5.0×10^{-3}	
Be	45 ± 5	67 ± 2	69.4 ± 10	8.1×10^{-5}	
B	210 ± 9	233 ± 4	212 ± 10	3.5×10^{-3}	
C	851 ± 29	760 ± 16	684 ± 27	1110	1300 ± 300
N	194 ± 8	208 ± 5	188 ± 6	231	230 ± 100
O	777 ± 28	707 ± 15	607 ± 28	1840	2300 ± 500
F	18.3 ± 1.3	17.0 ± 1.1	13.5 ± 2.3	0.078	0.093 (1.6)
Ne	112 ± 6	113 ± 3	100 ± 3	240	270 (1.7)
Na	27.3 ± 3.4	25.8 ± 1.1	21.3 ± 3.2	6	5.6 ± 0.9
Mg	143 ± 6	142 ± 4	125 ± 12	106	105 ± 3
Al	25.2 ± 3.0	28.2 ± 1.2	22.2 ± 3.2	8.5	8.4 ± 0.4
Si	100	100	100	100	100 ± 3
P	4.0 ± 0.7	5.3 ± 0.5	5.3 ± 1.6	0.65	0.96 ± 0.20
S	16.4 ± 1.2	23.1 ± 1.1	19.6 ± 0.9	50	45 ± 13
Cl	3.6 ± 0.5	6.4 ± 0.5	4.7 ± 0.4	0.47	0.47 (1.6)
A	6.3 ± 0.6	10.2 ± 0.7	8.2 ± 1.2	10.6	9.0 (1.7)
K	5.1 ± 0.6	7.2 ± 0.5	6.3 ± 0.4	0.35	0.36 ± 0.12
Ca	13.5 ± 1.0	16.1 ± 0.9	13.1 ± 1.2	6.25	6.2 ± 0.8
Sc	2.9 ± 0.5	4.5 ± 0.5	3.3 ± 1.1	0.003	0.0035 ± 0.0005
Ti	10.7 ± 0.9	10.2 ± 0.7	9.1 ± 0.9	0.24	0.27 ± 0.04
V	5.7 ± 0.6	6.7 ± 0.5	4.6 ± 0.3	0.025	0.026 ± 0.005
Cr	10.9 ± 1.0	11.8 ± 0.8	9.1 ± 0.8	1.27	1.30 ± 0.12
Mn	7.2 ± 1.2	8.2 ± 0.7	6.3 ± 0.4	0.93	0.79 ± 0.17
Fe	60.2 ± 3.2	69.8 ± 2.0	60.5 ± 7.6	90.0	88 ± 6
Co	0.2 ± 0.1		0.4 ± 0.2	0.22	0.21 ± 0.03
Ni	2.9 ± 0.4	3.7 ± 0.5	2.8 ± 0.6	4.78	4.8 ± 0.6
Cu			0.038 ± 0.006		0.052 (1.6)
Zn			0.035 ± 0.005		0.135 (1.6)

The cosmic ray data are taken from the review by Simpson (1983) who gives detailed references to the sources of these data. The data on the Solar System and Local Interstellar abundances are taken from the same review, supplemented by data from Cameron (1973).

sources, this would provide an important astrophysical clue to potential sites of acceleration. The relevant comparison of these abundance data is with the typical cosmic abundances of the elements and this is where we begin.

9.2.1 *The Solar System abundances of the elements*

The determination of the chemical abundances of the elements in different objects is one of the great goals of astronomical spectroscopy. The subject is vast and a gentle introduction to the subject is provided by Tayler in his excellent textbook *The origin of the chemical elements*. The problem is easily appreciated if one considers the enormous differences one would find if the elemental abundances of a representative sample of material dug up in the centre of Edinburgh is compared with those found on the surface of the Moon, the outer layers of the Sun or in a typical sample of interstellar gas. Every sample we consider has been influenced to a greater or lesser extent by its past history and the surface layers of the Earth are particularly unreliable because of the influence of the Earth's ecosystem. It is safer to use abundances derived from the photosphere of the Sun, from meteoric samples and estimates of chemical abundances in the interstellar gas.

The Sun

If the Sun were truly a black-body, we would obtain no information about the abundances of the chemical elements since the spectrum would be featureless. In the outer layers of the Sun, however, photons can escape without being reabsorbed and this means that the system is no longer in thermodynamic equilibrium. The fact that we observe absorption lines superimposed upon the underlying Planck spectrum is due to resonance absorption by ions, atoms and molecules in the Sun's photosphere. The problems of radiative transfer in the non-equilibrium physical conditions of the Solar photosphere are formidable but the great advantage of studying the Sun is that it is so bright that very faint lines indeed can be observed in its spectrum. The theory of the Solar atmosphere can then be used to determine relative abundances of the elements.

Meteorites

We have already described the importance of meteorites in determining the average cosmic ray flux over periods up to roughly the age of the Solar System (Section 6.3.2). The meteorites have the great advantage over studies of the Earth's composition that they have not undergone fractionation and other terrestrial geophysical phenomena and are believed to have chemical abundances similar to that of the primitive solar nebula. There are important abundance differences between different types of meteorite which probably reflect their differing origins. One class of meteorite, however, the chondritic meteorites, have chemical abundances which are similar to those found in the solar photosphere. The result is that the abundances in these meteorites can be used to fill in the abundances of some of the rarer elements in the periodic table. In Table 9.1, a set of abundances are given for the Solar System based upon studies of the solar atmosphere and

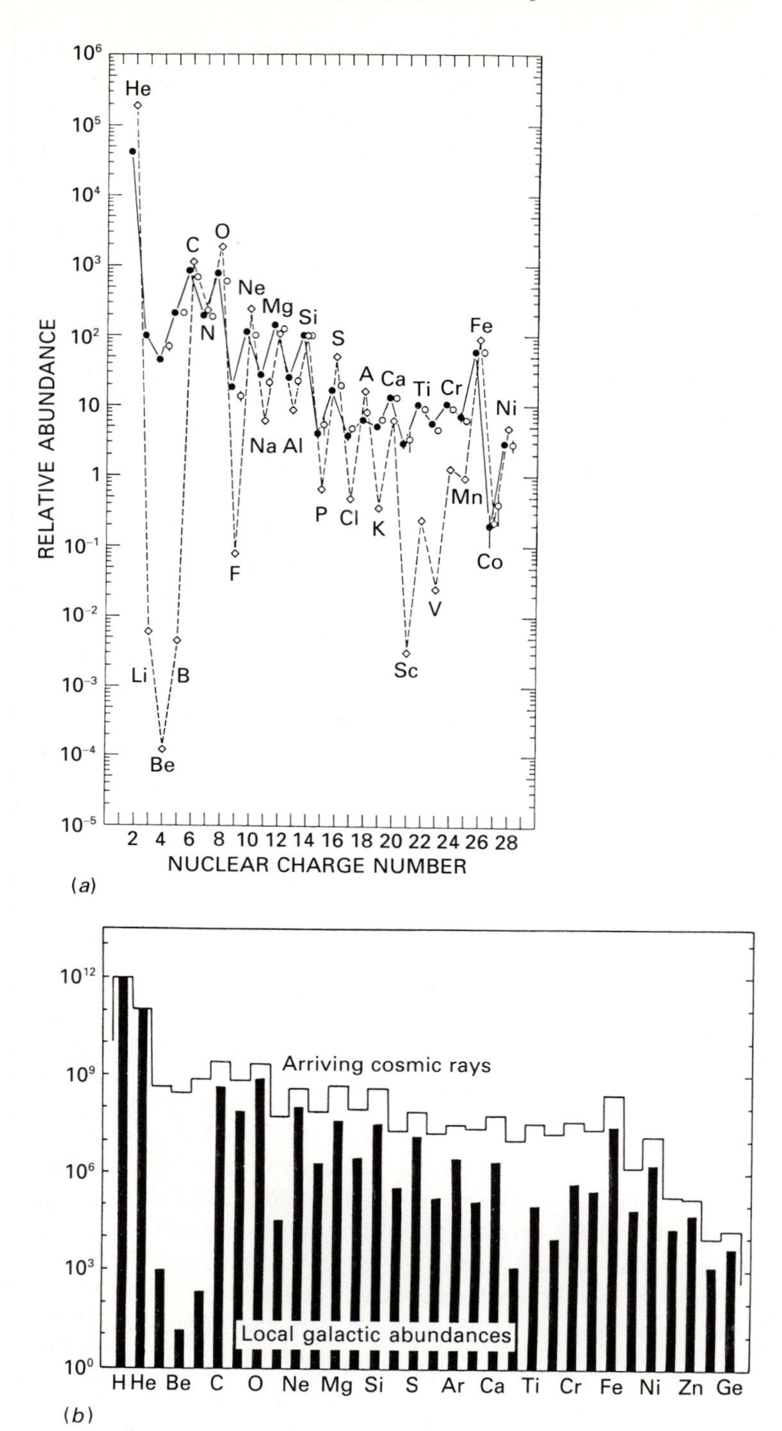

Figure 9.4. For legend see facing page

from meteorites. Also shown in that table are estimates by Meyer (1979) of local interstellar abundances. It can be seen that the values are in reasonable agreement.

What guarantee do we have that these abundances are typical of ordinary matter in the Universe? This question takes us deeply into the problem of the chemical evolution of our Galaxy. We cannot obtain nearly as refined information for the stars as we can for the Sun and only abundances for the commoner species can be determined with high accuracy. These studies show that, for stars like the Sun, there is not a great deal of variation in the chemical abundances. Our own Sun is only about 4.6×10^9 years old compared with the age of the Galaxy which is at least twice that age. This means that our Sun was formed from interstellar material which had already been rather thoroughly processed by previous generations of massive star formation. It is true that the chemical abundances of the very oldest stars in our Galaxy are deficient in heavy elements as compared to stars like the Sun and this is interpreted as meaning that they formed from material which had not been enriched by the products of stellar nucleosynthesis. The results of studies of the evolution of the chemical abundances of the elements show that the typical cosmic abundances of the elements are attained rather rapidly and this strongly suggests that on average the abundances found in stars like the Sun may well be fairly typical of the average cosmic abundances in the Galaxy at the present day. These ideas will become somewhat clearer once we have described the astrophysical background to the study of stars and stellar evolution in Volume 2, Chapter 14. For our present purposes, we will adopt the figures given in Table 9.1 as representing the typical chemical abundances of the matter in stars at the present epoch. In our studies we will not be interested in particularly subtle differences but rather in looking at global differences between the abundances observed in typical stellar matter and those found in the cosmic rays.

9.2.2 *The chemical abundances in the cosmic rays*

A wide range of data on the chemical abundances in the cosmic rays has been summarised by Simpson (1983) and Meyer (1985). Many of the very best data were obtained by the French–Danish experiment flown in the HEAO-C2 experiment and from balloon experiments in which large detector packages were flown at high altitude (see Section 7.2). Simpson's summary of the abundances of the elements

---

Figure 9.4. (*a*) The cosmic abundances of the elements in the cosmic rays found at the top of the Earth's atmosphere compared with the Solar System abundances, both given relative silicon. The solid circles are low energy data (70–280 MeV nucleon^{-1}), the open circles show a compilation of high energy measurements (1–2 GeV nucleon^{-1}) and the diamonds represent the Solar System abundances. These data are presented numerically in Table 9.1. The data have been normalised to [Si] = 100. (From J. A. Simpson (1983). *Ann. Rev. Nucl. Part. Sci.*, **33**, 326.)
(*b*) The same data as in (*a*) but shown in summary form and with the abundances shown relative to hydrogen (from N. Lund (1984). *Cosmic radiation in contemporary astrophysics*, ed. M. M. Shapiro, page 1, Dordrecht: D. Reidel Publishing Co.)

in the cosmic rays in different energy ranges is compared with the Solar System abundances and estimated abundances for the local interstellar medium in Table 9.1 (Simpson 1983). These data are displayed graphically in Fig. 9.4 which reveals the overall similarities and differences between the abundances in the cosmic rays and those of typical Solar System and interstellar matter. The following features are immediately apparent:

(i) the abundance peaks at carbon, nitrogen and oxygen and at the iron group are present both in the cosmic ray and Solar System abundances;
(ii) it can be seen that the odd–even effect in the relative stabilities of the nuclei according to atomic number known to be present in the Solar System abundances of the elements is also present in the cosmic rays but to a somewhat lesser degree.
(iii) the light elements, lithium, beryllium and boron are grossly overabundant in the cosmic rays relative to their Solar System abundances;
(iv) there is an excess abundance in the cosmic rays of elements with atomic and mass numbers just less than those of iron, i.e. elements with atomic numbers between about calcium and iron;
(v) there is an underabundance of hydrogen and helium in the cosmic rays relative to the heavy elements.

Although these differences are striking, it is remarkable that overall, the distribution of element abundances in the cosmic rays is not so different from those of typical Solar System abundances. Some of the differences, specifically points (iii) and (iv) listed above, can immediately be accounted for qualitatively as a result of *spallation*. The primary cosmic rays accelerated in their sources have to propagate through the interstellar medium to reach the Earth and in the process there are spallation collisions between the cosmic rays and the ambient interstellar gas (Section 5.2). The net result is that the common elements in the cosmic rays are chipped away and fragmented resulting in the production of nuclei with atomic and mass numbers just less than those of the common groups of elements. We will demonstrate in detail in Volume 2, Chapter 20 how the process of spallation can account for the observed abundances of the light elements, lithium, beryllium and boron as well as the elements lighter than iron in the cosmic radiation.

The abundances shown in Table 9.1 extend just beyond the iron peak. The HEAO-C3 and BUGS experiments described in Section 7.2 were specially designed to measure the abundances of elements beyond the iron peak and up to elements such as uranium. Interest in these very heavy nuclei was stimulated by some of the early work carried out using nuclear emulsions in which evidence for some particles with atomic numbers up to uranium were observed (Fig. 6.8). In some of the earliest experiments by Fowler and his group, it was suggested that some of the particles might belong to the actinide group of elements which would be of the greatest interest since many of these are likely to be radioactive nuclides with short lifetimes.

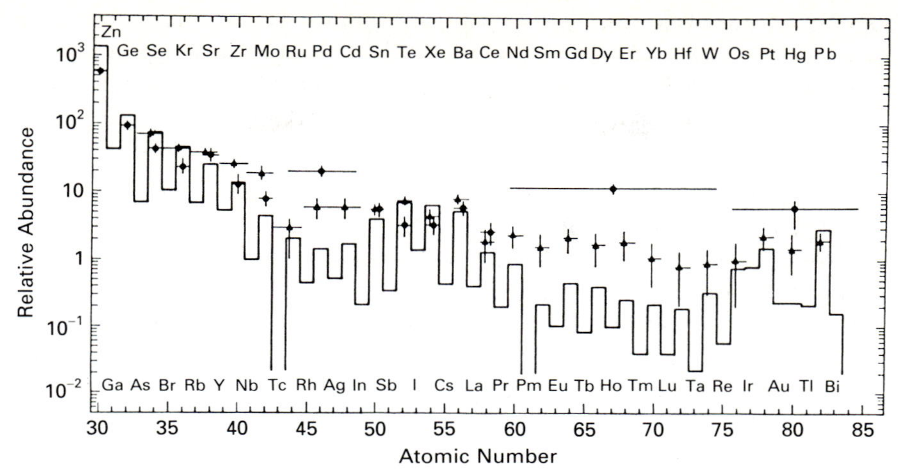

Figure 9.5. The abundances of the elements with very large atomic numbers in the cosmic rays. The abundances observed in the Ariel-VI observations are indicated by filled triangles and the HEAO-C3 observations by filled circles. The solid line indicates the cosmic abundances of the elements. In each case, the abundances are normalised to 10^6 iron nuclei. (J. P. Wefel (1988) *Genesis and propagation of cosmic rays*, eds. M. M. Shapiro and J. P. Wefel, page 1, D. Reidel Publishing Co, Dordrecht.)

There is now good agreement between these two experiments, a summary of the data being displayed in Fig. 9.5. The resolution of the experiments was not sufficient to resolve individual elements and so the data have been binned. In the case of the HEAO-C3 experiment, the bins shown in Fig. 9.5 are quite broad; in the case of the Ariel-VI data, the data are presented for even nuclei. In comparing the observations in Fig. 9.5, it should be recalled that the HEAO-C3 data are summed over broad bins (see also Fig. 7.7). The areas of excellent agreement are the abundances of heavy cosmic rays with atomic numbers in the range up to about 42, the peak at $50 < Z < 58$ and the high Z peak at $78 < Z < 84$. Between the peaks, the Ariel-VI observations show somewhat larger abundances of the elements in the cosmic rays as compared with the HEAO results but Meyer (1985) suggests that the differences are small if particles of the same energy per nucleon in the two experiments are compared. It can be seen in Fig. 9.5 that, between these peaks, the cosmic ray abundances are significantly greater than the Solar System abundances of these elements. Again, a natural explanation for this phenomenon is the spallation of elements in the peaks as they encounter the atoms and molecules in the interstellar gas.

The other remarkable feature of Fig. 9.5 is how the abundances of the heavy elements relative to iron are similar in both the cosmic rays and in the Solar System. There are important differences but the overall impression one obtains is that the cosmic ray particles must have been accelerated from material of quite similar chemical composition to the Solar System abundances of the elements.

One of the intriguing questions has been the presence or otherwise of members of the actinide group in the cosmic rays. There was little evidence for them in the

Table 9.2. *Isotope ratios of hydrogen and helium*

Isotope ratio	60 MeV nucleon^{-1}	80 MeV nucleon^{-1}	200 MeV nucleon^{-1}	Cosmic abundance
$^{2}H/^{1}H$		$(4.4 \pm 0.5) \times 10^{-2}$	$(5.7 \pm 0.5) \times 10^{-2}$	1.0×10^{-5}
$^{3}He/^{4}He$		$(9.5 \pm 1.5) \times 10^{-2}$	$(11.8 \pm 0.7) \times 10^{-2}$	3.0×10^{-5}
$^{2}H/^{4}He$	0.21 ± 0.09		0.31 ± 0.03	1.0×10^{-4}

All measurements except those at 60 MeV nucleon^{-1} were made near Solar minimum. The figures in the table are the *number ratios* of isotopes rather than their mass ratios. (From J. A. Simpson (1983). *Ann. Rev. Nucl. Part. Sci.*, page 357.)

HEAO-C3 experiment. In the Ariel-VI experiment, three candidates for particles with atomic number about 92 were found. The conclusion of both experiments is, however, that there is no large excess of actinides in the cosmic rays contrary to what was once thought to be the case.

An important analysis which we will take up in Volume 2, Chapter 20 is the estimation of the source abundances of the elements once the effects of spallation in the interstellar medium has been taken into account. Then, we will find a number of significant differences between the source abundances of the cosmic rays and those of the local interstellar medium. These similarities and differences will provide important clues concerning the origin of the cosmic rays.

9.2.3 *Isotopic abundances of cosmic rays*

In addition to overall chemical abundances, the development of detectors which are sensitive to the mass as well as the charge of the cosmic rays, means that isotopic abundances are available for a number of species. There are several reasons why these observations are of particular interest.

First of all, the very lightest stable elements, ^{1}H, ^{2}H, ^{3}He and ^{4}He, form a special group of isotopes. Only ^{1}H and ^{4}He are found with high abundances in the interstellar medium – it is widely accepted that most of the helium in the Universe was synthesised in the Hot Big Bang through the standard pp-chain as the matter and radiation cooled down through a temperature of about 10^{9} K – this process is described in all the standard text-books on cosmology. It is one of the remarkable features of the Hot Big Bang model that about 24% of helium by mass (corresponding to a helium-to-hydrogen ratio by number of about 0.08:1) is synthesised for a wide range of reasonable initial conditions. The abundances of the rarer isotopes, ^{2}H and ^{3}He, relative to hydrogen and helium are shown in Table 9.2. ^{2}H and ^{3}He are very fragile isotopes and they are destroyed rather than created in stars. They are, however, synthesised in significant abundances in the standard Hot Big Bang as by-products of the main pp-chain. The observed abundances of these elements in the interstellar gas can be accounted for in world models in which the baryon density is less than about one-tenth of the critical density required to close the Universe. The local interstellar abundances of these elements are

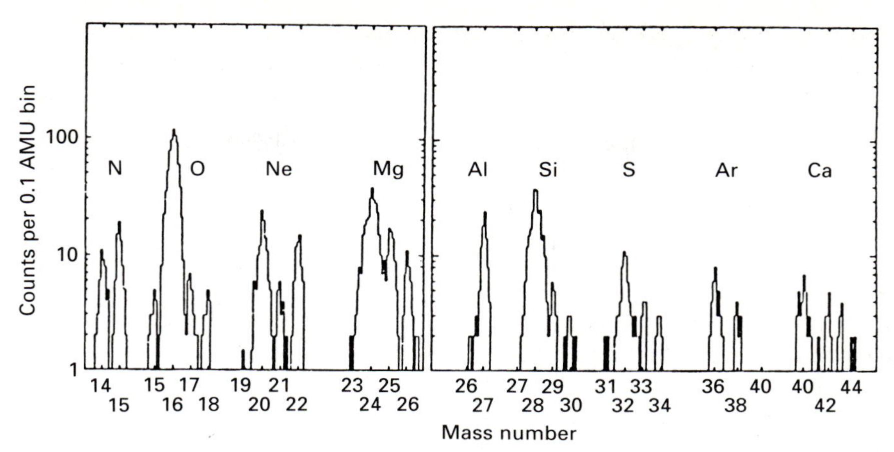

Figure 9.6. Examples of the determination of the isotopic abundances of the elements nitrogen, oxygen, neon, magnesium, aluminium, silicon, sulphur, argon and calcium. The charge resolutions in these observations ranges from 0.23 amu for oxygen to 0.29 amu for silicon. (W. R. Webber, J. C. Kish and D. A. Schrier (1983). *19th intl. cosmic ray conference*, La Jolla, USA, Vol. 2, page 88.)

compared with the cosmic ray abundances in Table 9.2. It is apparent that ^{2}H and ^{3}He are present in much greater abundances in the cosmic rays than they are in the interstellar medium. The cosmic ray abundances can be attributed to spallation reactions between the four species concerned. In fact, these elements can be considered as an independent check of the spallation models – their abundances are so much greater than those of the other elements that the spallation products of elements such as carbon, oxygen and nitrogen can contribute little to the relative isotopic abundances of hydrogen and helium.

A second important aspect of isotopic abundances is the fact that some of the species created in spallation reactions are radioactive and hence, if the production rates of the different isotopes of a given element are known, information can be obtained about the time that it has taken these samples to reach the Earth from their sources. The most famous of these 'cosmic rays clocks' is the isotope ^{10}Be which has a radioactive half-life of 1.5×10^6 years and so is a very useful discriminant for determining the typical lifetime of the spallation products in the vicinity of the Earth. We will take this story up in much more detail in Chapter 20.

A third aspect of isotopic abundances is related to the sources of the cosmic rays. Quite subtle tests of the origin of the cosmic rays are possible with the ability to determine the isotopic abundances of the heavy elements. It is found that the most common isotopes of the heavy elements are the same as the most common isotopes found in the Solar System and in the local interstellar medium but, in a number of cases, significantly greater abundances of relatively rare isotopes are found among the cosmic ray particles. Examples of these are shown in Fig. 9.6 in which the neon, magnesium and silicon isotopes have been clearly resolved. The advantage of using isotopic abundances is that the differential effects of spallation should be very much a second-order effect because the cross-sections for spallation should not be very different for isotopes of the same element. Thus, the isotopic ratios of a particular heavy element should be relatively insensitive to the spallation history of the species. In the example shown in Fig. 9.6, the ratio of $^{22}Ne/^{20}Ne$ is about four times greater than the value found in the Solar System. It is the case that for the elements which have been well studied, there is a greater abundance of neutron rich isotopes as compared with the Solar System abundances. For example, the isotopic abundances of $^{25}Mg/^{24}Mg$, $^{26}Mg/^{24}Mg$, $^{29}Mg/^{28}Mg$ and $^{30}Si/^{28}Si$ are each about 1.6 times greater in the cosmic rays than they are in the Solar System abundances.

These enrichments are important clues about the sites of acceleration of the cosmic ray particles. Evidently, a source is required which favours the production of neutron rich elements. We will have to investigate these features in much more detail once we have disentangled the effects of spallation upon the abundances of the elements.

9.3 The isotropy and energy density of cosmic rays

It is immediately apparent from Fig. 9.1 that the flux of cosmic rays with energies less than about 10 GeV must be significantly influenced by the process of solar modulation and hence information about the arrival directions of these cosmic rays at the Solar System is lost. In fact, only relatively high energy protons and nuclei penetrate to the vicinity of the Earth undeflected by the magnetic field in the interplanetary medium. As we will show in Chapter 11, a measure of the deflection suffered by a particle is the ratio of its radius of gyration in the interplanetary magnetic field to the scale of the Solar System. For a relativistic proton, the gyroradius is

$$r_g = 3 \times 10^9 \gamma (B/10^{-9} T) \text{ m} \tag{9.3}$$

where the magnetic field strength B is measured in tesla and $\gamma = (1 - v^2/c^2)^{-\frac{1}{2}}$ is the Lorentz factor. Therefore, adopting the local value of the magnetic field strength in the interplanetary medium $B = 10^{-9}$ T, relativistic protons with $\gamma = 10^3$ (i.e. energies of 10^{12} eV) have gyroradii which are 3×10^{12} m $= 20$ AU, i.e. 20 times the distance from the Sun to the Earth, corresponding roughly to the radius of the orbit of the planet Uranus about the Sun. Thus, particles with these energies and greater are likely to preserve information about their arrival directions at the Solar System when they arrive at the top of the Earth's atmosphere.

Observations of the arrival directions of high energy protons and nuclei can be undertaken using the Earth's atmosphere itself as a 'convertor'. As was shown in Chapter 5, very high energy protons and nuclei create pions in collisions with the nuclei of atoms and molecules in the upper layers of the atmosphere. The charged pions decay into muons which have half-lives such that they decay long before they reach the surface of the Earth unless they are highly relativistic, $\gamma \geqslant 20$ (see Section 5.4). This process is the basis of studies of the isotropy of cosmic rays using underground muon detectors. These experiments are designed to be sensitive to primary particles entering the atmosphere with energies of 10^{12} eV and greater.

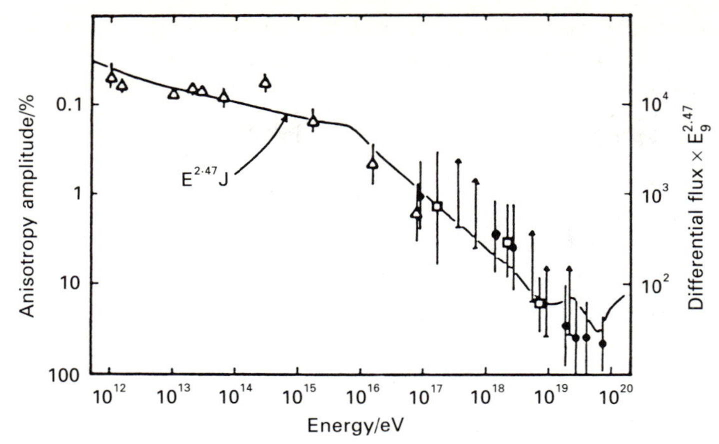

Figure 9.7. The amplitude of the anisotropy in the distribution of arrival directions of cosmic rays as a function of energy. In each case, a best fitting sine wave has been fitted to the data and the percentage amplitude of this harmonic measured. The solid line shows the shape of the differential spectrum of cosmic rays. (From A. M. Hillas (1984). *Ann. Rev. Astr. Astrophys.*, **22**, 425.)

These data are summarised together with data on much higher energy particles in Fig. 9.7 which is taken from Hillas (1984) (see also Wdowczyk and Wolfendale (1989)). It can be seen that, in the range 10^{13}–10^{14} eV, the distribution of arrival directions of high energy particles is remarkably uniform, anisotropy only being detected at the level less than 1 part in 10^3. Quantitatively, the amplitude of the best-fitting first harmonic function to the distribution of arrival directions of these cosmic rays corresponds to 0.06%. This is an important result for many aspects of the high energy astrophysics of the interstellar medium. The implication is that the net streaming velocity of the flux of high energy particles relative to the local frame of reference in our Galaxy is small. This, in turn, has implications for the diffusion of high energy particles in the interstellar medium and also for the escape of these particles from the Galaxy. We return to this important topic in Section 9.5 and Volume 2, Section 20.4.

We can now derive an estimate of the local energy density in high energy particles. It is immediately apparent that this calculation is strongly affected by assumptions about the degree to which the fluxes of particles are influenced by solar modulation. We obtain a lower limit by assuming that the flux of particles observed at the top of the atmosphere is representative of that present in the local interstellar medium. Although the spectrum of the cosmic rays extends to very high energies, there is little total energy in these because of the steepness of the energy spectrum of the particles, $\mathrm{d}N \propto E^{-2.6}\,\mathrm{d}E$. The maximum of the proton spectrum corresponds to about 2 protons $\mathrm{m^{-2}\,s^{-1}\,sr^{-1}\,MeV^{-1}}$ at an energy of about 1 GeV. Wdowczyk and Wolfendale (1989) show the total energy density in high energy particles with energies greater than different limiting energies. According to their estimates, the total energy density of cosmic rays with energies greater than 1 GeV is about 1 MeV $\mathrm{m^{-3}}$ (or 1 eV $\mathrm{cm^{-3}}$).

When the effects of solar modulation are taken into account, the local interstellar energy density is expected to be greater although, even when the effects of modulation are removed, the demodulated spectrum converges at energies less than about 1 GeV (see Fig. 9.1).

It is intriguing that this energy density is similar to that present in the interstellar magnetic field, $B^2/2\mu_0 \approx 0.2$ MeV m^{-3}. It is also remarkable that it is similar to the local energy density in starlight (about 0.3 MeV m^{-3}) and to the energy density of the Microwave Background Radiation (about 0.3 MeV m^{-3}). Some of these coincidences have real physical significance. Others are probably just genuine coincidences of no deep astrophysical significance. Opinions differ about this topic.

9.4 The highest energy cosmic rays and extensive air-showers

Evidence on the highest energy cosmic rays is provided by observations of extensive air-showers. Showers of cosmic rays were noted in some of the earliest air-shower experiments. Auger and his colleagues began systematic studies with separated detectors in the late 1930s and observed coincidences between two or more detectors separated by distances up to 300 m. They made the correct inference that these coincidences were associated with showers of secondary and higher-order cosmic rays triggered by the arrival of a single cosmic ray of very high energy at the top of the atmosphere. These showers were called *extensive air-showers*.

In these showers the *nuclear cascades* described in Section 5.4 are initiated by cosmic rays of such high energy that large fluxes of secondary, tertiary and higher-order products reach the ground without losing all their energy by interactions and ionisation losses in the atmosphere. A great deal of experimental and theoretical work was undertaken during the 1960s and 1970s to understand these showers and several very large cosmic air-shower arrays have been constructed specifically to detect the most energetic of these particles. The objectives were to measure the energy spectra, chemical composition and isotropy of the highest energy cosmic rays.

It is found empirically that the relative number of charged particles as a function of distance from the core of the shower is similar for showers of different sizes (Fig. 9.8). In practice, it is found that the distribution with distance from the core of the shower is not very different from the distributions expected from electron–photon cascades. Only very close to the core of the shower do the distributions vary. Sometimes, the cores are multiple and this has suggested that these showers are initiated by nuclei rather than by protons but there is no general agreement about this.

The mere fact that we observe 10^6 or more relativistic particles at ground level tells us that the initiating particle must have had very high energy. To estimate the energy of the incoming particle, we need a model of the overall development of the shower and an example of how this is done is provided by the results of a combined US–Japanese collaboration carried out at an altitude of 5200 m on Mount

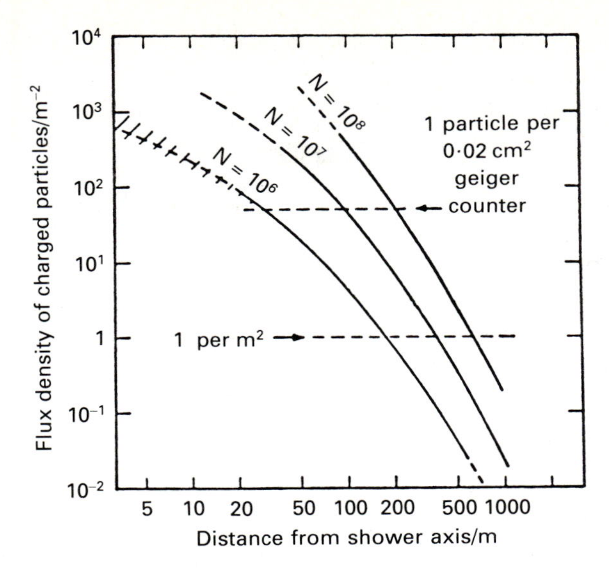

Figure 9.8. The distribution of charged particles as a function of distance from the axes of extensive air-showers of different sizes as described by the total number of particles in the shower. (From A. M. Hillas (1972). *Cosmic Rays*, page 87. Oxford: Pergamon Press.)

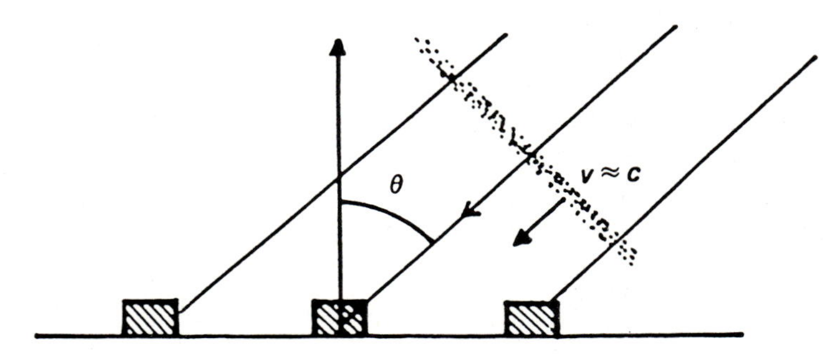

Figure 9.9. A diagram illustrating the arrival of an extensive air-shower at zenith angle θ.

Chacaltaya in Bolivia (for details, see Hillas (1972)). At this altitude, the detector array is about half-way through the atmosphere, the vertical column density above the detector being only about 5200 kg m^{-2}. On the site, detectors for particles of all types were distributed over a wide area and the numbers of extensive air-showers arriving at different zenith angles θ were noted, the angle θ being measured with respect to the vertical direction (Fig. 9.9). From the geometry of Fig. 9.9, it can be seen that the path-length of the shower through the atmosphere varies as 5200secθ kg m^{-2}. The angle θ is measured from the delay time between the triggering of the detectors at various points in the array. This technique works because the thickness of the shower is no more than a few metres. The thickness of the layer of electrons, which are the lightest particles, is less than 2 m and that

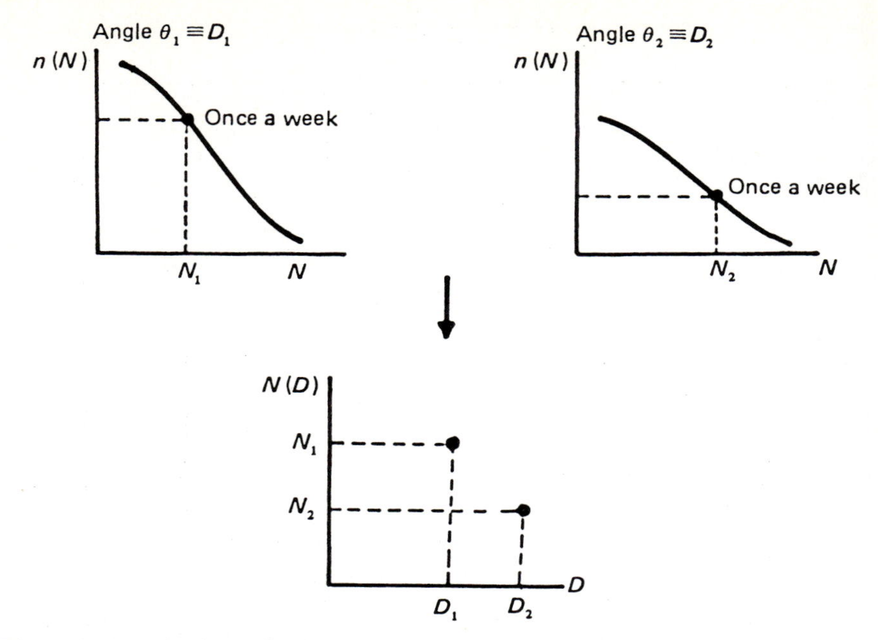

Figure 9.10. The determination of the development of extensive air-showers of different sizes through the atmosphere from observations at different zenith angles θ; $n(N)$ is the rate of occurrence of showers of N particles. A zenith angle θ is equivalent to a depth through the atmosphere of $5200 \sec\theta$ kg m^{-2}.

of the muons about 3–4 m. The hadrons, which are the most massive particles, trail behind the others at a distance of about a few metres. Thus, the particles constituting the showers may be thought of as arriving in 'pancakes'.

To interpret these results, it is assumed that showers initiated by particles of the same energy develop in the same way and that the rate of arrival of cosmic rays does not vary with time. We can then plot the frequency distribution of the sizes of the showers at different depths D through the atmosphere from those arriving at different zenith angles θ and obtain distributions of the form shown in Fig. 9.10. It is assumed that showers which have the same frequency of occurrence at different zenith angles were initiated by particles of the same energy and hence the development of the showers with depth through the atmosphere can be derived directly from the distributions observed at different zenith angles. This procedure provides an average picture of the development of showers for particles of a particular energy. Fig. 9.11 shows examples of the results of this type of analysis. On the diagram, the development of showers initiated by particles of two different energies are shown. The estimated strengths of the showers as observed at sea-level are also shown. Both of these showers reached maximum development at about 5200 kg m^{-2} through the atmosphere. In Fig. 9.11, the development at shorter path lengths through the atmosphere is derived from theoretical modelling.

More recently other techniques have been used to map out the development of the showers through the atmosphere (see e.g. Sokolsky (1989)). One approach is

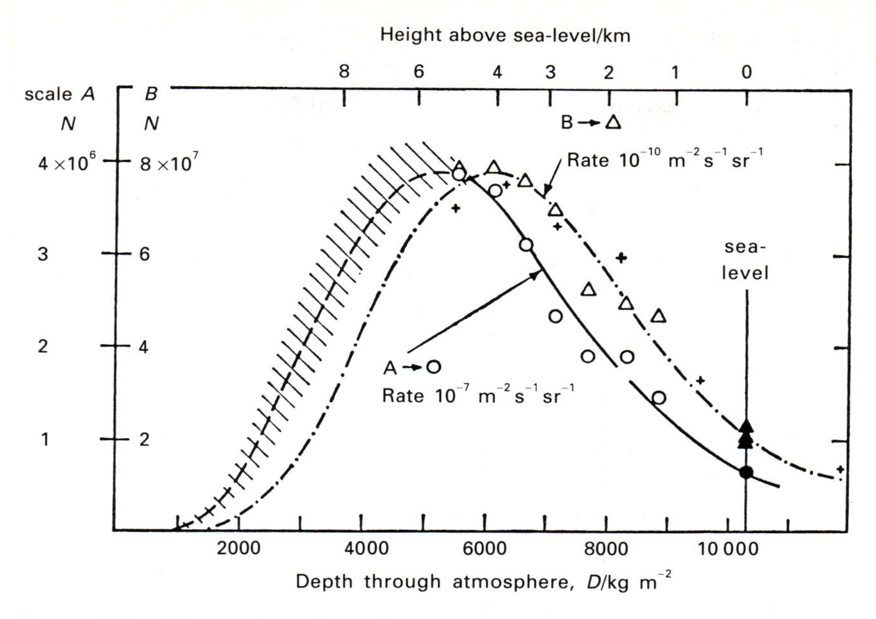

Figure 9.11. The number of particles N in showers which are detected at fixed rates A and B as a function of depth through the atmosphere from inclined showers at Chacaltaya and from sea-level laboratories. $N(D)$ shows approximately the growth and decay of showers of a particular energy. The continuation of the curves to path lenghs less than 5200 kg m^{-2} is based upon theory. (From A. M. Hillas (1972). *Cosmic rays*, page 89, Oxford: Pergamon Press.)

to measure the optical Cherenkov radiation emitted by the relativistic electrons created in the shower. Another clever optical method makes use of the fact that the relativistic electrons in the shower excite the molecules of the atmosphere and as a result they emit fluorescent radiation. The intensity of this radiation is proportional to the flux of electrons and so, by measuring the flux of fluorescent radiation, the development of the shower through the atmosphere can be determined. These observations are made with 'Fly's Eye' telescopes in which a number of optical telescopes observe the flux of optical emission at different zenith angles. If more than one Fly's Eye telescope is used separated by a few kilometres, the geometry of the shower can be determined. Fig. 9.12 shows the type of data which can be accumulated by this means. This new technique for measuring ultrahigh energy cosmic ray particles has already detected some of the highest energy particles known.

Since many of the nuclear interactions end up producing electromagnetic cascades, much of the energy of the shower ends up in the form of relativistic electrons. In Figs. 9.11 and 9.12 the fluxes are the numbers of relativistic electrons, each of which loses energy at a rate of 0.22 MeV kg^{-1} m^2, the mean energy loss rate for electrons in the ultrarelativistic regime. Therefore, we can estimate the total energy loss in the form of relativistic electrons by integrating the total area under the curve, since it registers all the relativisitic electron energy resulting from the

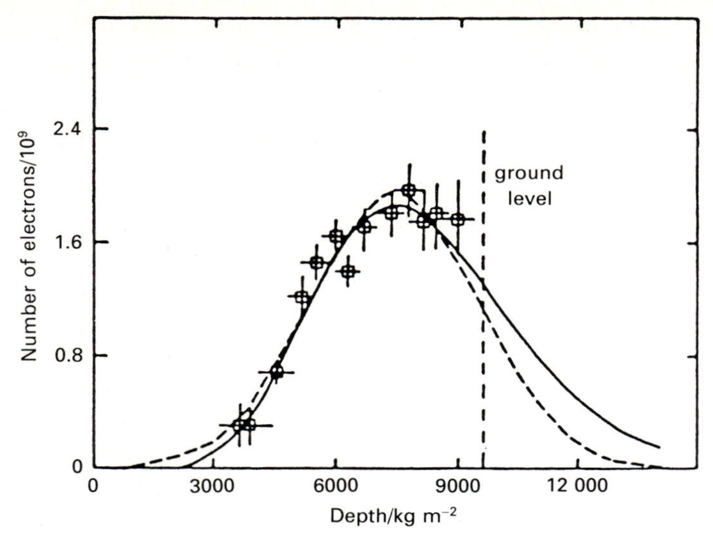

Figure 9.12. The determination of the development of an extensive air-shower through the atmosphere using the 'Fly's Eye' technique. The energy of the primary particle was 2×10^{18} eV and the angle of the shower from the vertical was 28°. (From A. A. Watson (1985). *19th intl. cosmic ray conference*, La Jolla, USA, Vol. 9, page 111.)

shower. Because most of the electron energy is expended in the relativistic regime, we obtain a good estimate of the total electron energy by simply taking the integral

$$\int 0.22 N_e(D)\,\mathrm{d}D \text{ MeV} \tag{9.4}$$

Let us now look in a little more detail at the energies of the two particles responsible for initiating the showers shown in Fig. 9.11 (Table 9.3). The following points can be noted.

(i) Most of the energy arriving at ground level is actually in the form of muons, although at maximum development the electrons contain most of the energy.
(ii) The total energy of the shower is proportional to the number of particles present at maximum development. A useful rule is that, to within about 25 % accuracy, the energy of the initiating particle may be found by assuming that there is 1.4 GeV of energy per particle present at shower maximum.
(iii) The more energetic the particle, the lower the maximum occurs in the atmosphere.

These showers have been modelled in considerable detail using computer simulations which include a very wide range of particle interactions and in which integrations are made over the appropriate probability distributions to estimate the characteristics of the showers such as size, lateral extent, composition as a function of initial energy, etc.

Table 9.3. *Energy content of two selected showers*

	Shower A	Shower B
Rate of arrival	$3\ m^{-2}\ sr^{-1}\ yr^{-1}$	$1/300\ m^{-2}\ sr^{-1}\ yr^{-1}$
Size of shower at maximum	3.9×10^6	7.8×10^7
Size of shower at sea-level	6.3×10^5	2.0×10^7
Ionisation above sea-level (eV) (from integration under curve)	4.4×10^{15}	8.8×10^{16}
Energy of remaining soft component at sea-level (eV)	0.14×10^{15}	0.4×10^{16}
Energy of nucleons and pions at sea-level (eV)	0.08×10^{15}	$0.2? \times 10^{16}$
Energy of muons at sea-level (eV)	0.64×10^{15}	1.3×10^{16}
Energy of neutrinos (estimated) at sea-level (eV)	0.33×10^{15}	0.6×10^{16}
Total energy (eV)	5.6×10^{15}	11.3×10^{16}

In order to detect the most energetic particles, very large arrays are required for two reasons – first, because they record the air-showers of the greatest lateral extent and, consequently, the particles of the highest energies and, second, because the most energetic particles are so rare that a large area detector is required to have a reasonable chance of detecting even a few of them each year. For example, the Yakutsk array in Siberia covered an area of 35 km^2 and the Giant Sydney Air-shower Array covered an area of 34 km^2.

9.5 Observations of the highest energy cosmic rays

A compilation of results is shown in Fig. 9.13 (from Watson, (1985); see also Hillas (1984) and Wdowczyk and Wolfendale (1989)). In this differential energy spectrum, corrections have been made to the raw data to account for the fact that the arrays are located at different depths through the atmosphere, have different collecting areas and so on. The experimenters responsible for each set of data have made their own energy calibrations and no attempt has been made to reduce the data to a common energy scale. Watson (1985) states that a systematic error in the energy scale of any of these experiments could be as much as 20% and the effect of such an error is indicated on Fig. 9.13. According to Watson, part of the scatter in the points could be due to this uncertainty in the calibration. The results are presented in the form of a differential energy distribution $j(E)$, multiplied by the energy of the particle E raised to the power 2.5, i.e. $j(E)E^{2.5}$. This presentation is used because the low energy spectrum of cosmic rays follows the law $j(E) \propto E^{-2.5}$ quite closely and therefore, if the high energy particles were to follow the same law, the function $j(E)E^{2.5}$ should be independent of energy E. This procedure of renormalising distributions which span many orders of magnitude is common in astronomy in which very wide ranges of parameters are often found.

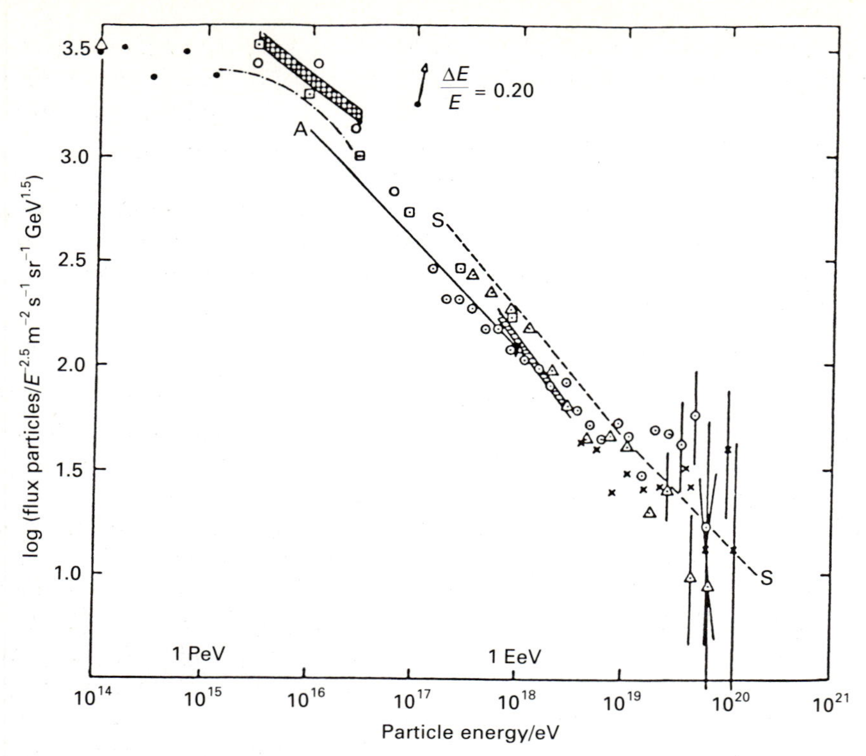

Figure 9.13. The differential energy spectrum of high energy cosmic rays presented in the form $N(E)E^{2.5}$. It is apparent that the spectrum is significantly steeper than that found at energies less than about 10^{14} eV. (From A. A. Watson (1985). *19th intl. cosmic ray conference*, La Jolla, USA, Vol. 9, page 111.)

This presentation is also a convenient way of displaying deviations from expectations and also of compressing all the information onto a single piece of normal graph paper.

Let us look at some of the features of this diagram.

(i) *The highest energy cosmic rays* It is immediately apparent from Fig. 9.13 that some cosmic rays of quite enormous energy have been detected; showers containing up to 10^{11} particles have been observed in the largest arrays and, using the rule that there is about 1.4 GeV per particle at maximum development, which occurs close to sea-level for the most energetic particles, the initiating particle must have had energy about 10^{20} eV, i.e. 16 J in one particle. In his review of 1985, Watson noted that 17 showers had been observed in which the inferred energies of the particles were greater than 10^{20} eV. His best estimate of the rate of arrival of particles with energies $E \geqslant 10^{20}$ eV was

$$I(E > 10^{20}\ \text{eV}) = \{3^{+2}_{-1}\} \times 10^{-16}\ \text{m}^{-2}\ \text{s}^{-1}\ \text{sr}^{-1}$$

or

$$\approx 1\ \text{km}^{-2}\ \text{sr}^{-1}\ \text{century}^{-1}$$

We will see that, if these results are correct, they have important implications for the origin of the very highest energy cosmic rays since they should interact with photons of the Microwave Background Radiation on a timescale significantly shorter than cosmological timescales (Volume 2, Chapters 20 and 26).

(ii) *The Spectrum* The differential spectrum shown in Fig. 9.13 spans the energy range 10^{14}–10^{20} eV. The first point to note is that there is good agreement among the experimenters about the slope of the energy spectrum in the energy range 10^{16}–10^{19} eV. Overall, the spectrum can be well represented in this energy range by the relation

$$j(E) = 2.1 \times 10^7 E^{-3.08}\ \mathrm{m^{-2}\ s^{-1}\ sr^{-1}\ GeV^{-1}} \tag{9.5}$$

where the energy E is measured in gigaelectron volts. This exponent of -3.08 should be compared with the value quoted for the energy range 1–10^3 GeV per nucleon of -2.5 to -2.7. There is therefore a steepening of the cosmic ray energy spectrum between the low and high energy range which occurs about 10^{15} eV.

The intermediate range of energies 10^{14}–10^{15} eV has been the subject of particular interest. The data of Fig. 9.13 suggest that the spectrum has slope -2.5 in this energy range and that this is somewhat flatter than the spectral slope observed at lower energies. Detailed analyses have suggested that there may be a 'bump' in the energy spectrum of the cosmic rays at about 10^{14}–10^{15} eV. The break in the overall spectrum at 10^{15} eV is often referred to as the 'knee' in the energy spectrum. The overall distribution of particle energies is displayed in Fig. 9.14.

Finally, at the very highest energies, $E \geqslant 10^{19}$ eV, there has been controversy about the precise shape of the energy spectrum. There is no question about the existence of these ultrahigh energy particles but the determination of the spectral shape is bedevilled by small statistics and the calibration of the energies of the particles. Taken literally, the data suggest that the energy spectrum of the very highest energy cosmic rays may flatten to a differential spectrum with slope -2.5 at energies greater than 2×10^{19} eV but it can be seen from Fig. 9.13 that there is considerable scatter in the data which would probably be consistent within the statistical uncertainties with an extrapolation of the lower energy spectrum to 10^{20} eV.

(iii) *Chemical composition* As mentioned above, the determination of the chemical composition of cosmic rays with energies greater than $E = 10^{14}$ eV is fraught with problems. The techniques described in Section 7.2 can only be used for particles with energies up to about 10^3 GeV nucleon^{-1}. To perform similar analyses for higher energy particles would require very much larger cosmic ray telescopes above the atmosphere, both to stop the energetic particles within the body of the detector and also to obtain a larger collecting area so that the chances of detecting the rare high energy particles are increased. The properties of the air-showers themselves have to be studied in order to obtain some information about the typical mass numbers of the particles initiating the showers. The observation of multi-cored showers has suggested that some of the primaries may be nuclei. Another approach involves measuring the Cherenkov radiation from the showers and estimating at what height the maximum development occurs as a function of

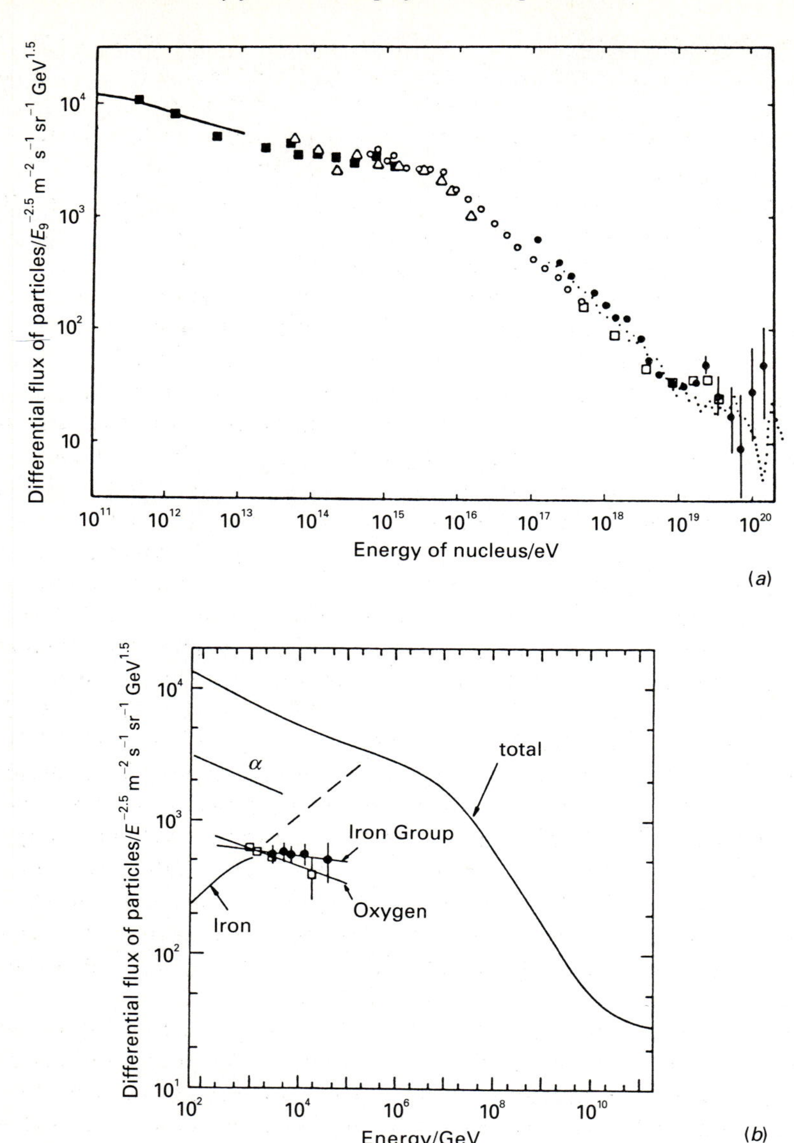

Figure 9.14. The overall spectrum of cosmic rays from 10^{11}–10^{20} eV. (*a*) The overall spectrum showing the observations at a wide range of particle energies. E_9 means energy measured in units 10^9 eV. (From A. M. Hillas (1984). *Ann Rev. Astr. Astrophys.*, **22**, 425.)
(*b*) A schematic representation of the overall spectrum, showing also the spectra of various elements which have somewhat flatter spectra than cosmic ray protons. (From J. Wdowczyk and A. W. Wolfendale (1989). *Ann. Rev. Nucl. Part. Sci.*, **39**, 43.)

the energy of the primary particles. If the primary particle of a given energy were an iron nucleus rather than a proton, the iron nucleus would have a lower Lorentz factor and therefore the secondary and subsequent generations of the electrons would have on average lower Lorentz factors and the maximum development would occur somewhat higher in the atmosphere than would the corresponding shower associated with a proton of the same energy. Unfortunately, there are internal disagreements in interpreting the experimental data and it is not clear whether the composition of the highest energy cosmic rays is similar to that at lower energies or whether there is a deficit of heavy nuclei relative to protons as suggested by some experiments. This is a key question which urgently needs resolution but the observations and their interpretation are not straightforward.

(iv) *Isotropy* In air-shower experiments, the incoming directions of the cosmic rays are determined and hence it is possible to estimate whether or not the arrival directions of the highest energy cosmic rays are isotropically distributed on the sky. The analysis is far from straightforward since there are many selection effects which could bias the observations. Data concerning the isotropy of the arrival directions of high energy cosmic rays are summarised in Fig. 9.7 which is due to Hillas (1984). In this diagram, the anisotropy in the arrival directions is found by taking the first 24 hr Fourier component of the intensity of cosmic rays on the sky. Hillas (1984) showed that there is now good agreement among the different observers about the magnitude of the anisotropy and its direction on the sky.

At energies less than about 10^{16} eV, an anisotropic flux of cosmic rays is observed with good statistical significance. In the energy range $10^{16} \leqslant E \leqslant 10^{19}$ eV, there are measurements and limits to the anisotropy. These data provide significant constraints on models of the volume within which the cosmic rays are confined. We will have a great deal to say about this topic later but we give here just one example of the type of question which can be asked. One interpretation of the break in the spectrum of the cosmic rays at 10^{15} eV is that it represents the energy at which cosmic rays can escape more freely from the Galaxy. In this case, one might expect the cosmic rays to be anisotropic above this energy if they were to originate within the Galaxy.

At the very highest energies, $E \geqslant 10^{19}$ eV, there is strong evidence for a large scale anisotropy in the distribution of arrival directions. According to Watson (1985), the anisotropy of all those cosmic rays with energies greater than 4×10^{19} eV amounts to 47 ± 18 % with a statistical significance of about 3 %. The direction of the maximum excess lies close to the direction of the Local Supercluster of galaxies. There is no evidence for these very high energy particles being concentrated towards the Galactic plane as might be expected if they were of Galactic origin. The very highest energy cosmic rays observed by the Sydney Giant Air-shower Array all came from a direction close to the north Galactic pole. This is suggestive evidence that the very highest energy particles may have an origin within the local supercluster.

9.6 The problems to be solved

It is useful to summarise the key astrophysical problems which have to be solved.

(i) The acceleration of particles to very high energies, $E \geqslant 10^{20}$ eV.
(ii) The nature of the acceleration processes which lead to the formation of a power-law spectrum of particle energies.
(iii) The origin of the high abundances of light elements such as lithium, beryllium and boron and of the heavy elements lighter than iron in the cosmic radiation and the variation of their relative abundances with energy.
(iv) The overall preservation of the universal abundances of the elements throughout the periodic table.
(v) The origin of the anisotropy in the distribution of cosmic rays.
(vi) Astrophysical sources of the cosmic rays and the processes of propagation from their sources to the Earth.

These are the most important astrophysical problems which have to be solved and they turn out to be related to a very wide range of problems in high energy astrophysics. We emphasise again the important point that, in studying the astrophysics of cosmic ray particles, we are dealing directly with the only high energy particles which originate in astronomical sources which are directly observable. As such, they have a unique contribution to make to high energy astrophysics.

10

The Solar Wind and its influence upon the local flux of cosmic rays

10.1 Introduction

The next problem we have to tackle is the propagation of high energy particles from the interstellar medium to the top of the atmosphere. A dominant theme of this topic is the dynamics of charged particles in magnetic fields which is a central subject for all high energy astrophysics. In the immediate vicinity of the Earth, the magnetic field distribution is dipolar and, as early as 1929, Clay found a correlation between the number of charged particles arriving at the surface of the Earth and geomagnetic latitude. This discovery stimulated research into the propagation of high energy particles in dipolar magnetic fields and it turns out that the Earth's magnetic field can be used to diagnose the properties of incoming high energy particles. Unfortunately, this particular story was overtaken by the realisation that the Earth's magnetic field is only dipolar rather close to the Earth. Beyond a few Earth radii, the magnetic field distribution is strongly influenced and distorted by the outflow of hot gas from the Sun, what is known as the *Solar Wind*. It turns out that the Solar Wind influences very strongly the observed flux of high energy particles with energies $E > 1$ GeV observed at the top of the atmosphere.

For these reasons, we discuss briefly in the Appendix to this chapter the dynamics of particles in dipole magnetic fields and emphasise in the chapter itself the plasma physics and magnetohydrodynamics of the interaction of the Solar Wind with the Earth's magnetic field structure. The physical processes discussed in this chapter and the next will reappear in their astrophysical context in the discussion of the physics of high energy phenomena in many different cosmic environments.

10.2 The Solar Wind

The photosphere of the Sun is at a temperature of about 5800 K as deduced from the absorption lines observed in the solar spectrum. Outside the surface

Figure 10.1. A High Altitude Observatory photograph of the solar corona taken during the total solar eclipse of 1970. (From G. W. Pneuman and F. Q. Orrall. *Physics of the Sun*, Volume 2, eds P. A. Sturrock, T. E. Holzer, D. M. Michalas and R. K. Ulrich, page 97, D. Reidel Publishing Co., Dordrecht, 1986.) The photograph was taken with a radially graded filter which enhances the brightness far from the Sun relative to that closer in. Notice that the hot corona is not circularly symmetric but that there are sectors in which there is little radiation. These are referred to as coronal holes, the largest of which is to the top left of the image.

layers of the Sun, however, there is observed to be high temperature gas in what is known as the *solar corona*. The corona is seen clearly during total eclipses of the Sun (Fig. 10.1) and its high temperature is inferred from the observation of very high excitation lines in its spectrum, for example, FeXIII, FeXIV, etc. which are often referred to as coronal lines. The heating mechanism for the coronal gas is still a matter of controversy but is almost certainly associated with the transport of energy by hydromagnetic waves or shock waves, the energy for which is generated in the surface layers of the Sun (see Chapter 12).

The physical situation led to one of the most beautiful predictions of modern astronomy. E. N. Parker wrote down the equations of hydrodynamics for the dynamics of a hot plasma in the gravitational field of the Sun and demonstrated that there must exist a general outflow of material from the Sun which was named the Solar Wind. We will not demonstrate this but refer the interested reader to the relevant sections of Parker's book *Interplanetary dynamical processes*. It is interesting that exactly the same set of equations describe stellar winds and

Table 10.1. *Typical parameters of the Solar Wind*

Particle velocity	~ 350 km s^{-1}
Particle flux	$\sim 1.5 \times 10^{12}$ m^{-2} s^{-1}
Particle concentration	$\sim 10^{7}$ m^{-3}
Energy of proton	~ 500 eV
Energy density of protons	$\sim 4 \times 10^{-10}$ J m^{-3}
Temperature	$\sim 10^{6}$ K

These figures refer to the normal Sun. In high speed streams, velocities up to 700–800 km s^{-1} are found and the particle concentrations are $\sim 5 \times 10^{6}$ m^{-3} so that the particle fluxes are more or less the same.

accretion flows, the latter being the energy sources in X-ray binary systems (see Volume 2, Chapter 16).

Parker's prediction was made in 1958. In 1959, one of the first Soviet satellites, Lunik II, measured a flux of particles emanating from the Sun and then, in 1962, the US Venus probe Mariner 2 measured continuously the number density of particles, their velocities and the magnetic field strength in the interplanetary medium between the Earth and Venus. At the distance of the Earth from the Sun, the unperturbed Solar Wind is characterised by the quantities given in Table 10.1 when there is little Solar activity. It should be emphasised that these are average figures and there is considerable variation about the mean values, particularly during periods of high solar activity. The average magnetic field strength in the Solar Wind is found to be about 5×10^{-9} T (5×10^{-5} G). The corresponding energy density in the Solar Wind magnetic field is

$$U_{\mathrm{mag}} = \frac{B^2}{2\mu_0} = 10^{-11}\ \mathrm{J\ m^{-3}}$$

Comparison with the figures quoted in Table 10.1 shows that most of the energy in the Solar Wind is carried by the bulk kinetic energy of the protons.

Our reasons for looking at the interaction of the Solar Wind with the Earth's magnetic field are multi-fold. First of all, there is convincing evidence that the level of solar activity strongly influences the flux of cosmic ray particles with energies about 1 GeV and less which we observe at the top of the atmosphere. Second, the Solar Wind and the environment of the Earth provide an ideal laboratory in which to study the physics of rarefied cosmic plasmas and these will prove to be important in studying many problems in high energy astrophysics involving high energy particles, plasmas and magnetic fields. Third, many of the processes observed in the interplanetary medium have direct analogues in astronomical systems but, in the case of the environment of the Earth, the particles and magnetic fields can be observed directly.

The study of the Solar Wind is only one part of the subject of solar–terrestrial relations which has become a discipline of the greatest importance. It has been established that the level of solar activity can influence the terrestrial environment in important ways and numerous space vehicles have been flown and are planned to increase this understanding.

10.3 Evidence for solar modulation

The evidence that the Solar Wind influences the local flux of cosmic rays is most dramatically illustrated by the inverse correlation between the intensity of the cosmic ray flux at the top of the Earth's atmosphere and the level of solar activity. A measure of the level of *solar activity* is the number of sun-spots observed on the surface of the Sun.

This is found to be correlated strongly with phenomena such as radio interference detected in long wavelength radio transmissions and the number of aurorae observed at high geomagnetic latitudes. The increase in Solar activity also has the effect of increasing the heating of the upper atmosphere of the Earth so that its scale height increases. This is important in determining the length of time satellites can remain in low Earth orbit since the higher the particle density encountered by the space probe, the more rapidly it is decelerated and spirals towards the Earth. The *solar cycle* is the 11 year period during which the sun-spot number and general solar activity rises from a low level to a maximum value and then returns to a low value.

The anticorrelation between solar activity, and the consequent turbulence in the plane of the ecliptic, and the flux of cosmic rays at the top of the Earth's atmosphere can be studied by comparing various measures of these three quantities. The cosmic ray flux can be studied at ground level using very stable counters which measure the fluxes of neutrons or muons at ground level. In the case of the muons, these are the secondary products produced by the flux of high energy cosmic rays incident upon the top of the atmosphere (see Section 5.5); in the case of the neutrons, these are tertiary or higher products observed at sea-level of the cosmic rays incident at the top of the atmosphere. The standard measure of the level of Solar activity is the *Zurich relative sun-spot number*. The number of sun-spots on the disc of the Sun is the standard measure of activity on the Sun and most of them are found between heliocentric latitudes $\pm 45°$. A measure of the degree of turbulence in the plane of the ecliptic plane in the vicinity of the Earth is provided by the *geomagnetic aa index* (see Mayaud (1990)). This index measures fluctuations in the magnetic field strength in the vicinity of the Earth and these result from the interaction of the Earth's dipole magnetic field with the Solar Wind as we will describe later. Fig. 10.2 shows the variations of these three quantities over the period 1954–80 (Shea and Smart 1985). The periods of solar minimum and maximum are indicated along the top of this diagram. The Mount Washington neutron monitor intensities have been used as a measure of the cosmic ray flux. This station has the advantage that it is located geographically at a point of the Earth's surface at which the geomagnetic cut-off is only 1.3 GV (see Appendix to this chapter) and hence this monitor provides a good measure of the cosmic ray flux incident at the top of the atmosphere.

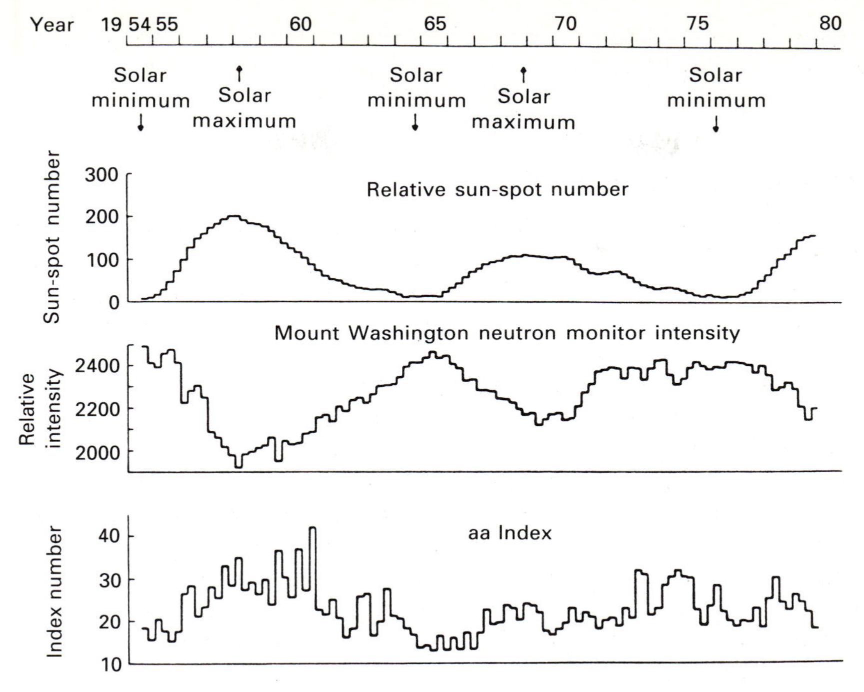

Figure 10.2. The correlation between the relative sun-spot number, the intensity of neutrons measured by the Mount Washington neutron monitor and the geomagnetic aa index. (From M. A. Shea and D. F. Smart (1985). *19th international cosmic ray conference*, La Jolla, USA, Vol. 4, 501.)

It can be seen from Fig. 10.2 that there is an excellent correlation between the relative sun-spot number and the aa geomagnetic index showing that the local degree of turbulence in the geomagnetic field is strongly correlated with the level of activity on the Sun. In addition there is a perfect anticorrelation between these quantities and the intensity of cosmic rays. Thus, the greater the level of solar activity, the more effective the Solar Wind is in preventing the interstellar flux of cosmic rays from reaching the Earth. This is the phenomenon known as *solar modulation*.

More evidence is provided by changes in the shape of the primary cosmic ray spectrum observed at the top of the atmosphere as a function of phase of the solar cycle. Fig. 10.3(*a*) shows the energy spectra of hydrogen and helium nuclei at different phases of the solar cycle. The top and bottom curves correspond to sun-spot minimum and maximum respectively whilst the middle curve corresponds to an intermediate level of sun-spot activity. Notice that for clarity of presentation, the curves for the protons have been displaced by a factor of 5. The influence of the Solar Wind in attenuating the flux of low energy cosmic rays is apparent. Webber and Lezniak (1974) also present these spectra in terms of the rigidity of the particles (Fig. 10.3(*b*)). As we will show in Section 11.1, the rigidity $R = pc/ze$ is a measure of the amount by which a particle of electric charge z and relativistic

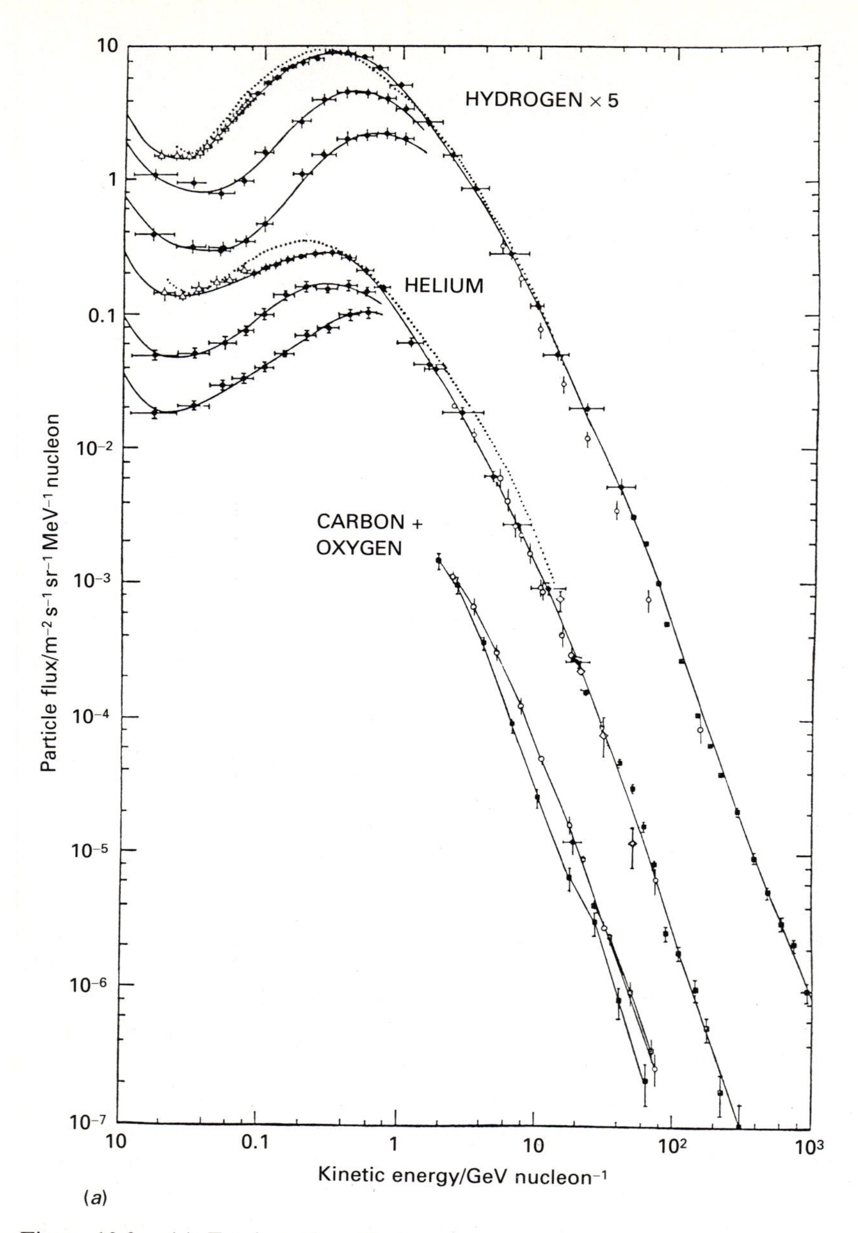

Figure 10.3. (*a*) For legend see facing page.

momentum $p = \gamma mv$ is deflected by a magnetic field. Thus, particles of the same rigidity should be deflected by the same amount and have similar trajectories in a given magnetic field configuration. It will turn out that it is often simplest to use the rigidity rather than the energy per nucleon as a measure of the degree of modulation of the cosmic rays.

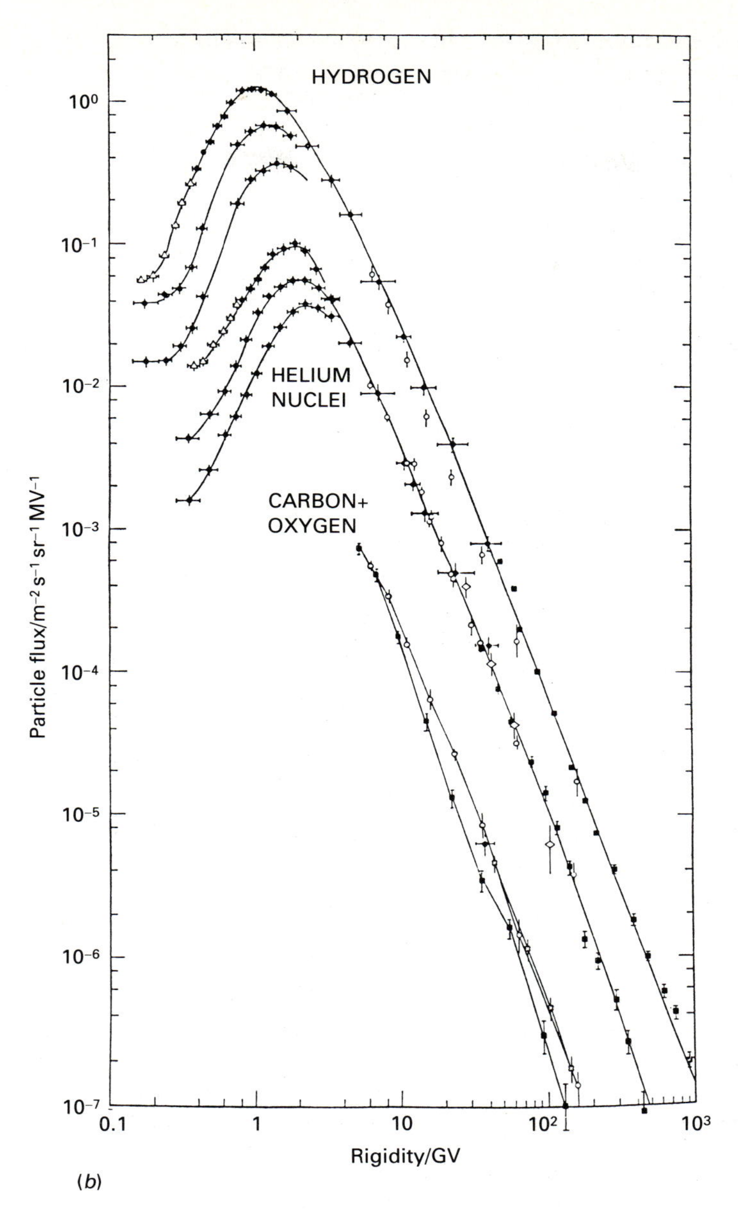

Figure 10.3. (*a*) The differential energy spectra of hydrogen and helium nuclei in the cosmic rays as a function of phase of the Solar cycle. In each set of spectra, the top curve shows the measurements at the time of minimum sun-spot number and the bottom curve at sun-spot maximum. The middle curve shows the spectra at intermediate levels of sun-spot number. The data were obtained in the period 1964–73. (From W. R. Webber and J. A. Lezniak (1974). *Astrophys. Sp. Sci.*, 30, 361.)
(*b*) The same data as in (*a*) but with the spectra plotted as differential rigidity spectra.

A third powerful method of investigating Solar modulation is to use the information provided by cosmic ray electron spectra. Unfortunately, we have to anticipate somewhat the future development of our story. Briefly, we obtain information about the cosmic ray electrons in two ways. First, direct measurements at the top of the atmosphere give us the modulated spectrum of the cosmic ray electrons (Fig. 9.3). Second, the same electrons in the interstellar medium emit synchrotron radiation in the radio waveband and, fortunately, the spectrum of the radio emission enables us to infer the energy spectra of the electrons in the interstellar medium outside the Solar System (Volume 2, Sections 18.2 and 18.3). By comparing these spectra, the degree of modulation can be determined directly as a function of the energy per nucleon (or rigidity) of the particles and the phase of the Solar cycle.

We therefore need to understand the structure of the magnetic field in the interplanetary medium and how it can influence the propagation of particles from interplanetary space to the Earth. First of all, we have to establish some key physical properties of the interplanetary plasma.

10.4 The electrical conductivity of a fully ionised plasma

The reason for introducing this topic here is that we must next build a model for the structure of the magnetic field in the interplanetary medium. To do this, we need to know the electrical conductivity (or resistivity) of the plasma since dissipative effects can have a major effect upon the dynamical behaviour of the magnetic field. This will become apparent in our discussions of Section 10.5.

We are all familiar with the phenomenon of electrical conductivity in metals which we can model by the Drude picture according to school physics or by phonon scattering according to university physics. Neither of these pictures is relevant to our present considerations because we are dealing with conduction in a fully ionised plasma. The picture is closer to that of the conduction of electricity in gases in which energy and momentum are transferred in particle collisions. This is straightforward in the case of a gas of molecules in which the particles are all neutral and the collision cross-section is roughly the same as the geometric cross-section of the molecules. In a plasma, however, the collisions are mediated through long-range electrostatic forces between the electrons and protons. The dynamics of a particle in the plasma are illustrated schematically in Fig. 10.4. We can define a collision between the particles in the plasma if we remember that, classically, a collision is just an interaction in which the particle loses all memory of its initial direction and starts off in a new random direction.

In a plasma, a charged particle feels a large number of little impulses, just as we worked out in the case of ionisation losses (Section 2.2) and the average of all these random impulses is to give no net sideways momentum to the particle. Statistically, however, just as in a random walk process, although there is no net sideways velocity $\langle \Delta v_\perp \rangle = 0$, the root mean square velocity $\langle \Delta v_\perp^2 \rangle_{\frac{1}{2}}$ is not zero, i.e. statistically, the particle acquires net sideways momentum by random scattering.

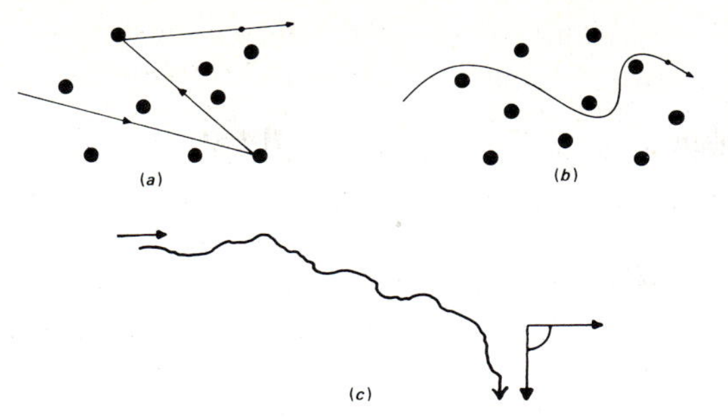

Figure 10.4. A schematic diagram illustrating the concept of a collision according to (*a*) the Drude model and (*b*) collisions mediated by long-range electrostatic forces. In (*c*) we illustrate how eventually a particle is deflected through 90° by the stochastic effect of a large number of distant encounters.

Notice that, if the process were truly continuous, there would never be any net velocity perpendicular to the line of flight but, because the impulses are given in discrete random packets, the net value of $\langle \Delta v_\perp^2 \rangle$ is not zero. Exactly as in the case of a random walk, if the rms sideways velocity acquired per second is $\langle \Delta v_\perp^2 \rangle^{\frac{1}{2}}$, then this velocity will become v after a time t_c

$$\langle \Delta v_\perp^2 \rangle t_c = v^2$$

We define this time t_c to be the time for a particle to make a collision. By this time, the perpendicular velocity is as large as the initial velocity and we can then deem the particle to have forgotten its initial direction. This calculation is quite straightforward if we make simplifiations such as the protons remaining stationary. The problem in the present case is that we are interested in the dynamics of the protons themselves since they carry all the momentum of the Solar Wind.

In the spirit of our previous calculations, let us work out a simple expression for the mean free path of a proton of velocity v interacting with the protons and electrons of the plasma. For simplicity, we assume that the protons and electrons of the plasma are stationary. Our analysis will indicate the origin of the dependences upon physical parameters which appear in the full calculation.

In a single collision, the proton receives an impulse perpendicular to its direction of motion

$$p = \frac{Ze^2}{2\pi\varepsilon_0 bv}$$

(see equation (2.1)). Hence,

$$\Delta v_\perp = \frac{Ze^2}{2\pi\varepsilon_0 bvm} \tag{10.1}$$

Our approach is to work out the mean square deviation of the particle's velocity about its initial direction. When the mean square velocity is v^2, the mean velocity squared, the particle is deflected through roughly 90° and a collision is deemed to have taken place. Using the same procedure as in Section 2.1, we find that

$$\langle \Delta v_{\perp}^2 \rangle = \int_{b_{\min}}^{b_{\max}} \left(\frac{Ze^2}{2\pi\varepsilon_0 bvm} \right)^2 2\pi b N v \, \mathrm{d}t$$

in one second

Therefore,

$$\langle \Delta v_{\perp}^2 \rangle = \frac{Z^2 e^4}{4\pi^2 \varepsilon_0^2 m^2 v} 2\pi N \ln \left(\frac{b_{\max}}{b_{\min}} \right)$$

$$= \frac{Z^2 e^4 N}{2\pi\varepsilon_0^2 m^2 v} \ln \Lambda \tag{10.2}$$

where $\Lambda = b_{\max}/b_{\min}$. Once again, we have encountered our old friend $\ln\Lambda$, a Gaunt factor. The old complications are present again. In the present instance, the maximum collision parameter $b_{\max}$ is the Debye length for a thermal plasma, $b_{\max} = (\varepsilon_0 kT/Ne^2)^{\frac{1}{2}}$; it will be recalled that this is the typical shielding distance of a proton in the plasma by the gas of oppositely charged electrons. The minimum collision parameter is again the closest distance of approach in the classical limit $b_{\min} = Ze^2/8\pi\varepsilon_0 m_e v^2$, the usual value for a low temperature plasma (see Section 2.2). Therefore,

$$t_c = v^2/\langle \Delta v_{\perp}^2 \rangle$$

$$= \frac{2\pi\varepsilon_0^2 m^2 v^3}{Z^2 e^4 N \ln \Lambda}$$

$$= \frac{2\pi\varepsilon_0^2 m^{\frac{1}{2}} (3kT)^{\frac{3}{2}}}{Z^2 e^4 N \ln \Lambda} \tag{10.3}$$

where we have taken the velocity v to be the typical thermal velocity of a proton in a plasma at temperature T, $\frac{1}{2}mv^2 = \frac{3}{2}kT$. There are considerable complications in the full theory because all the particles are in motion and there is a Maxwellian distribution of velocities. Spitzer (1962) give details of these results in his monograph *The physics of fully ionised gases* pp. 120–53.

Let us work out the typical relaxation time of a proton in the interplanetary medium. We can adapt the result (10.3) we have already derived and compare it with the result presented by Spitzer. He gives the relaxation time for a particle of mass number A and atomic number Z interacting with a gas of particles of the same species. In this case, we have to write $m = Am_p$ and the dependence upon charge becomes Z^4 rather than Z^2 because in our last sum we considered a proton interacting with the particles of a gas of atomic number Z. Then, we find

$$t_c = \frac{2\pi\varepsilon_0^2 m_p^{\frac{1}{2}} (3k)^{\frac{3}{2}}}{e^4} \left(\frac{T^{\frac{3}{2}} A^{\frac{1}{2}}}{NZ^4 \ln \Lambda} \right)$$

This is of identical form to that presented by Spitzer (his formula (5.26)) and the constant is the same to within 25%. Spitzer quotes the result

$$t_c = 11.4 \times 10^6 \frac{T^{\frac{3}{2}} A^{\frac{1}{2}}}{N Z^4 \ln \Lambda} \text{ s} \tag{10.4}$$

where T is the temperature in degrees Kelvin, N in particles m^{-3} and A is the mass number of the particles. The mean free path of the particle is just $\lambda = v t_c$. For the Solar Wind, we adopt $T = 10^6$ K, $A = 1$, $N = 5 \times 10^6$ m^{-3}, $Z = 1$ and $\ln \Lambda = 28$ (see Spitzer (1962) p. 128), Then, we find $\lambda = 3 \times 10^{13}$ m. This should be compared with the distance from the Earth to the Sun, 1.5×10^{11} m, the distance known as the astronomical unit (AU) We can therefore see that for the protons which carry all the momentum of the Solar Wind, the mean free path for electrostatic collisions is much larger than the Sun–Earth distance and therefore, for the purposes of describing the dynamics of the intermixed Solar Wind particles and magnetic field, we can consider the plasma to be collisionless with infinite conductivity.

In many astrophysical situations, we will encounter this same phenomenon that the mean free path is very long so that the plasma may be considered collisionless. The appropriate mean free path for any particular situation requires careful consideration and depends upon the physical problem being studied. For example, the effective mean free path may be related to the gyroradius of the particles (Section 10.8) or to the distance for scattering of electrons streaming along the magnetic field lines (Volume 2, Section 19.7).

10.5 Flux freezing

A distinctive feature of many of the plasmas we have to deal with in high energy astrophysics, and astronomy in general, is that they have very high electrical conductivities or, to put it another way, the particles have very long mean free paths. What simplifies many discussions is the fact that, in the limit of infinite conductivity, the magnetic field behaves as if it were frozen into the plasma. This phenomenon is known as *flux freezing*.

We can adopt two approaches. In one approach, we write down the equations of magnetohydrodynamics, take the limit of infinite electrical conductivity and then find the dynamics of the fields and the plasma. Unless one is fluent in magnetohydrodynamics, this approach does not give an immediate feel for what the answer means and so we will begin with a more physical approach in which we study the behaviour of the flux linkage of closed circuits in a fully ionised plasma when the circuits are moved or distorted.

First, we state the theorem we wish to prove. If we represent the magnetic field by magnetic lines of force, so that the number per unit area perpendicular to the lines is equal in magnitude to the magnetic field strength, then, when there are movements in the plasma, the magnetic field lines move and change their shape as though they were frozen into the plasma.

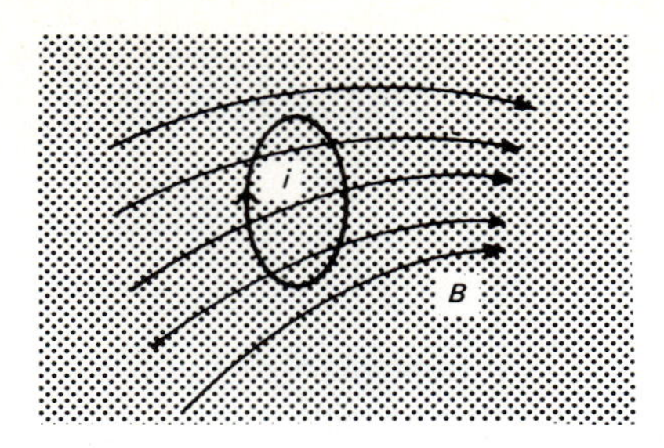

Figure 10.5.

10.5.1 *The physical approach*

We adopt the pleasant approach of Ratcliffe (1972) in his monograph *An introduction to the ionosphere and magnetosphere*. The analysis is performed in two parts. First, the changes in the magnetic flux linkage in a stationary current loop are studied when the magnetic field strength changes and then the effect of distorting the shape of the current loop is analysed.

We consider currents flowing in a plasma of infinite conductivity. Suppose we consider a little current loop to which no batteries are attached in the plasma (Fig. 10.5). Then, the emf induced in the circuit can only be due to the rate of change of magnetic flux through the circuit

$$\mathscr{E} = -\mathrm{d}\phi/\mathrm{d}t$$

The magnetic flux ϕ consists of two parts, one part due to the current in the loop itself ϕ_i and the other due to all external currents ϕ_{ex}. Let the inductance of the loop be L; then, by definition

$$\phi_i = Li \quad \phi = \phi_i + \phi_{\mathrm{ex}}$$

Now, if the external currents change so that ϕ_{ex} changes, then an emf is induced in the circuit and the resulting equation for the current induced in the wire is

$$L\,\mathrm{d}i/\mathrm{d}t + Ri = -\mathrm{d}\phi_{\mathrm{ex}}/\mathrm{d}t$$

We take the resistance of the coil to be zero, a superconducting loop, to model the case of a collisionless plasma. Then,

$$L\,\mathrm{d}i/\mathrm{d}t = -\mathrm{d}\phi_{\mathrm{ex}}/\mathrm{d}t = \mathrm{d}\phi_i/\mathrm{d}t$$

i.e.

$$\phi_i + \phi_{\mathrm{ex}} = \text{constant} \tag{10.5}$$

This means that, although ϕ_{ex} may change, by virtue of changing, it induces a current which exactly cancels out the decrease which might have been expected. This is a property of the fact that the current loop is supposed to be superconducting. It is not true if R is finite but is very closely so if R is just very, very small. Note that there is nothing inconsistent in assuming that there is a current i flowing without any emf being present initially. Because the current is flowing in a conductor of zero resistance, there is no way of dissipating the current.

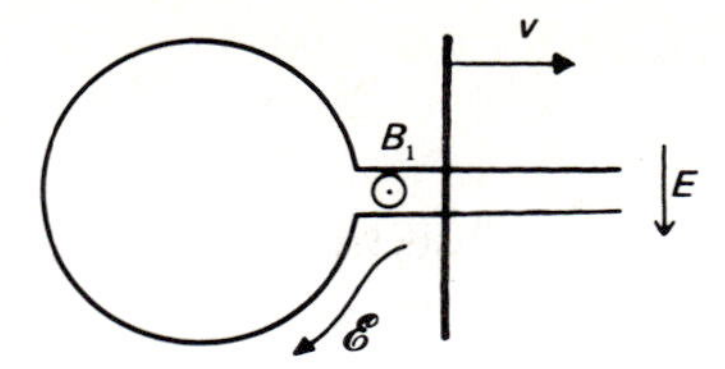

Figure 10.6.

A corollary of this proof is that, if the circuit is moved, the flux will also remain unchanged because, so far as the loop is concerned, it is only the external field which changes.

Finally, what happens if the loop changes shape? Let us consider the concrete example of the circuit shown in Fig. 10.6 which consists of a loop with parallel wires crossed by a conductor. We suppose that the entire circuit is made of superconducting material and that the field in the region of the parallel wires is B_1. Now let the conductor move down the wire a distance $\mathrm{d}x$ at a velocity v. The strength of the induced electric field is $|\mathbf{E}| = |\mathbf{v} \times \mathbf{B}| = vB_1$ in the sense shown in Fig. 10.6. The induced emf is

$$\mathscr{E} = El = vB_1 l$$

where l is the length of the conductor crossing the parallel wires. But $\mathscr{E} = -\mathrm{d}\phi/\mathrm{d}t$ and therefore the magnetic flux induced in the circuit is

$$\mathrm{d}\phi = (vB_1 l)\,\mathrm{d}t \text{ in the sense opposite to } B_1.$$

But, because the area is bigger, we enclose more magnetic flux. In fact, because all the changes are small,

$$\mathrm{d}\phi = B_1 l\,\mathrm{d}x = (B_1 lv)\,\mathrm{d}t \text{ in the same direction as } B_1.$$

Thus, the two effects cancel exactly and there is no change in the magnetic flux through the circuit after its shape has changed. Since the magnetic flux through the circuit is constant, $L_1 i_1 = L_2 i_2$, where the subscripts 1 and 2 refer to the values of L and i before and after the deformation of the circuit. Thus, the current flowing in the circuit changes as the loop deforms. Again, because there is no dissipation, the current simply maintains this new value. The emf produced while the loop is being distorted induces a further current in the loop which just 'stays around' because there is no means of dissipating it.

To express the result mathematically, if we choose any loop C in the plasma, and follow it as the shape changes due to motions in the plasma,

$$\int_S \mathbf{B} \cdot \mathrm{d}\mathbf{S} = \text{constant}$$

where $\mathrm{d}\mathbf{S}$ is the increment of surface area and S refers to the total surface area bounded by the loop C. If we take a little circular loop of wire and then let the plasma expand uniformly, the above result leads to $BA = B\pi r^2 = \text{constant}$, i.e.

$$B \propto r^{-2}$$

Thus, in a uniform expansion, the energy density of the magnetic field decreases as $B^2/2\mu_0 \propto r^{-4}$. This is the same result as is obtained for the adiabatic expansion of a gas for which the ratio of specific heats is $\gamma = \frac{4}{3}$. An ultrarelativistic gas has $\gamma = \frac{4}{3}$ (see Volume 2, Section 19.2) and this makes sense since such a gas behaves in many ways like a gas of massless particles and a magnetic field is an excellent example of a massless gas.

All these considerations are directly relevant to the interplanetary and interstellar plasma since these are collisionless plasmas and hence the field lines move with the plasma as if they were frozen in. This result is true for a very wide range of diffuse cosmic plasmas.

10.5.2 *The mathematical approach*

First, we write down the equations of magnetohydrodynamics. These consist of:

The equation of continuity

$$\frac{\partial \rho}{\partial t} + \nabla \cdot (\rho \mathbf{v}) = 0 \tag{10.6}$$

where ρ is the mass density and $\mathbf{v}$ is the velocity at a point in the fluid;

Force equation

$$\rho \mathrm{d}\mathbf{v}/\mathrm{d}t = -\nabla p + \mathbf{J} \times \mathbf{B} + \mathbf{F}_v + \rho \mathbf{g} \tag{10.7}$$

where p is the pressure, $\mathbf{J}$ is the current density, $\mathbf{B}$ is the magnetic flux density, $\mathbf{F}_v$ represents viscous forces and $\mathbf{g}$ is the gravitational acceleration. We note that $\mathrm{d}\mathbf{v}/\mathrm{d}t$ is the convective derivative, i.e. the force $\rho \mathrm{d}\mathbf{v}/\mathrm{d}t$ acts on a particular element of the fluid in the frame of reference which moves with the plasma. This is related to the partial derivative which describes change in a property of a fluid at a fixed point in space by

$$\frac{\mathrm{d}}{\mathrm{d}t} = \frac{\partial}{\partial t} + \mathbf{v} \cdot \nabla \tag{10.8}$$

Readers wishing a simple revision of these concepts may consult Longair (1984), Appendix to Chapter 5, pp. 105–11.

Maxwell's equations

We need the equations in the following form:

$$\nabla \times \mathbf{E} = -\partial \mathbf{B}/\partial t \tag{10.9}$$

$$\nabla \times \mathbf{B} = -\mu_0 \mathbf{J} \tag{10.10}$$

Notice that no displacement current $\mathbf{D}$ is included in the equation (10.10). This is because we are dealing with slowly varying phenomena, $v \ll c$. Therefore, there are no space charge effects present i.e. the particles of the plasma always have time to neutralise any charge imbalance on the scale of motion of the plasma. Finally, we have:

Ohm's law

$$\mathbf{J} = \sigma(\mathbf{E}+\mathbf{v}\times\mathbf{B}) \tag{10.11}$$

where σ is the electrical conductivity of the plasma. Now substituting (10.11) into (10.9)

$$\nabla\times(\mathbf{J}/\sigma-\mathbf{v}\times\mathbf{B}) = -\partial\mathbf{B}/\partial t \tag{10.12}$$

Now, putting (10.10) into (10.12)

$$\nabla\times\left(\frac{\nabla\times\mathbf{B}}{\sigma\mu_0}-\mathbf{v}\times\mathbf{B}\right) = -\partial\mathbf{B}/\partial t \tag{10.13}$$

Therefore

$$\frac{\partial\mathbf{B}}{\partial t} = \nabla\times(\mathbf{v}\times\mathbf{B})-\nabla\times(\nabla\times\mathbf{B})/\mu_0\sigma$$

We now use the identity $\nabla\times(\nabla\times\mathbf{B}) = \nabla(\nabla\cdot\mathbf{B})-\nabla^2\mathbf{B}$. Since $\nabla\cdot\mathbf{B} = 0$, we find

$$\partial\mathbf{B}/\partial t = \nabla\times(\mathbf{v}\times\mathbf{B})+\frac{1}{\sigma\mu_0}\nabla^2\mathbf{B} \tag{10.14}$$

The system of equations (10.6), (10.7), (10.8) and (10.14) form the basic equations of magnetohydrodynamics.

Let us look at two particular solutions. If the plasma is at rest, $\mathbf{v} = 0$,

$$\frac{\partial\mathbf{B}}{\partial t}-\frac{1}{\sigma\mu_0}\nabla^2\mathbf{B} = 0$$

This is just a standard diffusion equation. We can determine the time it takes the magnetic field to diffuse out of a region by the standard trick of writing $\partial\mathbf{B}/\partial t \sim B/\tau$ where τ is a characteristic diffusion time and $\nabla^2 B \approx B/L^2$, where L is the scale of the system i.e.

$$\frac{B}{\tau} \approx \frac{1}{\sigma\mu_0}\frac{B}{L^2}$$

$$\tau \approx \sigma\mu_0 L^2$$

This trick always works to order of magnitude. This result is important, for example, in the study of star formation if the gas cloud is permeated by a magnetic field. The time τ is that required for the field to diffuse out of a region of dimension L.

We are interested in the case of infinite conductivity $\sigma = \infty$ in which case equation (10.14) becomes

$$\partial \mathbf{B}/\partial t = \nabla \times (\mathbf{v} \times \mathbf{B})$$

If we consider any loop $\mathbf{S}$ in the plasma, there are two contributions to the change in $\mathbf{B}$ with time. First, there may be changes in the magnetic flux density due to external causes and second, there is an induced component of the flux density due to motion of the loop. The first contribution is simply

$$\int_S \frac{\mathrm{d}\mathbf{B}}{\mathrm{d}t} \cdot \mathrm{d}\mathbf{S}$$

The second contribution results from the fact that, because of the movement of the loop, there is an induced electric field $\mathbf{E} = \mathbf{v} \times \mathbf{B}$. Then, because $\nabla \times \mathbf{E} = -\partial \phi/\partial t$, there is an additional contribution to the total magnetic flux through the loop, i.e.

$$\int_S \frac{\mathrm{d}\mathbf{B}}{\mathrm{d}t} \cdot \mathrm{d}\mathbf{S} = -\int_S \nabla \times (\mathbf{v} \times \mathbf{B}) \cdot \mathrm{d}\mathbf{S}$$

Therefore, adding together both contributions, we obtain

$$\begin{aligned} \frac{\mathrm{d}}{\mathrm{d}t}\int_S \mathbf{B} \cdot \mathrm{d}\mathbf{S} &= \int_S \frac{\partial \mathbf{B}}{\partial t} \cdot \mathrm{d}\mathbf{S} - \int_S \nabla \times (\mathbf{v} \times \mathbf{B}) \cdot \mathrm{d}\mathbf{S} \\ &= \int_S (\partial \mathbf{B}/\partial t - \nabla \times (\mathbf{v} \times \mathbf{B})) \cdot \mathrm{d}\mathbf{S} \\ &= 0 \end{aligned}$$

i.e. the magnetic flux through the loop is constant, which is what we mean by *flux freezing*.

10.5.3 *Application to the Solar Wind*

Now let us apply this important result to the dynamics of the Solar Wind. The plasma and the magnetic field are strongly tied together and therefore the dynamics depend upon which of them has the greater energy (or mass) density. From Table 10.1 we observe that the kinetic energy of the protons is much greater than that of the magnetic field and therefore the magnetic field is dragged around by the particles.

The Sun rotates once every 26 days on its axis and the Solar Wind is released radially outwards with more or less constant radial velocity of the order 350 km s^{-1}. Therefore, with respect to the point on the Sun from which it was ejected, the Solar Wind traces out an Archimedian spiral. The particles are all tied to magnetic field lines rooted in the Sun and therefore the magnetic field in the Solar Wind takes up a spiral pattern. This is illustrated schematically in Fig. 10.7(a) which shows the dynamics of particles ejected at constant radial velocity from the Sun as it rotates. The dynamics are the same as those of a rotating garden sprinkler.

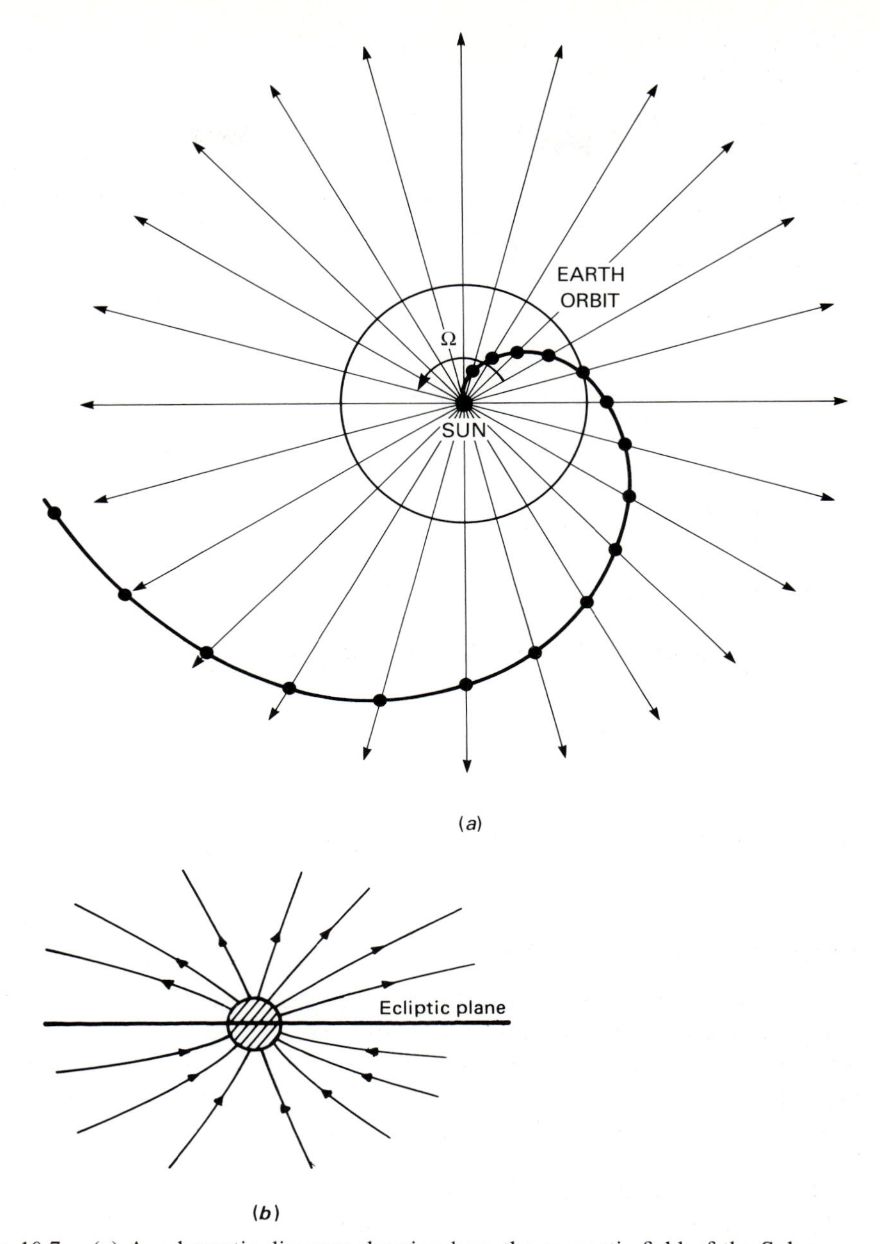

Figure 10.7. (*a*) A schematic diagram showing how the magnetic field of the Solar Wind takes up a spiral configuration. The plasma leaving the solar corona moves out more or less radially and the magnetic field is dragged with it. The diagram shows the dynamics of plasma associated with one field line while the Sun rotates through half of one rotation. At large distances, the spiral is Archimedean.
(*b*) A schematic diagram showing the structure of the magnetic field out of the plane. The magnetic field has opposite polarity on either side of the neutral sheet.

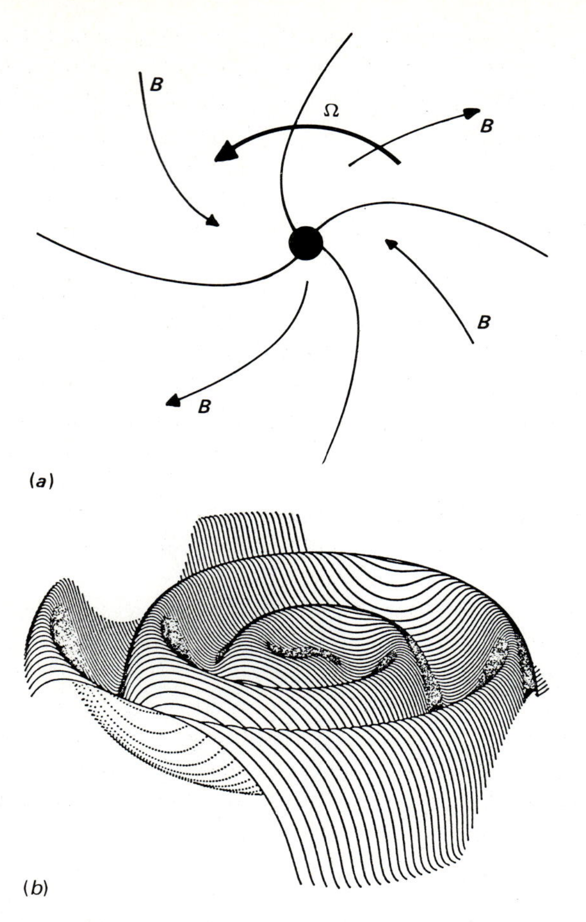

Figure 10.8. (*a*) A schematic diagram showing the sector structure observed in the magnetic field distribution in the Solar Wind when the field direction is measured close to the plane of the ecliptic. (After J. M. Wilcox and N. F. Ness (1965). *J. Geophys. Res.*, **70**, 5793.)
(*b*) A model of the wavy inclined neutral sheet (or heliospheric current sheet) which separates the opposite polarities of the magnetic field in the northern and southern halves of the Solar Wind. The sheet is as observed from 30° above the equatorial plane at a distance of 75 AU. The figure is 25 AU across. (From J. R. Jokipii and B. T. Thomas (1981). *Astrophys. J.*, **243**, 1115.)

Since the magnetic field of the Sun is dipolar, we would expect the polarity of the magnetic field to differ on either side of the plane of the ecliptic as illustrated in Figure 10.7(*b*). The magnetic field distribution in the interplanetary medium was found to have this basic spiral configuration by Wilcox and Ness (1965) but they also found that the magnetic field direction reversed so that, in some sectors, the magnetic field points inwards and in others outwards as indicated schematically in Fig. 10.8(*a*). It was subsequently noted, however, that the sector structure could equally well be accounted for if the basic north–south polarity of the Solar magnetic field is maintained and, in addition, the neutral sheet separating the two regions is inclined to the plane of the ecliptic. More recent measurements by the Pioneer-11 spacecraft as it travelled to northern solar latitudes between Jupiter and Saturn have confirmed this magnetic field configuration (for details of the observations, see Simpson (1989)). The wavy inclined neutral sheet has been modelled by Jokipii and Thomas (1981) (Fig. 10.8(*b*)). The observed sector structure is due to the inclined wavy neutral sheet passing by the observer twice per rotation of the Sun.

In addition to defining the basic structure of the magnetic field in the Solar Wind, the Voyager and Pioneer spacecraft have confirmed the tight wrapping of the spiral field beyond about 20–25 AU. Superimposed upon this basic pattern, there is a myriad of other phenomena. For example, the Solar Wind is not uniform over all latitudes and, in particular, at periods when there is a high level of solar activity, there are fast streams in the Solar Wind. These result in shock waves propagating outwards through the interplanetary medium which bring with them a wide variety of new phenomena in the plasma physics and magnetohydrodynamics of the Solar Wind. Simpson (1989) provides a clear introduction to these phenomena.

This outflow of material from the Sun also modifies the structure of the magnetic field of the Earth and the shape of the distorted magnetic dipole has been determined in detail from satellite studies. As we will show, the Solar Wind is highly supersonic when it encounters the Earth's magnetic field and hence a *shock front* forms and there is the characteristic 'stand-off' behaviour that one sees in front of blunt objects when they move supersonically. This is a topic of wide importance in many areas of astrophysics and one to which we now turn.

10.6 Shock waves

Shock waves are found ubiquitously in high energy astrophysics and play a key role in many different astrophysical environments. It is useful to derive some of their basic properties which we will find have application in as diverse fields as star formation in the spiral arms of galaxies, the high velocity outflows from young stars, extragalactic radio sources, compact galactic nuclei and our present concern, the interaction of the Solar Wind with the magnetic field of the Earth. The basic physics is set out in two classic texts, *Fluid mechanics* by Landau and Lifshitz (Chapter 9) and *Physics of shock waves and high-temperature hydrodynamic phenomena* by Zeldovich and Raizer.

It is a general property of perturbations in a gas that they are propagated away from their source at the speed of sound in the medium. It is therefore plain that, if a disturbance is propagated at a velocity greater than the speed of sound, the disturbance cannot behave like a sound wave at all. There is a discontinuity between the regions behind and ahead of the disturbance and the latter region can have no prior knowledge of its imminent arrival – the sound waves which would transmit the information are propagated at a speed less than that of the disturbance. These discontinuities are called *shock waves*. They commonly arise in

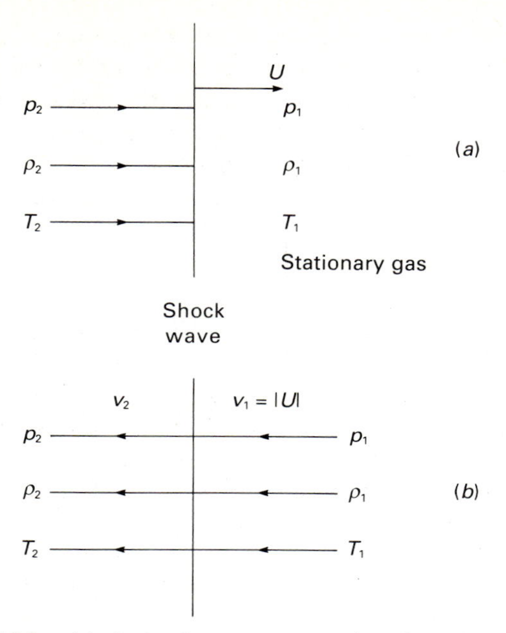

Figure 10.9. (*a*) A shock wave propagating through a stationary gas at a supersonic velocity U. The velocity U is supersonic with respect to the sound velocity in the stationary medium.
(*b*) The flow of gas through the shock front in the frame of reference in which the shock front is stationary.

situations such as explosions or where gases flow past obstacles at supersonic velocities or, equivalently, objects more supersonically through a gas. The basic phenomenon is the flow of gas at a supersonic velocity relative to the local velocity of sound. This is an enormous subject and our object here is to derive some useful relations for shock waves in a perfect gas.

10.6.1 *The basic properties of plane shock waves*

We assume that there is an abrupt discontinuity between the two regions of fluid flow. In the undisturbed region ahead of the shock wave, the gas is at rest with pressure p_1, density ρ_1 and temperature T_1. Behind the shock wave, the gas moves supersonically and its pressure, density and temperature are p_2, ρ_2 and T_2 respectively (Fig. 10.9(*b*)). It is convenient to transform to a reference frame moving at velocity U in which the shock wave is stationary (Fig. 10.9(*b*)). Then, the undisturbed gas flows towards the discontinuity at velocity $v_1 = |U|$ and, when it passes through it, its velocity becomes v_2 in the moving reference frame.

The behaviour of the gas on passing through the shock wave is described by a set of conservation relations. First, mass is conserved on passing through the discontinuity and hence

$$\rho_1 v_1 = \rho_2 v_2 \tag{10.15}$$

Second, the energy flux i.e. the energy passing per unit time through unit area parallel to v_1 is continuous. It is one of the standard results of fluid dynamics that the energy flux through a surface normal to the vector $\mathbf{v}$ is $\rho\mathbf{v}(\frac{1}{2}v^2+w)$ where w is the enthalpy per unit mass, $w = \varepsilon_m + pV$, ε_m is the internal energy per unit mass and V is the specific volume $V = \rho^{-1}$ i.e. the volume per unit mass. We consider only plane shock waves which are perpendicular to v_1 and v_2 and therefore the conservation of energy flux implies

$$\rho_1 v_1(\tfrac{1}{2}v_1^2 + w_1) = \rho_2 v_2(\tfrac{1}{2}v_2^2 + w_2) \tag{10.16}$$

Notice that it is the enthalpy per unit mass and not the energy per unit mass ε which appears in this relation. The reason for this is that, in addition to internal energy, work is done on any element of the fluid by the pressure forces in the fluid and this energy is available for doing useful work. Another way of looking at this relation is in terms of Bernoulli's equation of fluid mechanics in which the quantity $\frac{1}{2}v^2 + w = \frac{1}{2}v^2 + \varepsilon_m + p/\rho$ is conserved along streamlines which is the case for normal flow through the shock wave.

Finally, the momentum flux through the shock wave should be continuous. For the perpendicular shocks considered here, the momentum flux is given by $p+\rho v^2$ and hence

$$p_1 + \rho_1 v_1^2 = p_2 + \rho_2 v_2^2 \tag{10.17}$$

Notice that the pressure p, being the force per unit area, contributes to the momentum flux of the gas. Expressions (10.15), (10.16) and (10.17) are the three conservation equations which are often referred to as the *shock conditions*.

For simplicity, we will study the case of shock waves in a perfect gas for which the enthalpy is $w = \gamma pV/(\gamma-1)$ where γ is the ratio of specific heats and V again is the specific volume i.e. the volume of unit mass $V = \rho^{-1}$. Landau and Lifshitz show how it is possible to obtain many elegant results for this case. First of all, we define the mass flux per unit area $j = \rho_1 v_1 = \rho_2 v_2$. Then, from expression (10.17), which describes the conversation of momentum, we immediately find

$$j^2 = (p_2 - p_1)/(V_1 - V_2) \tag{10.18}$$

In addition, we obtain an expression for the velocity difference

$$v_1 - v_2 = j(V_1 - V_2) = [(p_2 - p_1)(V_1 - V_2)]^{\frac{1}{2}} \tag{10.19}$$

The next step is to find the ratio V_2/V_1 as a function of p_1 and p_2 for a perfect gas. We begin with the equation of conservation of energy flux (10.16) and substitute as follows:

$$w_1 + \tfrac{1}{2}v_1^2 = w_2 + \tfrac{1}{2}v_2^2$$
$$w_1 + \tfrac{1}{2}j^2V_1^2 = w_2 + \tfrac{1}{2}j^2V_2^2$$

Using expression (10.18), this expression reduces to

$$(w_1 - w_2) + \tfrac{1}{2}(V_1 + V_2)(p_2 - p_1) = 0 \tag{10.20}$$

We can now substitute the perfect gas expression, $w = \gamma p V/(\gamma-1)$ into the relation (10.20),

$$\frac{V_2}{V_1} = \frac{p_1(\gamma+1)+p_2(\gamma-1)}{p_1(\gamma-1)+p_2(\gamma+1)} \tag{10.21}$$

which gives the relation between the pressures and specific volumes on either side of the shock. We can immediately find the relation between T_2 and T_1 from the perfect gas law, $p_1 V_1/T_1 = p_2 V_2/T_2$.

$$\frac{T_2}{T_1} = \frac{p_2}{p_1}\frac{V_2}{V_1} = \frac{p_2}{p_1}\frac{p_1(\gamma+1)+p_2(\gamma-1)}{p_1(\gamma-1)+p_2(\gamma-1)} \tag{10.22}$$

Also, using expression (10.21), we can eliminate V_2 from the expression for the flux density j

$$j^2 = \frac{(\gamma-1)p_1+(\gamma+1)p_2}{2V_1} \tag{10.23}$$

From expression (10.23), we can find the velocities of the gas in front of and behind the shock

$$v_1^2 = j^2 V_1^2 = \frac{V_1}{2}[(\gamma-1)p_1+(\gamma+1)p_2] \tag{10.24}$$

$$v_2^2 = j^2 V_2^2 = \frac{V_2}{2}\frac{[p_1(\gamma+1)+p_2(\gamma-1)]^2}{p_1(\gamma-1)+p_2(\gamma+1)} \tag{10.25}$$

It is most convenient to write these results in terms of the *Mach number M* of the shock wave which is defined to be $M_1 = v_1/c_1$ where c_1 is the velocity of sound for the undisturbed gas, $c_1 = (\gamma p_1/\rho_1)^{\frac{1}{2}}$. Thus,

$$M_1^2 = v_1^2/(\gamma p_1/\rho_1) = v_1^2/\gamma p_1 V_1 \tag{10.26}$$

By simple substitution, we find the following results. From expressions (10.24) and (10.25), the *pressure ratio* is

$$\frac{p_2}{p_1} = \frac{2\gamma M_1^2-(\gamma-1)}{(\gamma+1)} \tag{10.27}$$

From expressions (10.24) and (10.25), we find the density ratio

$$\frac{\rho_2}{\rho_1} = \frac{v_1}{v_2} = \frac{(\gamma-1)p_1+(\gamma+1)p_2}{(\gamma+1)p_1+(\gamma-1)p_2}$$

$$= \frac{(\gamma+1)}{(\gamma-1)+2/M_1^2} \tag{10.28}$$

Finally, from expressions (10.22), (10.27) and (10.28), we find the temperature ratio

$$\frac{T_2}{T_1} = \frac{[2\gamma M_1^2-(\gamma-1)][2+(\gamma+1)M_1^2]}{(\gamma+1)^2 M_1^2} \tag{10.29}$$

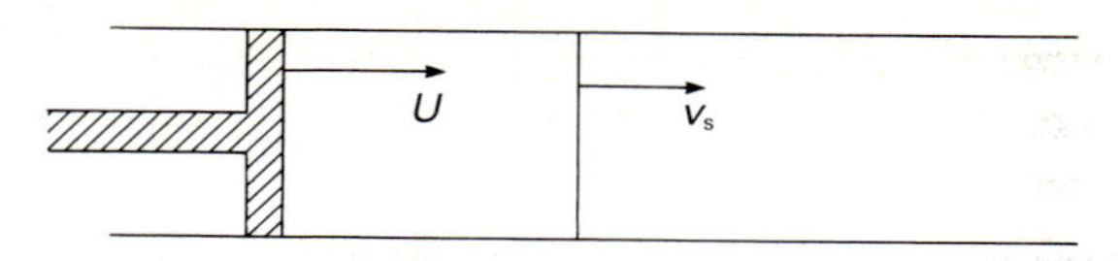

Figure 10.10. Illustrating the flow of gas in the case of a piston which moves at velocity U which is supersonic with respect to the velocity of sound in the stationary medium.

It is useful to look at these ratios in the limit of very strong shocks, $M_1 \gg 1$. Then,

$$p_2/p_1 = 2\gamma M_1^2/(\gamma+1) \tag{10.30}$$

$$\rho_2/\rho_1 = (\gamma+1)/(\gamma-1) \tag{10.31}$$

$$T_2/T_1 = 2\gamma(\gamma-1)M_1^2/(\gamma+1)^2 \tag{10.32}$$

These results show that in the limit of very strong shocks, the temperature and pressure can become arbitrarily large but the density ratio attains the finite value $(\gamma+1)/(\gamma-1)$. For example, a monatomic gas has $\gamma = \frac{5}{3}$ and hence $\rho_2/\rho_1 = 4$ in the limit of very strong shocks. These results demonstrate how efficiently strong shock waves can heat gas to very high temperatures and this is found to be the case in supernova explosions and supernova remnants. These are very useful results.

What exactly is happening in the shock front? It is apparent that the undisturbed gas is both heated and accelerated when it passes through the shock front and this is mediated by the atomic or molecular viscosity of the gas. It can be shown that the acceleration and heating of the gas takes place over a physical scale of the order of the mean free path of the atoms, molecules or ions of the gas. This makes physical sense because it is over this scale that energy and momentum can be transferred between gas molecules. Thus, the shock front is expected to be very narrow and the heating very strong over this short distance.

10.6.2 *The supersonic piston*

A common situation in high energy astrophysics is one in which an object is driven supersonically into a gas, or equivalently, a supersonic gas flows past a stationary object. A useful illustrative example is that of a piston driven supersonically into tube containing stationary gas (Fig. 10.10) (see Landau and Lifshitz (1987) p. 357). A shock wave forms ahead of the piston and the gas behind the shock moves at the velocity of the piston U. In the frame of reference of the shock front, which moves at some as yet unknown velocity v_s, the velocity of inflow of the stationary gas is $v_1 = |v_s|$ and the gas behind the shock moves at velocity v_2. As yet we do not know v_1 and v_2 but know that their difference $v_1 - v_2 = U$. As described by Landau and Lifshitz, the solution is a very neat one and involves a little elementary algebra.

First, using expression (10.18), we find

$$v_1 - v_2 = U = [(p_1 - p_2)(V_1 - V_2)]^{\frac{1}{2}} \tag{10.33}$$

Substituting for V_2 using equation (10.21) and squaring expression (10.33), the expression can be written in terms of the pressure ratio p_2/p_1,

$$\left(\frac{p_2}{p_1}\right)^2 - \left(\frac{p_2}{p_1}\right)\left[2+(\gamma+1)\frac{U^2}{2p_1 V_1}\right]+\left[1-\frac{(\gamma-1)U^2}{2p_1 V_1}\right] = 0 \qquad (10.34)$$

We can now write $\gamma p_1 V_1 = c_1^2$, where c_1 is the velocity of sound in the undisturbed medium, and solve for p_1/p_2

$$\frac{p_2}{p_1} = 1+\frac{\gamma(\gamma+1)U^2}{4c_1^2}+\frac{\gamma U}{c_1}\left[1+\frac{(\gamma+1)^2U^2}{16c_1^2}\right]^{\frac{1}{2}} \qquad (10.35)$$

The velocity $v_1 = |v_s|$ follows from expression (10.24)

$$v_1^2 = \frac{V_1}{2}[(\gamma-1)p_1+(\gamma+1)p_2] = \frac{c_1^2}{2\gamma}\left[(\gamma-1)+(\gamma+1)\frac{p_2}{p_1}\right] \qquad (10.36)$$

Some simple algebra shows that, substituting for p_2/p_1 using expression (10.35),

$$v_s = \frac{(\gamma+1)}{4}U+\left[c_1^2+\frac{(\gamma+1)^2U^2}{16}\right]^{\frac{1}{2}} \qquad (10.37)$$

This is a very neat result since it determines the thickness of the layer of shocked gas ahead of the piston for any supersonic velocity U. Let us look at the case of a very strong shock wave $U \gg c_1$. The expression (10.37) reduces to

$$v_s = (\gamma+1)U/2$$

Thus, the ratio of the position of the shock front to the position of the piston is $v_s/U = (\gamma+1)/2$. For a monatomic perfect gas $\gamma = \frac{5}{3}$ and hence $v_s/U = \frac{4}{3}$. Thus, all the gas which was originally in the tube between $x = 0$ and the position of the shock wave is squeezed into a smaller distance $(v_s - U)t$. It follows that the density increase over the undisturbed gas is $\rho_2/\rho_1 = v_s/(v_s - U) = (\gamma+1)/(\gamma-1)$, the same result we found in expression (10.31).

This simple calculation gives some feel for what is observed when supersonically moving gas encounters an obstacle or is ejected into a stationary gas. Ahead of the obstacle there is a shocked region which runs ahead of the advancing piston. This is what is expected to occur when a supernova ejects a sphere of hot gas into the interstellar medium. It also shows that there is a *stand-off* distance of a shock front from a blunt object placed in the flow and this is what occurs in the case of the flow of the Solar Wind past the Earth's magnetic dipole.

10.7 The Earth's magnetosphere

The Solar Wind is highly supersonic when it encounters the Earth and its magnetic field. To a rough approximation, the Earth and its associated magnetic field act as a spherical obstacle in the outflowing Solar Wind and, consequently, if this were a problem in standard gas dynamics, we would expect a stand-off shock to form in front of it, similar to those observed in hypersonic wind-tunnel

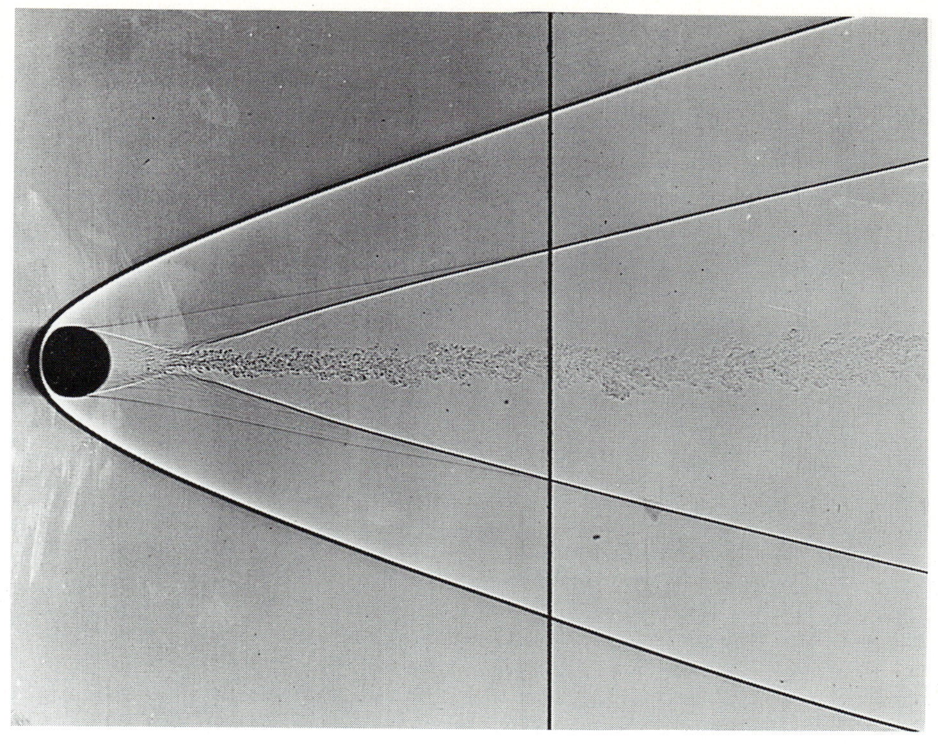

Figure 10.11. A laboratory experiment in which a solid sphere is placed in a supersonic flow. It can be seen that a stand-off shock is present around the obstacle. (From van Dycke, M. (1988). *An album of fluid flow* page 166, Parabolic Press.)

experiments (Fig. 10.11). The simple picture of a shocked zone in front of a piston developed in Section 10.6.2 is a crude picture of what might be expected. The important difference is that the gas can flow round the sides of the obstacle and so, while the shock wave is perpendicular at the equator, it becomes oblique round the edges of the object as shown in Fig. 10.11. In the case of oblique shocks, the component of flow velocity parallel to the shock wave is continuous whilst the normal component of the flow satisfies the shock conditions derived in Section 10.6.1. As a result, the streamlines are refracted on passing through the oblique shock. It should be noted that the velocity of the flow behind the shock can become supersonic if the shock wave is sufficiently oblique.

It is remarkable that, despite the many obvious differences between the case of a solid obstacle placed in a supersonic gas flow and the Solar Wind flowing past the Earth, the structures observed in the vicinity of the Earth can be very well described by classical gas dynamics. The magnetic field and particle distributions in the vicinity of the Earth have been well determined by space probe experiments and the results of these are shown schematically in Fig. 10.12.

There is a bow shock, similar to the bow shock in front of a solid object, at a stand-off distance of about $14R_E$ from the centre of the Earth in the direction of

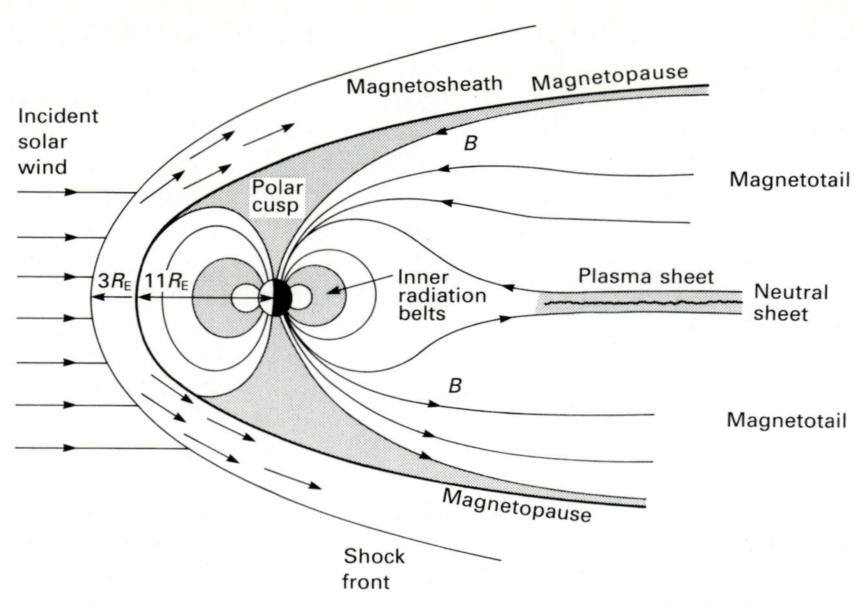

Figure 10.12. A schematic diagram showing the structure of the Earth's magnetosphere. The names of the various regions are shown.

incidence of the Solar Wind, where R_E is the radius of the Earth. Closer to the Earth, there is a boundary known as the *magnetopause* at a distance of about $11R_E$ which acts as the surface of the region within which the Earth's magnetic field is dynamically dominant. For the purpose of visualisation, the magnetopause may be thought of as the surface of a solid obstacle. The Solar Wind plasma flows past the Earth between the shock wave and the magnetopause. The whole region within the magnetopause is known as the *magnetosphere*, meaning the region in which the magnetic field of the Earth is the dominant dynamical influence. It is pleasant to note that the typically density enhancements across the shock wave are observed to be about a factor of 2–4, typical of the values expected for strong shocks in a monatomic gas. Of course, this is only a crude comparison because a full calculation would need to take account of the flow of the Solar Wind past the magnetopause.

The Earth's dipole magnetic field is strongly perturbed by the flow of the Solar Wind so that, although it can be well represented by a magnetic dipole close to the surface of the Earth, further away it is distorted as shown in Fig. 10.12. Perhaps the most significant distortion is the fact that the magnetic field-lines of the downstream side of the Earth are stretched out by the drag of the Solar Wind. The magnetospheric cavity is stretched out into a long cylindrical region which has radius about $25R_E$ at the distance of the Moon's orbit, i.e. at a distance of about $60R_E$. This region is known as the *magnetotail.* The magnetic field lines are oppositely directed on either side of the equatorial plane, those in the northern region heading towards the Earth, those in the southern region pointing away from the Earth. Between the two regions is a thick layer of hot plasma which is known as the *plasma sheet*. Because the magnetic field lines run in opposite directions on either side of the plasma sheet, there must be a surface of zero magnetic field separating the two regions which is known as the *neutral sheet*. Because the magnetic field changes sign through the neutral sheet, an induced electric current flows in the plasma sheet and hence particles can be accelerated in the vicinity of the neutral sheet. If the plasma moves in such a way as to bring together regions of oppositely directed magnetic field, the magnetic field lines can 'annihilate' transferring the magnetic field energy into particle energy by virtue of the electric fields created as the magnetic field strength changes. The Solar Wind particles flowing past the magnetotail are coupled into the magnetotail by instabilities acting at the magnetopause. The Kelvin–Helmholtz instability which results when one fluid streams past a stationary fluid enables Solar Wind particles to be entrained into the magnetosphere.

This picture of the Earth's magnetosphere suggests an explanation of the phenomena of *aurorae* observed at high geomagnetic latitudes. It can be seen from Fig. 10.12 that particles accelerated in the region of the magnetotail can drift along the magnetic field lines to high geomagnetic latitudes. Electrons with energies 0.5–20 keV entering the upper layers of the atmosphere at about 90–130 km can excite oxygen atoms producing the green 558 nm and red 630 nm red lines of oxygen characteristic of the aurorae.

From our perspective of learning about the behaviour of cosmic plasmas, there are a number of points of special interest. First of all, it should be noted that we have been able to use standard gas dynamics to understand the structure of the magnetosphere, despite the fact that the plasma is collisionless on the scale of an astronomical unit. The reason for this is the presence of the magnetic field which is frozen into the collisionless plasma. This is a very important aspect of plasma physics for many astrophysical plasmas. Despite the fact that the particles have very long mean free paths, the presence of even a very weak magnetic field ties the particles together. The fact that this works so well in the Earth's magnetosphere shows that this simplification can be used in many astrophysical environments.

Related to this point is the fact that there is a *shock wave discontinuity* at the boundary of the magnetosheath. As described in Section 10.6.1, it is expected that the thickness of the shock front should be of the same order as the mean free path of the particles. Again, the key point is the presence of the magnetic field. The magnetic field is frozen into the plasma and, as we will show in the next chapter, the particles of the plasma gyrate about the magnetic field direction at the gyrofrequency. The effective friction and viscosity needed to transfer momentum and energy through the shock wave are provided by the magnetic field tied to the particles of the plasma. The distance over which energy and momentum are transferred is, to order of magnitude, the gyroradius of a proton in the interplanetary magnetic field. The mechanism by which energy is transferred is probably through various forms of plasma wave interaction involving the magnetic field. This is a highly complex subject but is of the greatest interest for plasma physicists. The shock wave which bounds the magnetopause is one of the best examples known of a *collisionless shock wave*.

There is one final point of importance to be made. We have stated that the Solar Wind flows supersonically but we should clarify what we mean by this. In the case of a normal gas, the flow is supersonic if the velocity is greater than the local sound speed. For a gas at temperature T, this velocity is $c = (\gamma p/\rho)^{\frac{1}{2}} = (\gamma kT/\bar{m})^{\frac{1}{2}}$ where it is assumed that the gas is a fully ionised hydrogen plasma and $\bar{m}$ is the mean mass of the particles contributing to the pressure of the plasma. For a fully ionised hydrogen plasma in thermal equilibrium $\bar{m} = m_{\mathrm{H}}/2$ since the protons and electrons contribute equally to the thermal energy. The temperature of the Solar Wind is about 10^6 K and hence $v = 170$ km s^{-1}. Thus, the Solar Wind is supersonic with respect to its internal sound speed.

We have to ask, however, what the appropriate sound speed within the magnetosphere should be where the dynamics are dominated by the strength of the magnetic field. In this case, the appropriate sound speed is the *Alfven velocity* $v_{\mathrm{A}} = B/(\mu_0 \rho)^{\frac{1}{2}}$. The significance of this relation is that all sound speeds are roughly the square root of the ratio of the energy density of the system to its mass density $v \approx (\varepsilon/\rho)^{\frac{1}{2}}$ where ε is the energy density in the medium. Since the magnetosphere is magnetically dominated, $\varepsilon = B^2/2\mu_0$ and hence $v \approx B/(\mu_0 \rho)^{\frac{1}{2}}$. The exact answer is the Alfven velocity quoted above which is the speed at which hydromagnetic waves can be propagated in a magnetically dominated plasma. Inserting appropriate values for the magnetosphere, $B = 5$ nT, $n = 10^7$ m^{-3}, we find $v_{\mathrm{A}} = 35$ km s^{-1}. Thus, the flow of the Solar Wind is certainly highly supersonic with respect to the Alfven velocity within the magnetosphere. The important point of principle is the fact that, if any region of space is magnetically dominated, the appropriate sound speed is the Alfven velocity rather than the standard sound speed in the gas. Often, the flow of the Solar Wind is described as super-Alfvenic rather than supersonic.

Appendix The influence of the Earth's magnetic field

It is interesting to look at some of the early work on the propagation of cosmic rays in a dipole magnetic field. Until the discovery of the Solar Wind and the Earth's magnetosphere, it was natural to suppose that the magnetic field of the Earth is of dipole form and that it is through this field that the particles are propagated to the surface of the Earth. As early as 1929 there was evidence that the dipole component of the Earth's magnetic field influences the flux of cosmic radiation. In that year, Clay was travelling from Java to the Netherlands to calibrate his ionisation detectors at the bottom of a Dutch salt mine. During the voyage, he observed variations of the numbers of charged particles arriving at the Earth as a function of geomagnetic latitude, i.e. the polar angle λ measured with respect to the equatorial plane of the Earth's magnetic field (Fig. A10.1). We now understand that he was detecting high order products of primary cosmic rays arriving at the top of the atmosphere. By 1932, Compton showed that the intensity of the cosmic radiation follows closely lines of geomagnetic latitude. The problem of describing the dynamics of particles in a dipole magnetic field was first tackled by the Norwegian physicist Størmer in an attempt to explain the polar aurorae. Consequently, the theory is often referred to as Størmer theory and the orbits of the particles as Størmer orbits.

A10.1 Dynamics of charged particles in a dipole magnetic field

The problem is an entirely classical one. Given a particle approaching a magnetic dipole from infinity, what is its trajectory? The analysis is long but straightforward and results in some pleasant answers. Details of the analysis can be found in Appendix B of the book by Hopper (1964). It is shown in Section 11.1 that the equations of motion of a charged particle of rest mass m_0, charge ze and Lorentz factor $\gamma = (1 - v^2/c^2)^{-\frac{1}{2}}$ in a static magnetic field **B** reduce to

$$\gamma m_0 \frac{\mathrm{d}v}{\mathrm{d}t} = ze(\mathbf{v} \times \mathbf{B}) \tag{A10.1}$$

All we need now is the relevant magnetic field distribution, remembering that, as every schoolchild knows, a north pole points along magnetic lines of force towards the south pole of a magnet and so the north pole of the Earth's magnetic dipole is located close to the

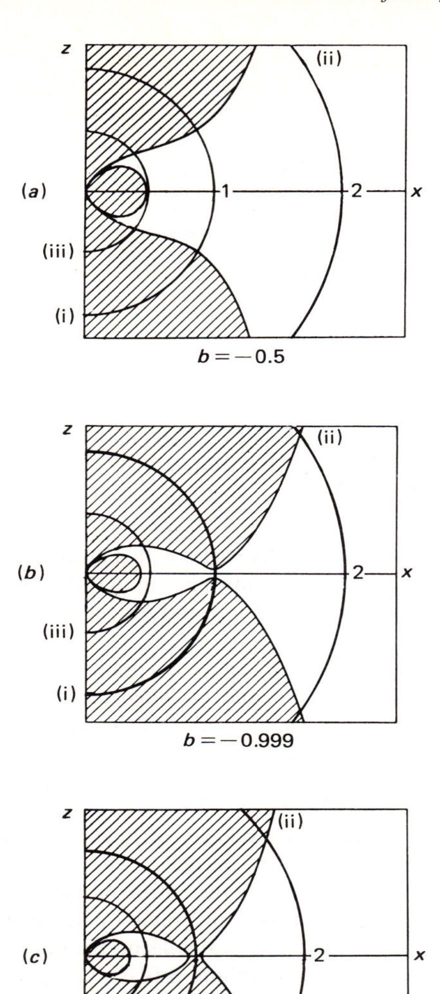

Figure A10.3. Diagrams illustrating the regions in polar coordinates which are accessible to high energy particles entering the Earth's magnetic field. The shaded regions have $|\sin\theta| > 1$ and consequently are forbidden. The values of the collision parameter b are given in Størmer units on the diagrams. The circumference of the Earth is shown on the diagrams for three values of the rigidity for protons: (i) $R = 59.6$ GeV/c; (ii) $4R$; (iii) $R/4$. The distance units are Størmer units, i.e. distances in units of $r_s = (ze\mu_0 M/4\pi p)^{\frac{1}{2}}$. (From Vallatra, M.S. 1961). *Cosmic rays I: Handbuch der Physik*, Vol. 61/1, 88. Berlin: Springer-Verlag).

in the meridian plane as defined in Fig. A10.2 and hence have $\theta = 0$. For these particles, we find from the expression (A10.4)

$$2b = -\cos^2\lambda/r$$

We are interested in particles which arrive at the surface of the Earth and hence, in terms of Størmer units, particles for which

$$r = \frac{r_E}{r_S} = \frac{r_E}{(ze\mu_0 M/4\pi p)^{\frac{1}{2}}} = \left(\frac{p}{59.6z \text{ GeV}}\right)^{\frac{1}{2}} = 1$$

Therefore, the critical value of b is

$$b = -\frac{1}{2}\frac{\cos^2\lambda}{(p/59.6z \text{ GeV})^{\frac{1}{2}}}$$

The necessary condition that particles reach the Earth if they have momenta less than 59.6 GeV c^{-1} is that they have collision parameters $b \geqslant -1$. Therefore, particles arriving vertically must satisfy the following relation

$$-\frac{1}{2}\frac{\cos^2\lambda}{(p/59.6z \text{ GeV})^{\frac{1}{2}}} \geqslant -1$$

i.e.

$$p \geqslant 14.9z \cos^4\lambda \text{ GeV } c^{-1}$$

Thus, only particles with momenta greater than a certain value can reach the Earth at different latitudes. Putting in some numbers, the geomagnetic cut-off momenta are

			Kinetic energy
$z = 1$	$\lambda = 0°$	$cp \geqslant 14.9$ GeV	14.0 GeV
	$\lambda = 40°$	$cp \geqslant 5.1$ GeV	4.3 GeV
	$\lambda = 60°$	$cp \geqslant 0.93$ GeV	0.48 GeV

Thus, at each point on the Earth's surface, there is a threshold energy below which the particles cannot reach the surface. The geomagnetic field acts as a filter for the particles of different energies and use can be made of this fact in balloon and space experiments in low Earth orbit to distinguish particles of different energies. Nowadays, maps of the geomagnetic cut-off can be made and there are kinks and irregularities in the cut-off energy which reflect irregularities in the magnetic field.

This analysis can be extended to obtain information about the energy distribution of the particles from the distribution of arrival directions. Particles only reach the Earth if they have $b \geqslant -1$, i.e. from expression (A10.4)

$$-\frac{r}{2}\sin\theta\cos\lambda - \frac{\cos^2\lambda}{2r} \geqslant -1$$

$$\sin\theta \leqslant \frac{2}{r\cos\lambda} - \frac{\cos\lambda}{r^2} \qquad \text{(A10.5)}$$

Therefore, when observations are made at a fixed geomagnetic latitude, we should only observe particles of a given energy within the range of angles defined by expression (A10.5). Notice that the directions of arrival should be opposite for protons and electrons. These expectations are in full agreement with observation.

The complete discussion of this problem becomes rather complicated and, if we require more detailed predictions, we have to follow particle trajectories and this can become a complex business. The major problem with this analysis is that, although the Earth's magnetic field resembles a dipole close to the surface, at distances greater than a few Earth radii, it is severely distorted by the flow of the Solar Wind past the Earth as described in the main body of this chapter.

11

The dynamics of charged particles in magnetic fields

We now know the basic magnetic field configuration through which high energy particles have to pass in order to reach the Earth. As they travel towards the Earth, they have to make their way upstream against the Solar Wind and into a converging configuration of magnetic field lines. This analysis again has applications far outside the immediate problem of understanding the degree to which the energy spectrum of the cosmic rays is influenced by the Solar Wind. The results are applicable very widely in high energy astrophysics but there is the great advantage in the case of the cosmic ray particles that the effects of the magnetic field can be observed directly. This subject can become one of daunting complexity but let us begin with some simple results of importance for much of our future discussion.

11.1 A uniform static magnetic field

We begin by writing down the equation of motion for a particle of rest mass m_0, charge ze and Lorentz factor $\gamma = (1 - v^2/c^2)^{-\frac{1}{2}}$ in a uniform static magnetic field $\mathbf{B}$.

$$\frac{\mathrm{d}}{\mathrm{d}t}(\gamma m_0 \mathbf{v}) = ze(\mathbf{v} \times \mathbf{B}) \tag{11.1}$$

We recall that the left-hand side of this equation can be expanded as follows:

$$m_0 \frac{\mathrm{d}}{\mathrm{d}t}(\gamma \mathbf{v}) = m_0 \gamma \frac{\mathrm{d}\mathbf{v}}{\mathrm{d}t} + m_0 \gamma^3 \mathbf{v} \frac{(\mathbf{v} \cdot \mathbf{a})}{c^2}$$

because the Lorentz factor γ should be written $\gamma = (1 - \mathbf{v} \cdot \mathbf{v}/c^2)^{-\frac{1}{2}}$. In a magnetic field, the three-acceleration $\mathbf{a} = \mathrm{d}\mathbf{v}/\mathrm{d}t$ is always perpendicular to $\mathbf{v}$ and consequently $\mathbf{v} \cdot \mathbf{a} = 0$. As a result,

$$\gamma m_0 \, \mathrm{d}\mathbf{v}/\mathrm{d}t = ze(\mathbf{v} \times \mathbf{B}) \tag{11.2}$$

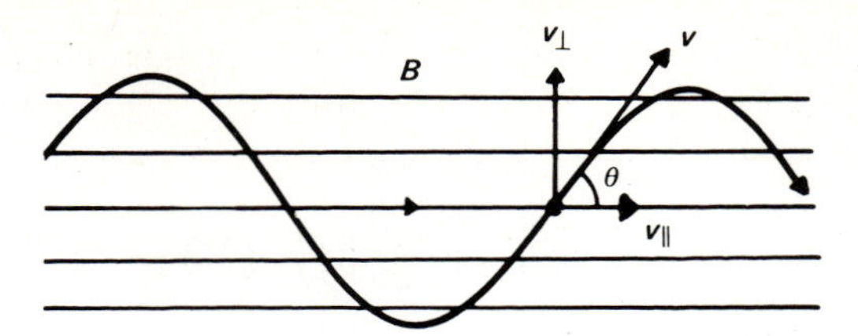

Figure 11.1. Illustrating the dynamics of a charged particle in a uniform magnetic field.

We now split **v** into components parallel and perpendicular to the uniform magnetic field, $v_\parallel$ and $v_\perp$ respectively (Fig. 11.1). The pitch angle θ of the particle's orbit is given by $\tan\theta = v_\perp/v_\parallel$ i.e. the angle between the vectors **v** and **B**. Since $v_\parallel$ is parallel to **B**, equation (11.2) tells us that there is no change in $v_\parallel$, *i.e.* $v_\parallel$ = constant. The acceleration is perpendicular to the magnetic field direction and to $v_\perp$.

$$\gamma m_0 \frac{d\mathbf{v}}{dt} = zev_\perp B(\mathbf{i}_v \times \mathbf{i}_B) = zevB \sin\theta \, (\mathbf{i}_v \times i_B)$$

where $\mathbf{i}_v$ and $\mathbf{i}_B$ are unit vectors in the directions of **v** and **B** respectively.

Thus, the particle's acceleration vector is perpendicular to the plane containing both the instantaneous velocity vector **v** and the direction of the magnetic field **B**. Because the magnetic field is uniform, this constant acceleration perpendicular to the instantaneous velocity vector results in circular motion about the magnetic field. Equating this acceleration to the centrifugal acceleration, we find

$$v_\perp^2/r = zevB\sin\theta/\gamma m_0$$

i.e.

$$r = \gamma m_0 v \sin\theta/zeB \qquad (11.3)$$

Thus, the motion of the particle consists of a constant velocity along the magnetic field direction and circular motion with radius r about it. This means that the particle moves in a *spiral path* with *constant pitch angle* θ. The radius r is known as the *gyroradius* of the particle. The angular frequency of the particle in its orbit ω_g is known as the *angular cyclotron* or *angular gyrofrequency* and is given by

$$\omega_g = v_\perp/r = zeB/\gamma m_0 \qquad (11.4)$$

The corresponding *gyrofrequency* ν_g, i.e. the number of times per second that the particle rotates about the magnetic field direction, is

$$\nu_g = \omega_g/2\pi = zeB/2\pi\gamma m_0 \qquad (11.5)$$

In the case of a non-relativistic particle, $\gamma = 1$ and hence $\nu_g = zeB/2\pi m_0$. A useful figure to remember is the non-relativistic gyrofrequency of an electron $\nu_g = eB/2\pi m_e = 28$ GHz T^{-1} where the magnetic field strength is measured in tesla; alternatively, $\nu_g = 2.8$ MHz G^{-1} for those not yet converted to teslas.

In this simple case, the axis of the particle's trajectory is parallel to the magnetic field direction and this axis is known as the *guiding centre* of the particle's motion, i.e. it is the mean direction of translation of the particle about which the gyration takes place. In more complicated magnetic field configurations, it is convenient to work out what the *guiding centre motion* is and this determines the general drift of particles in the field. Examples of this are discussed in the next section.

Let us rewrite the expression for the radius of the particle's path in the following form.

$$r = \frac{\gamma m_0 v}{ze}\frac{\sin\theta}{B} = \left(\frac{pc}{ze}\right)\frac{\sin\theta}{Bc} \qquad (11.6)$$

where p is the relativistic three-momentum of the particle. This means that, if we inject particles with the same value of pc/ze into a magnetic field $\mathbf{B}$ at the same pitch angle θ, they have exactly the same dynamical behaviour. Obviously, by extension, the result remains the same for any magnetic field configuration. The quantity pc/ze is called the *rigidity* or *magnetic rigidity* of the particle. Since pc has the dimensions of energy and e has the dimensions of charge, pc/ze has the dimensions of volts. A useful unit for practical purposes is gigavolts (GV). In cosmic rays studies, the energies of cosmic rays are often quoted in terms of their rigidities rather than their energies per nucleon. It is useful to compare the various ways of describing the energies of protons, carbon and iron nuclei with $\gamma = 2$ and 100 (Table 11.1).

11.2 A time-varying magnetic field

11.2.1 *Physical approach to the non-relativistic case*

In the configuration depicted in Fig. 11.1 the particle moves in a spiral path, the radius of which is $r = (\gamma m_0 v \sin\theta)/zeB$. An important case is that in which the magnetic field strength $\mathbf{B}$ varies slowly with time. By this we mean that the fractional change in the magnetic field strength $\Delta B/B$ changes very little in a single orbital period $T = \nu_g^{-1}$. Let us work out the non-relativistic version of this problem in a way which brings out the essential physics.

A particle gyrating about the magnetic field is equivalent to a current loop. The current is simply the rate at which charge passes a particular point in the loop per second and therefore $i = zev_\perp/2\pi r$. The area of the loop is $A = \pi r^2$ and therefore the magnetic moment of the loop is

$$\mu = iA = \frac{zev\sin\theta}{2\pi r}\pi r^2 = \frac{zev_\perp}{2}r$$

Now, in the non-relativistic limit, $r = m_0 v_\perp/zeB$ and therefore

$$\mu = \frac{m_0 v_\perp^2}{2B} = \frac{w_\perp}{B} \qquad (11.7)$$

where $w_\perp$ is the kinetic energy of the particle in the direction perpendicular to the guiding centre.

Now let there be a small change ΔB in the magnetic field strength B during one orbit. Then, an emf is induced in the loop because of the changing magnetic field

Table 11.1. *The properties of protons, carbon and iron nuclei having Lorentz factors $\gamma = 2$ and 100*

	Protons		Carbon nucleus		Iron nucleus	
Lorentz factor, γ	2	100	2	100	2	100
Velocity, v	$(\sqrt{3}/2)c$	$0.99995c$	$(\sqrt{3}/2)c$	$0.99995c$	$(\sqrt{3}/2)c$	$0.99995c$
Mass number, A	1	1	12	12	56	56
Atomic number, z	1	1	6	6	26	26
Rest mass energy, mc^2	1 GeV	1 GeV	12 GeV	12 GeV	56 GeV	56 GeV
Total energy, γmc^2	2 GeV	100 GeV	24 GeV	1200 GeV	112 GeV	5600 GeV
Kinetic energy, $(\gamma-1)mc^2$	1 GeV	99 GeV	12 GeV	1188 GeV	56 GeV	5544 GeV
Kinetic energy per nucleon	1 GeV	99 GeV	1 GeV	99 GeV	1 GeV	99 GeV
Momentum, $pc = (\gamma mv)c^a$	$\sqrt{3}$ GeV	99.995 GeV	20.8 GeV	1199.9 GeV	96.99 GeV	5599.7 GeV
Rigidity, pc/ze	$\sqrt{3}$ GV	99.995 GV	$2\sqrt{3}$ GV	199.99 GV	3.73 GV	215.4 GV

[a] To obtain the dimensions of GeV, the momentum has been multiplied by c, the velocity of light.

and so the particle in its orbit feels an acceleration. The work done on the particle per orbit by this emf is just

$$ze\mathscr{E} = ze\pi r^2 \frac{dB}{dt} = ze\pi r^2 \frac{\Delta B}{\Delta T}$$

where $\Delta T = 2\pi r / v_\perp$ is the period of one orbit. Therefore, the change in kinetic energy of the particle in one orbit is just

$$\Delta w_\perp = \frac{ze\pi r^2}{2\pi r} v_\perp \Delta B = \frac{m_0 v_\perp^2}{2B} \Delta B = \frac{w_\perp}{B} \Delta B$$

What is the change in the magnetic moment of the current loop?

$$\Delta\mu = \frac{\Delta w_\perp}{B} - \frac{w_\perp \Delta B}{B^2} = \frac{\Delta w_\perp}{B} - \frac{\Delta w_\perp}{B} = 0 \tag{11.8}$$

i.e. the magnetic moment of the particle in its orbit is an *invariant* provided the field is slowly-varying. There are other useful ways of expressing this important result. $\Delta\mu = 0$ is equivalent to

$$\Delta(w_\perp / B) = 0 \tag{11.9}$$

Since $w_\perp = p_\perp^2 / 2m_0$, this in turn is the same as

$$\Delta(p_\perp^2 / B) = 0 \tag{11.10}$$

This is a particularly interesting result because it accounts for the phenomenon of *magnetic mirroring*. If the particle moves into a region of converging magnetic field lines, B increases and therefore $p_\perp^2$ must also increase. However, the kinetic energy of the particle is constant because no work is done by the magnetic field and therefore the increase in $p_\perp^2$ must take place at the expense of the parallel component of the particle's motion $p_\parallel^2$. Now $p_\parallel^2$ goes to zero at the point where $p_\perp^2 = p^2$. Thus, the particle is reflected back along the magnetic field configuration (Fig. 11.2). This is what happens to particles trapped in the radiation belts when they approach the converging field lines towards the magnetic poles of the Earth (see Fig. 10.12).

Finally, since $r = m_0 v_\perp / zeB$ and $p_\perp^2 = (zerB)^2$, $\Delta(p_\perp^2 / B) = 0$ implies that

$$\Delta(Br^2) = 0 \tag{11.11}$$

i.e. the particle follows the guiding centre in such a way that the number of field lines within the particle's orbit is a constant (see Fig. 11.3). These expressions are all called the *first adiabatic invariant* of the particle's motion in a magnetic field.

These relations can all be derived from the *principle of adiabatic invariance*. It is also the best way of deriving the relativistic generalisations of these formulae. These are:

$$\left.\begin{aligned} \Delta(Br^2) &= 0 \qquad r = \gamma m_0 v_\perp / zeB \\ \Delta(p_\perp^2 / B) &= 0 \qquad p_\perp = \gamma m_0 v_\perp \\ \Delta(\gamma\mu) &= 0 \qquad \mu = \gamma m_0 v_\perp^2 / 2B \end{aligned}\right\} \tag{11.12}$$

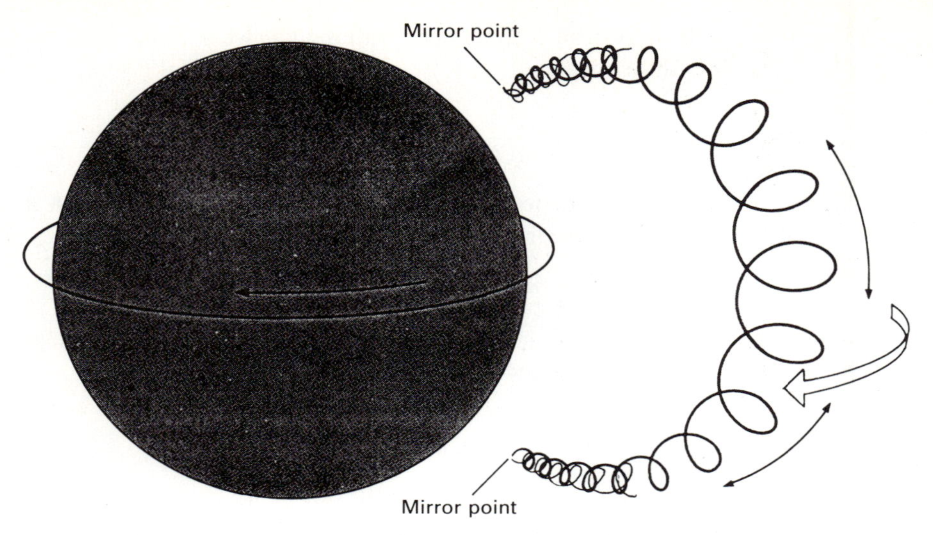

Figure 11.2. A schematic diagram illustrating the magnetic mirroring of charged particles in a dipole magnetic field. Trapping of this type accounts for the existence of the radiation belts in the Earth's ionosphere. (From H. Friedman (1986). *Sun and earth*, page 167, New York: Scientific American Books.)

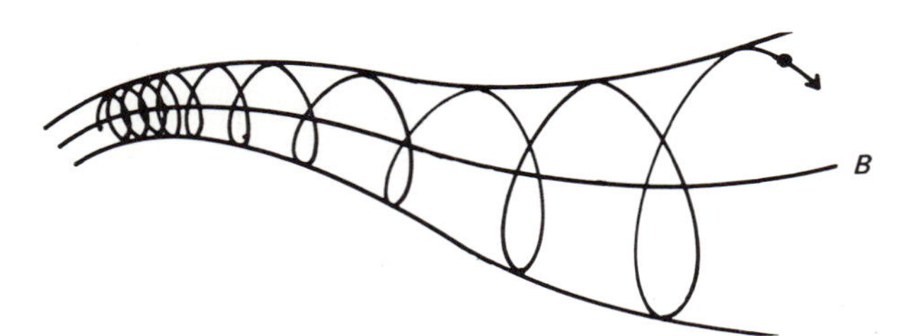

Figure 11.3. The dynamics of a charged particle in a slowly-varying magnetic field illustrating how the particle's guiding centre follows the mean magnetic field direction. The radius of curvature of the particle's path is such that a constant magnetic flux is enclosed by its orbit.

If you are happy to accept these results, then you may proceed to Section 11.3. If not, continue with the next subsection.

11.2.2 *Adiabatic invariant approach*

This procedure uses the Lagrangian formulation of classical dynamics and we follow Jackson's approach (1975, pp. 588–93). We start from the result of classical dynamics that if q_i and p_i are generalised canonical coordinates and momenta, then for each coordinate that is periodic, the action integral J_i defined by $J_i = \oint p_i \, dq_i$ is a constant for a given mechanical system with specified initial conditions. If the properties of the system change only slowly compared with the period of oscillation, the action integral is an invariant. Such a change is called an

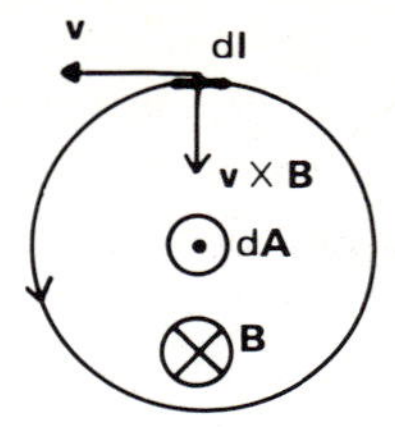

Figure 11.4. Illustrating how to find the sign of the increment of magnetic flux in evaluating the action integral J.

adiabatic change. This is exactly what we need to investigate the dynamics of a charged particle moving in a magnetic field.

The components of velocity and position perpendicular to the magnetic field direction are both periodic. Thus, the action integral is

$$J = \int \mathbf{P}_\perp \cdot \mathbf{dl} \qquad (11.13)$$

where $\mathbf{P}_\perp$ is the canonical momentum of the particle perpendicular to the magnetic field direction and $\mathbf{dl}$ is the line element along the circular path of the particle. The important result we need is that, for a charged particle in a magnetic field, the canonical momentum perpendicular to the field is

$$\mathbf{P}_\perp = \mathbf{p}_\perp + e\mathbf{A}$$

where $\mathbf{p}_\perp$ is the relativistic three-momentum of the particle perpendicular to $\mathbf{B}$ and $\mathbf{A}$ is the vector potential of the magnetic field, $\mathbf{B} = \nabla \times \mathbf{A}$. Therefore

$$\begin{aligned} J &= \oint_C \mathbf{P}_\perp \cdot \mathbf{dl} \\ &= \oint_C \mathbf{p}_\perp \cdot \mathbf{dl} + e \int_C \mathbf{A} \cdot \mathbf{dl} \\ &= \oint_C \gamma m_0 \mathbf{v}_\perp \cdot \mathbf{dl} + e\int_S \mathbf{B} \cdot \mathbf{dS} \\ &= 2\pi r \gamma m_0 v_\perp + e\int_S \mathbf{B} \cdot \mathbf{dS} \end{aligned} \qquad (11.14)$$

$\mathbf{dS}$ is the vector area associated with the line integral $\oint \mathbf{dl}$ around C. Let us look at the vector relations between $\mathbf{dl}$, $\mathbf{B}$ and $\mathbf{dS}$. If the $\mathbf{B}$ is directed into the paper and $\mathbf{v}$ has the direction shown in Fig. 11.4 the Lorentz force ($\mathbf{v} \times \mathbf{B}$) for a positively charged particle results in circular motion as shown. The vector area $\mathbf{dS}$ consequently points out of the paper, i.e. in the opposite direction to $\mathbf{B}$. Thus, the second term in equation (11.14) is negative. Therefore,

$$J = 2\pi r^2 \gamma m_0 \omega - e\pi r^2 B$$

But the angular gyrofrequency $\omega = eB/\gamma m_0$ and hence

$$J = e\pi r^2 B = eAB$$

where A is the area swept out by the particle. According to the above rule, this is a constant for changes in $\mathbf{B}$ which are slow compared with the period of the particle in its orbit, i.e.

$$\Delta(\pi r^2 B) = 0 \qquad (11.15)$$

This is the same result quoted in equation (11.11) and the other invariants follow immediately from the relations $r = \gamma m_0 v_\perp / eB$, $p_\perp = \gamma m_0 v_\perp$ and $\mu = \gamma m_0 v_\perp^2 / 2B$.

We could go on and work out the behaviour in more and more complicated cases – what happens when the particles are in regions where there is a magnetic field gradient, what is the effect of a gravitational field, etc? However, I think the point will now be clear that individual particles are tied to magnetic field lines and it takes a great deal to make them move across. For those interested in more details of the complexities which can arise, Northrop's monograph (1973) provides an excellent introduction.

We can see that these results for individual particles closely parallel those derived in Section 10.5 concerning flux freezing. They are, however, separate problems, although the treatments I have given make them look rather similar. The point is that the flux freezing argument is fundamentally a magneto-hydrodynamic process in which we treat the plasma as a perfectly conducting fluid. The treatment using individual particles, which I have just given, is a microscopic approach and to rationalise the two approaches one has to show that the equations of magnetohydrodynamics can be derived from the microscopic equations of motion. This is far from trivial (see the excellent chapter in Clemmow and Dougherty's *The electrodynamics of particles and plasmas* (1969) Chapter 11).

11.3 The diffusion of high energy particles in the interplanetary medium

An ideal environment in which to test the theory of the propagation of high energy particles in magnetic fields under conditions which approximate to typical cosmic conditions is the interplanetary medium. We have argued in Section 10.5.3 that the magnetic field configuration should be that of an Archimedean spiral and this has been confirmed by direct measurements of the interplanetary magnetic field. We therefore expect high energy particles to spiral in towards the Earth along the magnetic field lines against the outflowing Solar Wind. Unfortunately, this picture provides no explanation of the modulation of the flux of cosmic rays as a function of the level of Solar activity. The answer is that we have considered an idealised structure for the magnetic field in the Solar Wind. On top of the overall spiral pattern, there are small scale fluctuations in the field associated with the outflow of the Solar Wind. In addition, there are shocks associated with high velocity streams initiated by high energy events in the solar atmosphere such as solar flares. Thus, the particles have to make their way towards the Earth through a magnetic field structure which possesses both large and small-scale irregularities.

There is an ongoing debate about the relative importance of large and small-scale irregularities in determining the modulation of the flux of high energy particles. The 'conventional' model of solar modulation assumes that the process is associated with small-scale irregularities in the Solar Wind and that the diffusion of the particles along and across the magnetic field lines can be described by a diffusion tensor κ_{ij}. The basis for this model was described in a classic paper by

Parker (1965). According to McKibben (1986), it is generally believed that this approach provides a correct local description of the physics of the diffusion of high energy particles from the interstellar medium to the Earth but the constants of the diffusion equation may be a function of time and location within the cavity evacuated by the Solar Wind. Let us look at the relevant physics involved.

11.3.1 *The diffusion equation*

First of all, let us consider the simplest one-dimensional version of the diffusion problem. The flux of particles N through unit area in the positive x-direction is just

$$N = -\kappa \, \partial n/\partial x \tag{11.16}$$

where n is the number density of particles and κ is a scalar diffusion coefficient. Opposing the inward diffusion is the outward convection of the particles by the Solar Wind. The flux of particles associated with this flow is just nv where v is the velocity of the Solar Wind and hence the net flux of particles through unit area at any point is just

$$N = nv - \kappa \, \partial n/\partial x \tag{11.17}$$

Therefore, in the steady state, $N = 0$ and hence, assuming the velocity of the wind to be constant,

$$nv = \kappa \, \partial n/\partial x$$

$$n = n_0 \exp\left[\frac{v}{\kappa}(x-R)\right] \tag{11.18}$$

where n_0 is the number density of particles at distance R through the medium which may be taken to be the boundary between the interstellar medium and the Solar Wind. Thus, the number density of particles at the origin $x = 0$ is reduced by a factor $\exp(-vR/\kappa)$ as compared with that at distance R. We will find exponentials of this form occurring throughout the theory.

Now, in fact, we have to deal with a three-dimensional problem in which particles diffuse inwards towards the Sun from interstellar space. We therefore have to rewrite equation (11.16) in spherical polar coordinates. We also have to write the diffusion of particles in the i-direction in terms of a diffusion tensor κ_{ij} which describes the scattering of particles into the i-direction from the j-direction. We can therefore write

$$N_i = nv_i - \kappa_{ij} \, \partial n/\partial x_j \tag{11.19}$$

where we use the usual summation convention over repeated indices. Now we take the divergence of this equation and find

$$\frac{\partial N_i}{\partial x_i} = \frac{\partial}{\partial x_i}(nv_i) - \frac{\partial}{\partial x_i}\left(\kappa_{ij}\frac{\partial n}{\partial x_j}\right)$$

We can write the divergence of the flux of particles in terms of their number density n. By the usual procedure of considering the flow of particles into and out of an elementary volume $dx\,dy\,dz$, we can readily show that

$$\operatorname{div}\mathbf{N} = \partial N_i/\partial x_i = -\partial n/\partial t$$

Therefore, we can write the diffusion equation in the form

$$\frac{\partial n}{\partial t}+\frac{\partial}{\partial x_i}(nv_i)-\frac{\partial}{\partial x_i}\left(\kappa_{ij}\frac{\partial n}{\partial x_j}\right) = 0 \tag{11.20}$$

We can now write this equation explicitly in terms of spherical polar coordinates. We make the simplifying assumptions that the flow is spherically symmetric, that we can adopt a scalar diffusion tensor $\kappa_{ij} = \kappa$ and that the velocity is constant. Then, (see Longair (1984), page 57 if necessary)

$$\frac{\partial n}{\partial t}+\frac{v}{r^2}\frac{\partial}{\partial r}(r^2 n)-\frac{1}{r^2}\frac{\partial}{\partial r}\left(\kappa r^2\frac{\partial n}{\partial r}\right) = 0 \tag{11.21}$$

Notice that this is not the complete diffusion equation since we have not taken account of energy gains or losses of the particles. We will take this subject up in the next sub-section.

Again, it is useful to look at the steady-state solution, $\partial n/\partial t = 0$. Then,

$$\frac{\partial}{\partial r}(r^2 n) = \frac{\partial}{\partial r}\left(\frac{\kappa}{v}r^2\frac{\partial n}{\partial r}\right)$$

i.e.

$$\frac{\partial n}{\partial r} = \frac{v}{\kappa}n, \tag{11.22}$$

exactly the same result as in the one-dimensional case.

11.3.2 *Adiabatic deceleration of the high energy particles*

In the case of a spherical wind, there is one further effect which has to be taken into account. This is the fact that the high energy particles diffuse in from the interstellar medium but, at the same time, they are convected outwards in an expanding flow. They are therefore subject to 'adiabatic deceleration'. We now derive suitable forms for this energy-loss process.

Let us first derive the important results for a non-relativistic Maxwellian gas. The loss of internal energy U of the gas when it does work in expanding its volume by dV is

$$dU = -p\,dV$$

where p is the pressure of the gas. Now, for a perfect gas, $U = \frac{3}{2}nkTV$ and $p = nkT$ where n is the number density of particles and T is their temperature. Since the average energy of each particle is $\frac{3}{2}kT$, we find

$$dU = nV\,dE = -\tfrac{2}{3}nE\,dV$$

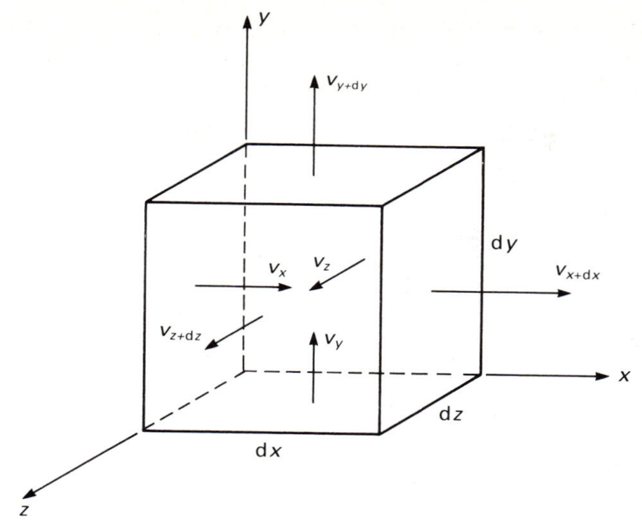

Figure 11.5.

Therefore setting the total number of particles $nV = N$, we find

$$\frac{dE}{dt} = -\frac{2}{3}\frac{nE}{N}\frac{dV}{dt} \tag{11.23}$$

Now dV/dt is just the rate of expansion of the volume V which is determined by the velocity field $\mathbf{v}(\mathbf{r})$. If we consider the volume of a cube of sides dx, dy, dz moving with the flow, we can go through the standard procedure for finding the change in volume (Fig. 11.5). We add together the changes in volume due to the differential velocities across each of the three pairs of faces of the cube i.e.

$$\frac{dV}{dt} = (v_{x+dx} - v_x)\,dy\,dz + (v_{y+dy} - v_y)\,dx\,dz + (v_{z+dz} - v_z)\,dx\,dy$$

Making the usual Taylor expansions, we find

$$\frac{dV}{dt} = \left(\frac{\partial v_x}{\partial x} + \frac{\partial v_y}{\partial y} + \frac{\partial v_z}{\partial z}\right) dx\,dy\,dz = (\nabla\cdot\mathbf{v})V \tag{11.24}$$

Substituting into (11.23), we find

$$\begin{aligned}\frac{dE}{dt} &= -\frac{2}{3}\frac{nV}{N}E(\nabla\cdot\mathbf{V})\\ &= -\tfrac{2}{3}(\nabla\cdot\mathbf{v})E\end{aligned} \tag{11.25}$$

This is the general expression for the energy loss rate due to adiabatic losses of a non-relativistic particle in an expanding flow. We can write this alternatively in terms of the momentum p of the particle, since $E = p^2/2m$, in which case

$$dp/dt = -\tfrac{1}{3}(\nabla\cdot\mathbf{v})p \tag{11.26}$$

The generalisation to the relativistic case is straightforward if we follow exactly the same procedure but use instead $U = 3nkTV$ and $p = \frac{1}{3}U$. It is straightforward to show that, in the ultrarelativistic limit,

$$\mathrm{d}E/\mathrm{d}t = -\tfrac{1}{3}(\nabla\cdot\mathbf{v})E \qquad (11.27)$$

$$\mathrm{d}p/\mathrm{d}t = -\tfrac{1}{3}(\nabla\cdot\mathbf{v})p \qquad (11.28)$$

Thus, in addition to diffusing in towards the Earth, the particles lose energy by being convected outwards in the expanding wind.

Before considering that case, let us show that this expression reduces to a possibly more familiar expression for the adiabatic expansion of a uniform gas sphere in both the relativistic and non-relativistic ases. Suppose the particles partake in the uniform expansion of a spherical volume which could be a supernova remnant or the Universe itself! In such a spherical expansion $v = v_0(r/r_0)$ where the outer radius r_0 of the volume expands at velocity v_0. Taking the divergence of $\mathbf{v}$ in spherical polar coordinates, we find

$$\nabla\cdot\mathbf{v} = \frac{1}{r\sin^2\theta}\left[\frac{\partial}{\partial r}(r^2\sin\theta\, v_r)\right] = 3\left(\frac{v_0}{r_0}\right)$$

Thus, for non-relativistic particles we find

$$\frac{\mathrm{d}E}{\mathrm{d}t} = -\frac{2v_0}{r_0}E$$

$$\frac{\mathrm{d}E}{\mathrm{d}r_0} = -\frac{2E}{r_0} \qquad E = E_0\left(\frac{r_0}{r}\right)^2 \qquad (11.29)$$

In the same way, for ultrarelativistic particles

$$\frac{\mathrm{d}E}{\mathrm{d}r_0} = -\frac{E}{r_0}$$

$$E = E_0\left(\frac{r_0}{r}\right) \qquad (11.30)$$

We will find that this relation describes the cooling of ultrarelativistic particles in the expansion of spherical objects and leads to the famous 'adiabatic loss' problem in non-thermal radio sources.

In the case of the Solar Wind, we can approximate the velocity field by $v = v_0$ constant, independent of radius from the Sun. In this case,

$$\nabla\cdot\mathbf{v} = 2v_0/r$$

and hence, substituting into the expression (11.25) for a non-relativistic gas,

$$\frac{1}{E}\frac{\mathrm{d}E}{\mathrm{d}t} = -\frac{4}{3}\frac{v_0}{r}$$

$$\frac{\mathrm{d}E}{E} = -\frac{4}{3}\frac{\mathrm{d}r}{r}$$

$$E = E_0\left(\frac{r_0}{r}\right)^{\frac{4}{3}} \qquad (11.31)$$

Similarly, for an ultrarelativistic gas, we find

$$\frac{1}{E}\frac{\mathrm{d}E}{\mathrm{d}t} = -\frac{2}{3}\frac{v_0}{r}$$

$$E = E_0\left(\frac{r_0}{r}\right)^{\frac{2}{3}} \tag{11.32}$$

This calculation shows that particles flowing out from the Sun suffer large energy losses because of adiabatic deceleration. In the case of relativistic particles diffusing in from the interstellar medium, the losses are not so severe since they diffuse at a velocity close to the velocity of light and there is not time for energy losses due to convection to have been too great by the time the particles reach the Earth.

Let us indicate the nature of this result in the following way. The solution (11.18) shows how the number density of particles is expected to decrease as they diffuse into the inner Solar System from the interstellar medium. For particles with energy ~ 1 GeV nucleon^{-1}, the value of $v_0 R/\kappa$ must be of order 1 to account for the amplitude of the modulation seen from solar maximum to solar minimum. In this case, we can work out the time the particle takes to reach the inner Solar System. From standard diffusion theory $\kappa = \frac{1}{3}\lambda v$ where λ is the mean free path and v the velocity of the particle which in our case can be taken to be the velocity of light. Then, the time to diffuse a distance r is just the time it takes to make N scatterings such that $r = N^{\frac{1}{2}}\lambda$, i.e.

$$t = N\lambda/v = r^2/3\kappa \tag{11.33}$$

Now the time for the particle to lose roughly half its energy is

$$T = E/(\mathrm{d}E/\mathrm{d}t) = 3r/2v_0$$

Taking $r = R$, we see that

$$\frac{T}{t} = \left(\frac{3r}{2v_0}\right)\bigg/\left(\frac{r^2}{3\kappa}\right) = \frac{9}{2}\frac{\kappa}{rv_0} \tag{11.34}$$

But we have argued that $\kappa/Rv_0 \sim 1$ and hence the particle does not, in fact, lose much of its energy during the time it takes to diffuse into the inner Solar System.

Thus, the adiabatic losses are relatively small for this case but need to be taken into account in a full treatment of this problem. To include the adiabatic loss term into the diffusion equation we have already derived in (11.21), we can use the same formulation we develop in Volume 2, Chapter 19 to describe the diffusion-loss equation for relativistic electrons. Anticipating the results of that section, we can show that it is necessary to include a term of the form

$$+\frac{\partial}{\partial E}[b(E)\,n(E)]$$

in the equation where $b(E) = -(\mathrm{d}E/\mathrm{d}t)$ is the energy loss rate of the particle. Thus, the full diffusion-loss equation is of the form

$$\frac{\partial n}{\partial t} + \frac{v}{r^2}\frac{\partial}{\partial r}(r^2 n) - \frac{2}{3}\frac{v}{r}\frac{\partial}{\partial E}(nE) - \frac{\kappa}{r^2}\frac{\partial}{\partial r}\left(r^2\frac{\partial n}{\partial r}\right) = 0 \tag{11.35}$$

This is the standard form of the equation used to describe the diffusion of high energy particles into the Solar System and was derived by Parker in his classic paper of 1965 (his equation (4)). If anisotropic diffusion across field lines is also taken into account, the equation can be written in the more general form

$$\frac{\partial n}{\partial t}+\frac{\partial}{\partial x_i}(v_i n)+\frac{\partial}{\partial E}[nb(E)]-\frac{\partial}{\partial x_i}\left(\kappa_{ij}\frac{\partial n}{\partial x_j}\right)=0 \qquad (11.36)$$

where the derivatives are in spherical polar coordinates and n describes the number density of particles with energies in the range E to $E+\mathrm{d}E$.

11.3.3 *Empirical diffusion–convection model for solar modulation*

Ideally, one would like to set about solving the general diffusion–convection equation (11.35) for solar modulation. This can be done but it is also useful to adopt a more empirical approach by deriving a model which can account for the observed effects of solar modulation. This model will act as a guide for the formulation of theoretical models of the physical processes involved in the solar modulation.

Figs 10.3 (*a*) and (*b*) show the types of data which have to be accounted for. We will discuss the results of modelling by Evenson *et al.* (1983) and Garcia-Munoz, Pyle and Simpson (1985) as an example of what can be achieved by this approach. The model makes the following assumptions. First of all, it is assumed that the modulation can be described by a steady-state model so that $\mathrm{d}n/\mathrm{d}t = 0$ and also that the modulation volume can be assumed to be spherically symmetric about the Sun. Drifts and non-stationary phenomena such as high velocity streams are neglected. The aim is to derive an empirical model which takes account of convection, diffusion and adiabatic deceleration. This is achieved by building the dependences upon energy, rigidity, location within the Solar Wind, etc. into the diffusion coefficient κ so that the number density of particles of a particular energy or rigidity is given by the simple solution of the diffusion equation (11.22) which can be written

$$n = n_0 \exp\left[\int_{r_0}^{r} \frac{V}{\kappa(r)}\,\mathrm{d}r\right] \qquad (11.37)$$

where V is the velocity of the Solar Wind and r_0 is the outer radius of the modulation volume. Thus, at any energy, if we assume that the velocity of the Solar Wind is constant, which is known to be a good approximation, the number density of particles is determined by an integral over the empirical diffusion coefficient. In addition, we know that we can write the diffusion coefficient in terms of the velocity of the particle v and its mean free path λ as $\kappa = \frac{1}{3}\lambda v$. Thus, the problem reduces to finding suitable forms for the mean free path of the particles as a function of energy or rigidity. We have argued that the dynamics of the particles in magnetic fields should be primarily a function of their rigidities since particles of the same rigidities should experience identical accelerations in any

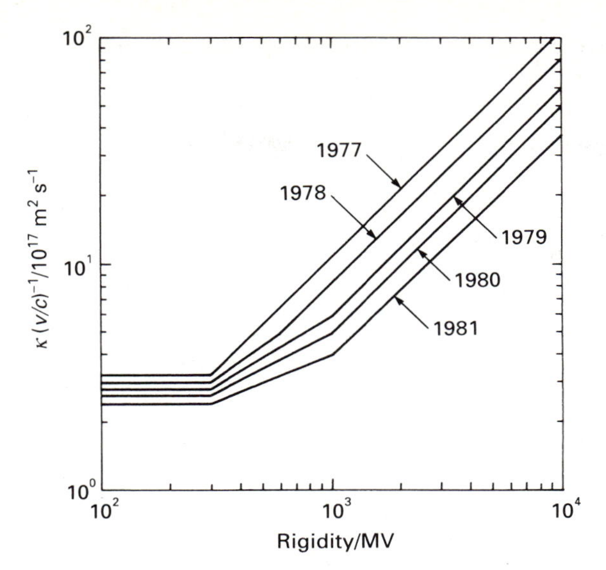

Figure 11.6. The rigidity dependence of the diffusion coefficient κ for the period 1977–81 derived from the empirical diffusion–convection model described in the text. (From M. Garcia-Munoz, K. R. Pyle and J. A. Simpson (1985). *19th International cosmic ray conference*, La Jolla, USA, Vol. 4, page 409.)

magnetic field configuration (see Section 11.1). In the approach of Evenson *et al.* (1983), it is assumed that the dependences of κ upon rigidity and distance from the Sun are mathematically separable. The effects of adiabatic deceleration are built into the dependence of the diffusion coefficient upon distance from the Sun r.

By analysing the variation of the energy spectra of protons and helium nuclei from solar minimum to solar maximum over the period 1977–81 as well as evidence on the modulation of high energy electrons by comparing the Galactic radio spectrum and local relativistic electron spectrum (see Section 9.1 and Volume 2, Sections 18.2 and 18.3), Evenson *et al.* (1983) and Garcia-Munoz *et al.* (1985) find that the diffusion coefficient κ can be satisfactorily described by the following forms:

$$\kappa(r) = \frac{v}{c} R^{\alpha} \kappa_0(r)$$

$$\kappa_0(r) = \kappa_0 \exp[(r-1)/29] \tag{11.38}$$

where $R = pc/ze$ is the rigidity of the particle and α is a spectral index which takes different values in different rigidity ranges; the distance r is measured in astronomical units. The exponential factor gives an empirical description of the effects of adiabatic deceleration. The adopted variations of κ_0 with rigidity are shown in Fig. 11.6 at different phases of the solar cycle. The authors emphasise that what is observable is the integral over the adopted form of the diffusion coefficient $\kappa(r)$ and hence a variety of different forms could be consistent with the experimentally observed spectra.

Fig. 11.6 may be interpreted as showing the variation of the mean free path of the high energy particles with rigidity within the Solar System. As expected, particles with the highest rigidities have the longest mean free paths and the particles with small rigidities have the shortest. The functions at different phases of the Solar cycle show how the mean free path of the particles change as the degree of turbulence in the interplanetary medium changes. We will attempt to explain these relations by elementary theory in the next sub-section.

According to the Evenson *et al.* and Garcia-Munoz *et al.*, these forms of diffusion coefficient can successfully account for essentially all the observations of electrons, protons and helium nuclei and their relative degrees of modulation over the period 1977–81 during which the degree of Solar activity changed markedly.

11.4 Diffusion coefficients for high energy particles scattered by magnetic irregularities

Let us make some simple calculations to demonstrate how high energy particles propagating in an irregular magnetic field are scattered and how order of magnitude estimates can be made of the diffusion coefficient κ. This is the type of calculation one would like to be able to make for the interstellar medium in our own and other galaxies but unfortunately we do not have access directly to the relevant experimental data about the spectrum of magnetic irregularities.

11.4.1 *The power spectrum of irregularities in the Solar Wind*

Fortunately, in the case of the Solar Wind, direct measurements of the spectrum of irregularities in the interplanetary magnetic field have been made through magnetometer observations from space probes. A good example of these types of measurement is provided by the Mariner 4 space probe which went on to take the first pictures of the Martian surface. The strength of the magnetic field was measured continuously throughout the flight from the Earth to Mars. We need to know the strength of the magnetic irregularities as a function of physical scale and so we adopt the standard procedure of taking the power spectrum of the magnetic field strength as a function of wavelength. To do this, we use the same procedure as in Section 3.3.5 – Parseval's Theorem enables us to find the power spectrum of the fluctuations in the magnetic field from the Fourier transform of the time variation of the interplanetary magnetic field strength as follows:

$$\int_{-\infty}^{\infty} B^2(t)\,\mathrm{d}t = \int_{-\infty}^{\infty} B^2(\omega)\,\mathrm{d}\omega \tag{11.39}$$

where $B(\omega)$ is the Fourier transform of the observed distribution of the magnetic field strength with time, $B(t)$. Note that we use the same definitions of the Fourier transform pairs as in Section 3.3.5. The frequencies have been referred to a frame of reference in which the Sun is stationary and therefore the corresponding

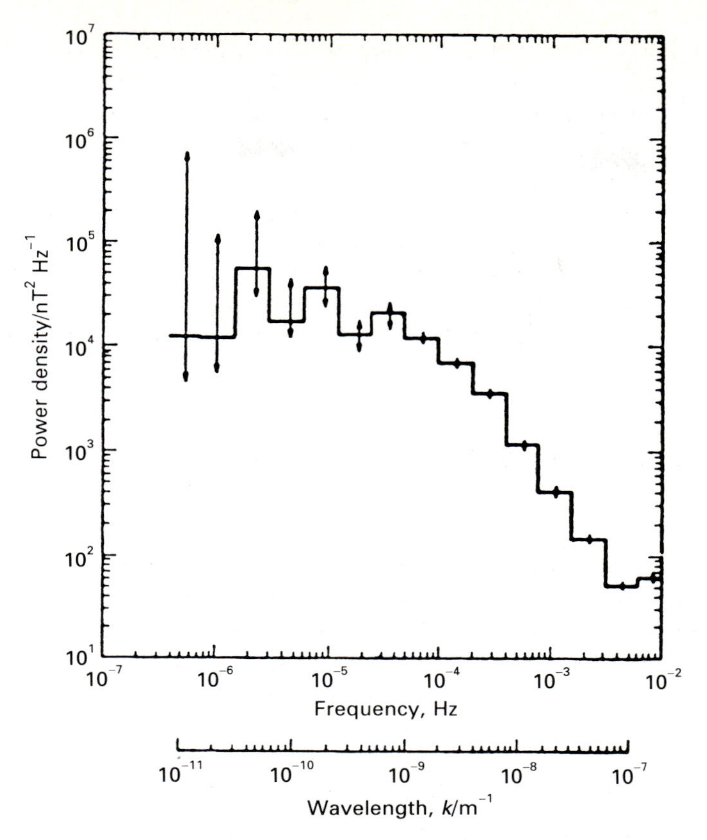

Figure 11.7. The power spectrum of the magnetic field energy density per unit frequency interval as measured by the magnetometers on board the Mariner 4 spacecraft. The strength of the magnetic field is measured in nanoteslas, 1 nT = 10^{-9} T which is the same as the traditional unit used in these studies, 10^{-5} G, which is known as one 'gamma' or γ. (From J. R. Jokipii (1973). *Ann. Rev. Astr. Astrophys.*, **11**, 1.)

physical scales are $\lambda = V/\nu$ where V is the velocity of the Solar Wind and ν the frequency of the Fourier component. In these early observations, the spectrum displayed in Fig. 11.7 could be described by the following relations

$$\left.\begin{array}{ll} B^2(\omega) \propto \omega^{-\frac{3}{2}} & \omega > \omega_c \\ \mathrm{B}^2(\omega) = \text{constant} & \omega < \omega_c \end{array}\right\} k_c = \frac{\omega_c}{V} \tag{11.40}$$

where the critical wavenumber at which the spectrum turns over is $k_c = 6 \times 10^{-10}$ m^{-1}. We now derive an estimate for the mean free path as a function of rigidity for high energy particles propagating through this irregular magnetic field.

We know that, if the high energy particles have gyroradii much smaller than the scale of the fluctuations in the magnetic field, the particles simply follow their guiding centre motion and only change their pitch angles as a result of conserving

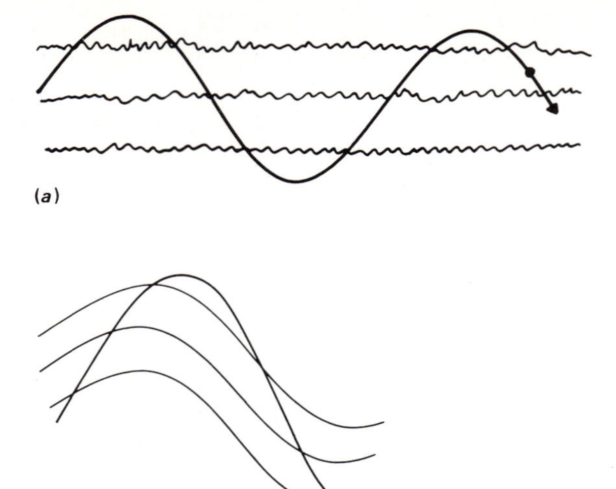

Figure 11.8. Illustrating the dynamics of a charged particle in a magnetic field, (*a*) when the irregularities in the magnetic field are on a scale which is much smaller than the gyroradius of the particle's orbit in the mean magnetic field; (*b*) when they are of the same order.

their adiabatic invariants (see Section 11.2). In the opposite limit in which the particles have gyroradii much greater than the scale of the fluctuations, the particles do not 'feel' the fine structure but move in orbits determined by the mean magnetic field which, in the case of the Solar Wind, is much greater in magnitude than the fluctuating component. Thus, it is only in the intermediate case in which the fluctuations have the same scale as the gyroradii of the particles that there is significant scattering. The process is illustrated schematically in Fig. 11.8 It can be seen from the figure that there is a significant change in the pitch angle of the particle in a single gyroradius and it is this random scattering process which can result in the scattering of the high energy particles.

These expectations are borne out by more detailed calculations. We can therefore associate a particular magnetic rigidity with each physical scale in the power spectrum of magnetic irregularities and it is these fluctuations which provide the most important scatterers for the particles in pitch angle. To illustrate that this idea is at least consistent with the observations of the power spectrum of the magnetic irregularities, let us work out the rigidity at which we would expect a break in the spectrum of the magnetic fluctuations to be reflected in the modulated spectrum of cosmic rays. First, we write down the gyroradius of the particle in terms of its magnetic rigidity

$$r_{\mathrm{g}} = \left(\frac{pc}{ze}\right)\frac{1}{Bc} = \frac{R}{Bc} \tag{11.41}$$

We now equate this value to the wavelength at which the break in the power spectrum of magnetic irregularities is observed. Taking $r = \lambda_c = 2 \times 10^9$ m and $B = 3$ nT, we find $R = 2$ GV. It can be seen that this value is remarkably close to the change in slope of the mean free path of the cosmic rays as determined by the empirical diffusion–convection model of Fig. 11.6. Let us now make an order of magnitude calculation for the mean free path of high energy particles in the interplanetary magnetic field to see if they are of the observed order.

11.4.2 *The diffusion coefficients for high energy particles in the Solar Wind*

The evaluation of the diffusion coefficients for high energy particles given the spectrum of magnetic irregularities is an interesting problem in the theory of stochastic processes and detailed calculations have been carried out by Jokipii (1971). In the spirit of our present discussion, we will indicate the essence of the calculation by an order of magnitude calculation.

The important assumption is that the magnetic field irregularities are random. The power spectrum of the magnetic field strength tells us how much energy there is in each Fourier component of the field and it is implicit in this procedure that the phases of the waves are random. What this means physically is that each particle 'feels' the influence of a particular field component for only about one wavelength before it encounters another wave with a random phase relative to the last wave. Our model of the diffusion process is therefore one in which the particle experiences any given wave for about one wavelength before it is scattered by another wave of random phase.

Now, in a single wavelength, the average inclination of the field lines from the mean field direction due to the magnetic irregularities is $\phi \approx B_1/B_0$ where B_0 is the strength of the mean magnetic field and B_1 is the amplitude of the random component. Therefore, the pitch angles of particles with gyroradii $r_g \approx \lambda$ must change by about this amount per wavelength. Let us look at the change in pitch angle from the point of view of the change in the guiding centre of the particle in one wavelength. The guiding centre is displaced by a distance $r \approx \phi r_g$ and this represents diffusion of the particles across the magnetic field lines as well as a change in their pitch angles. Thus, scattering by magnetic irregularities enables particles to diffuse across magnetic field lines.

In the next wavelength, the particle meets another wave of about the same energy density but the change in pitch angle is now random with respect to the previous wave and so the particle is scattered randomly in pitch angle. Therefore, to be scattered randomly through 1 rad, N scatterings are required where $N^{\frac{1}{2}}\phi = 1$. The distance necessary for scattering through 1 rad is thus $\lambda_{sc} \approx N\lambda \approx Nr_g \approx r_g\phi^{-2}$. This is the effective mean free path of the particle diffusing along the magnetic field in the Solar Wind. In this distance, the pitch angle of the particle has been changed by a large factor so that, just as in the case of a charged particle in a plasma considered in Section 10.4, the particle loses all memory of its initial pitch angle in this distance. This is our measure of the mean free path λ_{sc} of the particle in the interplanetary medium.

We can now combine this result with the spectrum of irregularities in the interplanetary magnetic field to work out the mean free path as a function of magnetic rigidity R. All we have to do is to work out the energy in the irregularities on a particular scale. The only thing to beware of is that the power spectrum of the irregularities is quoted in terms of the power per unit frequency interval. What we need is the energy density of the magnetic field on a particular physical scale λ. To order of magnitude, this is given by multiplying by the power spectrum by the angular frequency of the waves. Therefore, the energy density in the magnetic field on scale $\lambda = v/\nu$ is

$$B_1^2(\omega)\omega/2\mu_0$$

We can now insert appropriate power-law approximations for $B_1^2(\omega)$. In the low frequency range, $\omega < \omega_c$

$$B_1^2(\omega) = \text{constant} \tag{11.42}$$

To relate this scale to the rigidity of the resonating particles, we recall that $\nu = v/\lambda$ where $\lambda \approx r_g = R/Bc$. Consequently, as the frequency decreases, the rigidity of the particles with which the waves 'resonate' increases. Thus, for high energy particles having $R > R_c$,

$$\lambda_{sc} \approx r_g/\phi^2 = r_g B_0^2/B_1^2(\omega)\omega \propto r_g^2 \propto R^2$$

At high frequencies, $\omega > \omega_c$, in exactly the same way,

$$B_1^2(\omega) \propto \omega^{-\frac{3}{2}}$$

and the mean free path for low energy particles $R < R_c$ is

$$\lambda_{sc} \approx r_g B_0^2/B_1^2(\omega)\omega \propto R^{\frac{1}{2}}$$

Results of exactly the same form come out of the detailed calculations by Jokipii (1973), including, to order of magnitude, the values of the numerical constants in front of the relations. These predictions of the dependence of the mean free path of high energy particles on magnetic rigidity are not dissimilar from those found from the empirical diffusion–convection model shown in Fig. 11.6. It is true that the exact dependences have not been found but our analysis has been based upon early observations. The encouraging aspect of the analysis is that the observed qualitative dependence of mean free path upon rigidity can be derived from this physical model of the modulation process.

One interesting example of the application of calculations of this type is the scattering of high energy particles streaming through the interstellar medium. In that case, the scattering is due to waves generated by streaming instabilities of high energy particles in the interstellar medium. We will find that this is an important calculation when we study the diffusion of cosmic rays in interstellar space.

11.5 High energy particles in the outer heliosphere

A natural consequence of Solar modulation is that, as we move out through the Solar System, we would expect the flux of particles with energies 0.1–1 GeV to increase as interstellar space is approached. This has been demonstrated beautifully by a combination of observations made by the Voyager 1 and 2 and Pioneer 10 and 11 spacecraft, the space probes which explored the outer Solar System (Fig. 11.9). It will be recalled that the Voyager space probes were responsible for taking the magnificent close-up images of Jupiter and the outer planets as is indicated in Fig. 11.9. The payloads of the Pioneer missions include particle detectors which are sensitive to protons with energies $E > 67$ MeV. It can be seen from the shape of the energy spectra in Fig. 9.1 that this means that all the particles over the flat peak of the energy distribution are detected, the mean energy of the particles detected being about 1 GeV. Fig. 11.10 shows the percentage radial gradient of the particle flux with distance from the Sun measured by the Pioneer 10 spacecraft while it travelled between about 10 and 45 AU from the Sun. The fluxes of particles were normalised to the fluxes observed in the vicinity of the Earth by the IMP-8 spacecraft. The increase in particle flux with distance from the Sun corresponds to a fractional increase of between about 2–3% AU^{-1}.

Adopting the steady state diffusion–convection model for the propagation of the particles, we can estimate their average mean free path since

$$D\frac{\mathrm{d}n}{\mathrm{d}r} = nV$$

where n is the number density of the high energy particles and D is their diffusion coefficient which can be written in terms of the mean free path λ as $D = \frac{1}{3}\lambda v$ where v is the velocity of the particles between collisions. The fractional gradient of cosmic rays with distance is therefore

$$\frac{1}{n}\frac{\mathrm{d}n}{\mathrm{d}r} = \frac{3V}{\lambda v}$$

Taking $V = 300$ km s^{-1} and v to be the velocity of light for illustrative purposes, we find $\lambda \approx 0.1$ AU. This mean free path is not so different from the values derived from the theoretical model of Section 11.4.3. Let us select the wavelength of the turn-over in the power spectrum of magnetic irregularities, $\lambda = 2 \times 10^9$ m, $\nu = 3 \times 10^{-5}$ Hz. The power spectrum of irregularities of this wavelength corresponds to 10^{-14} T^2 Hz^{-1}. Assuming the mean field in the Solar Wind is 3×10^{-9} T, we find the mean free path $\lambda \approx (B_0^2/\nu B_\nu^2)r_g \approx 0.3$ AU.

One important point about these observations is that there is still a significant amount of modulation of the high energy particle flux even at a distance of 45 AU. from the Sun. Several authors have shown that, as a result, the region responsible for the modulation must extend to at least 60 AU from the Sun.

This same set of observations produced many other important results

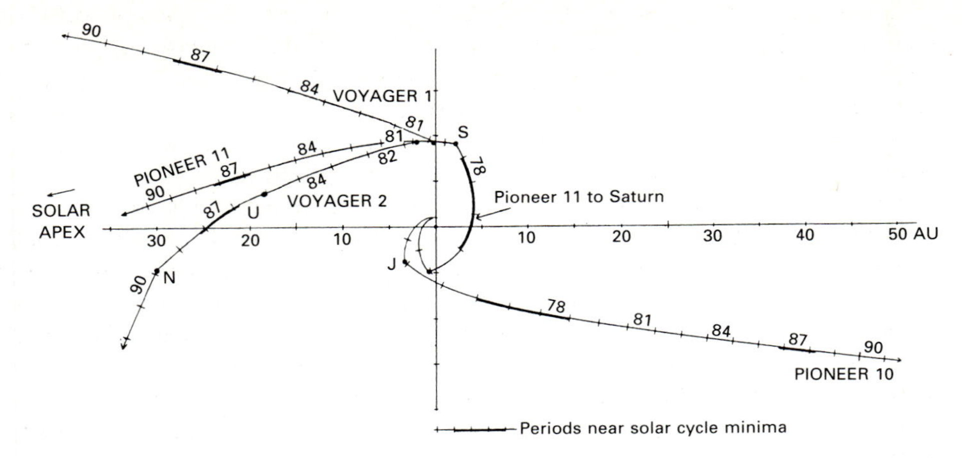

Figure 11.9. Trajectories projected onto the plane of the ecliptic of the space proves, Voyager 1 and 2 and Pioneer 10 and 11. The positions of the planets Jupiter (J), Saturn (S), Uranus (U) and Neptune (N) are shown when the different space craft had close encounters with them. Periods near the minima of the Solar cycle are indicated by heavy lines. (From J. A. Simpson (1989). *Adv. Sp. Res.*, **9**. No. 4, (4)5–(4)20.)

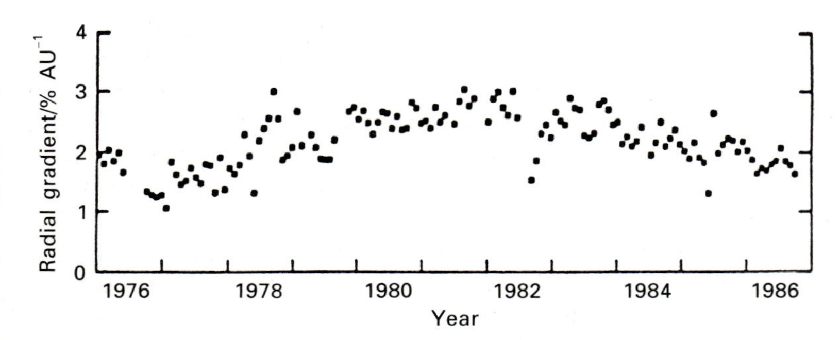

Figure 11.10. The percentage radial gradient of the number density of protons with energy $E \geqslant 67$ MeV normalised to the number density of protons with energies $E > 106$ MeV in the vicinity of the Earth as measured by the IMP-8 satellite. Each point is a 27-day average. (From J. A. Simpson (1989). *Adv. Sp. Res.*, **9**, No. 4, (4)11.)

concerning modulation processes in the interplanetary medium. For example, there are significant time variations in the degree of modulation in addition to the general correlation with the phase of the solar cycle. Some of these increases in modulation have been associated with increases in solar flare activity. In addition some of these changes in modulation have been observed both in the vicinity of the Earth and at much greater heliocentric distances. The cross-correlation of these changes has shown that the causes of the increase in modulation propagate at a velocity of about 300 km s^{-1}, i.e. roughly at the velocity of the Solar Wind.

Whilst the above arguments show that it is plausible that scattering by magnetic irregularities is a prime cause of the modulation of the flux of high energy particles entering the Solar System, it is likely that other factors play an important role in determining the local flux of high energy particles. These phenomena have been reviewed by McKibben (1986) and by Simpson (1989). As usual, this story can become as complicated as one desires and we will do no more than list some of the issues which are currently the subject of active investigation.

First of all, it is now apparent that the changes in modulation do not occur smoothly with the solar cycle. Rather, the modulation changes occur rather abruptly. These have been described as occurring 'in a series of steps separated by plateaus of nearly constant intensity' (McKibben 1986). The steps have been shown to move out through the interplanetary medium at the velocity of the Solar Wind. It has also been shown that the steps in the modulation are associated with high velocity transient flows and shocks in the Solar Wind and that, between these events, the Solar Wind structure appears to be determined by more-or-less steady state flows.

This strong association between shocks and increases in modulation has been confirmed by studies of the correlation of the magnetic field strength in the Solar Wind with intensity variations in the high energy particle fluxes and the local Solar Wind velocity (Figure 11.11). These data from the Voyager 2 experiment show that when the magnetic field strength increases due to the passage of shocks or other disturbances in the Solar Wind, the modulation of the high energy particle flux increases. Thus, it may well be that solar modulation is not a smooth continuous process but one in which the effects are associated with the superposition of successive regions of enhanced magnetic field which could be due to a variety of energetic phenomena. These regions have been referred to as 'merged interaction regions', to describe the amplification of the magnetic field which could be due to a variety of causes.

One of the major questions which can be addressed by these new data is the influence of the reversal of the Sun's magnetic polarity upon the process of modulation. If there were no small-scale irregularities in the magnetic field in the Solar Wind, the dynamics of charged particles would be determined entirely by the large-scale magnetic field distribution. We have given a qualitative description of recent studies of the magnetic field distribution in the Solar Wind in Section 10.7. It is an intriguing problem in the dynamics of charged particles in magnetic fields to work out the particle drifts which would be expected in the magnetic field structures suggested by Figs 10.7 and 10.8, especially when account has to be taken of the wavy inclined neutral sheet shown in Fig. 10.8(*b*). The results of calculations by Jokipii and his colleagues indicate that positively and negatively charged particles have completely opposite drifts in such a magnetic field structure. The nature of these particle drifts is indicated schematically in Fig. 11.12. Thus, when there is a reversal of the polarity of the Sun's magnetic field, as occurred in 1980, differences in the fluxes of positively and negatively charged high energy particles might be expected. This model has had some success in accounting for the increase in modulation with the tilt of the neutral current sheet but the relative magnitudes

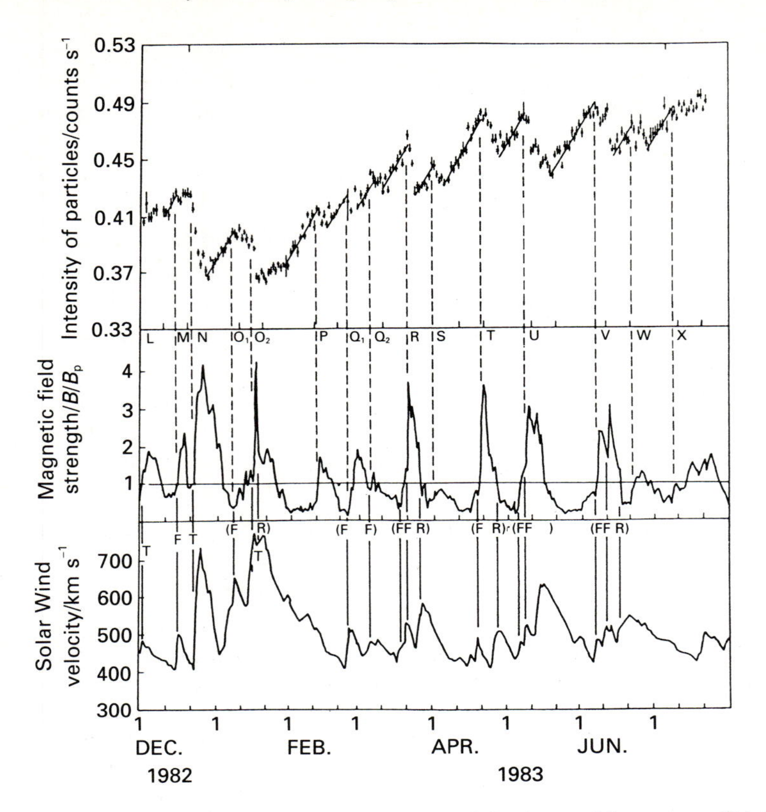

Figure 11.11. Voyager 2 observations of the integral intensity of high energy particles with energies greater than 75 MeV nucleon^{-1} compared with simultaneous measurements of the strength of the magnetic field and the velocity of the Solar Wind. It can be seen that there is a very strong correlation between decreases in the number density of particles and enhancements in the magnetic field intensity and the velocity of the Solar Wind. (From R. B. McKibben (1986). *The Sun and the heliosphere in three dimensions*, ed. R. G. Marsden, pages 361–74, Dordrecht: D. Reidel Publishing Co.)

of the radial intensity gradients for high energy particle nuclei before and after the solar polar reversal of 1980 were opposite to the predictions of this model.

The question of the relative importance of large-scale drifts as compared with diffusion in the conventional diffusion–convection model of solar modulation is of the greatest interest and will be the subject of intense study by the forthcoming Ulysses space mission which is a joint project of the ESA and NASA. The spacecraft will be launched into a trajectory which takes it out to Jupiter. Its trajectory has been chosen so that, as it sweeps past Jupiter, it will be deflected into an orbit which takes it back over the north pole of the Sun at high ecliptic latitudes. For the first time, it will be possible to measure directly the particle fluxes and magnetic field structure far out of the ecliptic plane. These observations will provide a wealth of information about the relative importance of drifts and scattering for high energy particles within the heliosphere.

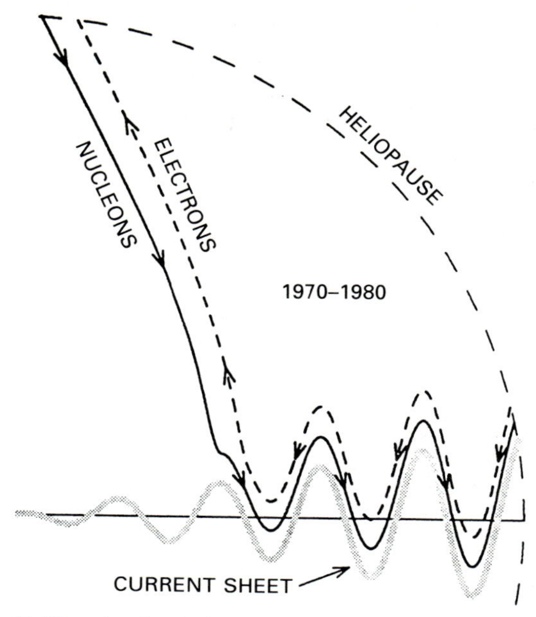

Figure 11.12. A schematic diagram illustrating the drifts expected for positively and negatively charged particles in the model for the heliosphere which includes the wavy current sheet. The drifts of the positively and negatively charged particles are expected to be in opposite directions. The drifts shown are expected when the north pole of the Sun has positive polarity, as was the case between 1970 and 1980. (From J. A. Simpson (1989). *Adv. Sp. Res.*, **9**, (4)13, after R. B. McKibben (1988). *The 6th Intl Solar Wind Conference*, eds V. J. Pizzo, T. E. Holzer and D. G. Sime, NCAR/TN-306, **2**, 615.)

11.6 Discussion

The above discussion provides an introduction to the many effects which can influence the propagation of high energy particles under conditions which must bear some resemblance to those found in other astrophysical environments. It is salutory to note the uncertainties and problems which arise, even when we are able to measure the particle fluxes and magnetic field strengths directly, circumstances which are excluded in any other astrophysical environment.

An example of the results of correcting for the effects of solar modulation upon the local flux of cosmic ray protons was shown in Fig. 9.1. This curve shows that the turn-over at low energies is largely due to the influence of solar modulation and that at energies less than about 1 GeV nucleon^{-1}, there is a considerable amount of energy present in high energy particles. Estimating the amount of modulation present at low energies is important for a number of astrophysical questions because this correction determines the total energy density of high energy particles in the Galactic disc. For example, this energy density directly influences the stability of the gaseous disc of the Galaxy and also the rate of heating of the interstellar gas by low energy cosmic rays. As discussed in Section 2.4.2, the ionisation losses of these low energy cosmic rays are important in heating cool gas clouds in which stars are forming and in determining the rate at which many chemical processes can proceed in cool molecular clouds. Unfortunately, the demodulation models are too uncertain to enable reliable estimates of the energy densities of the lowest energy cosmic rays to be made and astrophysicists invert the problem so that the heating rate due to these particles is determined from the requirements of the chemistry.

12

The high energy astrophysics of the Solar System

12.1 Introduction – a word of caution

This chapter is of an entirely different nature from the others in this volume. So far, we have tried to set out reasonably systematically the development of the tools needed to undertake studies in high energy astrophysics. This chapter is different in that I make no attempt to derive everything we will need. Rather, it is an essay which describes phenomena occurring within the Solar System which are of direct relevance to high energy processes occurring in Galactic and extragalactic astronomy.

The word of caution refers to the fact that these are areas which are normally far outside the interests of Galactic and extragalactic astronomers, including those of the present author. I therefore tread with some trepidation into what are for me uncharted areas. The story is, however, fascinating because, although there are important differences between physical conditions in the Solar System and, say, the environs of an active galactic nucleus, qualitatively similar phenomena are observed. For example, particles are accelerated to high energies in solar flares and they emit intense radio and X-ray emission. Similar phenomena are observed in active galaxies and quasars and it is an intriguing question whether or not these phenomena are related. The advantage of studying solar flares is that it is possible to study the acceleration of high energy electrons, protons and nuclei simulataneously and, in addition, the accelerated particles can be detected directly at the top of the Earth's atmosphere once they have travelled from their source in a solar flare to the Earth through the Solar Wind.

Some of the most important advances have resulted from observations with space missions dedicated to the study of the Sun, for example, the Skylab space laboratory launched in 1972, the Solar Maximum Mission (SMM) and the Japanese Hintoro mission. These have revealed high energy processes occurring on the surface of the Sun – for example, the emission of X and γ-rays from Solar flares as well as the products of nuclear interactions involving the high energy particles accelerated in flares. This chapter is intended to stimulate interest in these topics by astrophysicists whose interest normally only begin outside our own Solar System.

12.2 The atmosphere of the Sun and the solar corona

The structure of the atmosphere of the Sun and its corona is an excellent example of how wrong one's naive expectations can be. Intuitively, it might be expected that the atmosphere of the Sun would simply form the natural extension of its internal structure out to very large distances at which the envelope becomes highly rarefied. The internal structure of the Sun is now quite well understood, particularly as a result of the combination of theoretical studies of stellar structure with observations of solar oscillations, or helioseismology. Energy is generated within the core of the Sun principally by the pp-chain in which protons combine to form helium nuclei in the hot core of the Sun, its central temperature being about 1.6×10^7 K. These nuclear processes take place within the central 10% of the Sun by radius and are responsible for the total luminosity of the Sun which is 3.90×10^{26} W. The energy liberated in this process diffuses outwards through the Sun, the transport process being radiative diffusion in the inner 70% of the Sun (again by radius) and by convection in the outer 30%. The temperature gradient associated with the diffusion of energy outwards from the centre results in an effective temperature of about 5780 K at its visible surface. What we call the *atmosphere* of the Sun is the region from which photons can reach the observer on Earth without being scattered many times. This corresponds roughly to the layer at which the optical depth of the Sun has $\tau = 1$. We can think of the lower regions of the Sun's atmosphere as being a spherical shell from which the Sun's optical radiation was last scattered. This region is known as the *photosphere* and lies at a radius of 6.98×10^8 m from the centre.

Naively, one might think that that is the end of the story but this is very far from the case. Observations of the solar atmosphere have now been made at a wide range of wavelengths and, because it is such a bright star, the temperature and density distribution in and above the photosphere can be determined with considerable precision. It turns out that the temperature of the atmosphere does not continue to decrease outwards but becomes slightly hotter above the photosphere in a region known as the *chromosphere*. This region is so called because it can be clearly observed as a thin ring of coloured light surrounding the disc of the Sun during total eclipses by the Moon. Above the chromosphere, there is a narrow *transition region*, less than 100 km thick, which separates the chromosphere from the hot *corona* of the Sun. In the transition region, the temperature of the corona rapidly attains a value of about 3×10^5 K and then continues to increase outwards reaching a maximum value of about 1.5×10^6 K at a distance of about 3×10^8 m above the solar disc. These variations of density and temperature are shown in Fig. 12.1. Also shown in Fig. 12.2 are the ions of different elements which are sensitive probes of this temperature and density

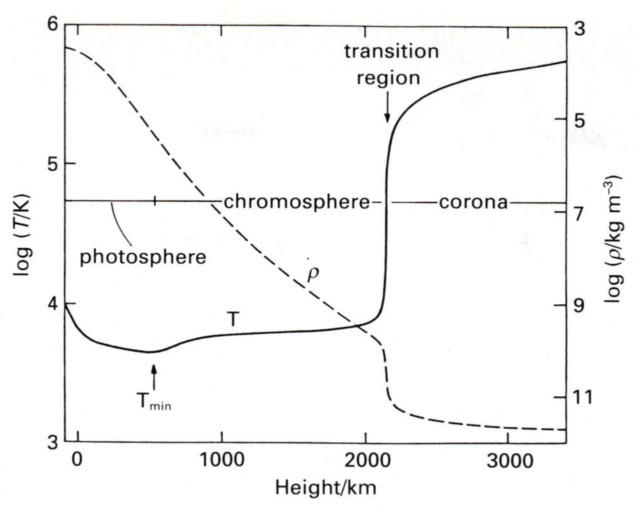

Figure 12.1. The variations of temperature and density as a function of height above the base of the Sun's photosphere. (From E. H. Avrett (1991). *The reference encyclopaedia of astronomy and astrophysics*.)

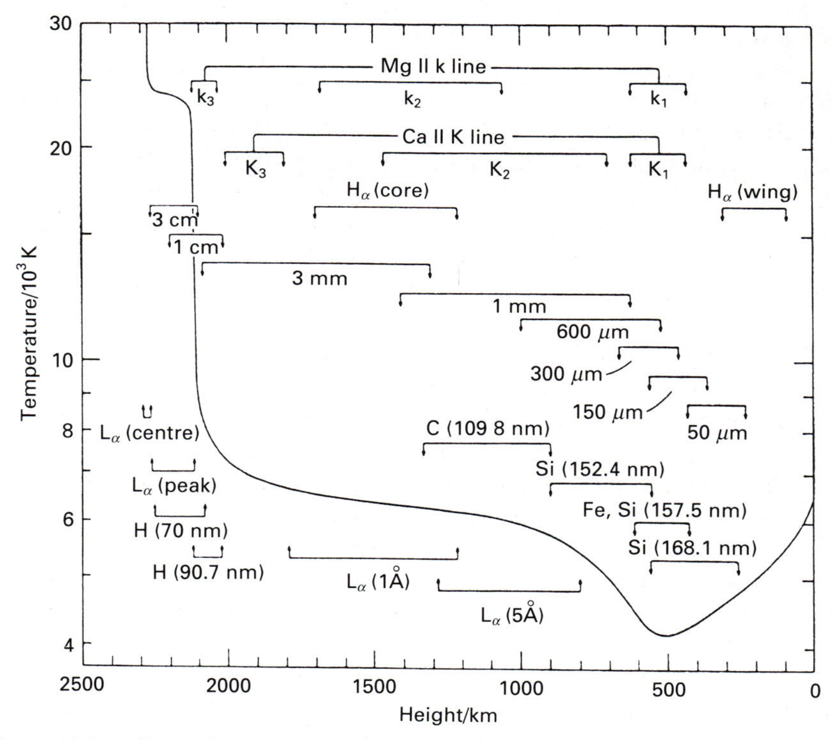

Figure 12.2. Illustrating the different atoms and ions which enable the density and temperature of the chromosphere and the lower regions of the corona to be determined. (From Athay, G. (1986). *Physics of the Sun*, Volume 2, eds P. A. Sturrock, T. E. Holzer, D. M. Michalas and R. K. Ulrich, pages 1–50, Dordrecht: D. Reidel Publishing Co.)

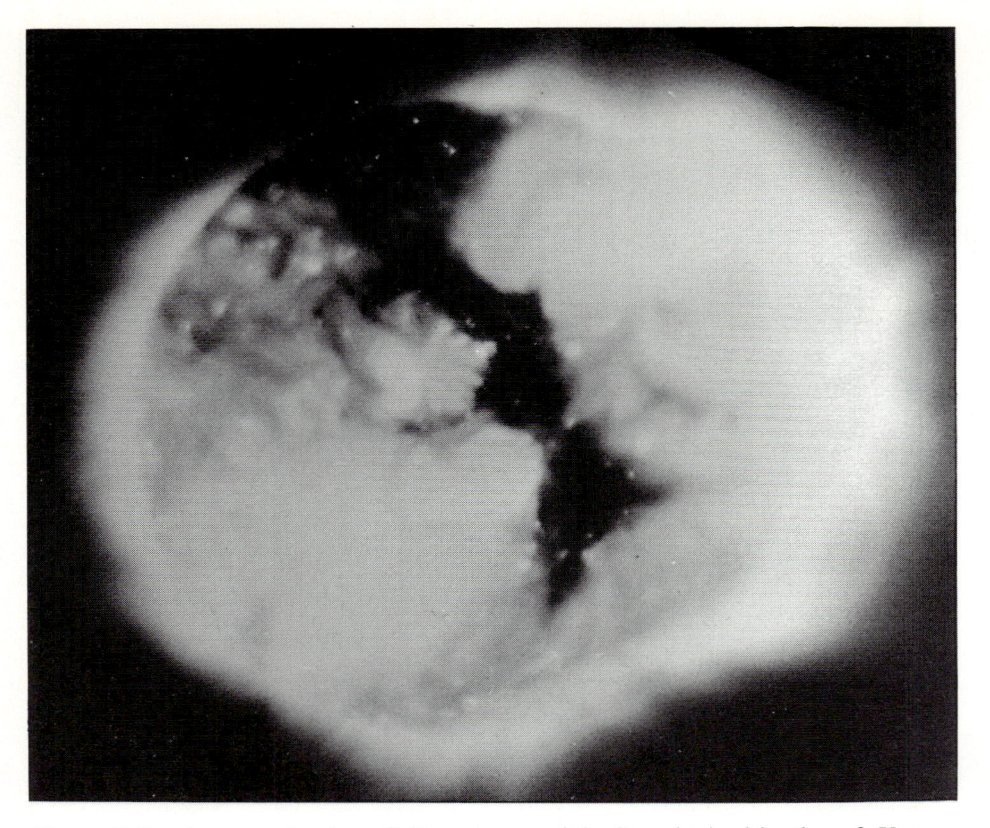

Figure 12.3. An example of a soft X-ray image of the Sun obtained by the soft X-ray telescope on board the Skylab space station. The X-ray telescopes were sensitive in the wavebands 0.2–3.2 nm (6.2–0.4 keV) and from 4.4 to 5.4 nm (0.28–0.23 keV). (Courtesy of NASA.)

structure. It should be noted, however, that this is only a very simple averaged model and in reality the whole atmosphere is highly non-uniform and dynamic.

Because the temperature of the corona reaches values exceeding 10^6 K, it is an intense emitter of soft X-rays. Magnificent images of the Sun in these wavebands were obtained by the soft X-ray telescopes on board the Skylab Observatory (Fig. 12.3) and these show the detailed structure of the hot gases in the lower region of the corona. It is now evident that qualitatively similar phenomena are found in essentially all classes of star. One of the most important discoveries of the Einstein X-ray Observatory was that all classes of main sequence star are X-ray emitters and it is natural to associate the emission with hot coronae, qualitatively similar to that observed about the Sun.

None of this was predicted by theory. We now understand that what was missing from the theory was the combined effect of convection and magnetic fields in the outer regions of the Sun. As noted above, the principal means of energy transport in the outer 30% of the Sun (by radius) is convection. The convection cells themselves do not extend into the photosphere but vestiges of the convective transport of energy are observed in the cellular granulation patterns which are present in images of the photosphere. The major complication is caused by the magnetic field which is frozen into the bulk convective motions of the material of the Sun. It is the combination of these convective motions and the Sun's magnetic field which is responsible for most of the active phenomena observed on the surface and in the atmosphere of the Sun. This leads us into the complexities of the magnetohydrodynamics and plasma physics of the solar atmosphere. This is an enormous subject and goes far beyond what we can hope to deal with in this text. The enthusiast is strongly recommended to study the book *Solar magnetohydrodynamics* by Priest (1982).

Deep inside the Sun, the magnetic field does not exert a large pressure compared with the thermal pressure of the gas. On the contrary, flux freezing ensures that the magnetic field is stretched and distorted by the convective motions in the outer layers of the Sun. This process is likely to be associated with the origin of the Sun's magnetic field through magnetic dynamo processes. However that field is generated, direct observations of the surface structure of the magnetic field show that, at the base of the atmosphere, the magnetic field is compressed into small flux tubes. As we progress up through the atmosphere, however, the gas pressure decreases relative to the magnetic pressure, the magnetic flux tubes expand and they can have a profound dynamical effect upon the structure of the corona.

To estimate the importance of the magnetic field dynamically, we can compare the thermal pressure of the hot gas at the base of the corona with the pressure of the magnetic field. A simple way of expressing this is in terms of the magnetic field strength which would exert the same pressure as the hot gas. If we write

$$\frac{p_{\mathrm{g}}}{p_{\mathrm{mag}}} = \frac{2nkT}{B^2/2\mu_0} \tag{12.1}$$

assuming that there are equal pressure contributions from the protons and electrons, then the ratio of pressures is unity when

$$B = B_{\mathrm{equ}} = (4\mu_0 nkT)^{\frac{1}{2}} \tag{12.2}$$

In Fig. 12.4, the variation of B_{equ} with distance through the photosphere and chromosphere is shown on the basis of the density and temperature distributions shown in Fig. 12.1. The strength of the magnetic field is difficult to measure directly above the photosphere. Photospheric magnetic fields are measured to be in the range 10^{-4}–10^{-3} T outside sun-spots and Solar flares. Within sun-spots, fields up to 0.3 T are found while coronal fields in flaring spot regions are indirectly inferred to be about $(5\text{–}50) \times 10^{-3}$ T from their microwave burst spectra. Thus, as we move out through the Solar atmosphere, magnetic forces are likely to be a dominant factor in determining the dynamics of the hot gas.

In general, it is now believed that most of the active phenomena observed on the Sun's surface and in its atmosphere are different manifestations of the dissipation

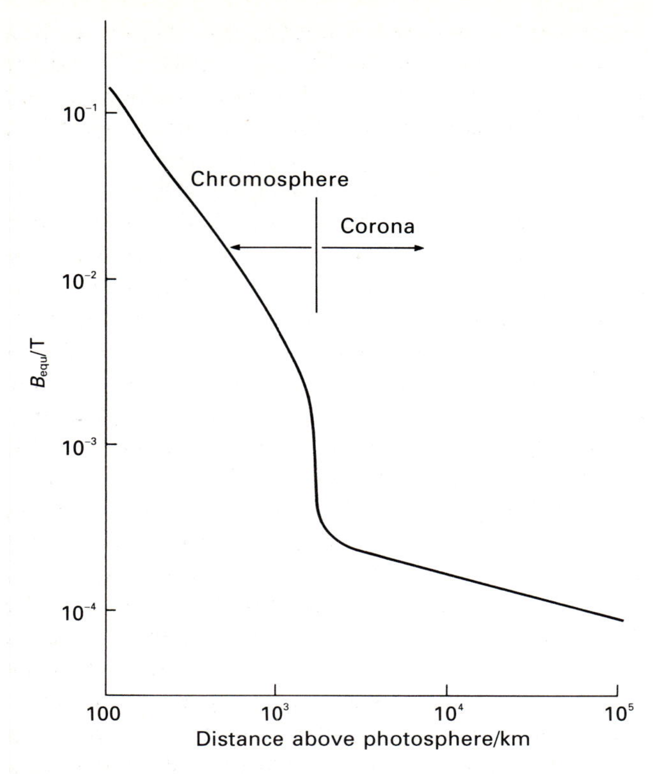

Figure 12.4. The variation of the magnetic flux density B_{equ} through the chromosphere and corona. If the magnetic flux density is greater than the value shown at any radius, the magnetic field will be dominant in determining the dynamics of the plasma.

of magnetic field energy. There is, however, uncertainty about the precise physical processes which are important in some of the most conspicuous physical phenomena. A key example is the problem of heating the solar corona to temperatures in excess of 10^6 K. Evidently, some form of mechanical heat pump is required to transfer energy from its source in the vicinity of the photosphere at a temperature of about 6000 K to values of about 10^6 K in the corona. The identification of the precise mechanical mechanism has proved elusive.

There is little problem in understanding how waves can be generated in the outer layers of the Sun. The convective motions inside the Sun generate waves which can transport energy outwards into the corona. The problem has been to identify an effective means by which energy can be dissipated in the corona. There is a vast literature on this topic (see e.g. Heyvaerts and Priest (1983) and the references therein). An example of the type of mechanism which may be important is the dissipation of shear Alfven waves by phase mixing in the steep Alfven velocity gradient in the Solar corona. In addition, the Kelvin–Helmholtz and tearing-mode instabilities may generate turbulence which assist in the dissipation of energy.

Another prime candidate is the creation and dissipation of singularities or current sheets in the magnetic field distribution (see Section 12.4.3). In this case, magnetic stresses are built up in field which can be dissipated at the reconnection rate (Heyvaerts and Priest 1984). This is a major problem for the understanding of the heating of stellar coronae in general.

The electrical conductivity of the coronal plasma is very high and, as a result, the process of magnetic flux freezing described in detail in Section 10.5 is very effective. This is excellent news from the point of view of explaining the dynamics of magnetic flux tubes since it can be assumed that the plasma and magnetic fields move together. Thus, in magnetically dominated regions, processes can be envisaged involving the dynamics of magnetic flux tubes which bear more than a passing resemblance to phenomena such as prominences seen in the solar corona. Indeed, the solar atmosphere is a gold-mine of phenomena for the magneto-hydrodynamicist and plasma physicist. Whenever energy is released in the Solar atmosphere, the gas is heated up causing expansion so that buoyancy effects can cause magnetic flux tubes to rise up through the atmosphere. If particles are accelerated in magnetic loops, they spiral along the flux tubes, much as is known to occur in the radiation belts surrounding the Earth, and these particles can deposit their energy in the denser plasma at the bases (or 'footpoints') of the magnetic loops. The gyrosynchrotron radiation of these particles can be observed at radio wavelengths as well as much higher brightness temperature phenomena associated with the coherent emission of these particles at very low frequencies.

Fig. 12.5 is a schematic diagram which pulls together a number of phenomena observed on the surface of the Sun. The central role of magnetic fields in producing such features can be appreciated. To the astrophysicist, these phenomena provide a unique opportunity for studying the behaviour of reasonably dense cosmic plasmas. Of these phenomena, the most important are the *solar flares* which we now study in a little more detail.

12.3 Solar flares

The most energetic events observed on the Sun are *solar flares*. They were first observed in 1859 by R. C. Carrington and R. Hodgson who noted independently an intense brightening of the intensity of the Sun in the vicinity of a complex group of sun-spots (Carrington 1859). In that particular case, the enhancement was what is now known as a 'white light' flare in which the continuum intensity is significantly enhanced over the background light of the Sun for a few minutes. Observations of these types of event remain rare but all flares can be observed in the optical waveband as enhancements of the Hα intensity. The Hα line in the Solar spectrum is one of the dark Fraunhofer lines which absorb the continuum emission from the Solar interior and flares are observed as reversals into emission of the dark lines. In large flares, the Hα intensity can exceed the intensity of the neighbouring continuum. The surface area of the Sun which can be covered by a flare can be as much as 10^9 km^2. These Hα flares have been studied

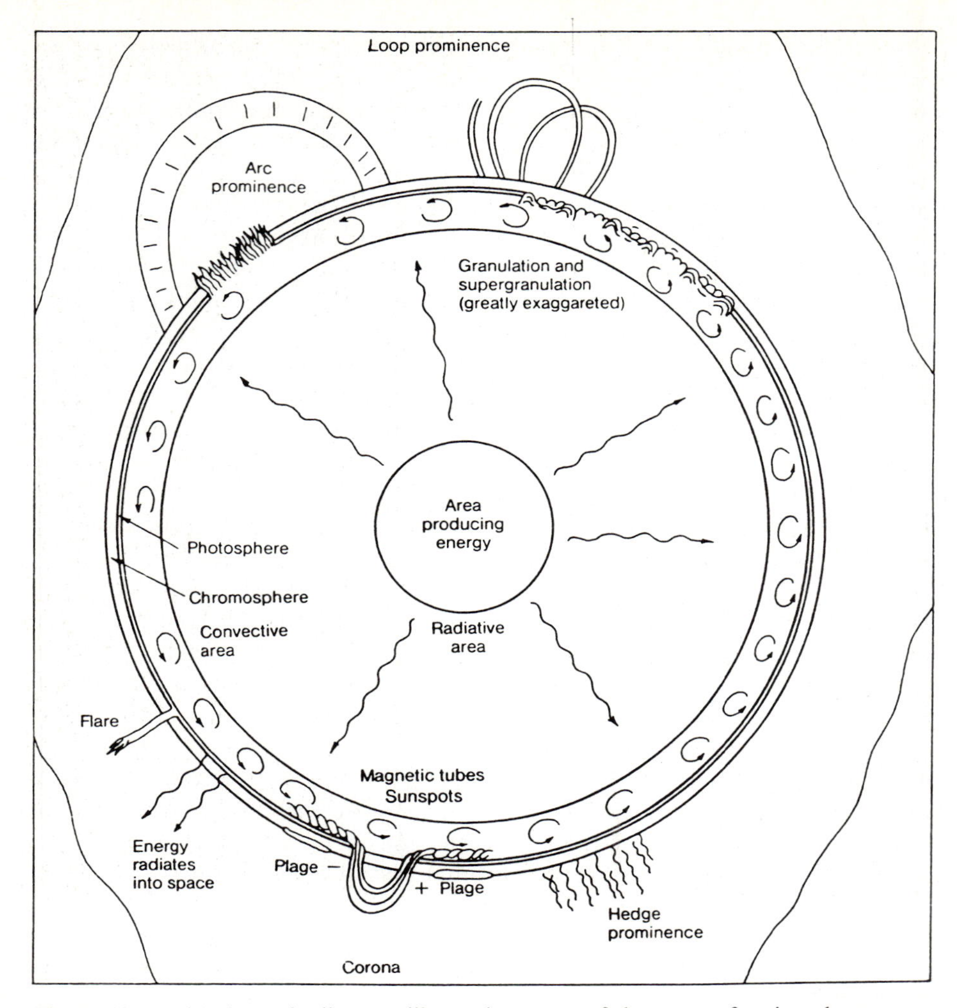

Figure 12.5. A schematic diagram illustrating some of the types of active phenomena observed on the surface of the Sun. (From H. Karttunen, P. Kroger, H. Oja, M. Pountanen and K. J. Donner (eds) (1987). *Fundamental astronomy*, page 278, Berlin: Springer-Verlag.)

extensively for many years. Sometimes the flaring takes the form of a 'double-ribbon' pattern and these are known as 'two-ribbon flares' (Fig. 12.6).

The study of solar flares has been revolutionised by the ability to make observations in the radio, ultraviolet, X and γ-ray wavebands. The following brief summary is based upon the monograph *The physics of solar flares* by Tandberg-Hanssen and Gordon Emslie (1988) and the volume *Solar flare magnetohydrodynamics* edited by Priest (1981), as well as volumes **118** (1988) and **121** (1989) of the journal *Solar Physics* which were devoted to the impact of space observations upon the study of solar flares.

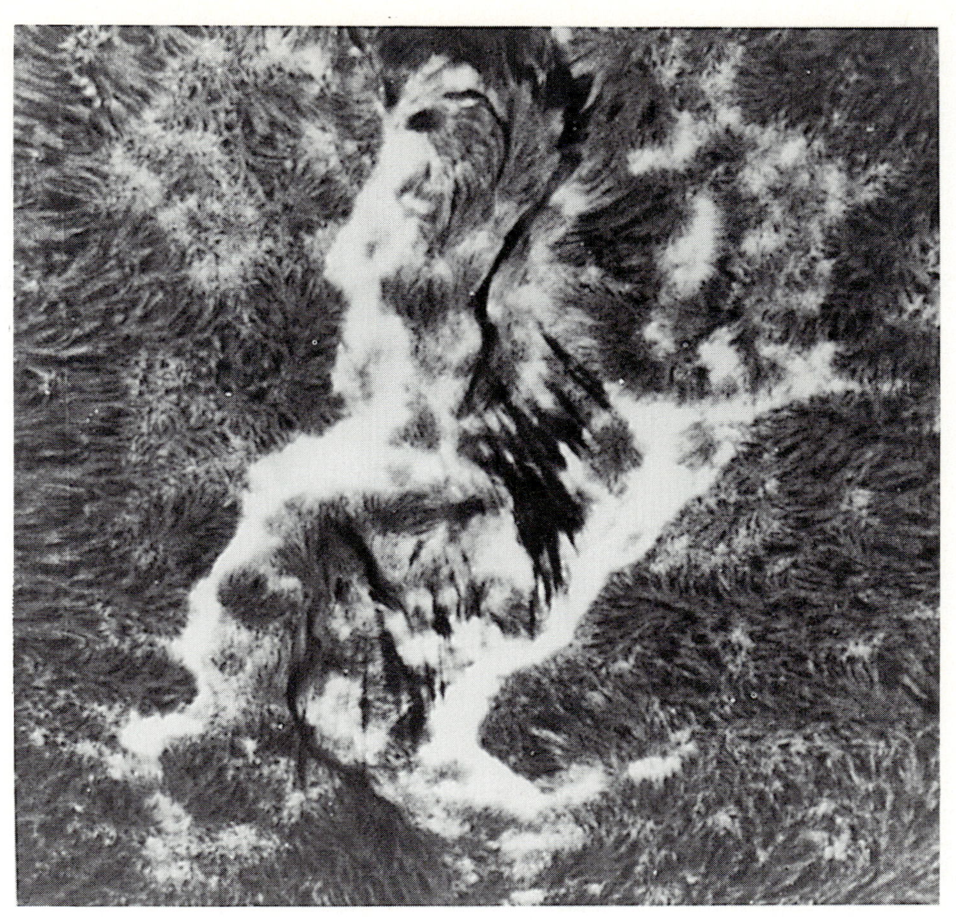

Figure 12.6. An example of a two-ribbon flare. (From E. Tandberg-Hanssen and A. Gordon Emslie (1988). *The physics of solar flares*, cover picture. Cambridge: Cambridge University Press.)

There is considerable variation in the time evolution of the luminosity of solar flares. Fig. 12.7(*a*) which is taken from Priest's review of 1981 summarises the various phases of the typical solar flare. If we were only interested in flares as intense emitters of Hα radiation, there would only be two phases – there would be the *flash phase* during which the Hα intensity increases to maximum intensity over a period of about 5 min followed by a *main phase* which lasts about an hour. The other wavebands, however, reveal a much more complex behaviour. In the soft X-ray waveband, corresponding to temperatures of about $(1\text{–}10) \times 10^6$ K, flares show both the flash and main phases but, in addition, there is a *pre-flare phase* which lasts for about 10–30 min before the start of the flash phase. Also, for the first 10–100 s after the onset of the flash phase, there is often an *impulsive phase* which is observed at radio microwave wavelengths, in the hard X-ray waveband and

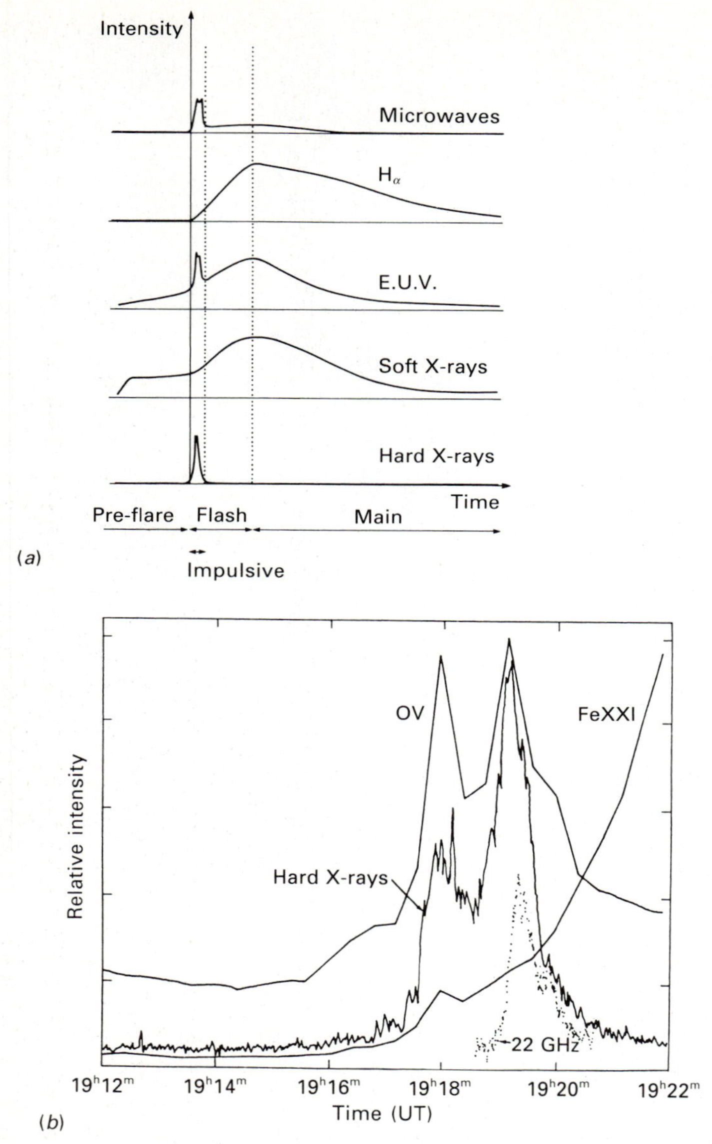

Figure 12.7. (*a*) A schematic diagram showing the time development of the luminosity of a solar flare in different wavebands. (After E. R. Priest (1981). *Solar flare magnetohydrodynamics*, ed. E. R. Priest, page 2, London: Gordon and Breach Science Publishers.)
(*b*) The early time development of the solar flare of 1 November 1980 showing the impulsive phase of the flare. (From E. Tandberg-Hannsen and A. Gordon Emslie (1988). *The physics of solar flares*, page 5, Cambridge: Cambridge University Press.)

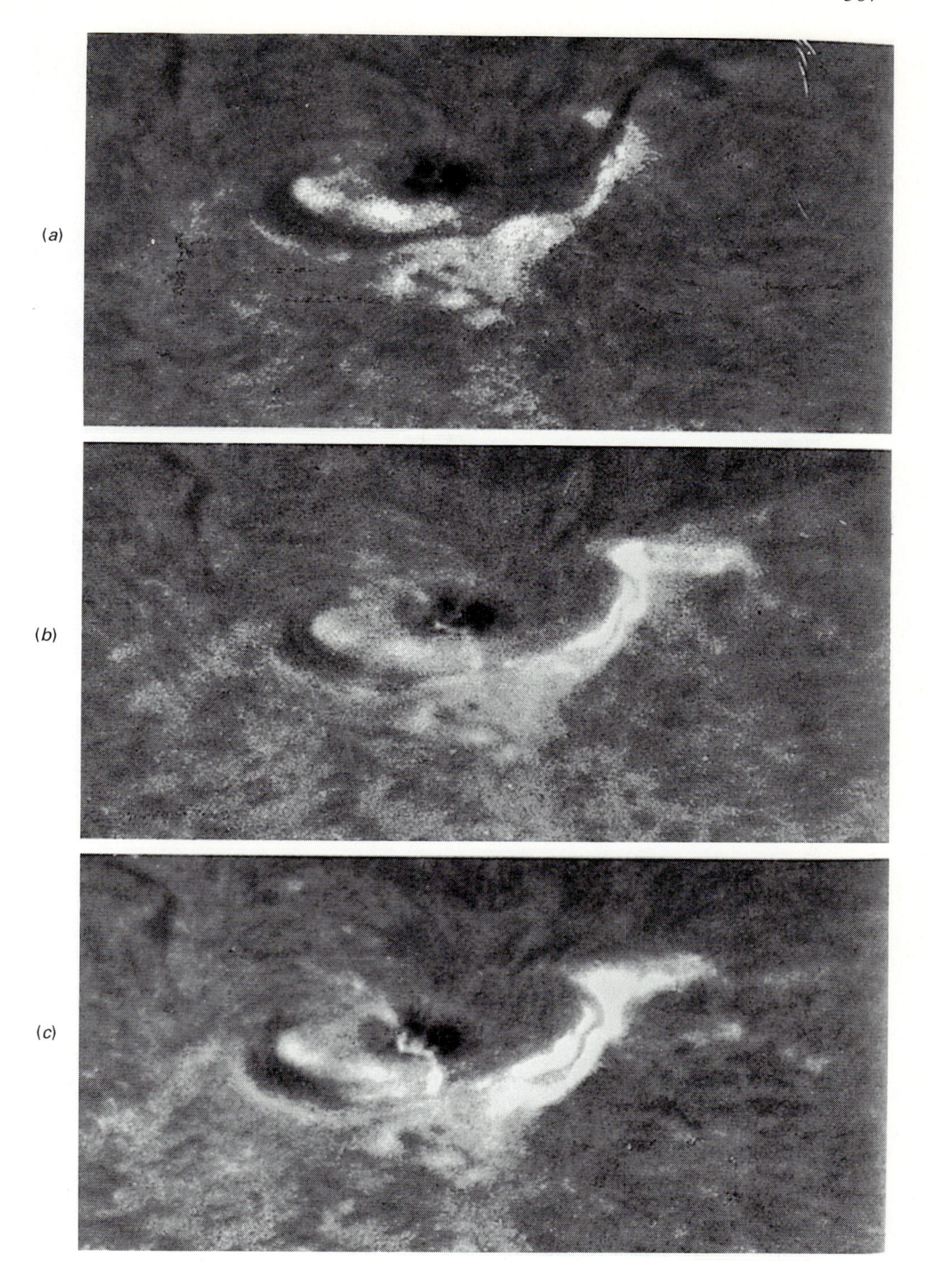

Figure 12.8. (*a*–*c*). For legend see page 369

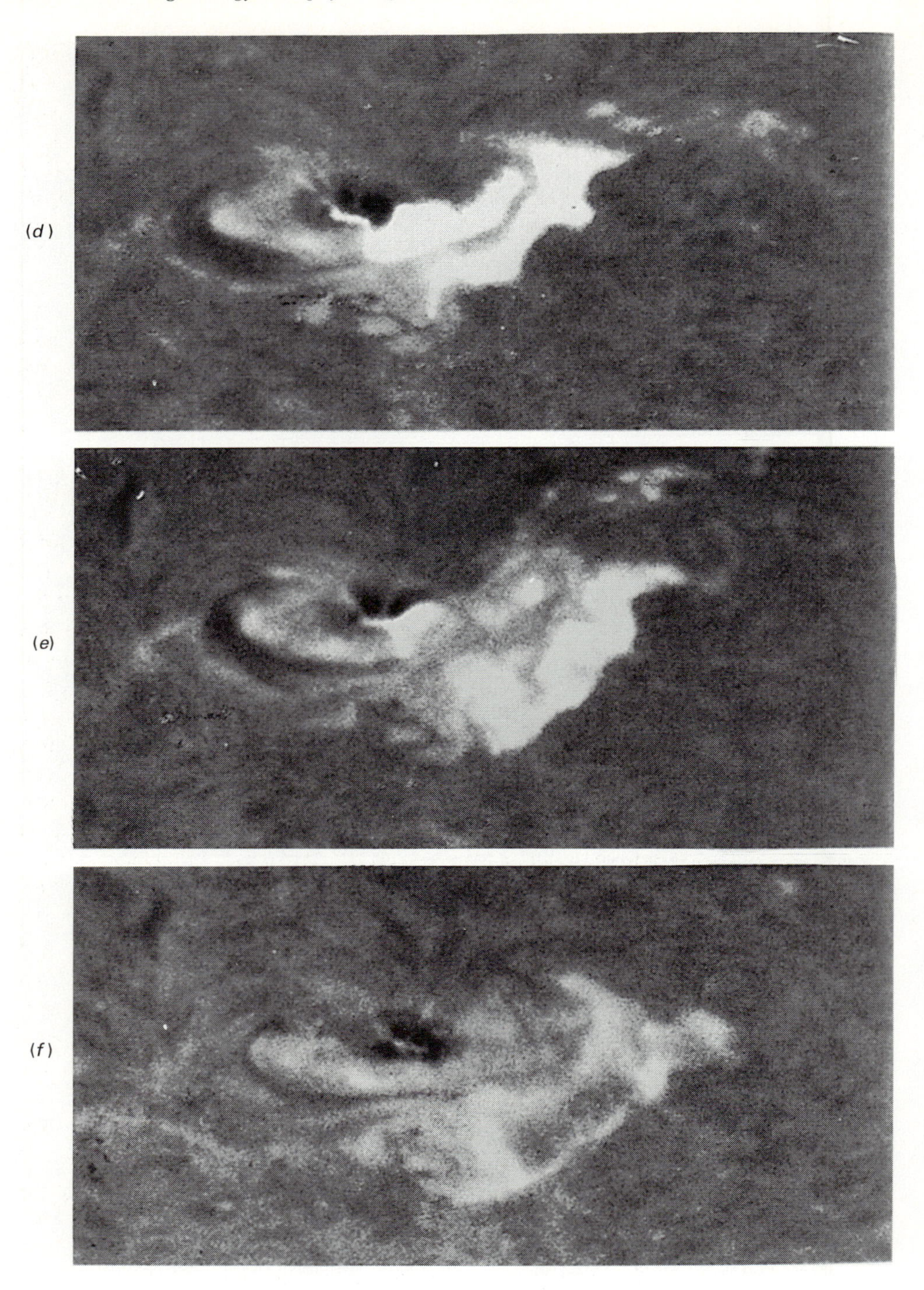

Figure 12.8. (*d–f*). For legend see page 369

sometimes in γ-ray lines. This impulsive phase of the flare is clearly seen in Fig. 12.7(*b*) which shows the early development of the strong flare of 1 November 1980 at radio, ultraviolet and hard X-ray wavelengths – the ultraviolet intensity is in fact the line emission of highly ionised oxygen (OV) and iron (FeXXI). It can be seen that there are rapid temporal variations of the intensity of the emission during the impulsive phase, the shortest variations occurring on timescales as short as 5–10 s. These intense emissions during the impulsive phase are attributed to the emission of high energy electrons accelerated on very short timescales in the solar flares themselves. In the strongest flares, there is sometimes an extended impulsive phase corresponding to multiple bursts of particle acceleration.

There is a wide range of luminosities among solar flares but there appear to be two basic types. The low luminosity flares appear to consist of a single magnetic flux tube which emits X-rays and remains essentially unchanged in structure and size throughout the solar flare. These are referred to as *simple loop flares* or *compact flares*. There seems to be a strong heat source which causes the whole of the flux tube to be raised to a high temperature.

Much more dramatic is the other class of solar flare – the *two-ribbon flare*. These are much larger than compact flares and take place in the vicinity of solar prominences. Solar prominences are those features observed at the limb of the Sun during total eclipses by the Moon and they can take a variety of forms. Typically, they take the form of long arches, to all appearances looking like magnetic flux tubes with their ends anchored in the sun (see Fig. 12.5). During total eclipses of the Sun, they are observed in emission but, when they are observed on the disc of the Sun, they absorb the continuum emission of the Sun and so are observed as dark bands, particularly when observed in the strong resonance line of Hα. When these dark filaments are located close to active regions on the Sun, for example, close to complex networks of the sun-spots, and they become unstable, the most violent and energetic flares are observed. During the flash phase, the Hα intensity increases on either side of the dark filament, giving rise to the characteristic 'two-ribbon' structure. The Hα emission increases to maximum intensity and then, during the main phase of the flare, the two ribbons move apart at a velocity of about 2–10 km s^{-1} (Fig. 12.8). In those strong solar flares which are associated with

---

Figure 12.8. The development of a strong solar flare as observed in the Hα line on 7 September 1973. The six images show the development of the solar flare over a period of less than one hour.
(*a*) 11^h35^m UT This is the pre-flare state of the region in which there is sun-spot activity and a dark filament which breaks a bright filament into two parts.
(*b*) 11^h42^m UT Bright spots of Hα begin to appear along the dark filament.
(*c*) 11^h50^m UT The flash phase of the flare. The bright zones begin to rise and lengthen.
(*d*) 12^h02^m UT The time of maximum development of the solar flare.
(*e*) 12^h32^m UT The flare is decaying. The two ribbons have moved apart.
(*f*) 14^h22^m UT The region has almost returned to a quiescent state.
(From Vial, J-C. (1988). *The Cambridge atlas of astronomy*, eds J. Audouze and G. Israel, page 40, Cambridge: Cambridge University Press.)

regions of intense solar activity and complex magnetic field configurations, the flare is accompanied by impulsive hard X-ray emission and the bursts of radio emission known as Type III solar bursts. Some of these large flares give rise to mass ejections into the Solar Wind at supersonic velicities so that shock waves are formed at the leading edges of these high velocity streams. More details of one possible classification of solar flares is given by Bai and Sturrock (1989).

According to Priest (1981), the energy released in flares can range from about 10^{22} J in small flares to about 3×10^{25} J in the largest flares. Much of this energy is released in the form of electromagnetic energy at wavelengths up to the soft X-ray waveband and in the form of mechanical energy as an interplanetary blast wave. A significant fraction of the energy is also released in the form of the high energy particles which are responsible for the hard X-ray emission although the exact values are somewhat uncertain and, in the case of the high energy electrons, are model-dependent.

Further important clues concerning the central importance of magnetic fields are provided by *sympathetic flares* in which flares are observed to begin one after another at different locations on the Sun's surface. Skylab observations have shown that different active regions are connected by arch or loop-like structures observed in soft X-rays. These features are characteristic of magnetic flux tubes and it is natural to assume that the disturbances are propagated along the magnetic flux tubes from one region to another. The images obtained at soft X-ray wavelengths have been crucial in showing that most of the activity on the Sun's surface can be associated with magnetic flux tubes.

An aspect of great interest for high energy astrophysics is the means by which the energy needed to power the Solar flares is generated in the Solar atmosphere and, in particular, how the energy can be released rapidly enough. Let us deal, first of all, with the energetics. The problem is to release an energy of 10^{25} J or more from the volume of the flare. The typical dimensions of a large flare are about $L \sim 3 \times 10^7$ m in diameter and about $H \sim 2 \times 10^7$ m in height so that the flare volume is about 2×10^{22} m^3 (Priest 1981). Following Tandberg-Hannsen and Gordon Emslie, we can estimate the available thermal energy in the quiescent chromosphere and corona using the column depths $\rho_{\rm col}$ and temperatures of these regions. In the chromosphere, the column depth is about $\rho_{\rm col} \sim 0.1$ kg m^{-2} and the temperature about 10^4 K whereas, in the corona, the values are $\rho_{\rm col} \sim 3 \times 10^{-5}$ kg m^{-2} at a temperature of about 3×10^6 K. The thermal energy of each component is therefore $\mathscr{E}_{\rm th} \approx 3\rho_{\rm col} kTL^2/m_{\rm H}$ assuming that the electrons and protons make equal contributions to the thermal energy of the gas. Therefore, the best case, that of the chromosphere, can provide only about $\mathscr{E}_{\rm th} \approx 2 \times 10^{22}$ J which is only enough to power the smallest flares and certainly not the most powerful examples for which at least a thousand times more energy is required. In any case, we have already argued that some form of mechanical heat pump is necessary to heat the material of the flare to a high temperature. Nuclear energy generation can be excluded because the temperatures in the solar atmosphere are not high enough and so this leaves only magnetic energy as a potential energy source.

If we take the volume of the flare to be $V = 3 \times 10^{22}$ m^3, the magnetic energy within the volume is

$$\mathscr{E} = V\frac{B^2}{2\mu_0} = 10^{28}B^2 \text{ J}.$$

Thus, if the magnetic field strength in the flare is 0.03–0.1 T, the available energy is almost 10^{25}–10^{26} J which is more than adequate for the most energetic flares, provided it can be used efficiently and provided it can be released fast enough. This magnetic field strength is of the same order of magnitude as the values estimated from observations of the microwave radio emission of flares during their early impulsive phases.

12.4 Magnetic fields in the solar atmosphere

The magnetohydrodynamics of the solar atmosphere is a subject of considerable complexity. The problem is a fully three-dimensional one in which magnetic flux tubes are stretched and sheared in the presence of convective and turbulent motions as well as having their own buoyancy in the Sun's gravitational field. These processes give rise to a myriad of topological possibilities for the behaviour of magnetic fields in the presence of turbulence. Let us outline some of the key concepts necessary to understand the basic physics of solar flares. Priest's text *Solar magnetohydrodynamics* provides a full discussion of these and other important phenomena.

12.4.1 *Magnetic flux freezing*

We have already treated the problem of magnetic flux freezing in the context of understanding the structure of the magnetic field in the Solar Wind. Let us repeat these calculations for the conditions found at the base of the corona. In Section 10.4, we worked out the mean free path of a proton in the Solar Wind. In the present instance, we are interested in the electrical conductivity of the plasma in the Sun's chromosphere and corona. We can derive an appropriate conductivity using the same tools described in Section 10.4 but now we are interested in the transfer of energy between electrons and protons by collisions – we recall from the discussion of that section that, in a plasma, a collision is the statistical exchange of energy between the electrons and the protons mediated by the electrostatic fields of the particles. Repeating the same order of magnitude calculation as in Section 10.4, we find that the collision time for an electron to transfer its energy to a gas of nuclei of charge Ze is

$$t_e = \frac{2\pi\varepsilon_0^2 m_e^{\frac{1}{2}}(3kT)^{\frac{3}{2}}}{Z^2e^4N \ln \Lambda} \tag{12.3}$$

where Λ is a Gaunt factor and N is the number density of nuclei of electric charge Z. We can now use the classical formula to work out the conductivity σ of the plasma, $\sigma = (n_e e^2/m_e \nu_c)$, where n_e is the number density of electrons and ν_c is the collision frequency $\nu_c = t_e^{-1}$. If the plasma is electrically neutral, then $n_e = ZN$ and hence

$$\sigma = \frac{n_e e^2}{m_e \nu_e} = \frac{2\pi\varepsilon_0^2(3kT)^{\frac{3}{2}}}{Ze^2 m_e^{\frac{1}{2}} \ln \Lambda} \tag{12.4}$$

A detailed description of the conductivity of a plasma is given by Spitzer (1962) who gives the following result:

$$\sigma = \frac{32\pi^{\frac{1}{2}}\varepsilon_0^2(2kT)^{\frac{3}{2}}}{Ze^2 m_e^{\frac{1}{2}} \ln \Lambda} = 2.63 \times 10^{-2} \frac{T^{\frac{3}{2}}}{Z \ln \Lambda} \text{ mho m}^{-1} \tag{12.5}$$

This formula neglects electron–electron interactions and Spitzer shows that, for a hydrogen plasma, this has the effect of decreasing the conductivity by a factor of 0.582. For our present purposes, it is adequate to use a Gaunt factor $\ln \Lambda = 10$ (see Spitzer (1962) page 128) and hence the expression $\sigma = 10^{-3} T^{\frac{3}{2}}$ mho m^{-1}. This is, in fact, a very high conductivity indeed and leads to the phenomenon of magnetic flux freezing in the Solar corona.

Let us demonstrate this in a slightly different way by returning to the equations of magnetohydrodynamics and in particular to equation (10.14).

$$\frac{\partial \mathbf{B}}{\partial t} = \nabla \times (\mathbf{v} \times \mathbf{B}) - \frac{1}{\sigma\mu_0} \nabla^2 \mathbf{B} \tag{10.14}$$

We recall that the condition for magnetic flux freezing is that the first term on the right hand side of the equation far exceeds the second. Suppose we are interested in phenomena on the scale L. Then, to order of magnitude, the ratio of the first to the second terms on the right hand side is just

$$R_m = \sigma\mu_0 \frac{\nabla \times (\mathbf{v} \times \mathbf{B})}{\nabla^2 \mathbf{B}} \sim \frac{(vB/L)}{(B/\sigma\mu_0 L^2)} \sim \sigma\mu_0 vL \tag{12.6}$$

where v is the velocity of the plasma. The quantity R_m is known as the *magnetic Reynolds' number* and is a measure of the importance of the phenomenon of magnetic flux freezing on the scale L. In the case of Solar flares, we can adopt the following values: $T = 2 \times 10^6$ K, $v = 20$ km s^{-1} and $L = 2 \times 10^7$ m. In this case, $R_m \approx 10^{12}$ so that it is a very secure assumption that, on the scale of the Solar flare, the magnetic field is frozen into the plasma.

This is the source of both good news and bad news. The good news is that we can assume that the magnetic field is effectively frozen into the plasma. The bad news is that it is only the finiteness of this same conductivity which allows there to be dissipation of the magnetic field energy. The timescale to dissipate the magnetic field energy into heat is just the time it takes to reduce the magnetic field to zero by ohmic dissipation which is represented by the last term in equation (10.14). This timescale is just

$$\tau = \frac{B}{\partial B/\partial t} \sim L^2\mu_0\sigma \tag{12.7}$$

Adopting the same values as above, we find that the timescale for the dissipation of energy is $\tau \sim 5\times10^{11}$ s, much greater than the typical duration of a Solar flare which is of the order of 10^3 s and very much greater than the timescale for the acceleration of high energy particles which is of the order of seconds. This is one of the biggest problems in the theory of solar flares and we will look at it in a little more detail in a moment.

12.4.2 *Magnetic buoyancy*

One of the very attractive pieces of physics which displays some of the important features of flux tubes is the concept of magnetic buoyancy. Following Tandberg-Hanssen and Gordon Emslie, suppose there is an isolated magnetic flux tube in a plane-parallel stratified atmosphere. The number density of protons in the atmosphere is n_o and that inside the flux tube is n_i. As in any buoyancy calculation, the atmosphere and the tube are in pressure balance and hence $p_o = p_i$. The buoyancy arises from the fact that the mass density inside the flux tube is less than that surrounding it and consequently, in the Sun's gravitational field, the lighter volume 'floats up' the potential gradient. For simplicitly, let us assume that the material inside and outside the flux tube are at the same temperature. Then, assuming that we are dealing with fully ionised gas, the electrons and ions each contribute a pressure nkT and therefore the equation of pressure balance becomes

$$2n_o kT = \frac{B^2}{2\mu_0} + 2n_i kT \tag{12.8}$$

and hence

$$n_i = n_o - \frac{B^2}{4\mu_0 kT}$$

The force acting upon the flux tube in the potential gradient is therefore

$$F = (n_o - n_i) m_H g V = \frac{B^2 m_H g V}{4\mu_0 kT}$$

where m_H is the mass of the hydrogen ion and V is the volume of the flux tube. Now, for an atmosphere in hydrostatic equilibrium, $dp/dx = -\rho g$ and since $p = 2\rho kT/m_H$, the scale height of the atmosphere H defined by $d\rho/\rho = dx/H$ is

$$H = 2kT/m_H g$$

Therefore,

$$F = B^2 V/2\mu_0 H$$

After the tube has risen a height H it has acquired a kinetic energy

$$\tfrac{1}{2}Mu^2 = \tfrac{1}{2}\rho V u^2 = FH = B^2 V/2\mu_0$$

because of the work done in accelerating the flux tube by the accelerating force F, The resulting velocity u is therefore

$$u = (B^2/\mu_0 \rho)^{\frac{1}{2}} \tag{12.9}$$

We recognise that this is just the local Alfven speed $v_A = (B^2/\mu_0 \rho)^{\frac{1}{2}}$. This is the rather elegant result we have been seeking – the flux tube rises up through the atmosphere at roughly the local Alfven speed. Since we expect the flux tube will be tied to the material of the solar atmosphere at its footpoints, we can see how it is very natural for flux tubes to develop into loop-like structures driven by the buoyancy of the magnetic field. Notice that this property of the buoyancy of magnetic flux tubes is a very general one in that it simply occurs when the matter density inside the tube is less than that outside and the system is located in a gravitational potential gradient. For example, a similar process applies to the magnetic fields confined to the plane of the Galaxy and in accretion discs. More details of these ideas and their application more generally are given by Parker (1979).

12.4.3 *Neutral current sheets and the reconnection of magnetic lines of force*

The problem identified in Section 12.4.1 was the means of tapping the large amounts of magnetic field energy present at the base of the corona, despite the fact that the conductivity of the plasma appears to be far too large to enable energy to be tapped efficiently. The considerations of the last two sub-sections suggest that the magnetic field structures can become remarkably complex. The magnetic flux tubes behave as though they are frozen into the plasma and are subject to a variety of forces. For example, the buoyancy of the flux tube makes it rise up through the atmosphere, the convective motions just below the photosphere are in the end responsible for energising the magnetic field and there are strong perturbations of the magnetic field distribution due to a wide variety of other disturbances seen on the surface of the Sun. Among the most important of these processes are the shearing and twisting of magnetic field lines, driven by motions in the plasma. These have a crucial effect in that they transfer energy from the bulk turbulent and convective motions in the plasma into magnetic field energy. In this way, as in other astrophysical situations, the strength of the magnetic field can be amplified. There is thus every reason to suppose that strong magnetic fields are generated at the base of the corona.

We have still not extracted any energy and this is where the concepts of the reconnection and annihilation of magnetic lines of force become important. In the Solar atmosphere, situations frequently occur in which the magnetic field lines in neighbouring flux tubes run in opposite directions. It is in such regions that there is the possibility of converting magnetic field energy into thermal energy if the

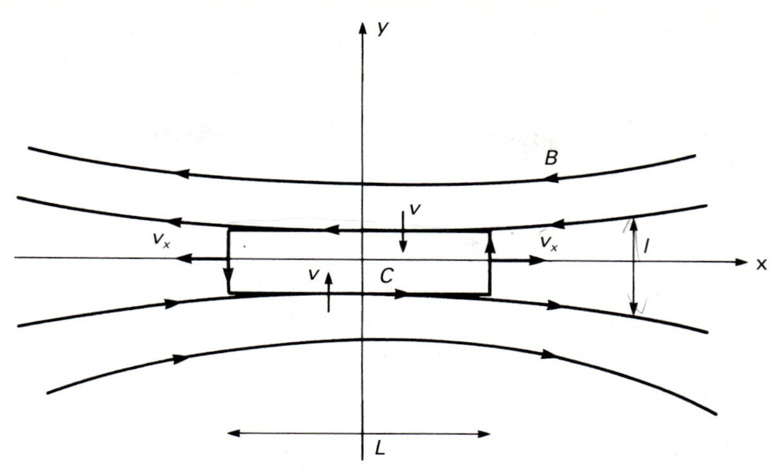

Figure 12.9. Illustrating the process of magnetic field line reconnection. The magnetic field reverses in direction along the x-axis leading to a large current density in the z-direction. The magnetic field is convected into the neutral sheet in the y-direction and this is balanced by the outflow of material along the positive and negative x-axes. The dimensions of the reconnection region are shown on the diagram. It is assumed that the geometry extends indefinitely in the $\pm z$-direction. (From E. Tandberg-Hannsen and A. Gordon Emslie (1988). *The physics of solar flares*, page 151, Cambridge: Cambridge University Press.)

regions of opposing magnetic field come close enough together and there is sufficiently high resistivity to dissipate the currents flowing in the *neutral sheet* between the oppositely directed field lines.

The simplest picture is illustrated in Fig. 12.9. The magnetic field reverses direction along the x-axis and the oppositely directed field lines are convected towards the x-axis at velocity v in the y-direction. The sheet is considered to be infinite in the z-direction. To preserve continuity, the inflow of plasma and magnetic field have to be balanced by outflow along the x-axis. It is a pleasant sum to work out the rate at which magnetic field energy is dissipated by ohmic losses in this system. First, we construct a closed loop path about the dissipation region and then apply Stokes' theorem to work out the current flowing through the loop. We write Stokes' theorem in the standard form

$$\int_S \mathbf{J}\cdot d\mathbf{S} = \frac{1}{\mu_0}\int_C \mathbf{B}\cdot d\mathbf{l} \tag{12.10}$$

where $\mathbf{J}$ is the current density passing through the loop and the integral on the right hand side is taken round the closed loop. Writing the expression in terms of the current density through the loop, we find to order of magnitude

$$lLJ \approx 2BL/\mu_0 \quad J \approx 2B/l\mu_0 \tag{12.11}$$

where l is the width of the loop and L its length. Note that this is no more than writing $\mathbf{J} = \text{curl}\,\mathbf{B}/\mu_0$ in the vicinity of the neutral sheet. Thus, as we make the value of l smaller and smaller, the current density J in the reconnection region becomes greater and greater so that, even if the conductivity of the region is very high, it would appear that there can be efficient ohmic losses in the neutral sheet provided the region in which the dissipation takes place is narrow enough. An absolute lower limit to the width of this region is set by the gyroradii of the particles in the field. If the resistivity of the plasma is $\eta = \sigma^{-1}$, the dissipation rate is $\eta J^2 = 4\eta B^2/l^2$ per unit volume.

What we have omitted, however, is the influence of the gas pressure in the neutral sheet. The plasma as well as the magnetic field is convected into the dissipation region and hence the material cannot be compressed indefinitely. The limiting pressure is set by the equality between the magnetic pressure and the thermal pressure of the gas. The solution is to alleviate the build up of pressure by allowing the material to flow out along the positive and negative x-axes. We can guess that this outflow velocity can only be the characteristic velocity at which the magnetoplasma can respond to the inflow and that is the Alfven speed. Let us demonstrate this slightly more properly than simply by assertion.

Neglecting magnetic forces, the equation of motion of the plasma along the x-axis is just

$$\rho\, \mathrm{d}v_x/\mathrm{d}t = -\partial p/\partial x \tag{12.12}$$

In the steady state $\partial v_x/\partial t = 0$ and, since $\mathrm{d}/\mathrm{d}t = \partial/\partial t + (\mathbf{v}\cdot\nabla)$, we can rewrite the equation in terms of Eulerian coordinates so that

$$\rho v_x \partial v_x/\partial x = -\partial p/\partial x \tag{12.13}$$

Now, integrating from $x = 0$ to $x = \pm\infty$,

$$p_\mathrm{i} - p_\mathrm{o} = \tfrac{1}{2}\rho v_x^2 \tag{12.14}$$

But the pressure difference is just that due to the magnetic pressure within the region of the neutral sheet and so

$$p_\mathrm{i} - p_\mathrm{o} = B_x^2/2\mu_0 \tag{12.15}$$

Therefore, the velocity of escape of the material along the x-axis is just of the order of the Alfven velocity $v_x \approx B/(\mu_0\,\rho)^{\frac{1}{2}} = v_\mathrm{A}$. By continuity, this outflow is balanced by the inflow along the y-axis and hence by mass conservation we find that the speed at which the material is convected into the dissipation region is $v = (l/L)v_\mathrm{A}$.

To complete the analysis, we can now equate the dissipation rate by ohmic losses to the rate at which magnetic energy is convected into the reconnection region i.e.

$$\int_V \eta J^2 \mathrm{d}V = \int_S \frac{B^2}{2\mu_0} v \mathrm{d}\underset{\sim}{S} \tag{12.16}$$

Therefore, per unit length in the z-direction, we find

$$\eta J^2(Ll) = \frac{B^2}{2\mu_0} 2vL \quad \eta J^2 l = \frac{B^2 v}{\mu_0} \tag{12.17}$$

But, from the expression (12.12), $J = 2B/\mu_0 l$ and hence

$$v = 4\eta/l\mu_0 \tag{12.18}$$

Notice that we can obtain the same result simply by setting the right hand side of equation (10.14) equal to zero – in a stationary (Eulerian) frame of reference, there is no variation in B at any point. Using expression (12.18) and the relation $v = (l/L)v_A$, we can solve for v and l

$$v^2 = \frac{4\eta}{\mu_0 L} v_A \quad l^2 = \frac{4\eta L}{\mu_0 v_A} \tag{12.19}$$

Notice that, in this model, the thickness of the reconnection region l has disappeared from the expression for v.

It is now convenient to introduce a 'longitudinal' Reynolds' number R_m for the neutral sheet in which the length scale L is the length of the neutral sheet and the velocity v the Alfven velocity v_A. We can therefore write

$$R_m = \sigma\mu_0 vL = \mu_0 v_A L/\eta$$

Therefore, we find that the convective or reconnection velocity v_r into the neutral sheet is just

$$v = v_r = \left(\frac{4\eta}{\eta_0 L} v_A\right)^{\frac{1}{2}} = 2v_A/R_m^{\frac{1}{2}} \tag{12.20}$$

and the thickness of the neutral sheet is

$$l = \left(\frac{4\eta L}{\mu_0 L}\right)^{\frac{1}{2}} = 2L/R_m^{\frac{1}{2}} \tag{12.21}$$

Since $R_m \gtrsim 10^{12}$ (see Section 12.4.1), it is clear that the velocity at which magnetic field lines are convected into the neutral sheet corresponds to 10^{-6} of the Alfven speed. The implication of this result is that this type of neutral sheet dissipates energy slowly. Notice, however, that dissipation in neutral sheets takes place much more rapidly than the diffusive dissipation of energy over a length scale L for which the timescale is $\tau_D \sim \sigma\mu_0 L^2$ (see equation (10.4)). The diffusive velocity is $v_D \sim L/\tau_D \sim 1/\sigma\mu_0 L \sim v_A/R_m$ which is less than the reconnection velocity by a factor of $R_m^{\frac{1}{2}}$. We can define a reconnection timescale τ_A associated with the neutral current sheet $\tau_R = l/v \sim L/v_A \sim \tau_A R_m^{\frac{1}{2}}$ where τ_A is the time it takes an Alfven wave to cross the neutral sheet.

Let us now see how well we are doing in terms of explaining the rate at which energy is released from the current sheet. The total amount of magnetic energy in the neutral sheet is $(B^2/2\mu_0)V$ where $V \sim L^2 l \sim L^3/R_m^{-\frac{1}{2}}$ and this is released over a timescale τ_R. Following Tandberg–Hanssen and Gordon Emslie, we adopt the following parameters for the neutral sheet: $L = 10^7$ m, $B = 0.03$ T, $n = 10^{16}$ m^{-3} and $T = 2 \times 10^6$ K, which are representative values for pre-flare conditions. The Alfven velocity is then 6×10^6 m s^{-1}, the Reynolds' number $R_m \approx 10^{14}$, the

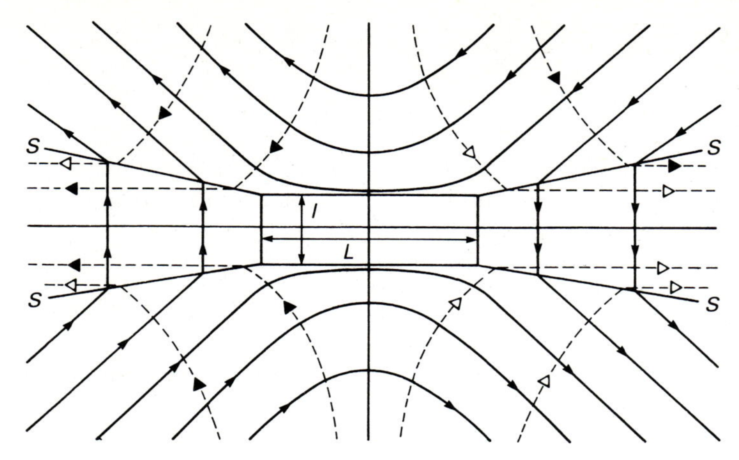

Figure 12.10. The geometry of reconnection according to Petschek (1964). The solid lines represent the magnetic field lines and the dashed lines the streamlines of the plasma flow. The standing shock waves are labelled S. It can be seen that the magnetic field lines do indeed reconnect in this picture. (From E. Tandberg-Hanssen and A. Gordon Emslie (1988). *The physics of solar flares*, page 155, Cambridge: Cambridge University Press.)

reconnection timescale about 1 s, the energy in the neutral sheet about 3×10^{16} J and hence the luminosity of the current sheet is $L_f \sim 3 \times 10^{16}$ W. The energy within the volume of the flare is $E \sim 3 \times 10^{23}$ J, not a particularly energetic flare, and therefore this energy would be liberated over a timescale $\tau \sim E/L_f \sim 10^7$ s, corresponding to months rather than hours or the much shorter timescales associated with the impulsive phases of large flares. Evidently, the energy of the pre-flare region is not liberated rapidly enough by a factor which can be as large as 10^8.

It was pointed out by Petschek (1964) that the dissipation rate can be increased if standing shock waves form on either side of the neutral sheet, creating the geometry shown in Fig. 12.10. The magnetic field lines genuinely reconnect as shown in the sketch. According to Petschek's analysis, the reconnection velocity can be as large as $v_A/\ln R_m$. The structure of these neutral sheets and their associated shock waves requires careful attention to the detailed microphysics and goes far beyond what can be covered here. Priest and Forbes (1986) have generalised the models for the reconnection of magnetic field lines in neutral sheets and shown that the reconnection velocity can almost be as large as the Alfven velocity v_A. They stress that the reconnection velocity is critically dependent upon the boundary conditions.

There are a number of ways in which the energy release is modified within the neutral current sheet. Among the most important of these considerations is the question of the stability of the current sheet. It turns out that the current sheet is susceptible to what is known as the *tearing mode instability*. In this instability, the neutral current sheet becomes unstable and, rather than the plasma flowing uniformly into the current sheet, the sheet breaks up into a number of X and O-

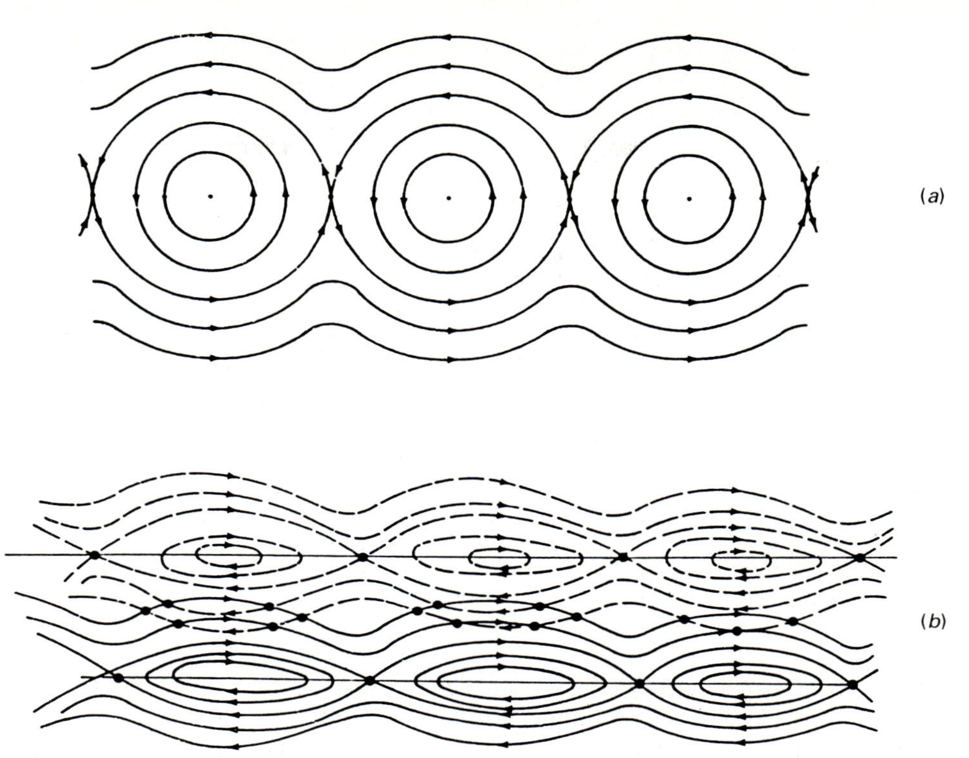

Figure 12.11. (*a*) Illustrating the formation of magnetic islands and O and X-type neutral points as a result of the development of the tearing mode instability in a neutral current sheet. (*b*) Illustrating the overlapping of magnetic reconnection regions in which tearing mode instabilities have developed. This process increases the number of magnetic neutral points and enhances the rate of magnetic field dissipation in the region. (From E. Tandberg-Hansen and A. Gordon Emslie (1988). *The physics of solar flares*, pages 156–7, Cambridge: Cambridge University Press.)

neutral points as illustrated in Fig. 12.11 (*a*). Thus, the current sheet is converted into a layer of current filaments. The flow pattern is different from that in the simple neutral current sheet. The magnetic islands collapse and dissipate energy with a much smaller length scale than that of the current sheet itself. The effect of the instability is not necessarily to enhance the reconnection rate but rather it makes the process impulsive and bursty. In addition to the simple tearing mode instability, Spicer has pointed out that these magnetic reconnection regions may overlap so as to produce many more X and O-neutral points (Fig. 12.11 (*b*)).

In addition to these instabilities, the resistivity of the plasma may be enhanced because of the phenomenon of *anomalous resistivity*. An important possibility is that the resistivity of the plasma is significantly increased because of the presence of waves or turbulence in the plasma. The effect of these waves is to move the particles of the plasma in a coherent fashion so that an individual electron interacts with the collective influence of a large number of particles rather than with a single particle. An example of the type of plasma instability which could have this effect in the neutral sheet is the *ion-acoustic instability*. This instability occurs when the drift velocity of the plasma exceeds the ion sound speed (or ion acoustic speed) $c_i = (kT_e/m_p)^{\frac{1}{2}}$. This condition is likely to be satisfied in the neutral current sheets in Solar flares.

Insofar as there is a consensus in this complex area, it is agreed that much more than simple neutral sheets are needed to account for the energy release observed in Solar flares. Whilst reconnection of lines of force can explain the basic process of energy release, more realistic models need to take account of the detailed physical processes taking place in the vicinity of neutral sheets. Detailed numerical modelling of these processes has been carried by Priest and his coworkers (see e.g. Forbes, Malherbe and Priest (1989), Priest and Forbes (1990)). The basic conclusion is that it may be possible to account for the most energetic flares, both their total energy release and their high luminosities during their impulsive phases. It is a salutory story for astrophysicists in that, even in what might appear to be a relatively straightforward problem in Solar magnetohydrodynamics, the precise means by which a solar flare generates its energy is far from trivial and by no means established with certainty.

12.4.4 *Models for solar flares*

There have been many attempts to model solar flares but we describe only two of the most popular field configurations which seem to account for most of the essential geometric features. The key point is to conceive of plausible magnetic field configurations in which neutral sheets appear naturally.

Sturrock (1968) proposed that the solar flare consists of a 'helmet and streamer' configuration (Fig. 12.12(*a*)). A neutral sheet is formed in the streamer where the magnetic field lines run in opposite directions and terminate in a Y-type neutral point. The field lines below the neutral point are closed. Reconnection begins at the Y-type neutral point which becomes the seat of a source of strong heating and possibly of particle acceleration. One of the attractions of the neutral sheet models is that strong electric fields can be induced in the neutral sheets where the strong magnetic fields are convected into the region in which the field is dissipated. These electric fields can contribute to the acceleration of high energy particles in the Solar flare. The strong magnetic energy conversion in the vicinity of the Y-type neutral point is the basic energy source for all the other phenomena observed in the flare. The hot gas and high energy particles stream down the magnetic loops below the reconnection region and deposit their energy in the denser regions of the solar atmosphere where they are responsible for the production of Hα ribbons. As further-out field lines are convected into the reconnection region, the Hα ribbons at the footpoints move outwards on either side of the neutral sheet producing the characteristic 'two-ribbon' flare pattern. The field lines in the 'streamer' above the Y-type neutral point are open and are responsible for the ejection of streams of particles into the interplanetary medium.

Another popular model is known as the *emerging flux model* due to Heyvaerts, Priest and Rust (1977) (Fig. 12.12(*b*)). This model was motivated by the fact that

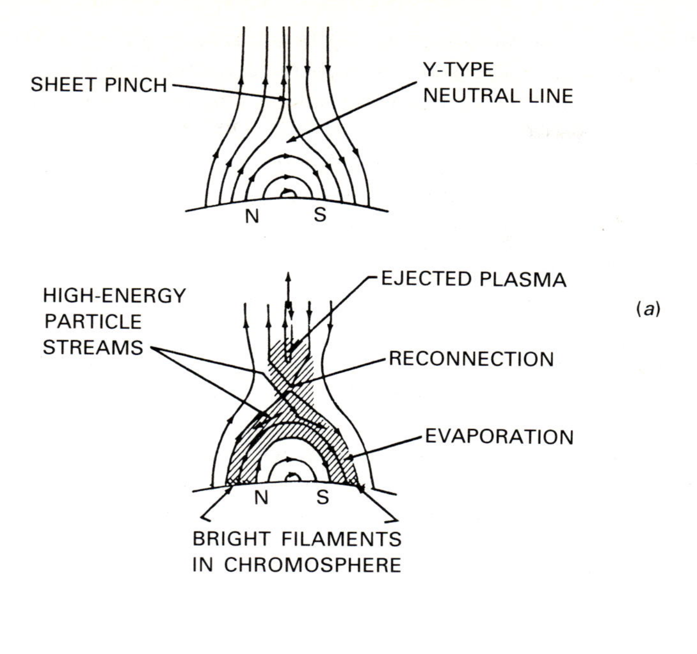

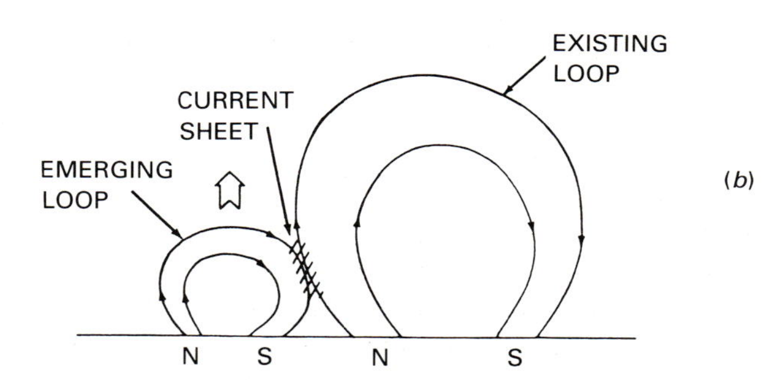

Figure 12.12. (*a*) The magnetic field geometry of a solar flare proposed by Sturrock (1968). The origin of the various phenomena observed during solar flares are indicated on the diagrams. (*b*) A schematic diagram of the emerging flux model of solar flares of Heyvaerts, Priest and Rust (1977). The emerging flux tube collides with a pre-existing flux tube, creating a neutral sheet. In this case, reconnection is a 'driven' process but relaxes to a steady state during the main phase. (From E. Tandberg-Hanssen and A. Gordon Emslie (1988), *The physics of solar flares*, pages 162–3, Cambridge: Cambridge University Press.)

emergent flux tubes are often observed just prior to the development of a two-ribbon flare. The idea is that the emerging flux tube rises up through the atmosphere driven by buoyancy and collides with a larger pre-existing magnetic flux loop. Reconnection of the magnetic lines of force takes place at the neutral sheet which is formed between the two flux tubes. In this case, the reconnection is driven by the buoyancy of the emerging flux tube. The magnetic field distribution

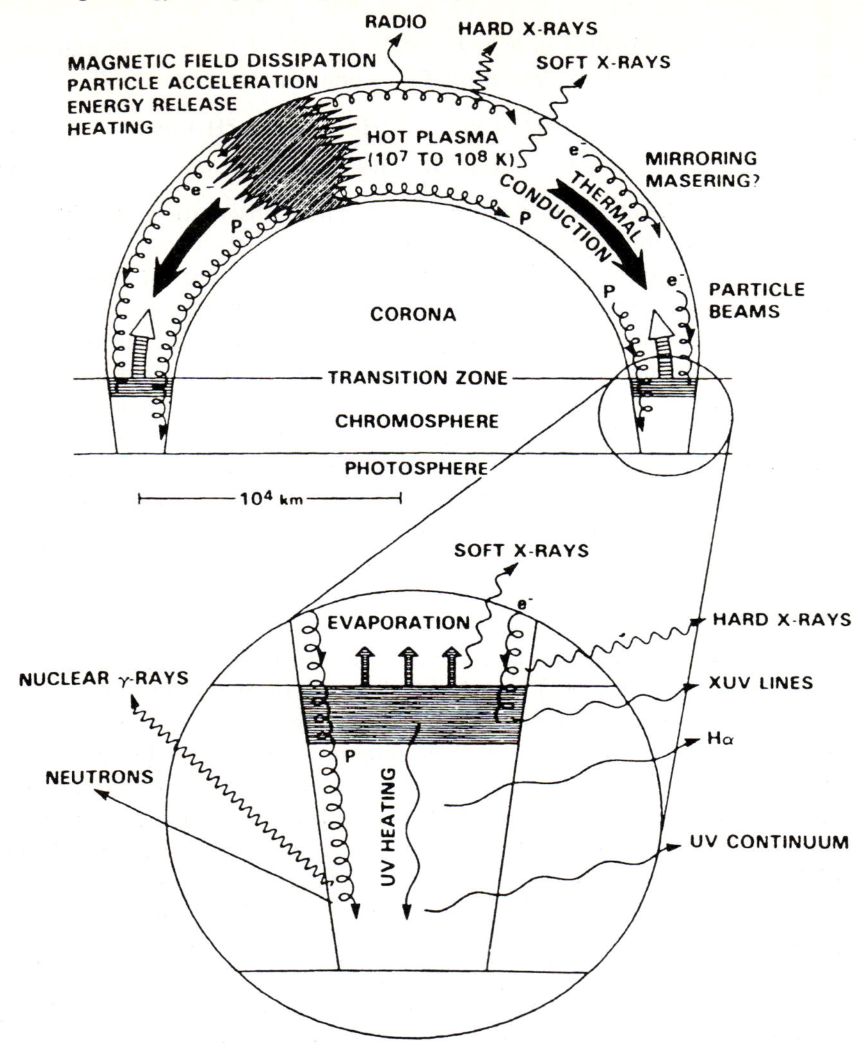

Figure 12.13. A possible model for a solar flare which is consistent with a wide range of observations (from J. B. Gurman (ed.) (1987). *NASA's Solar Maximum Mission: A new look at the Sun*. Greenbelt, Maryland: NASA Goddard Space Flight Center Publications). This picture shows the simplest geometry of a magnetic reconnection region located in a magnetic loop. The energy released in the magnetic reconnection process results in very rapid acceleration of electrons, protons and nuclei as well as the strong heating of a superhot flare 'kernel'. Hard X-rays, γ-rays, microwave radiation and neutrons are produced during this impulsive phase of the flare. These particles and plasma propagate along the magnetic field lines and interact with the material of the chromosphere at the footpoints of the loops where the particle beams are thermalised producing extreme ultraviolet radiation and Hα bursts. The energy of the beam may not be removed sufficiently rapidly by radiation and so explosive 'evaporation' of the chromospheric material occurs. This material is heated to temperatures of the order of 10^6 K or greater and is responsible for the soft X-rays. This gas gradually fills the magnetic loop during the gradual phase of the flare.

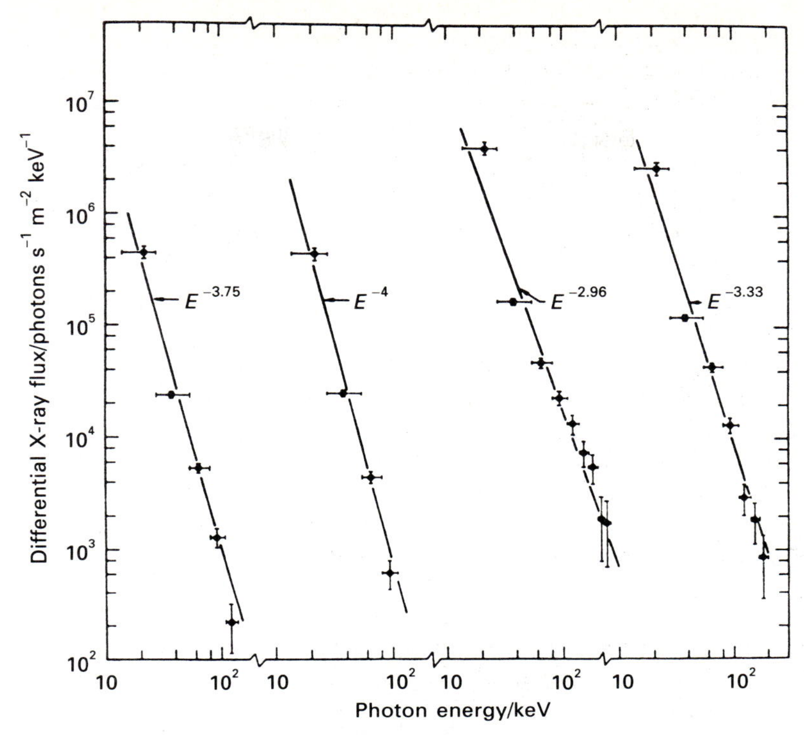

Figure 12.14. Hard X-ray spectra of four different events observed by the OSO-5 Solar observatory. The spectra can be well approximated by power laws over the energy range 10–100 keV. The horizontal bars represent the channel widths of the detectors. (From E. Tandberg-Hansen and A. Gordon Emslie (1988). *The physics of solar flares*, page 173, Cambridge: Cambridge University Press: after Kane, S. R., Frost, K. J. and Donnelly, R. F. (1979). *Astrophys. J.*, 234, 669.)

In a simple approximation, the spectrum is flat up to a frequency ω_{max} which corresponds to an electron of energy E_0 giving up all its energy in a single collision, $\hbar\omega = \frac{1}{2}m_e v^2 = E_0$. Therefore, we can find the emission spectrum for optically thin non-thermal non-relativistic bremsstrahlung by integrating over the energy distribution of all energies of electrons which can contribute to the intensity at frequency ω. We find

$$I(\omega) = \text{constants} \int_{E_0}^{\infty} N(E) E^{-\frac{1}{2}} \mathrm{d}E \tag{12.23}$$

If we assume that the spectrum of the electrons is of power-law form, $N(E)\,\mathrm{d}E \propto E^{-x}\,\mathrm{d}E$, the emission spectrum is $I(\omega) \propto \omega^{-(x-\frac{1}{2})}$. If we express this in terms of the number of photons per unit energy range, we divine by $\hbar\omega$ and hence the photon spectrum is

$$N(\omega) \propto \omega^{-(x+\frac{1}{2})} \tag{12.24}$$

However, we are interested in the case in which the electrons lose all their energy in colliding with the solar chromosphere. If we suppose that there is a steady inflow of electrons and that they lose all their energy in the emission region, then the emission spectrum of the electrons is modified because most of their energy is dissipated as heat through the ionisation loss process described in Chapter 2. Fortunately there is a simple way of treating this problem which we will use extensively in our discussion of the origin of cosmic rays in volume 2, Section 19.4. To quote a result derived in that volume, the steady state spectrum of electrons under continuous energy losses of the form, $-\mathrm{d}E/\mathrm{d}t = b(E)$ is given by the solution of

$$\frac{\mathrm{d}}{\mathrm{d}E}[b(E)n(E)] = Q(E) \tag{12.25}$$

where $Q(E)$ is the injection energy spectrum which can be assumed to be of the form κE^{-x}. Now, in the non-relativistic regime, the dominant energy loss for the electrons is ionisation losses as described by equation (2.10). The key point about this equation is the dependence upon the energy of the electron which is $\mathrm{d}E/\mathrm{d}t \propto v^{-1} \propto E^{-\frac{1}{2}}$ in the non-relativistic regime. We can now solve for the steady state spectrum of the electrons in the emission region and find that

$$N(E) \propto E^{-(x-\frac{3}{2})} \tag{12.26}$$

i.e. the effect of the energy losses is to flatten the energy spectrum of the electrons – this makes complete sense since the lowest energy electrons lose their energy fastest by ionisation losses. We can now enter this relation into the energy spectrum of the electrons as above and we find that the spectrum of the X-ray emission is 1.5 powers of ω flatter than was found in the previous example in which the particles lose only a small fraction of their energy in the source region,

$$N(\omega) \propto \omega^{-(x-1)} \tag{12.27}$$

This is exactly the same result found by Brown (1971) in his seminal paper. A much more general treatment of the processes involved in heating by fluxes of non-thermal electrons, including the hydrodynamics of the motion of the plasma along the field lines is given by Brown and Emslie (1989).

A problem with the thick target model and all non-thermal models, such as those involving a non-thermal population of electrons trapped in the corona, is that most of the energy of the non-thermal electrons is not radiated away as bremsstrahlung but is dissipated as heat and therefore large fluxes of high energy electrons are needed. An alternative possibility is that the electrons have a roughly Maxwellian energy distribution and so lose much more of their energy by radiation and conduction. The problem is then to heat up the flux of electrons to high enough energies to produce intense fluxes of hard X-rays. In one viable model, known as the 'dissipative thermal model' (Brown, Melrose and Spicer 1979), the streams of high energy particles are prevented from escaping from their place of origin by streaming instabilities associated with the ion-acoustic instability. In addition to restricting the flow of the bulk of the particles to the ion sound speed, the ion-acoustic waves enhance the resistivity of the plasma, increasing the heating of the hot gas. The apparent power-law distribution of the X-ray emission from the solar flare is attributed to the sum of a number of thermal spectra at different temperatures. The advantage of this model is that, because the electron distribution is close to local collisional equilibrium, more of the thermal energy of the electrons is used in producing bremsstrahlung as compared with the thick target model or indeed any non-thermal model.

There is an interesting relation between the spectra of the hard X-ray emission of the high energy electrons and their gyrosynchrotron emission. We will have an enormous amount to say about the emission of high energy electrons in Volume 2, Chapter 18. *Gyrosynchrotron radiation* is the radiation of mildly relativistic electrons gyrating in a magnetic field. In many areas of interest to astronomers, the electrons are of very high energy indeed, $E \gg m_e c^2$, and then this radiation process is simply known as *synchrotron radiation*. Fortunately, the same rules for relating the energy spectra of the electrons to the emitted radio spectrum applies to gyrosynchrotron radiation and synchrotron radiation. As we will show in some detail in Volume 2, Chapter 18, if the energy spectrum of the electrons is $N(E) \propto E^{-x}$, the spectrum of gyrosynchrotron radiation is $I(\nu) \propto \nu^{-\alpha_R} \propto \nu^{-(x-1)/2}$. On the other hand, the hard X-ray emission due to this spectrum of electrons is $N_X(\varepsilon) \propto \varepsilon^{\alpha_X} \propto \varepsilon^{-(x+\frac{1}{2})}$. Therefore, we see that the relation between the spectral indices in the radio and hard X-ray wavebands is

$$\alpha_R = 0.5\alpha_X - 0.75 \qquad (12.28)$$

According to Heyvaerts (1981), this relation is quite a good representation of the spectra of coincident hard X-ray and radio bursts observed during the impulsive phase of the flare. Notice that it does not matter which model is correct in producing the radio and X-ray emission – the only important thing is that the same electron spectrum is responsible for both emissions. There is one important caveat about this simple result. It has been assumed that the gyrosynchrotron radiation is optically thin and the analysis has also neglected the strong directional dependence of the radiation process. More details of these calculations as applied to solar flares are given by Crannell, Dulk, Kosagi and Magun (1988).

γ-ray lines

The other aspect of the spectra of the hard X-rays and γ-rays from solar flares is the presence of γ-ray lines, an example of which was shown in Fig. 5.6. Strong nuclear emission lines due to carbon, oxygen, neon, magnesium, silicon and iron are present in that spectrum and this enables the abundances of these elements in the solar flares to be determined. The excitation is due to collisions between high energy particles accelerated in the flare, principally protons and ^{4}He nuclei, and the ambient cold matter. According to Ramaty, Dennis and Emslie (1988), the intensities of the lines of these elements are in reasonable agreement with the expectations of models in which the excitation of the γ-ray lines is attributed to collisions between the accelerated ions and typical material in the solar atmosphere and in which it is assumed that the cross-sections for the high energy interactions are those measured in the laboratory. The success of this model is important because it enables a number of other possible models for the production of the nuclear emission lines to be excluded – for example, thermonuclear processes.

The neutron capture line at 2.223 MeV is produced by the following interaction between a hydrogen nucleus and a slow neutron

$$p + n \rightarrow {}^{2}H + \gamma \tag{12.29}$$

The neutrons themselves are produced by spallation through reactions such as those shown in (12.22) between high energy protons and the ambient material of the Solar atmosphere and have to be slowed down to thermal energies before neutron capture by the protons can take place efficiently – this results from the strong dependence of the neutron capture cross-section upon energy. As a consequence, this line is created quite deep in the Solar atmosphere. High energy neutrons can be created in spallation interactions of high energy helium nuclei with the ambient gas and these have been observed in the vicinity of the Earth, immediately following Solar flares. Timing of these events has shown that the high energy neutrons must have been created at the same time as the high energy electrons responsible for the hard X-ray and γ-ray emission. A detailed study of the implications of studies of neutrons in flares, the observation of the decay products of neutron decay in the interplanetary medium and direct observations of Solar neutrons at the top of the Earth's atmosphere is given by Chupp (1988).

12.5.2 *Solar energetic particles*

One of the most interesting aspects of the high energy astrophysics of solar flares for the extra-Solar System astronomer is the fact that the chemical abundances of the particles accelerated in Solar flares can be detected directly by observations made with cosmic ray telescopes on board space platforms. The energy spectra and chemical abundances of the particles turn out to have many properties in common with the high energy particles present in the interstellar medium and presumably those accelerated in exotic objects such as supernova explosions and active galactic nuclei. Guzik (1988) has reviewed many aspects of the study of Solar energetic particles relevant to our story.

One of the first problems to be addressed is the fact that the solar energetic particles arrive at the Earth having traversed the interplanetary medium from their source in the solar flare to the top of the Earth's atmosphere. They are therefore subject to the complete range of interplanetary phenomena which can strongly influence their propagation to the Earth and, in particular, modify the injection energy spectrum of the particles. In addition to the processes of Solar modulation described in Chapters 10 and 11, interplanetary shock waves can significantly modify the energy spectra of the particles. Therefore, care must be taken to select those samples for which there is evidence that the particles have travelled along simply-connected paths from the flares to the Earth.

The energies of particles detected on Earth range from about one to several hundred MeV nucleon^{-1}. Some examples of the spectra of protons are shown in Fig. 12.15. It can be seen that there is considerable variation in the forms of the

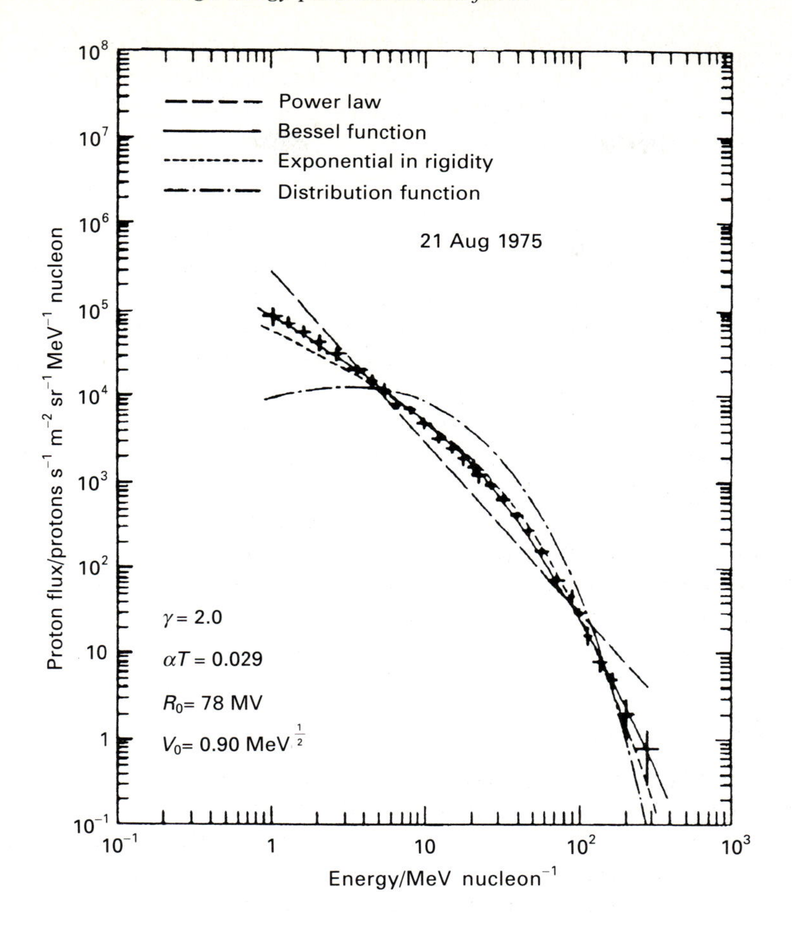

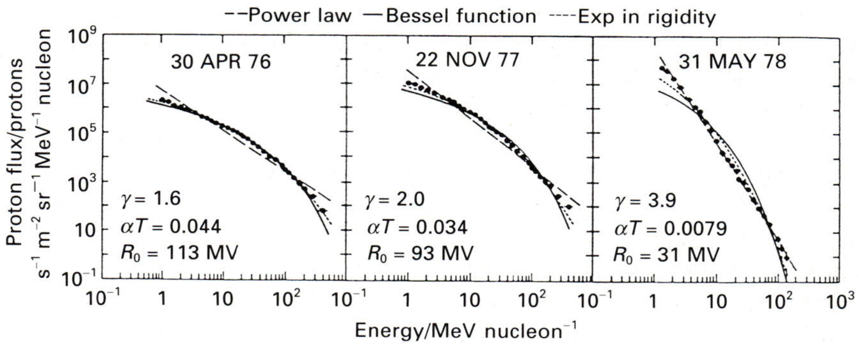

Figure 12.15. Examples of the energy spectra of protons accelerated in solar flares as observed at the top of the atmosphere by the IMP 7/8 satellites. (From R. E. McGuire and T. T. von Rosenvinge (1984). *Adv. Space Res.*, **4**, No. 2–3, 117.) The spectra have been fitted to various simple forms of spectra, including a power-law fit, a Bessel function fit, and exponential fits in rigidity and velocity. The authors describe the techniques involved in deriving these spectra which are believed to be unaffected by the effects of propagation from their source in the flare to the top of the atmosphere.

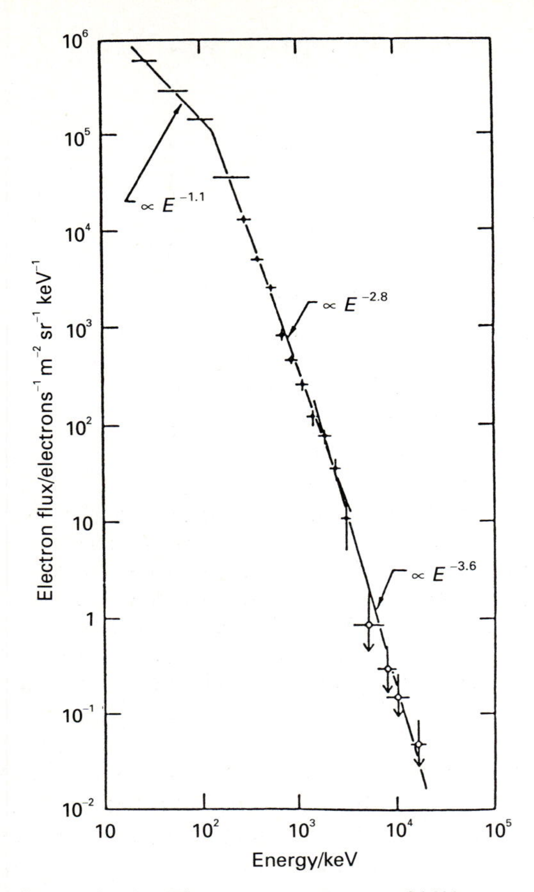

Figure 12.16. The energy spectrum of high energy electrons from the flare of 7 September 1973 as observed by the IMP 6/7 satellites. There is a change in slope of the spectrum at about 100 keV, the index being $\gamma \approx 1.1$ at low energies and about 2.8 at higher energies. (From Guzik, T. G. (1988). *Solar Physics*, **118**, 185.)

spectra, although they are qualitatively of roughly the same form. The figures show various fits to the spectra of the events, the adopted forms including power-laws, Bessel function fits and exponential fits in rigidity and velocity (or momentum). According to McGuire and von Rosenvinge (1984), Bessel functions provide generally the best fits, although there are examples in which power-law or double power-law fits can provide better agreement with the observations. The attraction of the Bessel function fits is that they are among some of the simpler solutions for the expected spectrum of particles accelerated by random scattering in flares. If the spectra are described in terms of power laws, spectral indices $\gamma \approx 1.5$–4 seem to be required. The spectra are more poorly defined for elements heavier than helium.

The spectrum of solar energetic electrons is shown in Fig. 12.16. The spectral index in the energy range 10–100 keV is about 2.8 with some evidence for a steepening above 100 keV and possibly an even steeper spectrum above about

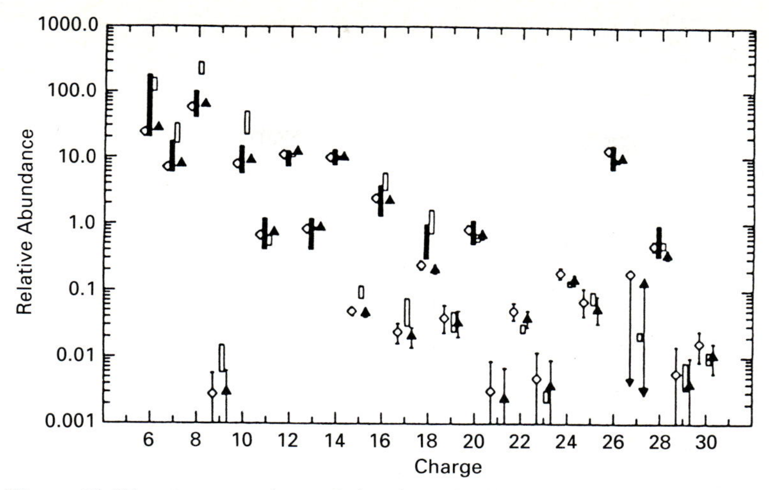

Figure 12.17. A comparison of the element abundances in the solar energetic particles with other abundance measurements. The average observed abundances for solar energetic particles are shown as filled triangles. These have been converted to source abundances in the corona by Breneman and Stone (1985) and these are shown as open diamonds. The abundances in the solar corona from EUV and X-ray spectroscopic observations are shown as filled boxes and the cosmic (or photospheric) abundances of the elements are displayed as open boxes. The normalisation is to silicon ($Z = 14$). (From Guzik, T. G. (1988). *Solar Physics*, **118**, 185.)

3 MeV. According to Guzik (1988), this form of spectrum is consistent with observations of the hard X-ray spectra and radio emission associated with the impulsive phase of solar flares although this statement is clearly model-dependent and considerable variations in the spectral indices are observed. It is therefore likely that this form of spectrum represents the spectrum of electrons accelerated in the flare.

Of particular interest are the element abundances of the solar energetic particles. The best data are derived from observations of particles with energies in the range 10–40 MeV nucleon^{-1} although data are available outside this energy range. The interesting comparisons are between these element abundances, those in the local interstellar medium and also those of the solar corona which have been derived from analyses of extreme ultraviolet and X-ray spectrographic studies (Meyer 1985). The results presented by Guzik (1988) are shown in Fig. 12.17 in which the three different abundance measurements are compared as a function of the atomic number (or charge) of the heavy elements. The solar energetic particle abundances have been found by averaging over many flares. Significant variations in the abundances are found in some flares, for example, those in which there are enhancements in the heavy ion abundances ('heavy-ion flares').

It is interesting that there is reasonable agreement between the coronal and high energy particle abundances but there are significant differences between these and the local cosmic, or interstellar, abundances. The disagreement is particularly significant for carbon, nitrogen, oxygen, neon, phosphorus, sulphur and argon. It

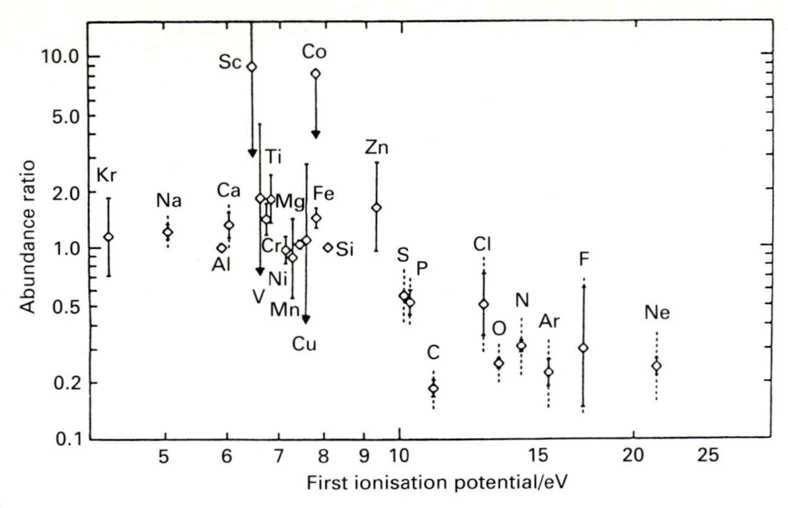

Figure 12.18. The ratio of the solar energetic particle abundances to the local cosmic abundances plotted as a function of the first ionisation potential of each element. The normalisation is to silicon. The dashed error bars include the uncertainties in the local galactic abundances. (From Guzik, T. G. (1988). *Solar Physics*, **118**, 185.)

was noted as long ago as the early 1970s, that this discrepancy can be ordered by the first ionisation potential of the elements involved. The ratios of the element abundances in the Solar energetic particles to the cosmic abundances are shown in Fig. 12.18, ordered by first ionisation potential. It can be seen that the elements showing a deficit in the Solar energetic particle fluxes have first ionisation potential greater than about 9.5 eV. It is of the greatest interest that exactly the same trend is apparent in the abundances of high energy particles in the local interstellar medium (Fig. 12.19).

The interpretation of these data is not at all clear. If the 'transition energy' of 8–12 eV is interpreted as a measure of the temperature of the region in which the particles are accelerated, a temperature of about 8×10^5 K is found, typical of the very base of the corona or the top of the chromosphere rather than the photosphere. However, if the particles were accelerated in these regions, the elemental abundances would be strongly modified by the effects of spallation as the particles escape into the interplanetary medium. Furthermore, measurements of the charge state of the elements observed in the solar energetic particles are more consistent with coronal temperatures rather than with values less than 10^5 K. Finally, it has been found that the element abundances in the Solar Wind itself are similar to those of the solar energetic particles. These arguments suggest that the correlation of element deficiency with first ionisation potential may not be associated with the mechanism by which the particles are injected or accelerated but rather may simply reflect intrinsic differences between the composition of the photosphere and the corona. We will return to this topic in Chapter 19 of Volume 2 in the context of the origin of Galactic cosmic rays.

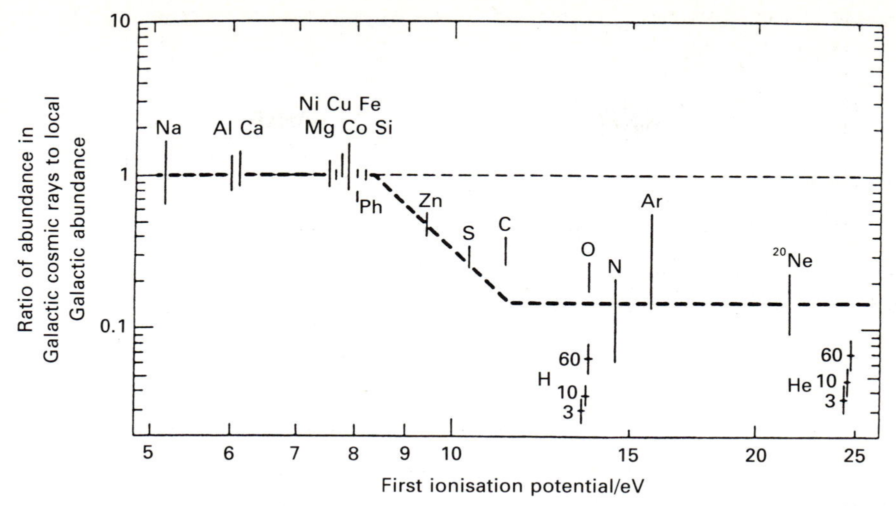

Figure 12.19. The same plot as in Fig. 12.18 but for the abundances of Galactic cosmic rays (from J-P. Meyer (1985). *19th Intl. Cosmic Rays Conference*, La Jolla, USA, Vol. 9, page 161.)

12.6 The acceleration of particles in solar flares

This section will be a severe disappointment to those astronomers who might look to solar flares for guidance in solving the problems of accelerating high energy particles in Galactic and extragalactic systems. This remains among the most difficult problems of high energy astrophysics and we will devote the whole of Chapter 21 of Volume 2 to these problems. The interested reader should study the impressive review article by Heyvaerts (1981) for details of the enormous amount of work which has been done in trying to understand the processes by which particles can be accelerated to very high energies in solar flares.

We will leave the bulk of the discussion to Chapter 21 and concentrate here upon the one important aspect in which the problems of particle acceleration in solar flares may differ from the mechanisms discussed in high energy astrophysics. In contrast to the situation in most astrophysical problems, we know that the principal source of energy is the reconnection of magnetic field lines and we can be certain that, in these regions, there are strong electric fields generated because of the facts that curl $\mathbf{B}$ and $\partial \mathbf{B}/\mathrm{d}t$ are not zero. The question is how effective these processes are in creating electric fields which can accelerate charged particles.

12.6.1 *DC electric field acceleration*

Let us suppose that a DC electric field E is created in the reconnection region. We can then write down the equation of motion for a charged particle in this field, taking account of the fact the particle will collide with other particles in the plasma in the usual stochastic way (see Sections 10.4 and 12.4.1). The collision frequency ν_c for electrostatic collisions can be found from the expression (12.5). We can therefore write the equation of motion for an electron

$$m_e \, \mathrm{d}\mathbf{v}/\mathrm{d}t = -e\mathbf{E} - \nu_c m_e \mathbf{v} \tag{12.30}$$

For simplicity, we consider fields parallel to the velocity of the electron and then we can write

$$m_e \, \mathrm{d}v/\mathrm{d}t = eE - \nu_c m_e v \tag{12.31}$$

We now require an expression for the collision frequency ν_c. We can adopt the expression (10.3) for the case of electron–electron collisions by setting $m = m_e$, $N = n_e$ and $Z = 1$. We therefore find that

$$\nu_c = \frac{e^4 n_e \ln \Lambda}{2\pi\varepsilon_0^2 m_e^2 v^3} = \frac{e^4 n_e \ln \Lambda}{2\pi\varepsilon_0^2 m_e^{\frac{1}{2}} (3kT)^{\frac{3}{2}}} \tag{12.32}$$

In the last equality of expression (12.32), we have set the mean squared velocity of the electrons equal to $3kT/m_e$ for the case in which the velocity distribution of the electrons is Maxwellian at temperature T. It can be seen that, because of the v^{-3} dependence of the collision frequency upon the velocity of the electron, once the electron's velocity becomes greater than a critical velocity v_c, the effect of collisions becomes less and less important with increasing velocity and hence the electrons are accelerated without any impediment under the influence of the electric field. This process is known as *electron runaway*. For velocities less than the critical velocity, the particles are not accelerated. The critical velocity is found by setting the right hand side of equation (12.31) equal to zero.

$$v_c = \left(\frac{e^3 n_e \ln \Lambda}{2\pi\varepsilon_0^2 m_e E}\right)^{\frac{1}{2}} \tag{12.33}$$

Correspondingly, for a thermal plasma, there is a critical electric field associated with this process which is known as the *Dreicer field* which we can write

$$E_D = \frac{e^3 n_e \ln \Lambda}{6\pi\varepsilon_0^2 kT} \tag{12.34}$$

at which *all* the electrons runaway. We can write this relation in terms of the Debye length of the plasma $\lambda_D = (\varepsilon_0 kT/ne^2)^{\frac{1}{2}}$ in which case

$$E_D = \frac{e \ln \Lambda}{6\pi\varepsilon_0 \lambda_D^2} \tag{12.35}$$

A remarkably similar result is quoted by Heyvaerts (1981), considering the simplifications we have adopted in deriving the collision frequency. Putting in the values of the numerical constants, we find

$$E_D = 2 \times 10^{-13} \frac{n_e \ln \Lambda}{T} \ \mathrm{V\,m^{-1}} \tag{12.36}$$

where n_e is measured in electrons $\mathrm{m^{-3}}$ and T in degrees Kelvin.

Let us put in some values which may be appropriate for the conditions found in Solar flares. We use the values adopted in Section 12.4.3. For the values used in that section, the Dreicer field is $E_D = 10^{-2}$ V m^{-1}. We can estimate the strength of the electric field in the neutral sheet by writing $E \sim v_R B \approx 0.02$ V m^{-1}. It can be seen that this process may well be important in the neutral sheets in solar flares.

There are, however, a number of problems. First of all, there is a limit to the current which can be associated with the accelerated beam because the beam of particles itself will create a magnetic field which cannot be greater than the initial magnetic field strength. The result is that a return current is induced which tends to neutralise the effect of the electric field. Second, the beam of accelerated particles will be subject to a variety of streaming instabilities. In particular, the beam may excite ion sound waves which will enhance the resistivity of the plasma. If only a small fraction of the electrons have velocities which exceed the critical velocity, these may be accelerated without exciting plasma instabilities. This process may also be important as part of the stochastic acceleration of particles. Heyvaerts (1981) gives a comprehensive description of the plasma instabilities which are likely to be associated with electron runaway. The significance of this mechanism of particle acceleration in solar flares has not been established.

12.6.2 *Double layers*

Another possible means of creating a potential gradient in which particles can be accelerated was proposed by Alfven and Carlqvist (1967). If a large enough current is present along a magnetic flux tube and a density fluctuation in the current arises, then it is possible that a potential gradient is set up along the flux tube. The type of configuration envisaged is shown in Fig. 12.20. It is called a double layer because it consists of two sheets of opposite charge density which act somewhat like the plates of a capacitor. Such a configuration is contrary to the usual impression that a plasma cannot tolerate charge accumulation but it *can* if it is strongly enough driven. The idea is that protons are accelerated down the potential gradient from left to right and electrons are accelerated from right to left. In fact, analysis of this configuration using the methods developed by Bernstein, Green and Kruskal (1957) has shown that self-consistent solutions of this form can be found. One can envisage the acceleration of the particles to take place in a similar manner to that discussed in Section 12.6.1. The applicability of this process to solar flares has been analysed by Heyvaerts (1981) to whom the reader is referred for further details. He concludes that there are three basic problems with the model – first, large currents are necessary to sustain the field configurations; second, whilst it can be shown that self-consistent charge and current distribution can be found, instabilities associated with the streams of accelerated particles are certain to limit the utility of the process; and third, the formation of these regions is not believed to be possible.

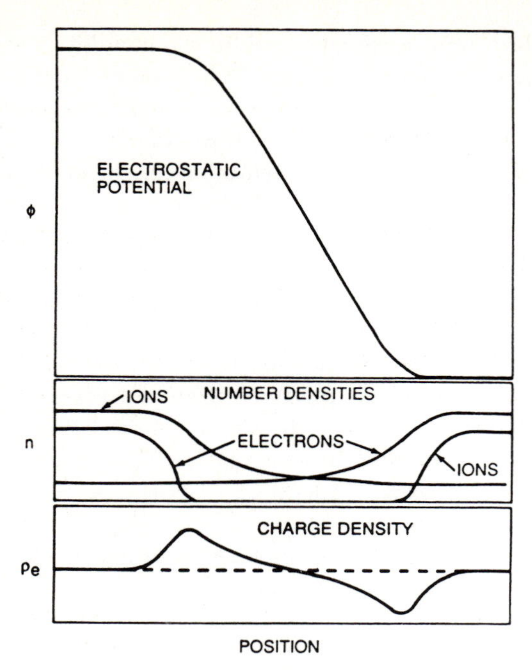

Figure 12.20. A schematic diagram of a double layer. Charged particles are accelerated when they find themselves in the electrostatic potential, the ions being accelerated down the potential and electrons up the potential. (From E. Tandberg-Hanssen and A. Gordon Emslie (1988). *The physics of solar flares*, page 167, Cambridge: Cambridge University Press.)

12.6.3 *Other processes*

In addition to these processes specific to flares, there are several other processes which are of importance more generally in astrophysics and which are almost certainly important in solar flares. Among these, the most important are first and second order Fermi acceleration but we will delay discussion of these topics to Volume 2, Chapter 21.

12.7 Particle acceleration in other regions in the Solar System

Almost certainly, solar flares are the most relevant phenomena in the Solar System for the high energy astrophysics of Galactic and extragalactic systems. It is, however, worthwhile noting other phenomena within the Solar System which may have some bearing on these topics.

The Earth's magnetosphere

We described in some detail the overall structure of the Earth's magnetosphere in Section 10.7. There are two footnotes to that story which we should bear in mind for the considerations of Volume 2, Chapter 21. First, it has been established by *in situ* experiments that particles are accelerated at the shock wave at the interface between the magnetosphere itself and the incident Solar Wind. This is an excellent example of particle acceleration in a collisionless shock wave and in Chapter 21 we will study how successfully the theory can account for the observations. This process is of the greatest importance for all astrophysical environments in which shock waves occur.

The second point concerns the magnetohydrodynamics of the magnetosphere. Space observations have now delineated the current structures present throughout the system so that models of the magnetic field distribution and its dynamics can be compared in detail with the theory. An example of this is the process of reconnection of lines of force in the neutral sheet which lies within the magnetotail and which stretches to many Earth radii. The processes of particle and magnetic field convection into this neutral current sheet and the instabilities associated with it have now been observed. These processes are of considerable interest for astrophysicists.

Io

Of all the discoveries made during the Voyager missions to the outer planets, none are more remarkable than the images of Io, the satellite of Jupiter closest to the planet itself. Whole books have been written about this remarkable object and its relation to Jupiter. Io does not have an internal heat source but the tidal forces due to its proximity to Jupiter are so strong that the mechanical stresses acting on its interior provide a strong heat source for the satellite. The combination of the tidal distortion of Io's shape and the dissipation of mechanical energy in its interior results in a great deal of volcanic activity on Io. The images of volcanos on Io taken by the Voyager space probes are among the most beautiful discoveries of these missions. The net result is that heavy elements, in particular sulphur and oxygen, are ejected into the atmosphere and environment of Io where they can be ionised by the Sun's ultraviolet radiation. Io therefore leaves behind a trail of ions and this is the origin of the ion torus which has been observed about Jupiter. The torus has the same radius as Io's orbit about Jupiter and lies in the magnetic equatorial plane of Jupiter. The result of all this activity is that Io has a highly ionised atmosphere and therefore can be considered a highly conducting sphere moving through Jupiter's strong magnetic field. It is little wonder that this gives rise to a number of remarkable plasma phenomena.

This is not the place to go into the details of these phenomena but one aspect is worthy of particular note. Jupiter is the strongest radio emitter of all the planets. In the wavelength range 100 MHz–10 GHz, the emission is the synchrotron radiation of high energy electrons trapped in the radiation belts about the planet. Presumably, these electrons are accelerated in the complex current systems in Jupiter's magnetosphere. There is certainly no lack of electrons because of the continual injection of atoms, ions and molecules from Io.

Equally remarkable is the radio emission at longer wavelengths, $\nu \sim 10$ MHz, which has a very steep spectrum and has never been observed above about

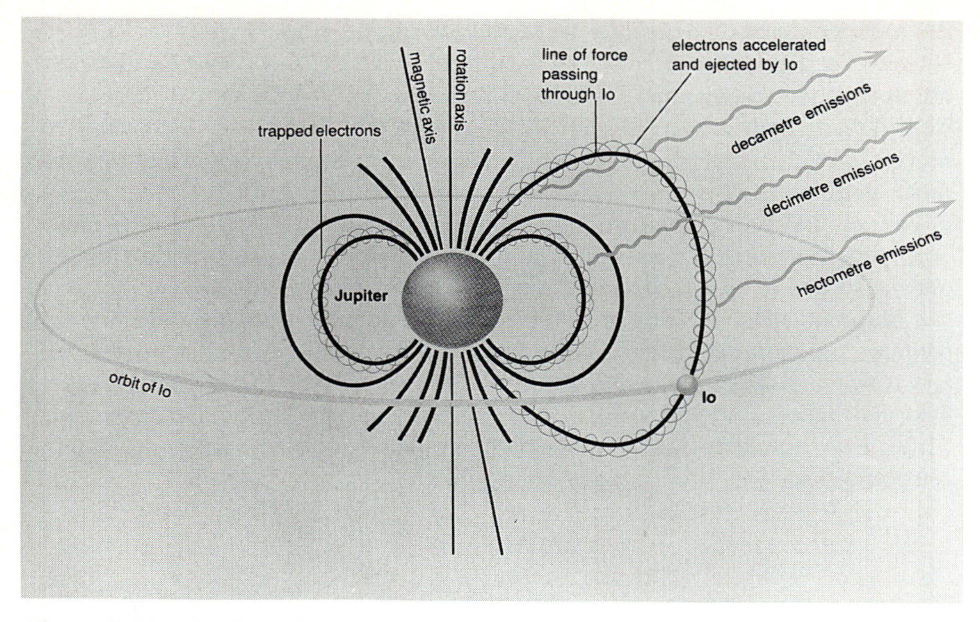

Figure 12.21. A schematic diagram illustrating the origins of the radio emissions from Jupiter. The decimetre radiation ($\lambda \sim 0.1$–1 m; $\nu \sim 0.3$–3 GHz) is synchrotron radiation from Jupiter's radiation belts. The decametre radiation ($\lambda \sim 10$–100 m; $\nu \sim 3$–30 MHz) is gyroradiation associated with particles tied to the magnetic lines of force passing through Io. The hectometre radiation ($\lambda \sim 100$–1000 m; $\nu \sim 0.3$–3 MHz) is associated with various coherent plasma processes and has been observed from space probes since radiation of this wavelength cannot penetrate the Earth's ionosphere. (From A. Boischot (1988). *The Cambridge atlas of astronomy*, eds J. Audouze and G. Israel, page 167, Cambridge: Cambridge University Press.)

40 MHz. The remarkable feature of this radiation is that it is highly variable but is strongly correlated with the position of Io in its orbit and with specific locations on the surface of the planet. Almost certainly, this radiation is the gyroradiation of electrons spiralling along the magentic field lines which join Io to Jupiter. The highest frequency radiation originates closest to the planet in the region of strongest magnetic field. In addition, the radiation has to be strongly beamed so that the radio bursts are only observed when Io and Jupiter are in the same relative positions. A sketch of some of these phenomena is included in Fig. 12.21.

These phenomena bear more than a passing resemblance to the observation of gyro-radiation features in the X-ray spectrum of X-ray binaries and possibly to the emission mechanisms in pulsars.

12.8 Conclusion

I had originally hoped that this final chapter of Volume 1 would have provided clues to physical processes which will be important in understanding high energy astrophysical objects such as supernovae, pulsars, quasars, radio galaxies and so on, the objects which will be the prime subject of Volume 2. In fact, many of the problems are just as complex as those found in Galactic and extragalactic systems and, generally, much more data are available. It is striking how many of the topics arising in this chapter will recur in Volume 2 – the acceleration of charged particles, the amplification of magnetic fields in different astrophysical environments, shock waves, the shaping of emission spectra by different energy loss processes for electrons, and so on. The links are clear and, in my view, their study will repay astrophysicists the effort needed to break into these fields.

References

Chapter 1

Further Reading

More details of the history of cosmic ray physics can be found in the following volumes:

Hayakawa, S. (1969). *Cosmic ray physics.* New York: Interscience.
Hillas, A. M. (1972). *Cosmic rays.* Oxford: Pergamon Press.
Sekido, Y. and Elliot, H. (eds) (1985). *Early history of cosmic ray studies.* Dordrecht: D. Reidel Publishing Co.

More details of the early history of atomic and nuclear physics can be found in:

Boorse, H. A., Motz, L. and Weaver, J. H. (1989). *The atomic scientists: A biographical history.* New York: John Wiley and Sons Inc.
Close, F., Marten, M. and Sutton, C. (1987). *The particle explosion.* Oxford: Oxford University Press.
Pais, A. (1987). *Inward bound.* Oxford: Oxford University Press.

Histories of the development of different branches of astronomy can be found in the following volumes:

Field, G. B. and Chaisson, E. J. (1985). *The invisible universe.* Boston: Birkhauser Boston Inc.
Henbest, N. and Marten, M. (1983). *The new astronomy.* Cambridge: Cambridge University Press.
Kellermann, K. and Sheets, B. (eds) (1983). *Serenditipitous discoveries in radio astronomy.* Green Bank: NRAO Publications.
Kondo, Y. (ed.) (1987). *Exploring the Universe with the IUE satellite.* Dordrecht: D. Reidel Publishing Co.
Learner, R. (1981). *Astronomy through the telescope.* London: Evans Brothers Ltd.
Ramana Murthy, P. V. and Wolfendale, A. W. (1986). *Gamma-ray astronomy.* Cambridge: Cambridge University Press.
Sullivan III, W. T. (ed.) (1985). *The early years of radio astronomy.* Cambridge: Cambridge University Press.
Tucker, W. and Giacconi, R. (1985). *The X-ray universe.* Cambridge: Harvard University Press.

For the conventions used in special relativity throughout this book, see:

Rindler, W. (1977). *Essential relativity*. New York: van Norstrand-Reinhold.
Longair, M. S. (1984). *Theoretical concepts in physics*, pages 261–75. Cambridge: Cambridge University Press.

References

Bothe, W. and Kolhorster, W. (1929). *Zeitschr fur Physik*, **56**, 571. (Translation: Hillas, A. M. (1972). *Cosmic rays*, p. 161. Oxford: Pergamon Press.)
Hess, V. (1912). *Physik. Zeitschr.*, **13**, 1084–91. (Translation: Hillas, A. M. (1972). *Cosmic rays*, p. 146, Oxford: Pergamon Press.)
Longair, M. S. (1984). *Theoretical concepts in physics*, p. 37–52. Cambridge: Cambridge University Press.
Wilson, C. T. R. (1901). *Proc. Roy. Soc.* **68**, 151.

Chapter 2

Further reading

There are many books which describe ionisation losses. The sources which I have mainly used are:

Enge, H. A. (1966). *Introduction to nuclear physics*. London: Addison-Wesley.
Hillas, A. M. (1972). *Cosmic rays*. Oxford: Pergamon Press.
Jackson, J. D. (1975). *Classical electrodynamics*. New York: Wiley and Sons, Inc.

The reader interested in following up the application of these ideas to the process of dynamical friction in stellar systems can consult:

Binney, J. and Tremaine, S. (1987). *Galactic dynamics*. Princeton: Princeton University Press.

Chapter 3

Further reading

The basic results on the radiation of accelerated charged particles are described in the standard textbooks. The ones which I refer to most are:

Clemmow, P. C. and Dougherty, J. P. (1969). *Electrodynamics of particles and plasmas*. London: Addison-Wesley.
Jackson, J. D. (1975). *Classical electrodynamics*. New York: Wiley and Sons, Inc.
Landau, L. D. and Lifshitz, E. M. (1975). *Course of theoretical physics*, Volume 2: *The classical theory of fields*. Oxford: Pergamon Press.

Bremsstrahlung is analysed in detail by Jackson (1975) and also by:

Rybicki, G. B. and Lightman, A. P. (1979). *Radiative processes in astrophysics*. New York: Wiley and Sons, Inc.
Blumenthal, G. R. and Gould, R. J. (1970). *Rev. Mod. Phys.* **42**, 237.

Results relevant to cosmic ray electrons are presented by:

Enge, H. A. (1966). *Introduction to nuclear physics*, London: Addison-Wesley.

Gaunt factors applicable to a wide range of astrophysical conditions are given by: Karzas, W. J. and Latter, R. (1961). *Astrophys. J. Suppl.* **6**, 167.

References

Abramovitz, M. and Stegun, I. A. (1965). *Handbook of mathematical functions*. New York: Dover Publications.

Gradshteyn, I. S. and Ryzhik, I. M. (1980). *Tables of integrals, series and products*. New York: Dover Publications.

Leighton, R. B. (1959). *Principles of modern physics*. London: McGraw-Hill.

Longair, M. S. (1984). *Theoretical concepts in physics*, pages 191–4. Cambridge: Cambridge University Press.

Chapter 4

Further reading

The processes discussed in this chapter are described in a number of books and articles. I have made most use of the following texts and articles:

Blumenthal, G. R. and Gould, R. J. (1970). *Rev. Mod. Phys.*, **42**, 237.

Chupp, E. L. (1976). *Gamma-ray astronomy*. Dordrecht: D. Reidel Publishing Co.

Clemmow, P. C. and Doughty, J. P. (1969). *Electrodynamics of particles and plasmas*. London: Addison-Wesley.

Enge, H. A. (1966). *Introduction to nuclear physics*. London: Addison-Wesley.

Jackson, J. D. (1975). *Classical electrodynamics*. New York: John Wiley and Sons, Inc.

Leighton, R. B. (1959). *Principles of modern physics*. London: McGraw-Hill.

Pozdnyakov, L. A., Sobol, I. M. and Sunyaev, R. A. (1983). *Astrophysics and space physics reviews*, **2**, 263. *Soviet Scientific Reviews*: Harwood Academic Publishers.

Ramana Murthy, P. V. and Wolfendale, A. W. (1986). *Gamma-ray astronomy*. Cambridge: Cambridge University Press.

Rybicki, G. B. and Lightman, A. P. (1979). *Radiative processes in astrophysics*. New York: John Wiley and Sons, Inc.

References

Birkinshaw, M. (1990). *The cosmic microwave background: 25 years later*. (eds) N. Mandolesi and N. Vittorio, p. 77. Dordrecht: Kluwer Academic Publishers.

Karzas, W. J. and Latter, R. (1961). *Astrophys. J. Suppl.* **6**, 167.

Kompaneets, A. S. (1956). *Zh. Eksp. Teor. Fiz*, **31**, 876. (English translation: *Sov. Phys. JETP*, **4**, 730, 1957.)

Longair, M. S. (1984). *Theoretical concepts in physics*, pages 226–34. Cambridge: Cambridge University Press.

Rossi, B. and Greisen, K. (1941). *Phys. Rev.* D1, 2252.

Heitler, W. (1954). *The Quantum Theory of Radiation*, pp. 204–211. Oxford: Clarendon Press.

Sunyaev, R. A. (1980). *Soviet Astronomy Letters*, **6**, 213.

Sunyaev, R. A. and Zeldovich, R. A. (1970). *Astrophys. Space Sci.*, **7**, 20.

Chapter 5

Further reading

The relevant aspects of nuclear physics are treated in many of the standard textbooks. The following contain much useful information:

Enge, H. A. (1966). *Introduction to nuclear physics*. London: Adison-Wesley.
Gaisser, T. K. (1990). *Cosmic rays and particle physics*. Cambridge: Cambridge University Press.
Hayakawa, S. (1969). *Cosmic ray physics*. New York: John Wiley and Sons, Inc.
Hillas, A. M. (1972). *Cosmic rays*. Oxford: Pergamon Press.
Leighton, R. B. (1959). *Principles of modern physics*. London: McGraw-Hill.
Wolfendale, A. W. (1973). *Cosmic rays at ground level*. London: Institute of Physics.

References

Arnett, W. D., Bahcall, J. N., Kirshner, R. P. and Woosley, S. E. (1989). *Ann. Rev. Astr. Astrophys.*, **27**, 629.
Damon, P. E., Lerman, J. C. and Long, A. (1978). *Ann. Rev. Earth Planet. Sci.*, **6**, 457.
Lingenfelter, R. E. and Ramaty, R. (1970). *Radiocarbon variations and absolute chronology*. Nobel Symposium No. 12, (ed.) I. U. Olsson, pp. 513–37, New York: John Wiley and Sons Inc.
Meyer, J-P. (1985). *19th Intl. Cosmic Ray Conference*, La Jolla, USA, Volume 9, page 141.
Murphy, R. J., Forrest, D. J., Ramaty, R. and Kozlovsky, B. (1985). *19th Intl. Cosmic Ray Conference*, La Jolla, USA, Volume 4, page 253.
Ramaty, R. and Lingenfelter, R. E. (1979). *Nature*, **278**, 127.
Silberberg, R., Tsao, C. H. and Letaw, J. R. (1985). *Astrophys. J. Suppl.*, **58**, 873.
Tsao, C. H. and Silberberg, R. (1979). *Proc. 16th Intl. Cosmic Ray Conference*, Kyoto, Vol. 2, page 202.
Webber, W. R., Kish, J. C. and Schrier, D. A. (1900a, b, c, d). *Phys. Rev. C.*, **41**, 520, 533, 547, 566.

Chapter 6

Further reading

I have found the following texts particularly useful in preparing this chapter:

Appenzeller, I. (1989). *Evolution of galaxies: Astronomical observations* (eds) I. Appenzeller, H. J. Habing and P. Lena, pp. 299–350. Berlin: Springer-Verlag.
Dodd, R. T. (1981). *Meteorites: A petrologic-chemical synthesis*. Cambridge: Cambridge University Press.
Enge, H. A. (1966). *Introduction to nuclear physics*. London: Addison-Wesley.
Fleischer, R. L., Price, P. B. and Walker, R. M. (1975). *Nuclear tracks in solids*. Berkeley: University of California Press.
Giacconi, R., Gursky, H. and van Speybroeck (1968). *Ann. Rev. Astr. Astrophys.*, **6**, 373.
Hillier, R. (1981). *Gamma-ray astronomy*. Oxford: Clarendon Press.
Kerridge, J. F. and Matthews, M. S. (eds) (1988). *Meteorites and the early Solar System*. Tucson: The University of Arizona Press.
Lal, D. (1972). *Sp. Sci. Rev.*, **14**, 3.
Lal, D. (1977). *Phil. Trans. R. Soc. London*, Series A, **285**, 69.
Powell, C. F., Fowler, P. H. and Perkins, D. H. (1959). *The study of elementary particles by the photographic method*. Oxford: Pergamon Press.
Price, P. B. and Fleischer, R. L. (1971). *Ann. Rev. Nucl. Sci.*, **21**, 295.
Reedy, R. C., Arnold, J. R. and Lal, D. (1983). *Ann. Rev. Nucl. Part. Sci.* **33**, 505.
Shapiro, M. M. (1958). In *Nuclear instrumentation – II*, page 342, Volume 45 of *Handbuch der Physik*. Berlin: Springer-Verlag.

Wasson, J. T. (1985). *Meteorites: Their record of early Solar System history*, New York: W. H. Freeman and Co.

Chapter 7

Further reading

I have made referemce to a wide range of sources for the material described in this chapter. The following are the principal sources of information:

Cosmic ray telescopes
Simpson, J. A. (1983). *Ann. Rev. Nucl. Part. Phys.*, **33**, 323.
X-ray telescopes
Peterson, L. E. (1975). *Ann. Rev. Astr. Astrophys.*, **13**, 423.
Tucker, W. and Giacconi, R. (1985). *The X-ray universe*. Cambridge: Harvard University Press.
γ-ray telescopes
Chupp, E. L. (1976). *Gamma-ray astronomy*. Dordrecht: D. Reidel Publishing Co.
Hillier, R. (1981). *Gamma-ray astronomy*. Oxford: Clarendon Press.
Ramana Murthy, P. V. and Wolfendale, A. W. (1986). *Gamma-ray astronomy*. Cambridge: Cambridge University Press.
Neutrino telescopes
Bahcall, J. N. (1989). *Neutrino astrophysics*. Cambridge: Cambridge University Press.

More information about many of these space projects is contained in:

Davies, J. K. (1988). *Satellite astronomy*. Chichester: Ellis Horwood Limited.

References

Peacock, A., Taylor, B. and Ellwood (1990). *Adv. Space Res.*, **20**(2), 273.

Chapter 8

Further reading

The information contained in this chapter is derived from a huge range of sources. The principal references are:

Appenzeller, I. (1989). *Evolution of galaxies: astronomical observations*, eds I. Appenzeller, H. J. Habing and P. Lena, pages 299–350. Berlin: Springer-Verlag.
Born, M. and Wolf. E. (1980). *The principles of optics*, 6th edition. Oxford: Pergamon Press.
Bracewell, R. (1965). *The Fourier transform and its applications*. New York: McGraw-Hill.
Clark, B. (1986). *Synthesis imaging*, eds. R. A. Perley, F. R. Schwab and A. H. Bridle, page 1. Green Bank: NRAO Publications.
Davies, J. K. (1988). *Satellite astronomy*. Chichester: Ellis Horwood Limited.
Eccles, M. J., Sim, E. M. and Tritton, K. P. (1983). *Low level light detectors in astronomy*. Cambridge: Cambridge University Press.
Felli, M. and Spencer, R. E. (1989). *Very long baseline interferometry – techniques and applications*. Dordrecht: D. Reidel Publishing Co.
Kitchin, C. R. (1984). *Astrophysical techniques*. Bristol: Adam Hilger.
Kraus, J. D. (1986). *Radio astronomy*, 2nd edition. Powell: Cygnus-Quasar Books.
Labeyrie, A. (1978). *Ann. Rev. Astr. Astrophys.*, **16**, 81.
Lena, P. (1988). *Observational astrophysics*. Berlin: Springer-Verlag.

Longair, M. S. (1984). *Theoretical concepts in physics*. Cambridge: Cambridge University Press.
Longair, M. S. (1989). *Alice and the Space telescope*. Baltimore: Johns Hopkins University Press.
McLean, I. S. (1989). *Electronic and computer-aided astronomy*. Chichester: Ellis-Horwood Limited.
Rohlfs (1986). *The tools of radio astronomy*. Berlin: Springer-Verlag.
Schroeder, D. J. (1987). *Astronomical optics*. San Diego: Academic Press.
Walker, G. (1987). *Astronomical observtions – an optical perspective*. Cambridge: Cambridge University Press.
Wolstencroft, R. D. and Burton, W. B. (1988). *Millimetre and submillimetre astronomy*. Dordrecht: Kluwer Academic Publishers.

References

Bowen, I. S. (1964). *Astron. J.*, **69**, 816.
Code, A. D. (1973). *Ann. Rev. Astr. Astrophys.*, **11**, 239.
Weigelt, G. (1989). *Evolution of galaxies: Astronomical observations*, eds I. Appenzeller, H. J. Habing and P. Lena, pp. 283. Berlin: Springer-Verlag.

Chapter 9

Further reading

Every two years there is a very large *International Conference on Cosmic Rays* at which all the most recent results are presented. The proceedings of the conferences are published in a number of volumes very soon after the conference. Of particular value are the volumes containing the *rapporteur* talks which summarise all the papers presented in different subject areas and which can be strongly recommended.

Hillas, A. M. (1972). *Cosmic rays*. Oxford: Pergamon Press.
Hillas, A. M. (1984). *Ann. Rev. Astr. Astrophys*, **22**, 425.
Shapiro, M. M. (ed.) (1983). *Composition and origin of cosmic rays*. Dordrecht: D. Reidel Publishing Co.
Shapiro, M. M. (ed.) (1984). *Cosmic radiation in contemporary astrophysics*. Dordrecht: D. Reidel Publishing Co.
Shapiro, M. M. and Wefel, J. P. (1988). *Genesis and propagation of cosmic rays*. Dordrecht: D. Reidel Publishing Co.
Simpson, J. A. (1983). *Ann. Rev. Astr. Astrophys.*, **33**, 323–81.
Sokolsky, P. (1989). *Introduction to ultra-high energy cosmic ray physics*. Redwood City, California: Addison-Wesley Publishing Co.
Tayler, R. J. (1972). *The origin of the chemical elements*. London: Wykeham Publications (London) Ltd.
Wdowczyk, J. and Wolfendale, A. W. (1989). *Ann. Rev. Nucl. Part. Sci.*, **39**, 43.

References

Cameron, A. G. W. (1973). *Space Sci. Rev.*, **15**, 121.
Cameron, A. G. W. (1982). *Essays in nuclear astrophysics*, eds C. Barnes, R. N. Clayton and D. N. Schramm, page 23. Cambridge: Cambridge University Press.
Meyer, J. P. (1979). *16th Intl. conf. cosmic rays*, Kyoto, Volume 2., page 115.
Meyer, J. P. (1985). *19th Intl. conf. cosmic rays*, La Jolla, Volume 9, page 141.

Watson, A. A. (1985). *19th Intl. conf. cosmic rays*, La Jolla, Volume 9, page 111.
Webber, W. R. (1983). *Compositions and origin of cosmic rays*, ed. M. M. Shapiro, page 83, Dordrecht: D. Reidel Publishing Co.

Chapter 10

Further reading

Hopper, V. D. (1964). *Cosmic radiation and high energy interactions*: Appendix B. London: Academic Press.
Jackson, J. D. (1975). *Classical electrodynamics*. New York: John Wiley and Sons, Inc.
Kaunde, Y. and Slavin, J. A. (1986). *Solar wind–magnetosphere coupling*. Dordrecht: D. Reidel Publishing Co.
Landau, L. D. and Lifshitz, E. M. (1987). *Course of theoretical physics*, Volume 6: *Fluid mechanics*. Oxford: Pergamon Press.
Longair, M. S. (1984). *Theoretical concepts in physics*, Cambridge: Cambridge University Press.
Marsden, R. G. (ed.) (1986). *The Sun and the heliosphere in three dimensions*. Dordrecht: D. Reidel Publishing Co.
Parker, E. N. (1963). *Interplanetary dynamical processes*. New York: Interscience Publishers.
Ratcliffe, J. A. (1972). *An introduction to the ionosphere and magnetosphere*. Cambridge: Cambridge University Press.
Spitzer, L. (1962). *Physics of fully ionised gases*. New York: Interscience Publishers.
Størmer, C. (1964). *The polar aurorae*. Oxford: Oxford University Press.
Sturrock, P. A., Holzer, T. E., Michalas, D. M. and Ulrich, R. K. (1986). *Physics of the Sun*. Volumes 1–3. Dordrecht: D. Reidel Publishing Co.
Zeldovich, Ya. B. and Raizer, Yu. P. (1966, 1967). *Physics of shock waves and high-temperature hydrodynamic phenomena*. New York: Academic Press.

References

Jokipii, J. R. and Thomas, B. T. (1981). *Astrophys. J.*, **243**, 1115.
Mayaud, P. N. (1980). *Derivation, meaning and use of geomagnetic indices*. Geophys. Monograph 22. Washington: AGU Publications.
Shea, M. A. and Smart, D. F. (1985). *19th Intl. conf. cosmic rays*, La Jolla, USA, Volume 4, page 501.
Simpson, J. A. (1989). *Adv. Space Res.*, **9**, No. 4, 5.
Webber, W. R. and Lezniak, J. A. (1974). *Astrophys. Space Sci.*, **30**, 361.
Wilcox, J. M. and Ness, N. F. (1965). *J. Geophys. Res.*, **70**, 5793.

Chapter 11

Further reading

Chen, F. C. (1984). *Introduction to plasma physics and controlled fusion*, Volume 1. *Plasma physics*. New York: Plenum Press.
Clemmow, P. C. W. and Dougherty, J. P. (1969). *The electrodynamics of particles and plasmas*. London: Addison-Wesley.
Jackson, J. D. (1975). *Classical electrodynamics*. New York: John Wiley and Sons, Inc.
Longair, M. S. (1984). *Theoretical concepts in physics*. Cambridge: Cambridge University Press.

Northrop, T. G. (1973). *Adiabatic motion of charged particles*. New York: Interscience Publishers.
Simpson, J. A. (1989). *Adv. Space Res.*, **9**, No. 4, 5.

References

Evenson, P., Garcia-Munoz, M., Meyer, P., Pyle, K. R. and Simpson, J. A. (1983). *Astrophys. J. Letters*, **275**, L15.
Garcia-Munoz, M., Pyle, K. R. and Simpson, J. A. (1985). *19th Intl. Conf. Cosmic Rays*, La Jolla, Volume 4, page 409.
Jokipii, J. R. (1971). *Rev. Geophys. and Sp. Sci.*, **9**, 27.
Jokipii, J. R. (1973). *Ann. Rev. Astr. Astrophys.*, **11**, 1.
McKibben, R. B. (1986). *The Sun and the heliosphere in three dimensions*, ed. R. G. Marsden, pp. 361–374. Dordrecht: D. Reidel Publishing Co.
Parker, E. N. (1965). *Planet. Space Sci.*, **13**, 9.

Chapter 12

Further reading

The books which I have found most useful in preparing this chapter are:

Audouze, J. and Israel, G. (eds) (1988). *The Cambridge atlas of astronomy*. Cambridge: Cambridge University Press.
Dessler, A. J. (ed.) (1983). *Physics of the Jovian magnetosphere*. Cambridge: Cambridge University Press.
Melrose, D. B. (1986). *Instabilities in space and laboratory plasmas*. Cambridge: Cambridge University Press.
Parker, E. N. (1979). *Cosmical magnetic fields*. Oxford: Clarendon Press.
Priest, E. R. (1982). *Solar magnetohydrodynamics*. Dordrecht: D. Reidel Publishing Co.
Priest, E. R. (ed.) (1981). *Solar flare magnetohydrodymamics*. London: Gordon and Breach Science Publishers.
Rucker, H. O., Bauer, S. J. and Pedersen, B. M. (eds) (1988). *Planetary radio emissions II*. Vienna: Verlag der Osterreichischen Akademie der Wissenschaften.
Stix, M. (1989) *The Sun – An introduction*. Berlin: Springer-Verlag.
Sturrock, P. A., Holzer, T. E., Michalas, D. M. and Ulrich, R. K. (eds) (1986). *Physics of the Sun*, Volumes 1–3. Dordrecht, D. Reidel Publishing Co.
Tandberg-Hanssen, E. and Gordon Emslie, A. (1988). *The physics of Solar flares*. Cambridge: Cambridge University Press.
Zirin, H. (1988). *Astrophysics of the Sun*. Cambridge: Cambridge University Press.
I have also made extensive use of Volumes 118 (1988) and 121 (1989) of the journal *Solar Physics*.

References

Alfven, H. and Carlqvist, P. (1967). *Solar Physics*, **1**, 220.
Bai, T. and Sturrock, P. A. (1989). *Ann. Rev. Astr. Astrophys.*, **27**, 421.
Bernstein, I. B., Green, J. M. and Kruskal, M. P. (1957). *Phys. Rev.*, **108**, 546.
Breneman, H. H. and Stone, E. C. (1985). *Astrophys. J.*, **299**, L57.
Brown, J. C. (1971). *Solar Physics*, **18**, 489.
Brown, J. C. and Emslie, A. G. (1989). *Astrophys. J.*, **339**, 1123.
Brown, J. C., Melrose, D. B. and Spicer, D. S. (1979). *Astrophys. J.*, **228**, 592.

Carrington, R. C. (1859). *Mon. Not R. Astr. Soc.*, **20**, 13.
Chupp, E. L. (1988). *Solar Physics*, **118**, 137.
Crannell, C. J., Dulk, G. A., Kosagi, T. and Magun, A. (1988). *Solar Physics*, **118**, 155.
Forbes, T. G., Malherbe, J. M. and Priest, E. R. (1989). *Solar Physics*, **120**, 285.
Guzik, T. G. (1988). *Solar Physics*, **118**, 185.
Heyvaerts, J. (1981). *Solar flare magnetohydrodynamics*, ed. E. R. Priest, page 429. London: Gordon and Breach
Heyvaerts, J. and Priest, E. R. (1983). *Astr. Astrophys.*, **117**, 220.
Heyvaerts, J. and Priest, E. R. (1984). *Astr. Astrophys.*, **137**, 63.
Heyvaerts, J., Priest, E. R. and Rust, D. M. (1977). *Astrophys. J.*, **216**, 123.
McGuire, R. E. and von Rosenvinge, T. T. (1984). *Adv. Space. Res.*, **4**, No. 2–3, 117.
Meyer, J. P. (1985). *19th Intl. conf. cosmic rays*, La Jolla, Volume 9, page 141.
Petschek, H. E. (1964). *AAS-NASA Symposium on the physics of solar flares*, NASA Special Publ. No. 50, p. 425.
Priest, E. R. and Forbes, T. G. (1986). *J. Geophys. Res.*, **91**, 5579.
Priest, E. R. and Forbes, T. G. (1990). *Solar Physics*, **126**, 319.
Ramaty, R., Dennis, B. R. and Emslie, A. G. (1988). *Solar Physics*, **118**, 17.
Spitzer, L. (1962). *Physics of fully ionised gases*. New York: Interscience Publishers.
Sturrock, P. A. (1968). *Structure and development of solar active regions*, ed. K. O. Kiepenheuer, p. 471. Dordrecht: D. Reidel Publishing Co.

Index

Items shown in bold type are physical processes or topics which are frequently encountered in High Energy Astrophysics. The page references in bold type indicate where the main discussions of these items are to be found.